Textbook of
Organic
Name Reactions

W0193326

Textbook of
Organic
Name Reactions

Sushil K Kashaw
M Pharm PhD MBA

Department of Pharmaceutical Sciences
Dr Harisingh Gour University (A Central University)
Sagar (MP)

Vikash K Mishra
M Pharm

Department of Pharmaceutical Sciences
Dr Harisingh Gour University (A Central University)
Sagar (MP)

CBSPD

CBS Publishers & Distributors Pvt Ltd

New Delhi • Bengaluru • Chennai • Kochi • Kolkata • Lucknow • Mumbai
Hyderabad • Jharkhand • Nagpur • Patna • Pune • Uttarakhand

Disclaimer

Science and technology are constantly changing fields. New research and experience broaden the scope of information and knowledge. The authors have tried their best in giving information available to them while preparing the material for this book. Although, all efforts have been made to ensure optimum accuracy of the material, yet it is quite possible some errors might have been left uncorrected. The publisher, the printer and the authors will not be held responsible for any inadvertent errors, omissions or inaccuracies.

Textbook of
Organic Name Reactions

ISBN: 978-81-239-2466-3

Copyright © Authors and Publisher

First Edition: 2015
 Reprint: 2023

All rights reserved. No part of this book may be reproduced or transmitted in any form or by any means, electronic or mechanical, including photocopying, recording, or any information storage and retrieval system without permission, in writing, from the authors and the publisher

Published by **Satish Kumar Jain** and produced by **Varun Jain** for

CBS Publishers & Distributors Pvt Ltd

4819/XI Prahlad Street, 24 Ansari Road, Daryaganj, New Delhi 110 002, India
Ph: 011-23289259, 23266861, 23266867   Website: www.cbspd.com
Fax: 011-23243014   e-mail: delhi@cbspd.com

Corporate Office: 204 FIE, Industrial Area, Patparganj, Delhi 110 092, India
Ph: 011-4934 4934   Fax: 011-4934 4935   e-mail: publishing@cbspd.com;
    publicity@cbspd.com

Branches

- **Bengaluru:** Seema House 2975, 17th Cross, KR Road, Banasankari 2nd Stage, Bengaluru 560 070, Karnataka, India
  Ph: +91-80-26771678/79   Fax: +91-80-26771680   e-mail: bangalore@cbspd.com
- **Chennai:** 7, Subbaraya Street, Shenoy Nagar, Chennai 600 030, Tamil Nadu, India
  Ph: +91-44-26680620, 26681266   Fax: +91-44-42032115   e-mail: chennai@cbspd.com
- **Kochi:** 42/1325, 1326, Power House Road, Opp KSEB, Power House, Ernakulum Kochi 682 018, Kerala, India
  Ph: +91-484-4059061-65,67   Fax: +91-484-4059065   e-mail: kochi@cbspd.com
- **Kolkata:** 147, Hind Ceramics Compound, 1st Floor, Nilgunj Road, Belghoria, Kolkata-700056, West Bengal, India
  Ph: +033-25633055, 033-25633056   e-mail: kolkata@cbspd.com
- **Lucknow:** Basement, Khushnuma Complex, 7 Meerabai Marg (Behind Jawahar Bhawan), Lucknow-226001, UP, India
  Ph: +0522-4000032   e-mail: tiwari.lucknow@cbspd.com
- **Mumbai:** PWD Shed, Gala no 25/26, Ramchandra Bhatt Marg, Next to JJ Hospital Gate no. 2, Opp. Union Bank of India,
  Noorbaug, Mumbai-400009, Maharashtra, India
  Ph: 022-66661880/89   e-mail: mumbai@cbspd.com

Representatives

- Hyderabad   0-9885175004
- Patna   0-9334159340
- Jharkhand   0-9811541605
- Pune   0-9923910676
- Nagpur   0-9421945513
- Uttarakhand   0-9716462459

Printed at Glorious Printers, Jhilmil Industrial Area, Delhi, India

to

Sir Dr. Harisingh Gour
(Founder of Dr. Harisingh Gour University)

and

our beloved parents

Smt Rajrani and   Smt Gayatri Mishra and
CSM/H Captain Kundan Lal (Retd.)   Shri Murlidhar Mishra

Preface

Writing a textbook at any level is always a challenge. In organic chemistry, exciting new discoveries are made at an ever-increasing pace. Organic chemistry is a lively and growing scientific discipline that touches a gigantic number of scientific areas. We thought to make *Textbook of Organic Name Reactions* because of their implications in the synthesis of many drugs and new chemical entities. Named reactions still are an important element of organic chemistry, and a detailed knowledge of such reactions is essential for the chemist. The scientific content behind the name is of great importance, and the names themselves are used as short expressions to ease spoken as well as written communication in organic chemistry. Furthermore, named reactions are a great support for learning the principles of organic chemistry. This text will provide the students an insight and the tools needed to address the tremendous challenges and the opportunities in the field of chemistry and pharmacy.

The book contains around five hundred name reactions and presents thoroughly updated matter on the name reactions with which we hope the reader will be highly benefited. This book will complement the knowledge of many organic chemistry textbooks. It is suitable for easy reading, learning and understanding the concept of the name reactions. Reactions with more implications have given more preference and have been dealt in detail. The book will serve its purpose to provide the service to the students of pharmacy and organic chemistry at UG, PG and PhD level. A student who has completed a course based on this book should be able to approach the literature directly, with a sound knowledge of modern basic organic chemistry.

The organization of matter in the book starts with introductory description of the reaction, chemical reaction and important applications. Basic mechanism of reaction along with considerations of reactivity and orientation is also discussed for important name reactions. These examples are not aimed at a complete treatment of every aspect of a particular reaction but are rather drawn from a didactic point of view. The reactions are arranged in alphabetical order. More emphasis is provided to the main reaction rather than the side reactions. Students preparing for qualifying examinations will find this book most suitable.

Special thanks to Prof NK Jain, Prof SP Vyas, Prof DV Kohli, Prof RK Agrawal, Prof SK Jain, Prof Asmita Gajbhiye, and other learned faculty members of the department for their time to time inspiration. In addition, we are indebted to Dr AK Saxena, Prof P Mishra, Dr Anshuman Dixit and Dr Alok Pal Jain, for their critical suggestions. Prof Varsha Kashaw and Ms Mitali Mishra, need special thanks for the proper proofreading of the book.

Authors are grateful to Prof Varsha Kashaw, Yash, Akshay, Ms Mitali Mishra and Krishna, for their constant support and patience during the preparation of the manuscript. We wish to thank CBSP&D for giving us opportunity to write this book. We gratefully acknowledge the blessings of our friends, students and well wishers.

Sushil K Kashaw
Vikash K Mishra

Contents

A

1. ACETOACETIC ESTER SYNTHESIS

Acetoacetic ester synthesis refers to the synthesis of α-substituted acetic acid esters or substituted acetones from acetoacetic ester; for this the ethyl acetoacetate is treated with a strong base, followed by alkylation and subsequent deacetylation or decarboxylation.

The acetoacetic ester synthesis is almost exactly like the malonic ester synthesis—involves a 1,3-dicarbonyl compound that is easily alkylated, and which also decarboxylates. But instead of making substituted acetic acid derivatives, this reaction makes substituted acetones.

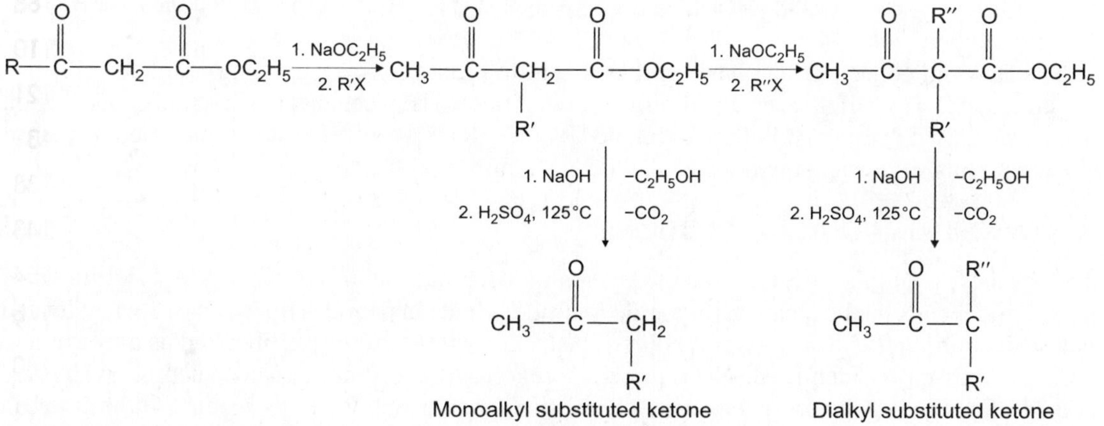

1. $NaOC_2H_5$
2. R′X
1. $NaOC_2H_5$
2. R″X
1. NaOH
2. H_2SO_4, 125°C
$-C_2H_5OH$
$-CO_2$
Monoalkyl substituted ketone
Dialkyl substituted ketone

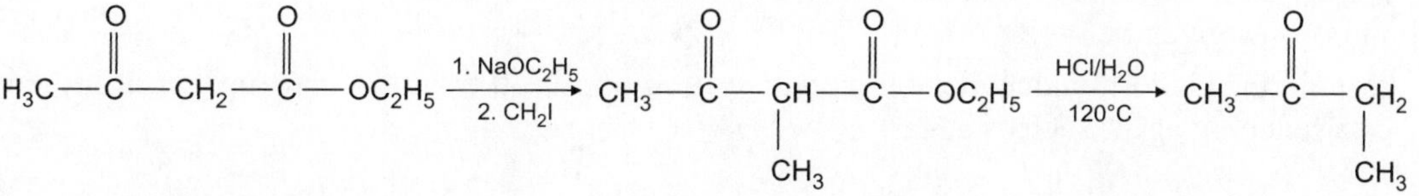

1. $NaOC_2H_5$
2. CH_3I
HCl/H_2O
120°C

The above reaction is a valuable method of preparing ketones called the acetoacetic ester synthesis of ketones. In the reaction mechanism, first the strong base deprotonates the dicarbonyl α-carbon which subsequently undergoes nucleophilic substitution. The protons between the carbonyl groups are particularly acidic. They can be deprotonated with alkoxide (e.g. NaOEt), or sodium hydride, to yield the stabilized anion. The carbon then undergoes nucleophilic substitution. When heated with aqueous acid, the newly alkylated ester is hydrolyzed to a β-keto acid, which is decarboxylated to form a methyl ketone.

2. ACYLOIN CONDENSATION

Acyloin condensation is a bimolecular reductive coupling of two carboxylic esters in an inert solvent such as ether, xylene using metallic sodium to yield an α-hydroxyketone (acyloin).

This reaction is favoured when R is an alkyl, also the trimethylchlorosilane has been found to improve the yield of this reaction. If the reaction is carried out in the presence of a proton donor, such as alcohol, simple reduction of the ester to the alcohol takes place (Bouveault-Blanc reduction).

This reaction has a great utility to synthesize cyclic ketones of intermediates to a large ring from diesters with long hydrocarbon chain between two ester groups. Cyclic acyloins are formed from appropriate substrates containing two ester groups. This reaction is found useful for the formation of rings, where the yield depends on ring size. The yields are found 60–95% for rings of 10–20 members.

The Benzoin condensation considered as a variant produces corresponding products with aryl substituents (R = aryl), although under different conditions. The mechanism of acyloin condensation is not known with certainty, but the diketone (RCOCOR) is assumed to be an intermediate, since small amounts of it can sometimes be isolated as a minor product.

3. AKABORI AMINO ACID REACTIONS

The Akabori reaction, devised about 60 years ago for the identification of the C-terminus of peptides, depends on the cleavage of amide (peptide) bonds in peptides by hydrazine when heated in a sealed tube at 125°C for several hours. The carboxy (C)-terminus is liberated as a free amino acid and thus easily identified. Now, there are several Akabori amino acid reactions which are named after Shiro Akabori, a Japanese chemist. Akabori reaction refers to the synthesis of an aldehyde, an α-amino aldehyde, or a primary amine from α-amino acid under different reaction conditions as given below.

1. α-amino acids form aldehydes by oxidative decomposition by molecular oxygen in presence of a reducing agent such as sugar:

$$R-\underset{\underset{COOH}{|}}{\overset{\overset{NH_2}{|}}{CH}} \xrightarrow[\text{sugar}]{O_2} RCHO + CO_2 + NH_3$$

2. α-amino acids and esters undergo reduction with sodium amalgam and ethanolic hydrogen chloride to the corresponding α-amino aldehydes:

$$R-\underset{\underset{COOC_2H_5}{|}}{\overset{\overset{NH_2}{|}}{CH}} \xrightarrow[\text{HCl/}C_2H_5OH]{NaHg} R-\underset{\underset{CHO}{|}}{\overset{\overset{NH_2}{|}}{CH}}$$

3. Formation of primary amines by heating mixtures of aromatic aldehydes (e.g. benzaldehyde) and amino acids.

4. ALDOL REACTION

The aldol reaction involves an addition reaction between the α-carbon of an enolizable aldehyde or ketone and the carbonyl group of another aldehyde or ketone yielding a β-hydroxy carbonyl compound which is traditionally known as the aldol. The reaction is catalyzed either by a base or sometimes by an acid but the most common form of the reaction uses a base. A simple case is addition of an enolate to an **ald**ehyde to afford an alco**hol**, thus the name **aldol**.

An enolizable aldehyde (or ketone) is a molecule that contains one or more alpha hydrogens. The carbon atom next to the functional group, i.e. adjacent to the C = O, is known as α-carbon (α-C) and hydrogens on this carbon are known as the α-hydrogens (α-H).

Enolizable ketones

Enolizable aldehydes

If the product (β-*hydroxy carbonyl* compound) undergoes subsequent dehydration, it yields an α, β-*unsaturated carbonyl* compound. The entire process (addition and dehydration steps) is known as aldol condensation.

The aldol reaction and the subsequent dehydration step are reversible and for a good yield of the desired product, the equilibrium of this reversible reaction might have to be shifted well to the right by appropriate reaction conditions. In aldol reaction the enolizable carbonyl compound after deprotonation at the α-carbon by base (e.g. alkali hydroxide) form a resonance stabilized enolate anion.

The hydroxide ion is not strong enough to convert substantially all of an aldehyde or ketone molecule to the corresponding enolate ion, that is, the equilibrium lies well to the left, subsequently, enough enolate ion will not be there for the reaction to proceed. Stronger bases, such as alkoxides (RO⁻) or amides (R₂N⁻), are commonly used now for a satisfactory yield.

Nature of solvent is another important factor because, for example, water and alcohol type protic solvent are found to react with the enolate anion and shift the equilibrium to the left.

Since both aldeydes and ketones can be the substrates for the reaction, there is the possibility to pair them by several ways as given below:

1. Reaction between two molecules of the same aldehyde.

Acetaldehyde
2 moles

beta-hydroxybutyrylaldehyde
(3-Hydroxybutanal)

2. Reaction between two molecules of the same ketone.

Acetone
2 moles

4-Hydroxy-4-methyl-2-pentanone

3. Reaction between two different aldehydes.

3-Phenyl-2-propenal

4. Reaction between two different ketones.
5. Reaction between an aldehyde and a ketone.

Among the above stated pairing possibilities, the aldol reaction between two molecules of the same aldehyde is generally quite feasible. Certainly, the aldehyde must possess enolizable hydrogen. On the other hand, the equilibrium lies well to the left in the analogous reaction of ketones, i.e. reaction between two molecules of the same ketone which can be shifted by allowing the reaction to proceed in a Soxhlet extractor or by treatment with basic Al₂O₃. There can be other provisions also such as the use of an amide base in aprotic solvents.

The *crossed aldol reactions*, which are also called *Claisen-Schmidt reactions*, leads to a mixture of products. From a mixture of two different aldehydes, each with α-hydrogens, four different aldols can be formed. The crossed aldol is successfully accomplished using amide bases in aprotic solvent. For reaction between two different ketones similar considerations do apply in addition to the unfavorable equilibrium mentioned above, which is why such reactions are seldom attempted.

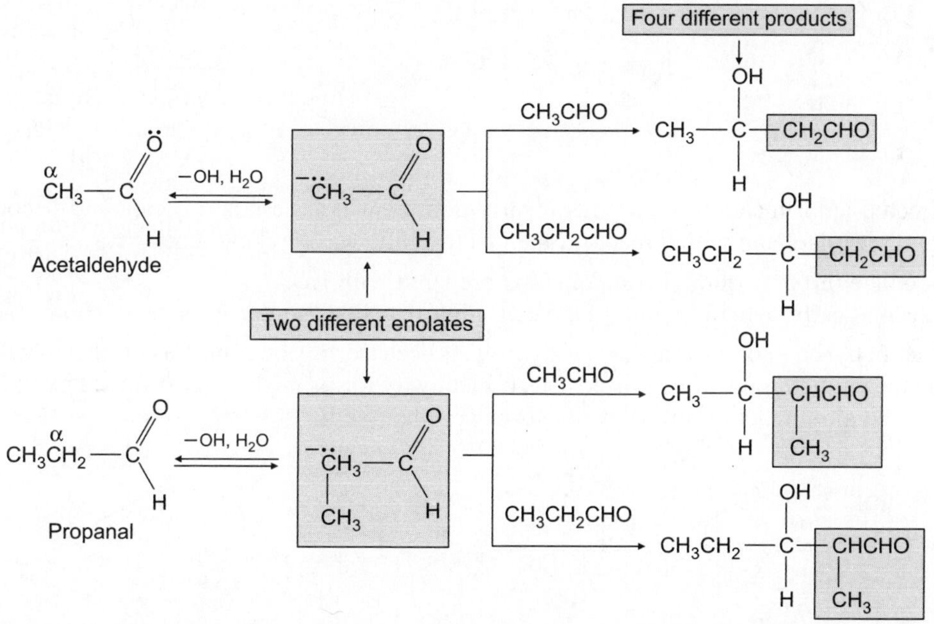

However, with only one non-enolizable aldehyde, i.e. if only one of the two aldehydes has an α-hydrogen, only two aldols are possible.

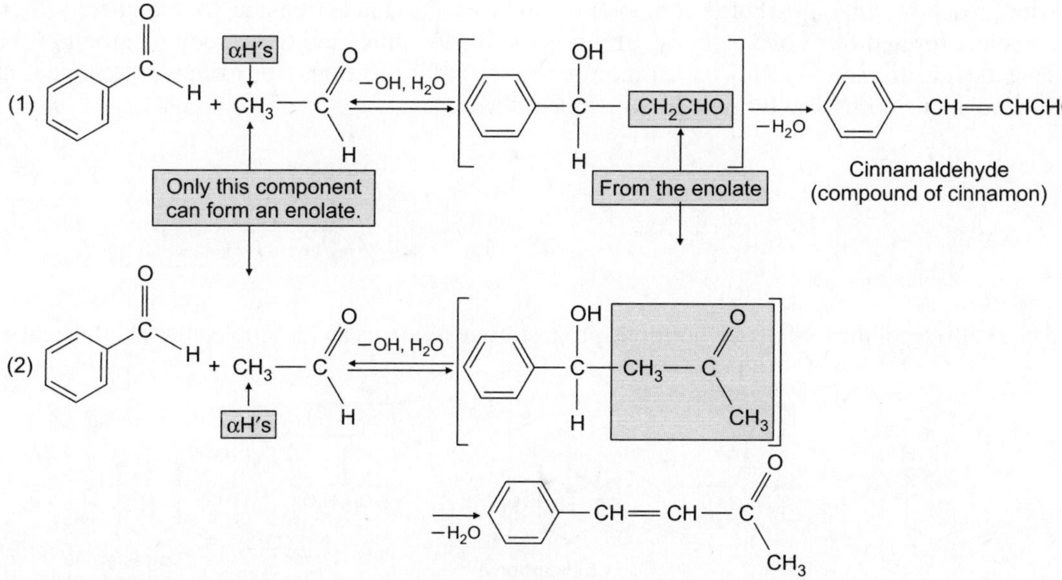

Aldol reaction between an aldehyde and a ketone is also called the Claisen-Schmidt reaction and is usually feasible with hydroxide or alkoxides bases in protic solvents.

The addition of the α-carbon of an aldehyde to the carbonyl carbon of the ketone can be achieved by using an amine instead of an aldehyde, and lithium diisopropylamide as the base. This is known as a directed aldol reaction. By this approach, unsymmetrical ketones can be made to react regioselectively.

A directed aldol reaction is one that clearly defines which carbonyl compound becomes the nucleophilic enolate and which reacts at the electrophilic carbonyl carbon:

1. The enolate of one carbonyl component is prepared with LDA.
2. The second carbonyl compound (the electrophile) is added to this enolate.

Both carbonyl components can have α-hydrogens because only one enolate is prepared with LDA. When an unsymmetrical ketone is used, LDA selectively forms the less substituted kinetic enolate.

A directed aldol reaction in the synthesis of periplanone B

deprotonation

nucleophilic addition of the enolate to the aldehyde carbonyl

periplanone B
sex pheromone of the female American cockroach

Aldol reactions with dicarbonyl compounds can be used to make five-and six-membered rings. The enolate formed from one carbonyl group is the nucleophile, and the carbonyl carbon of the other is the electrophile. This is called intramolecular aldol reactions, for example, treatment of 2,5-hexadienone with base forms a five-membered ring.

2,5-hexandedione

re-draw

NaOEt
EtOH

new C—C σ and π bonds

The synthesis of the female sex hormone progesterone involves an intramolecular aldol reaction:

Step [1]

1. O₃
2. Zn, H₂O

1,5-dicarbonyl compound

Step [2]

−OH, H₂O

Progesterone

Ozone oxidatively cleaves the C=C

Intermolecular aldol reaction forms the six-membered ring

There are several other analogous reactions to the aldol reaction, which also involve addition of enolate species to a carbonyl compound. Such reactions are often called aldol-type reactions but the term aldol reaction is reserved for the reaction of aldehydes and ketones.

5. ALGAR-FLYNN-OYAMADA (AFO) REACTION

Algar-Flynn-Oyamada reaction is the synthesis of flavonols via the intermediate dihydroflavonols made by oxidation of o-hydroxyphenyl styryl ketone (chalcones) using hydrogen peroxide. Chalcones undergo oxidative cyclization to form a flavonol in the reaction.

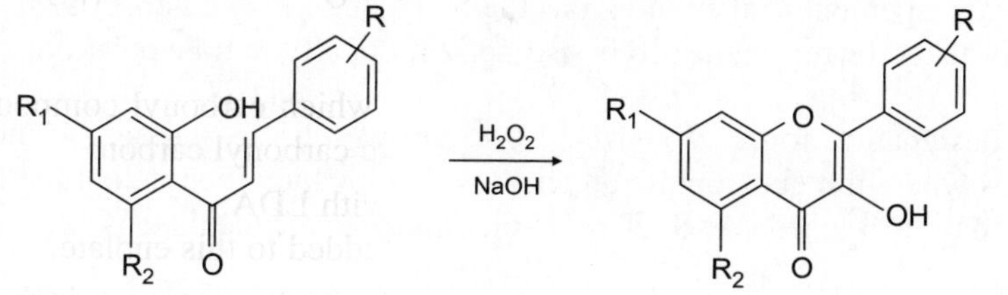

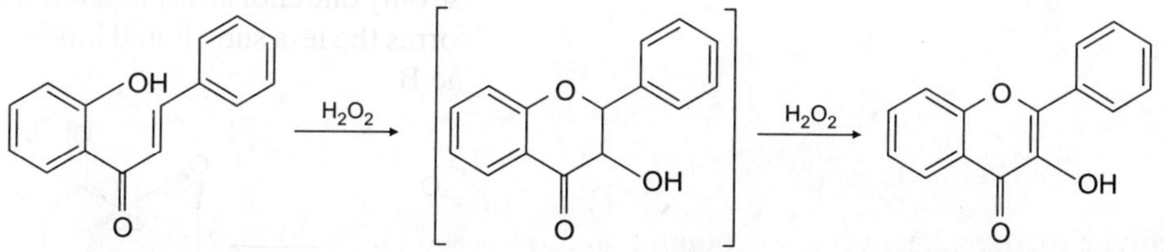

It has been found that a variety of 2-hydroxychalcones with methoxy groups in different positions in the two aromatic nuclei smoothly undergo the AFO oxidation. This reaction is applicable for the preparation of flavones and flavonols.

6. ALLAN-ROBINSON REACTION

The Allan-Robinson reaction involves the formation of flavones or isoflavones by condensing o-hydroxyaryl ketones with anhydrides of aromatic acids and their sodium salts. The structurally disparate flavonoids and isoflavones have been synthesized using this reaction.

If aliphatic anhydrides are used in the reaction, coumarins can also be formed (Kostanecki acylation).

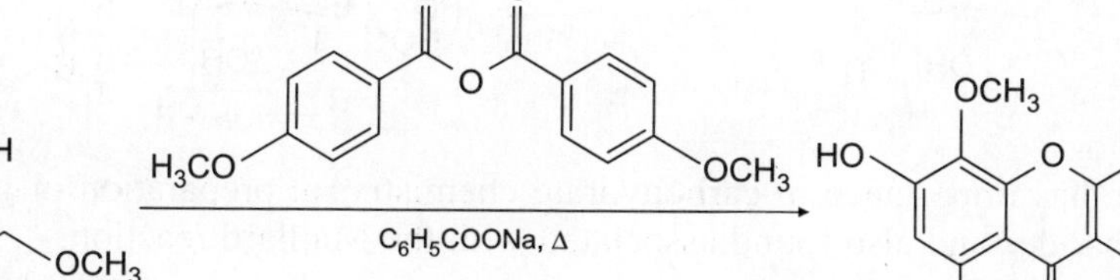

7. ALLYLIC REARRANGEMENTS

An allylic rearrangement involves migration of double bond in a three carbon allylic system to the next carbon atom on treatment with nucleophiles.

Such a rearrangement is often encountered in nucleophilic substitutions. Depending on the reaction conditions it may occur either under S_N1 or S_N2 conditions. Allylic system is very liable to rearrangement owing to the very plausible resonance.

When reaction conditions favour a S_N1 reaction mechanism the intermediate is a carbocation. Alternatively, under S_N2 conditions the nucleophile attack directly at the allylic position, displacing the leaving group in a single step.

The allylic rearrangement is also known as the S_N1' or S_N2' reaction. Alternatively, the allylic rearrangement is also referred as the allylic isomerization, allylic transposition, and/or allylic 1,3-rearrangement.

The IUPAC designation for S_N1' is $1/D_N+3/A_N$, in which the numbers 1 and 3 represent the relative positions where the nucleophile attacks and from which the nucleofuge leaves. Similarly, $3/1/A_ND_N$ designation is used for S_N2' reactions and the mechanism is a second-order allylic rearrangement.

The allylic rearrangement is also common in allyl-based Grignard reagents:

8. AMADORI REARRANGEMENT

Amadori rearrangement involves acid or base catalyzed conversion of N-glycosides of aldoses to N-glycosides of the corresponding ketoses. More generally it is known as the conversion of aldimines into ketoamines.

This reaction has importance in carbohydrate chemistry, in preparation of glycoproteins and aminopolysaccharides and also found associated with the Maillard reaction.

9. ARENS-VAN DORP SYNTHESIS

Arens-van Dorp synthesis involves preparation of alkoxyethynyl alcohols from ketones and ethoxyacetylene. In this synthesis, the ethoxyacetylene is converted to a Grignard reagent and then reacted with a ketone.

In the **Isler modification**, the tedious preparation of ethoxyacetylene is replaced by treating β-chlorovinyl ether with lithium amide to yield lithium ethoxyacetylene, which is then condensed with the ketone.

10. ARNDT-EISTERT SYNTHESIS

This reaction involves conversion of an acyl halide to a carboxylic acid with one additional carbon.

The diazomethane carbon adds to the acid halide to give an α-diazoketone. In the next step, treatment with water or silver salt initiates a rearrangement with the R-group migrating to produce an isocyanate, which readily hydrolyses to give a carboxylic acid.

If an alcohol (R′-OH) is used instead of water, the ester RCH_2COOR' is isolated directly. In analogous reactions with ammonia or amines (R′NH$_2$), the corresponding amides are accessible.

The reaction is of wide scope. The R group may be alkyl or aryl and may contain many functional groups including unsaturation. Therefore, the synthesis is useful for the synthesis of both, aliphatic and aromatic acids; however the substrate molecule should not contain other functional groups that can react with diazomethane. Reaction with other diazoalkanes (i.e. R′CHN$_2$) provides RCHR′COOH. The reaction has also been used successfully for ring contraction of cyclic diazo ketones.

The mechanism of the reaction is assumed to involve formation of a carbene. It is the divalent carbon that has the open sextet and to which the migrating group brings its electron pair. This reaction allows conversion of carboxylic acids into the next higher homolog and the reaction sequence is considered to be the best method for the extension of a carbon chain by one carbon atom in cases where a carboxylic acid is available.

11. AUWERS SYNTHESIS

This reaction was first reported by Karl von Auwers in 1908. Auwers reaction involves expansion of coumarones to flavonols by treatment of 2-bromo-2-(α-bromobenzyl)coumarones with alcoholic alkali.

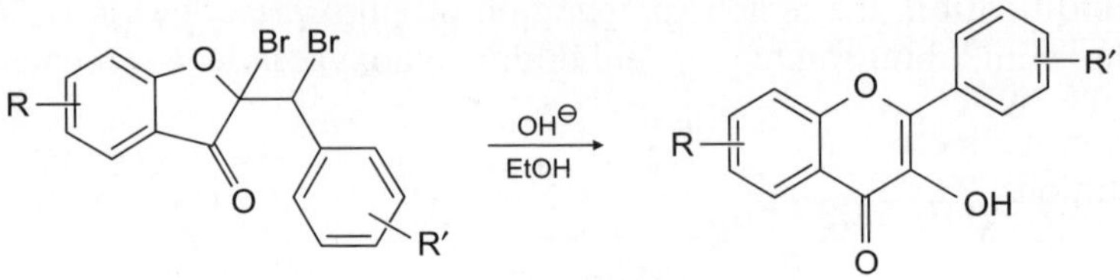

Synthesis of 2-Chloroflavonol

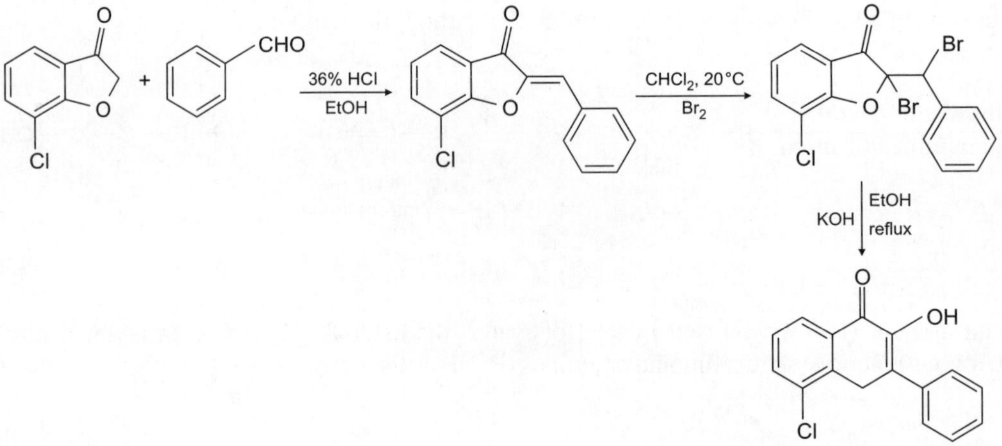

12. BAEYER-DREWSON INDIGO SYNTHESIS

In 1865, the German chemist Johann Friedrich Wilhelm Adolf von Baeyer began working with indigo. In 1880, his work resulted in the first synthesis of indigo.

Baeyer-Drewson indigo synthesis involves reacting o-nitrobenzaldehyde with acetone under highly basic conditions. **The reaction is an aldol condensation type reaction.**

This reaction works well for small scale reactions and is not used today in industry for producing large quantities of indigo.

13. BAEYER-VILLIGER REARRANGEMENT

The transformation of ketones into esters or of cyclic ketones into lactones by peracids was discovered as early as 1899 by Adolf von Baeyer and Victor Villiger, when they were working on the ring cleavage of cyclic ketones (terpene derivatives).

When ketones are treated with peroxyacids (RCO_3H), or peroxyacetic acid, or with other peroxy compounds in the presence of acid catalysts, carboxylic esters are obtained by "insertion" of oxygen. This "oxyinsertion" is called the Baeyer-Villiger oxidation or Baeyer-Villiger rearrangement. When cyclic ketones are used in the reaction, lactones are obtained as products.

11

Cyclopentanone → (RCO₃H) → Tetrahydro-pyran-2-one

Cyclohexanone → (CF_3COOH / CH_2CH_2) → Oxepan-2-one (Lactone)

A wide range of peroxy reagents can be used to carry out this rearrangement such as: meta-chloroperoxybenzoic acid (mCPBA), peroxyacetic acid (PAA), or peroxytrifluoroacetic acid (TFPAA), hydrogen peroxide/BF_3, etc.

This reaction has been applied in the synthesis of esters and lactones. The Baeyer-Villiger reaction is often applied to cyclic ketones to yield lactones. It is of synthetic use for shortening carbon chains, hydroxylating aromatic rings, converting carbocycles to heterocycles, and opening cyclic arrays to prepare functionalised chains and/or rings.

Besides being a method of producing cyclic and acyclic esters, the reaction, upon hydrolysis of these esters either as a separate step or *in situ*, may result in carboxylic acids and alcohols as products. Similar oxidation of aldehydes to the corresponding formate esters or their hydrolysis products also belong to this class of reactions.

When catalyzed by acid:

ester + water ⇌ (H^+) alcohol + carboxylic acid

ester + alkali hydroxide ⇌ alcohol + alkali salt of carboxylic acid

R_1—$COOR_2$ + NaOH ⇌ R_1—COONa + R_2—OH

The reaction mechanism of this oxidative cleavage starts with an initial addition of the peroxy group to the carbonyl carbon to give a tetrahedral intermediate also called the Criegee like intermediate. Then in next step, one of the groups attached to carbonyl carbon is migrated on to the electron deficient oxygen atom in a concerted step with loss of a carboxylic acid, which is the rate determining step.

If the migrating carbon is chiral, the stereochemistry is retained. The most electron-rich alkyl group (more substituted carbon) migrates first.

The general migration order:

tertiary alkyl > cyclohexyl > secondary alkyl > benzyl > phenyl > primary alkyl > methyl >> H.

For substituted aryls:

$$p\text{-MeO-Ar} > p\text{-Me-Ar} > p\text{-Cl-Ar} > p\text{-Br-Ar} > p\text{-MeOAr} > p\text{-O}_2\text{N-Ar}$$

In case of unsymmetrical ketones, the migrating group is usually the one that can best stabilize positive charge. Thus, cyclic ketones produce lactones. On the other hand, aldehydes usually react with migration of the hydrogen to yield the carboxylic acid.

Biocatalytic BV oxidation

The Baeyer-Villiger oxidation can be performed by biocatalysis with a modern variant of the reaction the so-called **Baeyer-Villiger monooxygenase** or **BVMO**. This variant allows oxidation under mild conditions in good yields, with one stereoisomer being formed predominantly.

Apart from this, the transition-metalcatalyzed Baeyer-Villiger reaction has also been reported.

14. BAKER-VENKATARAMAN REARRANGEMENT

The base induced rearrangement of aromatic ortho-acyloxyketones to the corresponding aromatic β-diketones (1,3-diketones) is known as the Baker-Venakataraman rearrangement. This reaction proceeds through the formation of an enolate, followed by intramolecular acyl transfer.

The most commonly used bases for the reaction are: KOH, potassium tert-butoxide in DMSO. Na metal in toluene, sodium or potassium hydride, pyridine, and triphenylmethylsodium. The reaction was named after Wilson Baker and Krishnasami Venkataraman.

β-diketones are important synthetic intermediates, and are widely used for synthesis of chromones, flavones, isoflavones, and coumarins.

15. BAMBERGER REARRANGEMENT

The aryl hydroxylamines when treated with strong aqueous acid (e.g. in aqueous sulfuric acid) rearrange to aminophenols. This reaction is called the Bamberger rearrangement and is named after the German chemist Eugen Bamberger. The rearrangement is intermolecular.

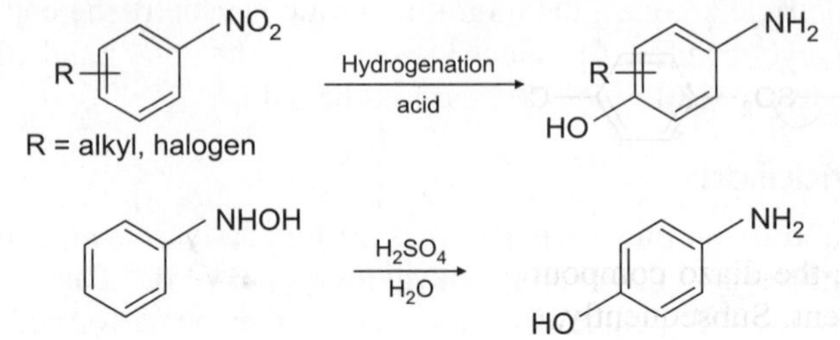

R = alkyl, halogen

The reaction of the Bamberger rearrangement starts with monoprotonation of N-phenyl-hydroxylamine and proceeds with formation of the nitrenium ion, which can react with nucleophiles (H_2O) to yield 4-aminophenol.

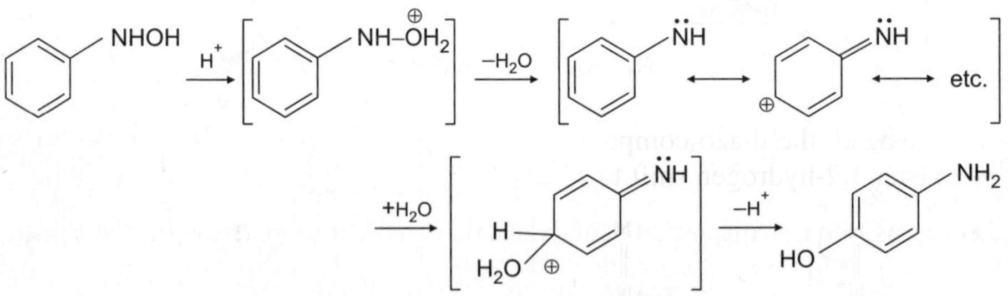

16. BAMFORD-STEVENS REACTION; SHAPIRO REACTION

This reaction describes the transformation of ketones and aldehydes into alkenes through the base-promoted decomposition of p-toluenesulfonylhydrazones (tosyl hydrazones) of the corresponding ketones or aldehydes. Thus, the reaction may be used to effect the overall transformation of a ketone to an alkene. It is named after William Randall Bamford Thomas Stevens.

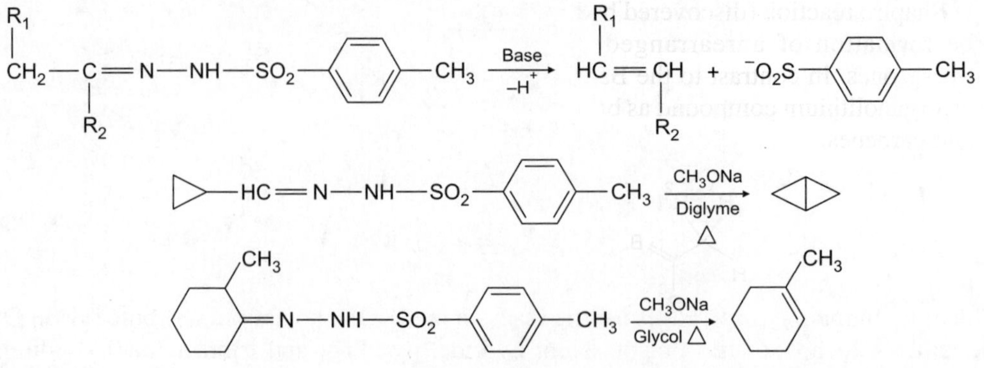

The bases used for the reaction are NaOMe, NaH, LiH, $NaNH_2$, etc. It has been observed that the reaction is strongly influenced by the type of base and the nature of solvent. The usage of aprotic solvent gives predominantly Z-alkenes, while protic solvent gives a mixture of E- and Z-alkenes. The more substituted alkene is formed as the thermodynamic product.

Reaction of tosyl hydrazone with a strong base initially leads to a diazo compound, where the intermediate diazo compound may be isolated. Subsequent, the reaction with protic or aprotic solvents strongly influences the outcome of the reaction.

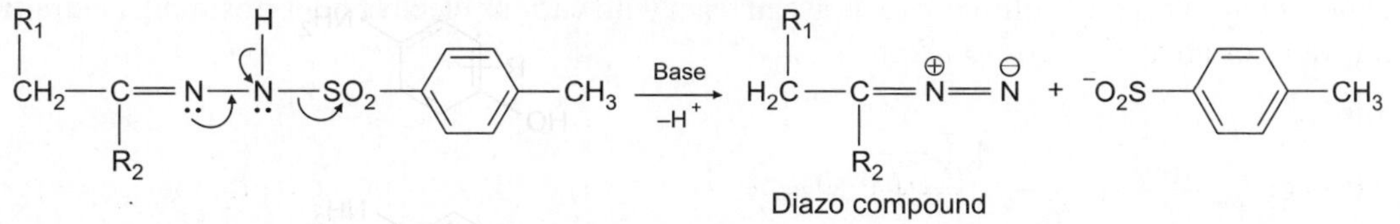

Diazo compound

In protic solvents, the diazo compound decomposes to the carbenium ion by abstraction of a proton from the solvent. Subsequently, the alkene is formed through the loss of dinitrogen and a proton by carbenium ion.

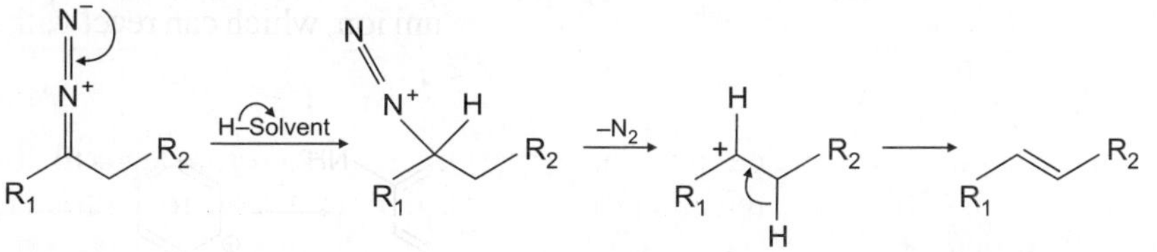

On the other hand, the diazo compound in aprotic solvent decomposes to the carbene which undergoes a faster 1,2-hydrogen shift to furnish an alkene.

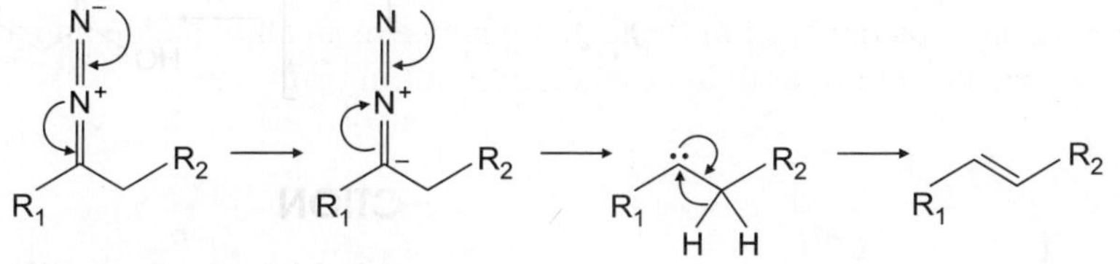

If an organolithium is used as the base, the reaction follows another mechanism without occurrence of carbenium ions and carbenes.

The Shapiro reaction, a variation of the Bamford-Stevens reaction, is the base-induced reaction of tosylhydrazones to afford alkenes. This reaction is carried out with two equivalents of an organolithium compound.

The Shapiro reaction (discovered by Robert H. Shapiro in 1975) is a more promising procedure for the formation of unrearranged alkenes (generally the less substituted isomers) from tosylhydrazones. In contrast to the Bamford-Stevens reaction, the Shapiro reaction is carried out with an organolithium compound as base. This reaction proceeds without occurrence of carbenium ions and carbenes.

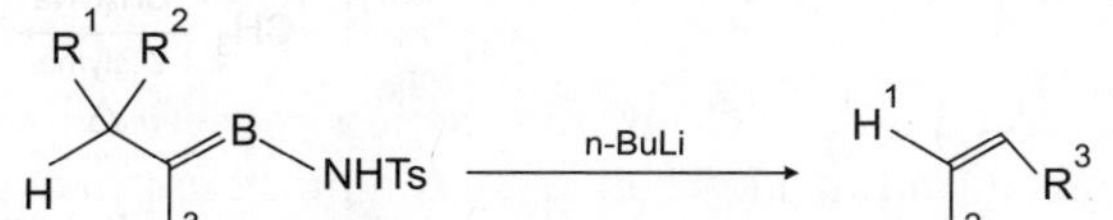

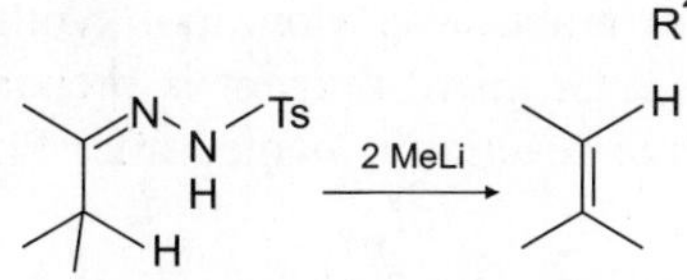

In Shapiro reaction, a ketone or an aldehyde is reacted first with p-toluenesulfonylhydrazide to form a p-toluenesulfonylhydrazone (or tosylhydrazone). Then, two equivalents of an organolithium compound abstract the proton from the hydrazone and leads to a carbanion. The carbanion undergoes an elimination reaction to create the carbon-carbon double bond and formation of a diazonium anion. The resulting diazonium anion losses neutral nitrogen molecule which result in an organolithium. This organolithium carbon can react with various electrophiles or simply neutralized with water or an acid to afford alkene.

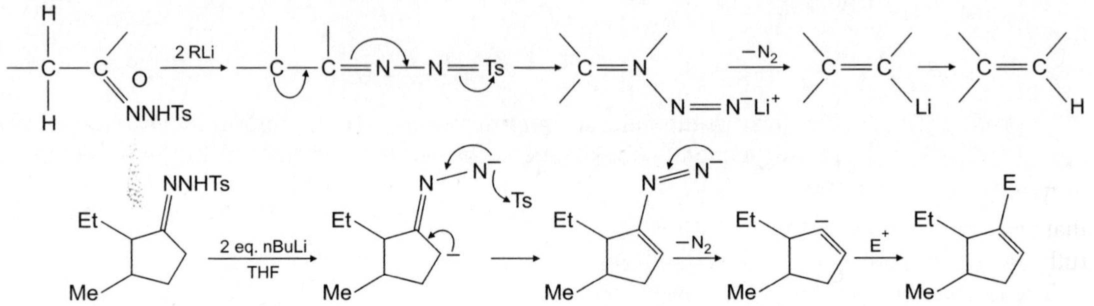

The dianion in Shapiro reaction does not tend to rearrange and therefore leads to non-rearranged alkenes in good yields. However, the Shapiro reaction does not lead to high stereoselectivity between the *E-Z*-isomers.

17. BARBIER COUPLING REACTION

It involves preparation of alcohols from the reaction between an alkyl halide and a carbonyl group in the presence of magnesium, aluminium, zinc, indium, tin or its salts.

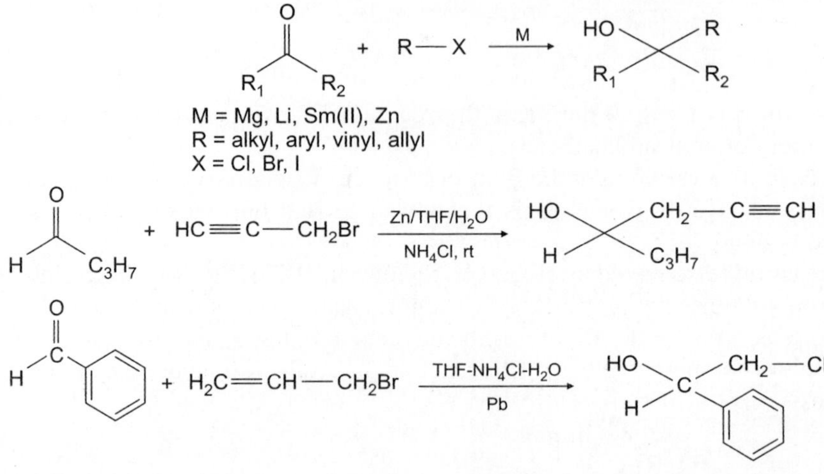

M = Mg, Li, Sm(II), Zn
R = alkyl, aryl, vinyl, allyl
X = Cl, Br, I

The Barbier coupling reaction is a Grignard reaction carried out *in situ,* where a carbonyl compound, magnesium metal, and an alkyl halide are combined *in situ* to give a substituted alcohol product. Therefore, the Barbier reaction is a one-pot synthesis carried out without preliminary formation of RMgX, whereas a Grignard reagent is prepared separately before addition of the carbonyl compound. The Barbier reaction is named after Philippe Barbier.

18. BARBIER-WIELAND DEGRADATION

Barbier-Wieland degradation describes stepwise degradation of aliphatic carboxylic acid by oxidative cleavage leading to chain-shortening, which finally results in next lower homologs.

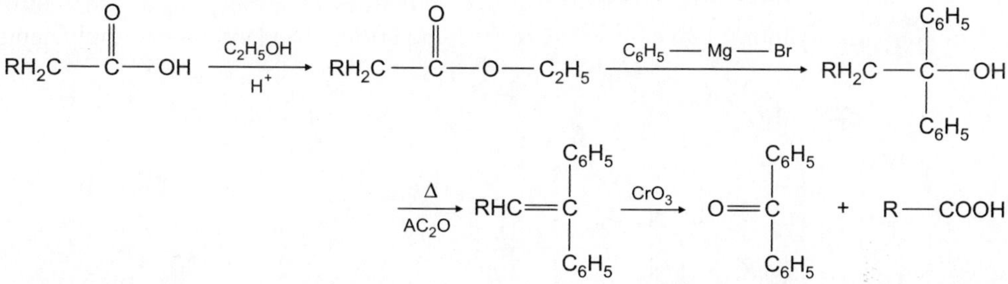

The reaction proceeds via four steps which comprise conversion of ester into a tertiary alcohol that is dehydrated with acetic anhydride, and the olefin oxidized with chromic acid ($NaIO_4$–ruthenium tetroxide has also been used) to a lower homologous carboxylic acid.

The Barbier-Wieland degradation is particularly useful in the degradation of steroid side chains and side chains of other cyclic compounds.

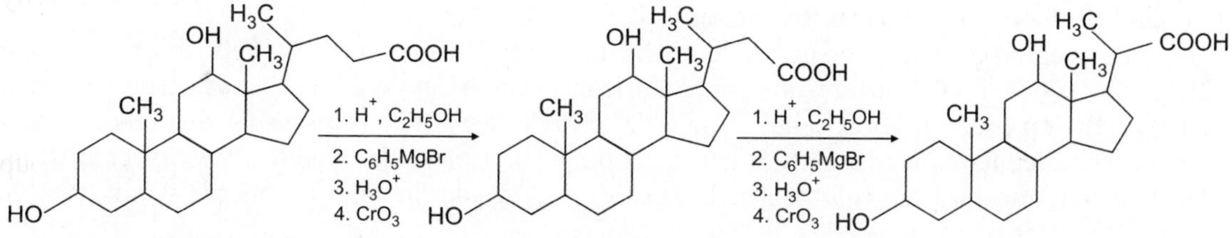

The inverse procedure is the Arndt-Eistert synthesis, where an acid is converted into acyl halide and reacts with diazomethane to give the higher homolog.

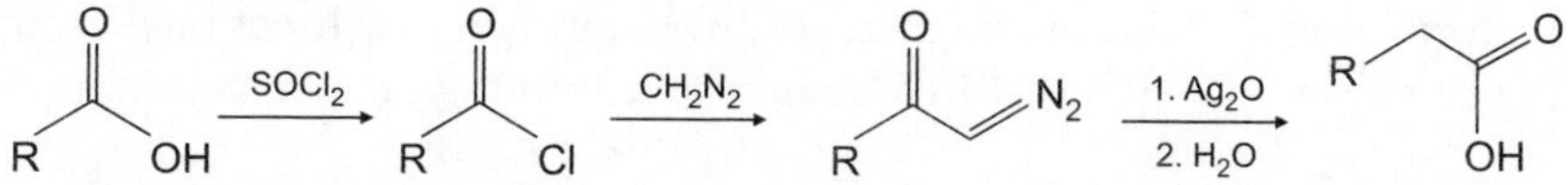

19. BART REACTION; SCHELLER MODIFICATION

Bart reaction involves reaction between aromatic diazonium compounds with alkali arsenites in the presence of cupric salts or powdered silver or copper to form arylarsenic acids.

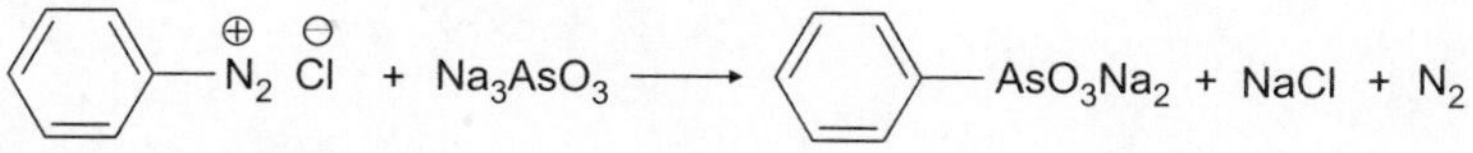

The Scheller modification involves the reaction between aromatic diazonium and arsenious chloride in the presence of trace amounts of cuprous chloride.

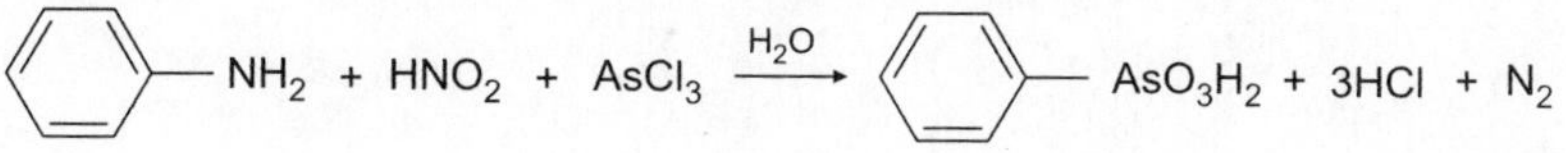

20. BARTON DECARBOXYLATION

Barton decarboxylation is a radical decarboxylation of organic acids to generate alkanes via the corresponding thiocarbonyl derivatives of the carboxylic acids. This reaction was named after its developer, the British chemist Sir Derek Barton.

Barton ester

The Barton decarboxylation process is a two-step process: in first step the carboxylic acid is first converted to a thiohydroxamate ester (Barton ester), which in the next step is heated in presence of a radical initiator and a suitable hydrogen donor (such as tri-n-butyltin hydride [$HSnBu_3$], t-butylmercaptan [t-BuSH], phenylselenol [PhSeH], and tri(trimethylsilyl)silane [$(Me_3Si)_3SiH$]) to complete the reductive decarboxylation.

Aromatic carboxylic acids do not undergo this reaction. The procedure is applicable to aliphatic carboxylic acids with the best results generally from primary and secondary acids. The reductive decarboxylation product of especially hindered tertiary carboxylic acids using this method may sometimes be contaminated with the corresponding alkyl chloride, which arises by competing chlorine atom transfer from chloroform. In these cases, the addition of a stronger H-donor, such as tertbutyl thiol, is recommended.

The reaction mechanism was proposed to involve homolytic cleavage, generally initiated by light or heat, leading to a 2-pyridylthiyl radical (PyS) and an acyloxy radical (RCOO). The alkyl radical (R), generated by radical decarboxylation, reacts with a radical-trapping species X–Y, resulting in the formation of R–X, the trapped product bearing a new functional group. The radical Y attacks the sulfur atom of the thiohydroxamate ester, forming an S-Y bond.

Initiation

Propagation

This sequence of reactions has been used to remove a carboxylic acid group and to replace it by another functional group, thereby generating new bonds such as carbon-carbon, carbon-oxygen, carbon-sulfur, carbon-selenium and carbon-halogen bonds.

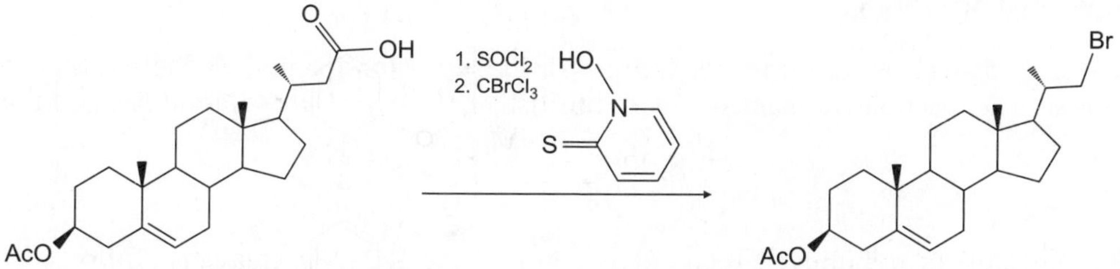

21. BARTON DEOXYGENATION (BARTON-MCCOMBIE REACTION)

Barton-McCombie reaction describes deoxygenation of alcohols via their thiocarbonyl derivatives by means of radical scission. The overall process is reduction of the ROH unit to RH. This reaction was named after the British chemists Sir Derek Harold Richard Barton and Stuart W. McCombie.

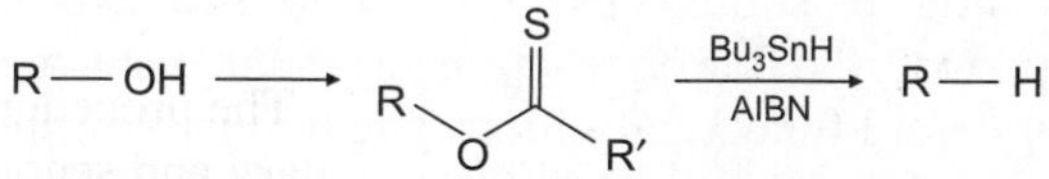

R = Alkyl, R′ = H, CH₃, SCH₃, OCH₃Ph, Oph, Imidazolyl

In the reaction mechanism, the alcohol is first converted to the thiocarbonyl derivative.

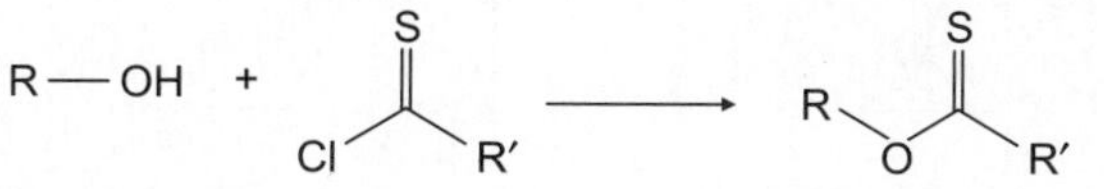

Reaction of tributylstannane with AIBN decomposes it into a tributylstannyl radical.

$$Bu_3Sn\text{---}H \xrightarrow{AIBN} Bu_3Sn\bullet + H\bullet$$

Overall reaction mechanism is given below:

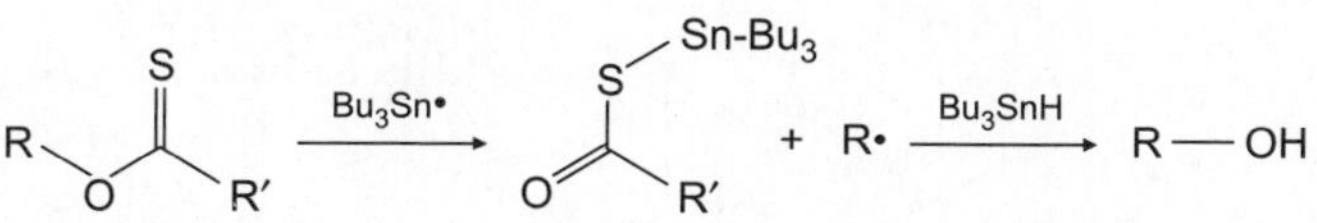

Formation of the very stable S-Sn bonds mainly drives the reaction.

22. BARTON OLEFIN SYNTHESIS (BARTON-KELLOGG REACTION)

Barton olefin synthesis involves coupling of two ketones (between a ketone and a thioketone) into alkenes. In this reaction, the steric constraints are gradually introduced via sequential processes involving a 1,3-dipolar cycloaddition to form five-membered Δ^3-1,3,4-thiadiazoline, followed by the nitrogen elimination to three-membered episulfide, and finally the sulfur extrusion to afford the alkenes.

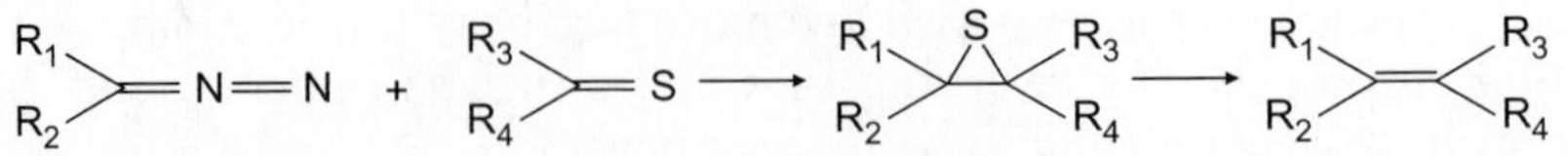

23. BARTON REACTION

Barton reaction involves transformation of nitrous acid esters into γ-oximino alcohol under photolytic conditions. The reaction was named after the British chemist Sir Derek Harold Richard Barton.

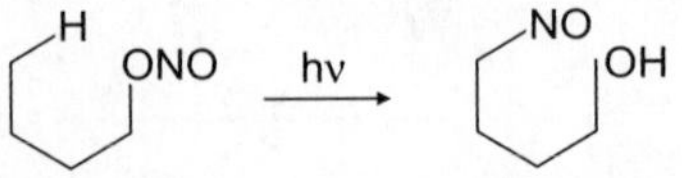

The mechanism is believed to involve a homolytic RO–NO cleavage, followed by an intramolecular hydrogen abstraction and free radical recombination.

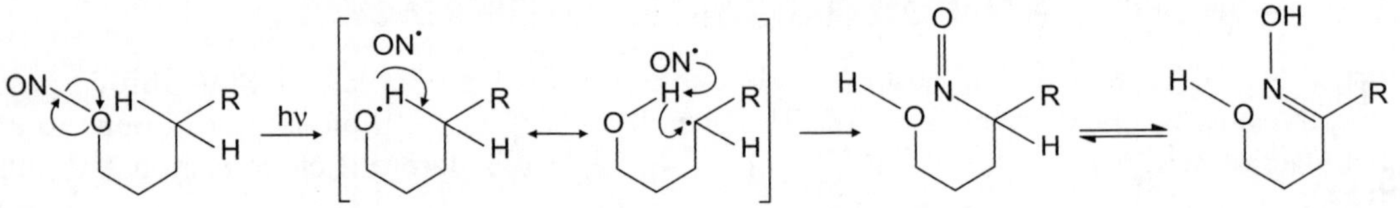

Having the unique potential of siteselective C-H bond cleavage at the d-position via a 1,5-radical translocation from O to C, the Barton reaction has found widespread application in synthesis processes, which includes steroid functionalisation. The required nitrite esters for the reaction are prepared from the corresponding alcohols and nitrosyl chloride.

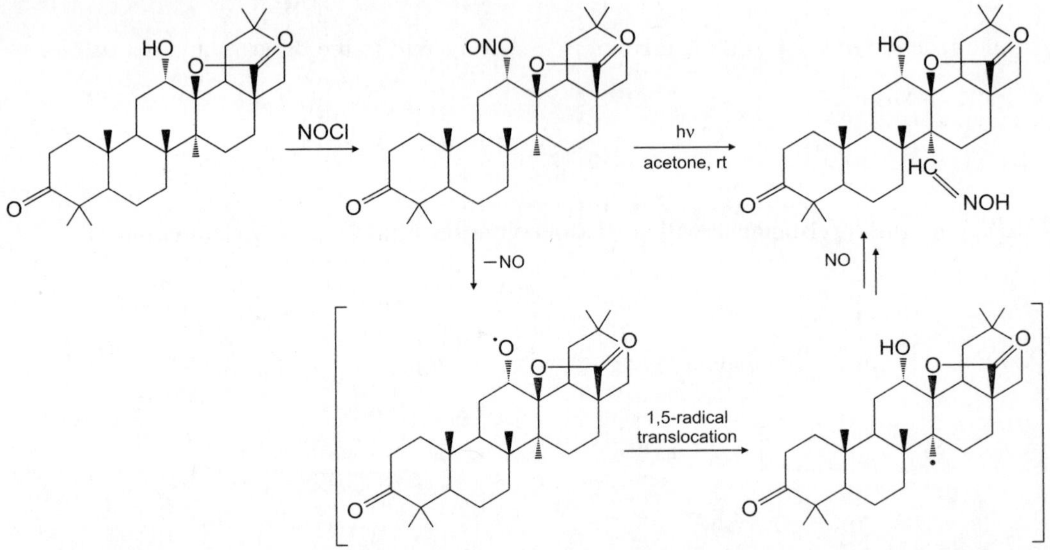

A related reaction is the Hofmann-Löffler reaction or Hofmann-Loeffler-Freytag reaction involving haloamines (N-chloroamines) proceeds by a similar mechanism and used for the synthesis of pyrrolidines.

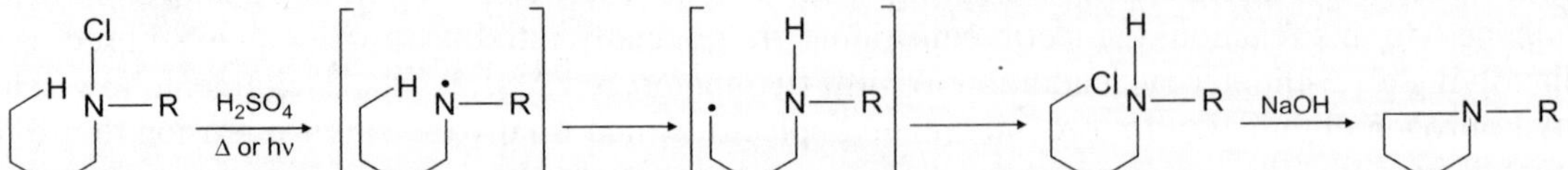

24. BARTON-ZARD REACTION OR BARTON-ZARD PYRROLE SYNTHESIS

Barton-Zard synthesis method of pyrrole synthesis involves synthesis of 2-substituted pyrroles (2-pyrrole-carboxylates or 2-sulfonyl pyrroles) by the base induced cyclisation reaction between α,β-unsaturated nitroalkene and α-isocyanoacetate. **A non-ionic base, such as 1,8 diazabicyclo [5.4.0]undec-7-ene (DBU) and guanidine is used in this reaction.**

R$_1$ = H, alkyl, aryl
R$_2$ = H, alkyl
R$_3$ = Me, Et, t-Bu
Base = KOt-Bu, DBU, guanidine bases

This method is often complicated because of poor yield and unwanted byproducts.

25. BAUDISCH REACTION

The reaction was discovered by Oskar Baudisch in 1939. The synthesis of o-nitrosophenols from the reaction in which a solution containing aqueous hydroxylamine hydrochloride and hydrogen peroxide reacts with benzene (or its derivatives), or phenol, with the assistance of copper(II) salts to give o-nitrosophenols is known as the **Baudisch reaction**.

26. BAYLIS-HILLMAN REACTION

The Baylis-Hillman reaction can be defined as a coupling reaction between an alkene activated by an electron withdrawing group often an acrylic ester and an electrophile (e.g. an aldehyde or ketone) under Lewis base (e.g. tertiary amine) catalysis.

R = aryl, alkyl, heteroaryl; R' = H, COOR, alkyl
X = O, NCOOR, NTs, NSO$_2$Ph
EWG (electron withdrawing group) =
COR, CHO, CN, COOR, PO(OEt)$_2$, SO$_2$Ph, SO$_3$Ph, SOPh

Apart from the tertiary amine the diazabicyclo[2.2.2]octane (DABCO), quinuclidin-3-one, quinuclidin-3-ol (3-QDL), tri-n-butylphosphine1 and diethylaluminium iodide can also be employed as catalysts for the reaction.

27. BECHAMP REDUCTION

It involves reduction of aromatic nitro compounds to the corresponding amines using iron and hydrochloric acid. The main shortcomings of this reaction include the slow reaction rate and costly steam distillation. The reaction was first described by Antoine Bechamp in 1854.

$$ArNO_2 + 2Fe + 6HCl \longrightarrow ArNH_2 + 2H_2O + 2FeCl_3$$

Bechamp reduction involves the use of stoichiometric amounts of finely divided iron metal and water in the presence of small amount of acid; although tin, zinc, and aluminium can also be employed.

This Bechamp reduction process suffers from several disadvantages:

- Quantity of formation of Fe-FeO sludge is very large and cannot be recycled, creating a serious effluent disposal problem.
- Fe-FeO sludge always contains the adsorbed reaction product and is difficult to filter and having a serious dumping problem.
- Erosion of reactor takes place due to Fe particles.
- Formation of side products.
- Difficult to separate the final product from the reaction mass.

Additionally, this method cannot be used for the reduction of a single nitro group in a polynitro compound, nor can it be used on substrates harmed by acid media (e.g. some ethers and thioethers), or containing additional substituents prone to being reduced (e.g. cyano, azo).

28. BECKMANN REARRANGEMENT

More than a century ago, Ernst Otto Beckmann first carried out in 1886 the conversion of an oxime into an amide. Since then, this reaction has been called after his name, the Beckmann rearrangement. The acid catalyzed rearrangement of oximes to the corresponding N-substituted amide, known as the Beckmann rearrangement.

Similar to the Hofmann and Schmidt reactions and the Curtius rearrangement, an electropositive nitrogen is formed that initiates an alkyl migration. The reaction proceeds by protonation of the oxime hydroxyl, followed by migration of the alkyl substituent *"trans"* to nitrogen. Electropositive nitrogen is formed in the reaction that initiates an alkyl migration. The reaction accomplishes in one stroke, i.e. the 1,2 shift of alkyl group (R) onto electron deficient nitrogen and the cleavage of N-O bond occurs simultaneously or both the cleavage of a carbon-carbon bond and the formation of a carbon nitrogen bond. Alkyl group which is 'anti' to the –OH group on nitrogen always undergoes 1, 2 shift.

Protic acids, Lewis acids, anhydrides, and acid halides, e.g. conc.H_2SO_4, HCl, PCl_5, PCl_3, $SOCl_2$, ZnO, SiO_2, PPA (poly phosphoric acid), etc. are commonly employed in Beckmann rearrangement which promote efficiently this rearrangement. Reaction is mostly applicable for ketoximes while aldoximes being less reactive are seldom used in the reaction. However, synthesis of isoquinoline from cinnamaldoxime is an example of Beckmann rearrangement where an aldoxime is involved.

Cinnamaldoxime
(an aldoxime)

Isoquinoline

This rearrangement represents a powerful method particularly for manufacturing ε-caprolactam in the chemical industry.

Cyclohexanone   Cyclohexanoxime   Caprolactam

It is also utilized for the synthesis of azithromycin. The reaction generally requires high reaction temperature and strongly acidic and dehydrating media. Thus, the reaction can lead to large amounts of by products and preclude its application to sensitive substrates.

The Semmler-Wolff- Schroeter reaction or chemically known as *oxime aromatization* is a peculiar deviation from the Beckmann rearrangement in that the oximes of α, β-unsaturated alicyclic ketones from the expected Beckmann rearrangement, producing instead aromatic amines in the presence of the Beckmann mixture [$(CH_3CO)_2O + CH_3COOH + HCl$].

29. BENARY REACTION

It provides an effective method for the synthesis of α,β-unsaturated carbonyl compounds from β-(N,N-dialkylamino)-vinyl ketone and Grignard reagent by 1,4-addition followed by hydrolysis of the reaction intermediate and elimination of the dialkylated amine. The Benary reaction was discovered in 1931 by Erich Benary. α,β-unsaturated aldehydes and α, β-unsaturated esters as well as poly-unsaturated ketones and aldehydes can also be formed by this reaction.

30. BENKESER REDUCTION

It involves catalytic hydrogenation of polycyclic aromatic hydrocarbons, especially naphthalene using lithium or calcium metal and in low molecular weight amines. It has been found that low molecular weight monoamines, particularly methylamine and ethylamine, are excellent media for reduction with lithium.

$$R = CH_3, CH_2CH_3, CH_2CH_2CH_3, CH_2CH_2NH_2$$

This reaction is a modification of the Birch reduction. Reduction in low molecular weight amines (in the absence of alcohol additives) furnishes more extensively reduced products than are obtained under Birch conditions (metal, NH_3, ROH). Apart from this, the Birch reduction with sodium is difficult when run on kilogram scale, whereas the Benkeser reduction gives better yield of products. With appropriate reaction conditions, the extent of reduction and selectivity can be controlled in the Benkeser reduction.

31. BENZIDINE REARRANGEMENT (SEMIDINE REARRANGEMENT)

When hydrazobenzene is treated with acids, it rearranges to give 4,4'-diaminobiphenyl (benzidine). This reaction is called **benzidine rearrangement** and is a general reaction for the N,N'-diarylhydrazines. Usually 4,4'-diaminobiaryl (benzidine) is the major product in the rearrangement, but if the hydrazobenzene contains a para substituent, then the favoured product is p-aminodiphenylamine and such rearrangement is known as the **semidine rearrangement**.

Benzidine Rearrangement

(Hydrazobenzene)
N,N'-diphenylhydrazine

(Benzidine)
Biphenyl-4,4'-diamine

Semidine Rearrangement

(p-aminodiphenylamine)
N-phenylbenzene-1,4-diamine

The benzidine rearrangement to form benzidine is due to sigmatropic shift ([5,5]-shift) and is concerted in nature. Usually in benzidine rearrangement the major product is the 4,4'-diaminobiaryl, but four other products (as given below) 2,4'-diaminobiaryl, 2,2'-diaminobiaryl and the o- and p-arylaminoanilines may also be produced in minor amounts.

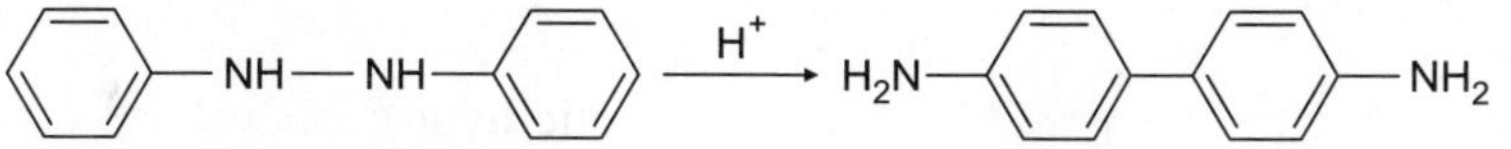

N,N'-diphenylhydrazine

Biphenyl-4,4'-diamine

Major product

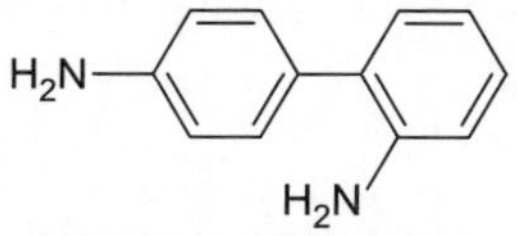

Biphenyl-2,4'-diamine

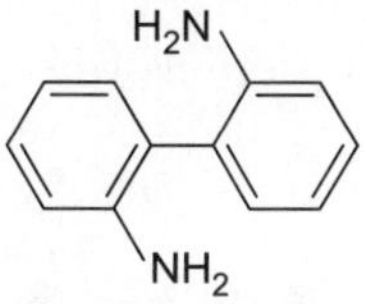

Biphenyl-2,2'-diamine

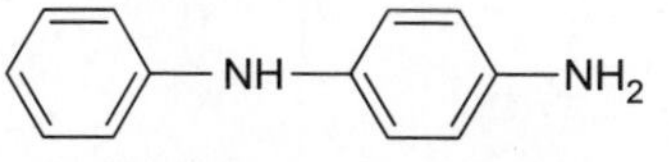

N-phenylbenzene-1,4-diamine

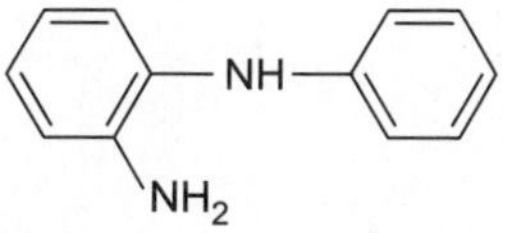

N-phenylbenzene-1,2-diamine

This reaction is of limited synthetic importance because of the many side products and low yields.

32. BENZIL-BENZILIC ACID REARRANGEMENT OR BENZILIC ACID REARRANGEMENT

Benzil-Benzilic acid rearrangement discovered in 1938 by Justus Liebig and it is one of the first C-C bond migration reaction. It involves base-induced rearrangement reaction of an α-diketone yielding an α-hydroxycarboxylic acid.

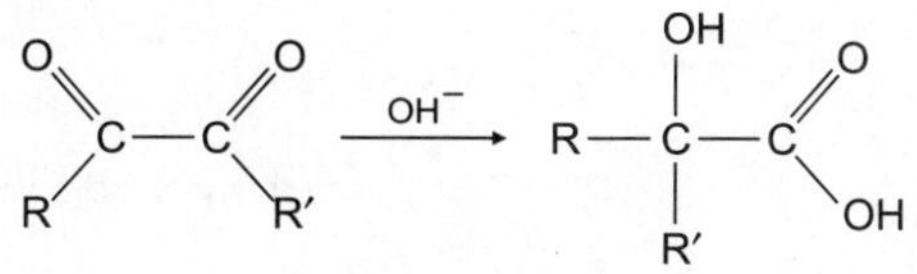

The name comes from the well known example of this type of conversion of benzyl into benzilic acid.

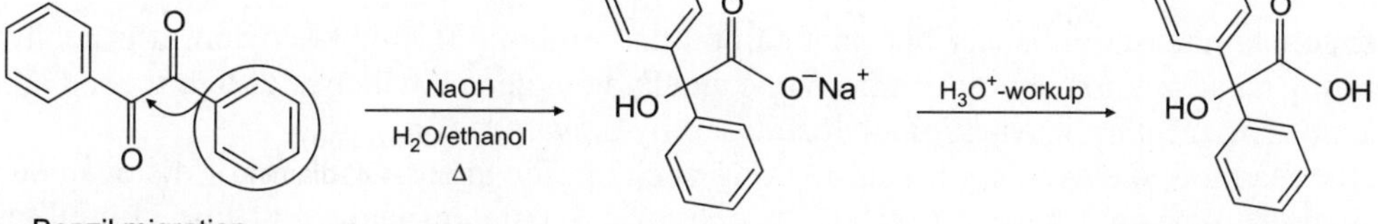

Benzil migration

Benzilic acid

The diketones which do not have enolizable protons give the best yields in this reaction. The benzilic acid rearrangement of cyclic diketones leads to ring contraction product:

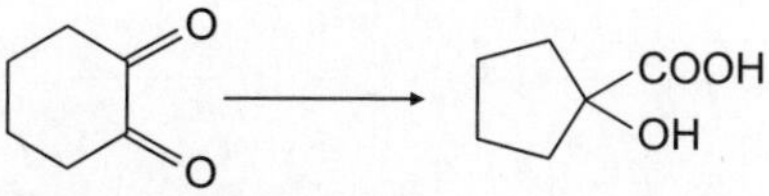

The *benzilic ester rearrangement*, a variant to the reaction, use an alkoxide instead of hydroxide as nucleophile gives the corresponding ester directly. The alkoxide such as OEt^- or $OCHMe_2^-$ are readily oxidized and reduce the benzil to benzoin.

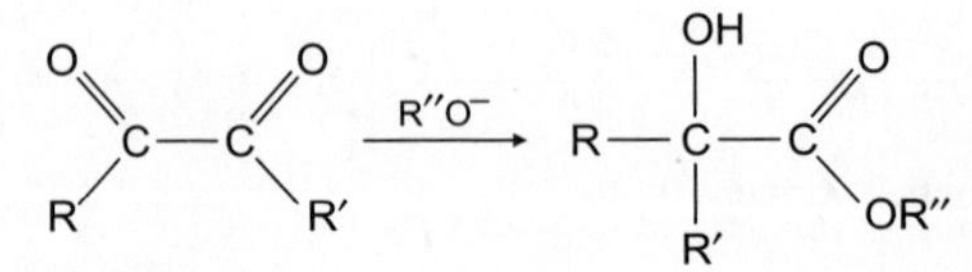

Another variation of this reaction known so far is the so-called **D-Homo rearrangement of steroids**. Occurs in certain steroids where a cyclopentane ring expands to a cyclohexane ring with added base.

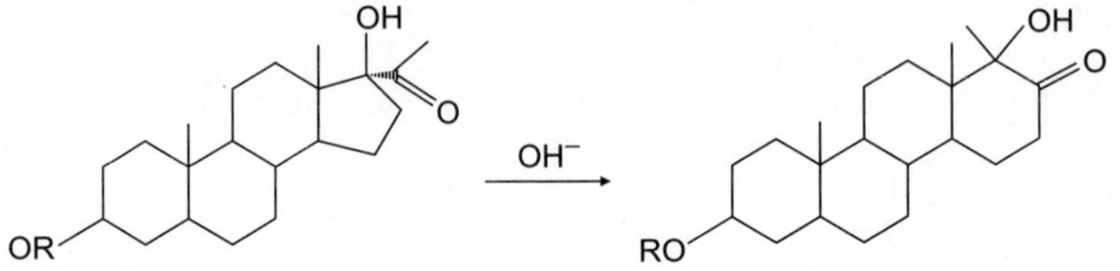

33. BENZOIN CONDENSATION

Two molecules of an aldehyde condense in presence of cyanide ions (CN^-) to give α-hydroxy ketone is known as the benzoin condensation.

The reaction was developed in 1832 by Justus von Liebig and Friederich Woehler during their research on bitter almond oil. Later in the 1830s, the catalytic version of the reaction was developed by Nikolay Zinin, and further in 1903 the reaction mechanism was proposed by A.J. Lapworth.

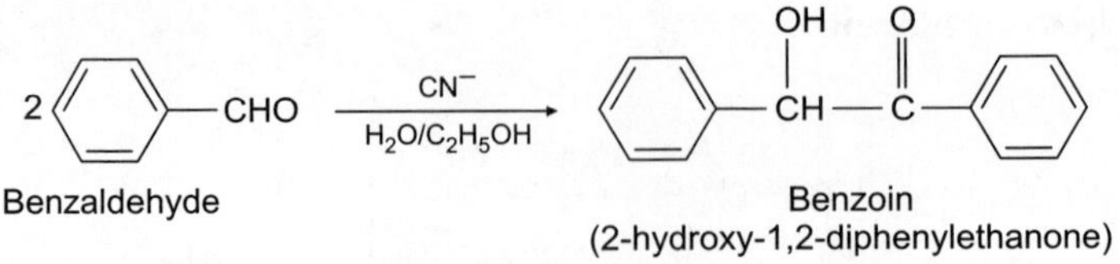

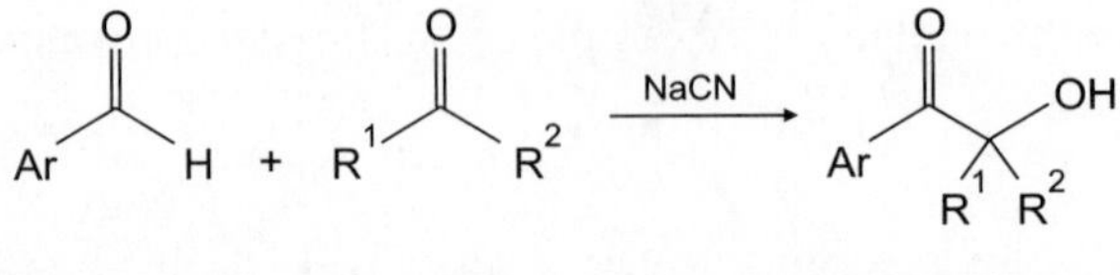

Benzoin condensation is an important carbon-carbon bond forming reaction. It is achieved by generating an acyl anion equivalent from one aldehyde molecule which adds to a second aldehyde molecule. The reaction is traditionally catalysed by a cyanide ion.

In the reaction, the cyanide acts as a strong nucleophilic agent. Addition of the cyanide ion to the aldehyde carbonyl forms a stable cyanohydrins. This carbanion, in turn, functions as a nucleophile and attacks a second molecule of benzaldehyde. This condensation intermediate experiences proton transfer and loses the cyanide ion producing benzoin.

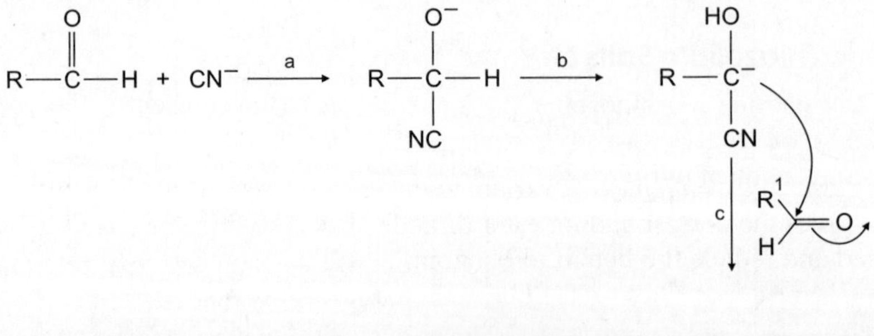

The uniquely successful role of cyanide ion in catalyzing benzoin reaction is due to its four qualities, namely:

  (i) High nucleophilic activity (stage a),
 (ii) Facilitating the proton transfer (stage b),
(iii) Ability to stabilize negative charge in active aldehyde intermediate **N** (stage c), and
(iv) Ability to depart finally (stage e).

In principle, any chemical entity that incorporates all these four features should be capable of bringing about benzoin condensation. Compared to cyanide ion, the **OH⁻ ion is a poor leaving group.** At the end of the mechanism, whatever attacked the aldehyde needs to be able to leave. A hydroxide is a horrible leaving group, whereas the cyanide is kicked with relative ease.

A reaction completely analogous to benzoin condensation occurs in our body, which however neither involves cyanohydrin intermediate nor is catalysed by cyanide ion. It is catalysed by the thiazolium moiety of the co-enzyme thiamine pyrophosphate (TPP).

Vitamin B$_1$ (Thiamine)

Thiamine pyrophosphate (TPP)

Thiozole ring

Pyrophosphate group

Pyrimidine ring

Vitamin B$_1$ and Thiazolium Salts as Catalysts

Vitamin B$_1$, i.e. thiamine pyrophosphate, TPP, is a co-enzyme present in our body, and other living organisms. TPP catalyses several reactions that include decarboxylation of pyruvic acid to acetaldehyde, conversion of pyruvic acid to acetoin, transfer of 2-carbongroup from sedoheptalose-phosphate to glyceraldehyde-3-phosphate to produce xylose-5-phospate (an acyloin reaction), which involve acyl ion or its equivalent intermediate as in cyanide catalysed benzoin reaction.

The benzoin condensastion is an example of "ompoulung"—a case in which the usual reactive polarities have been reversed. The carbonyl carbon, usually electrophilic, has been made nucleophilic. One might imagine that this reaction has two major problems:

(i) Competing aldol condensation

(ii) Use of the nasty reagent, cyanide

Nowadays, thiamine hydrochloride is being used rather than cyanide ion. Apparently, the thiamine-catalyzed reaction is a bit slower, but then the catalyst is edible.

- Benzoin is mainly used as a precursor to benzil, which is a photoinitiator.
- *A **photoinitiator** is any chemical compound that decomposes into free radicals when exposed to light.
- Benzoins can be converted into 1,2-diphenyl amino ketones or 1,2-diphenyl amino alcohols, which have shown tumor-necrotizing activities.
- The Benzoin condensation is a coupling reaction between two aldehydes that allows the preparation of α-hydroxyketones. The first method was only suitable for the conversion of aromatic aldehydes.

34. BERGIUS PROCESS

The Bergius process involves formation of petroleum-like liquid hydrocarbons for use as synthetic fuel by hydrogenation of coal at high temperatures and pressures with or without catalysts.It was first developed by Friedrich Bergius in 1913.

35. BERGMAN CYCLIZATION OR BERGMAN REACTION

The Bergman cyclization involves thermal or photochemical cycloaromatization of enediynes in the presence of a suitable H • donor (i.e. 1, 4-cyclohexadiene) to effect the construction of substituted arenes. This reaction is named after Robert George Bergman (an American chemist).

The reaction proceeds by a thermal procession or pyrolysis at temperature around 200°C, forming a short-lived and very reactive 1,4-benzenediyl diradical species which reacts with a hydrogen donor (H •) to give the corresponding arenes. However, despite the fact that both double and triple bonds display rich photoreactivities, comparatively fewer reports on the photochemistry of enediynes exist in the literature.

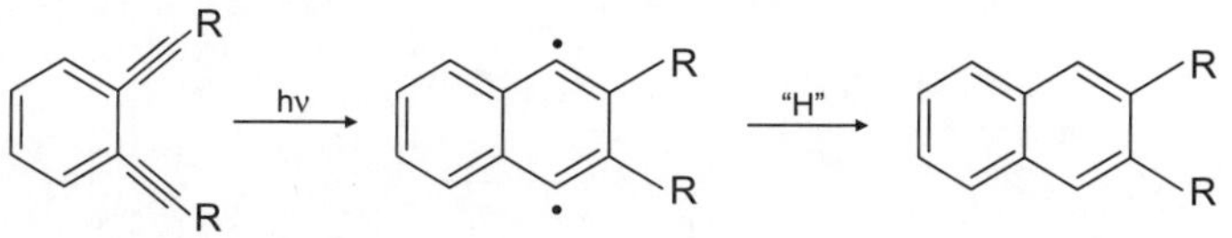

36. BERGMANN AZLACTONE PEPTIDE SYNTHESIS

Bergmann azlactone peptide synthesis involves synthesis of peptides by aminolysis of azlactones from α-amino acids or corresponding esters. In this reaction, an acetylated amino acid and an aldehyde are converted into an azlactone (5-oxazolone) with an alkylene side chain. Reaction of azalactone with a second amino acid results in ring opening and formation of an acylated unsaturated dipeptide, followed by catalytic hydrogenation and finally hydrolysis to give dipeptide.

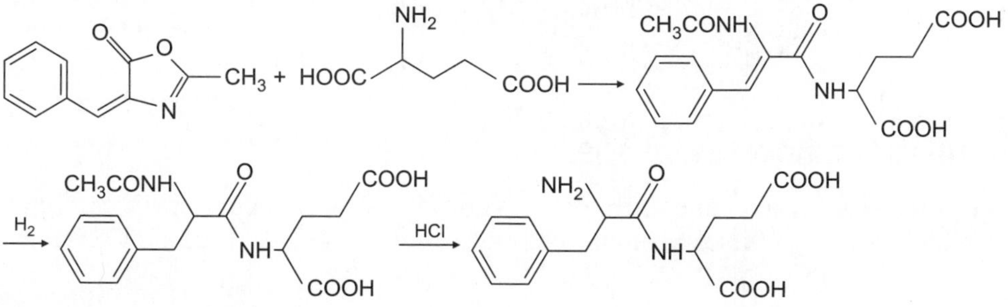

37. BERGMANN DEGRADATION

The Bergmann degradation is a series of chemical reactions designed to remove a C-terminal amino acid residue (single amino acid) from the carboxylic acid end of a peptide and leaves the rest of the peptide in the form of its amide. It involves stepwise degradation of polypeptides involving benzoylation, conversion to azides which undergoes Curtius rearrangement in the presence of benzyl alcohol to give benzyl carbamate which undergo catalytic hydrogenation and hydrolysis to the amide of the degraded peptide.

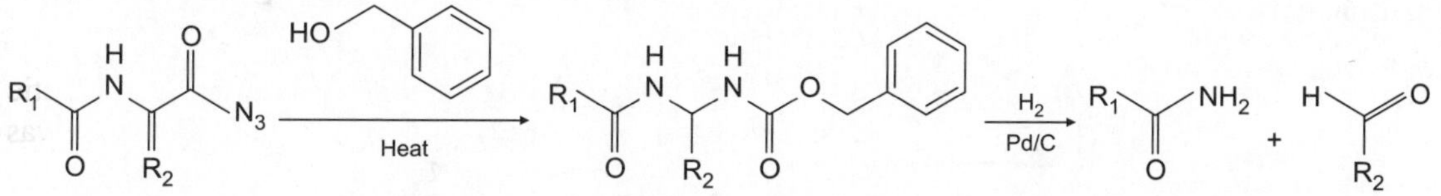

The reaction is useful in the elucidation of peptide sequences via carboxyl terminal (or C-terminal).

38. BERGMANN-ZERVAS CARBOBENZOXY METHOD

The method was first described by Max Bergmann and Leonidas Zerwas in 1932. This method is used to protect an amine functional group from electrophilic attacks in various organic reactions especially in peptide synthesis.

The synthesis of peptide by this method involves conversion of the amino acid into its N-carbobenzoxy derivative with benzyl chloroformate, activation of the carboxyl group, formation of peptide with the second amino acid, and deprotection via hydrogenolysis.

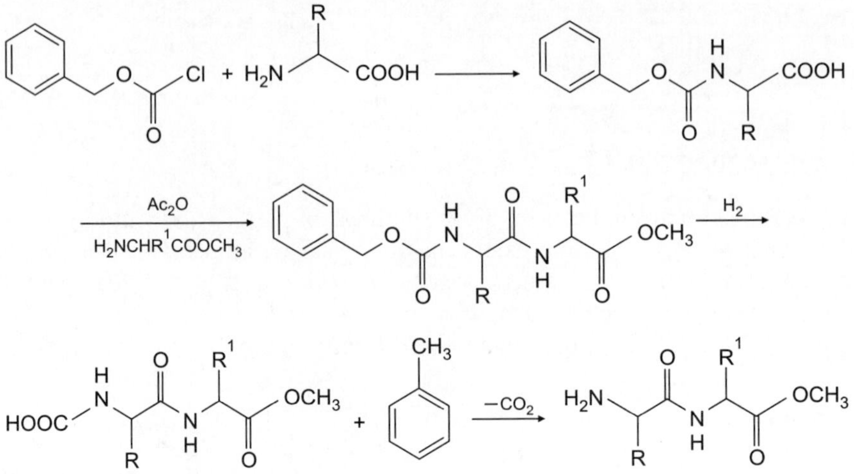

39. BERNTHSEN ACRIDINE SYNTHESIS

Bernthsen synthesis involves the chemical reaction of a diphenylamine with carboxylic acid (or acid anhydride) in the presence of zinc chloride resulting in the formation of acridine.

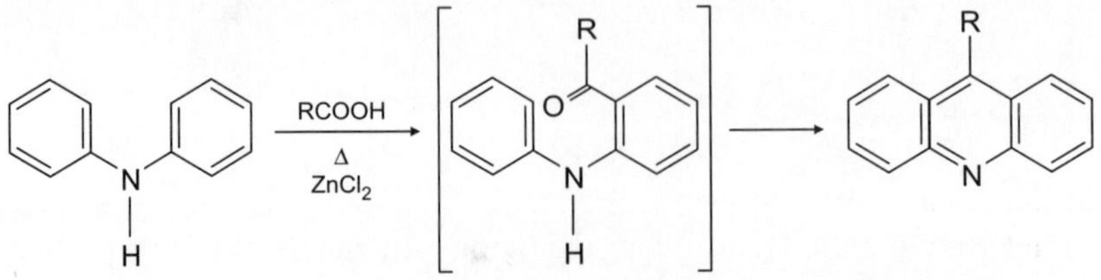

40. BETTI REACTION

The Betti reaction is a special case of the Mannich reaction leading to the formation of α-aminobenzylphenols by reaction of aromatic aldehydes, primary aromatic or heterocyclic amines and phenols.

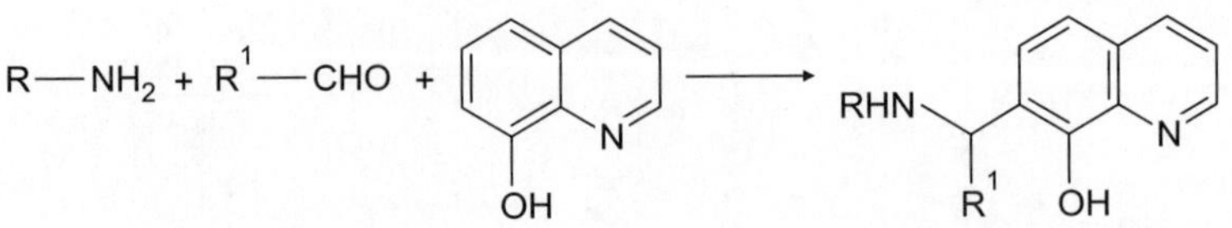

R, R^1 = Aryl, heterocyclic

41. BIGINELLI REACTION

The combination of an aldehyde, β-keto ester and urea under acid catalysis to give a dihydro-pyrimidine was first reported by Pietro Biginelli in 1893, referred to as the Biginelli reaction.

This acid-catalyzed, three-component reaction between an aldehyde, a β-keto ester and urea constitutes a rapid and facile synthesis of dihydropyrimidones, which are interesting compounds with a potential for pharmaceutical application.

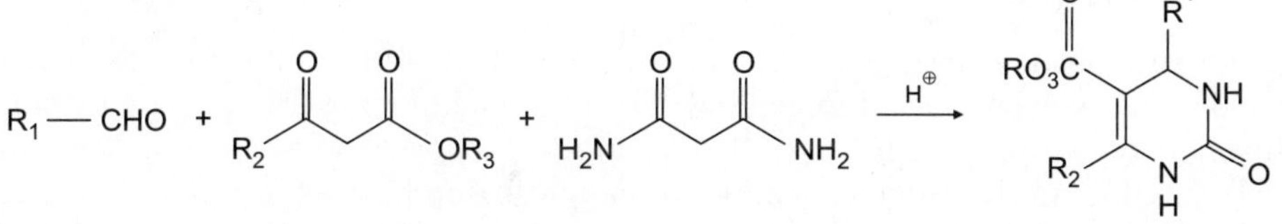

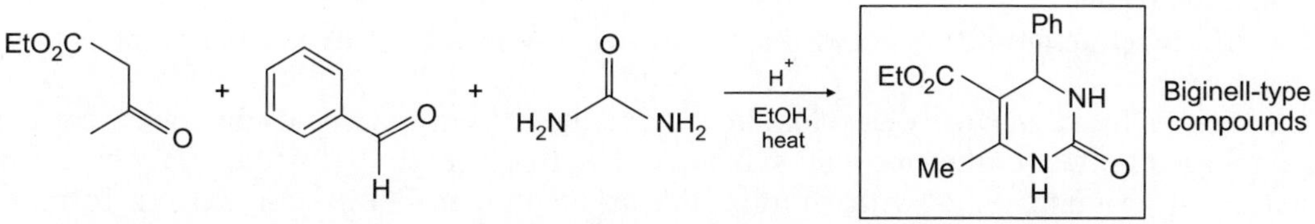

First, urea condenses with the aldehyde to produce an imine intermediate upon elimination of water. This intermediate then reacts with the enol form of the β-dicarbonyl compound to give an intermediate ureide, which cyclizes intramolecularly with loss of water to give 3,4-dihydro-pyrimidinones.

The reaction can be catalyzed by Brønsted acids and/or by Lewis acids such as copper(II) trifluoroacetate hydrate and boron trifluoride. Dihydropyrimidinones are interesting compounds widely used in the pharmaceutical industry as calcium channel blockers and antihypertensive agents.

Certain modifications to the Biginelli reaction were brought about by the need for better yields such as Atwal and co-workers introduced a modification to the original Biginelli reaction that affords high product yields and the preparation of previously inaccessible dihydropyrimidines. The Atwal modification involves reaction of preformed unsaturated keto esters with a protected urea.

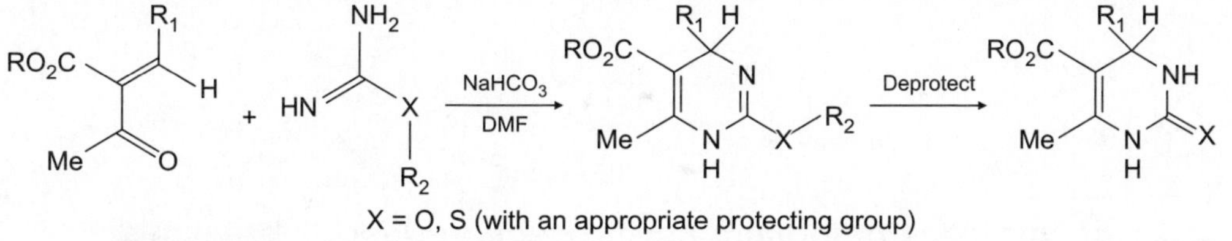

X = O, S (with an appropriate protecting group)

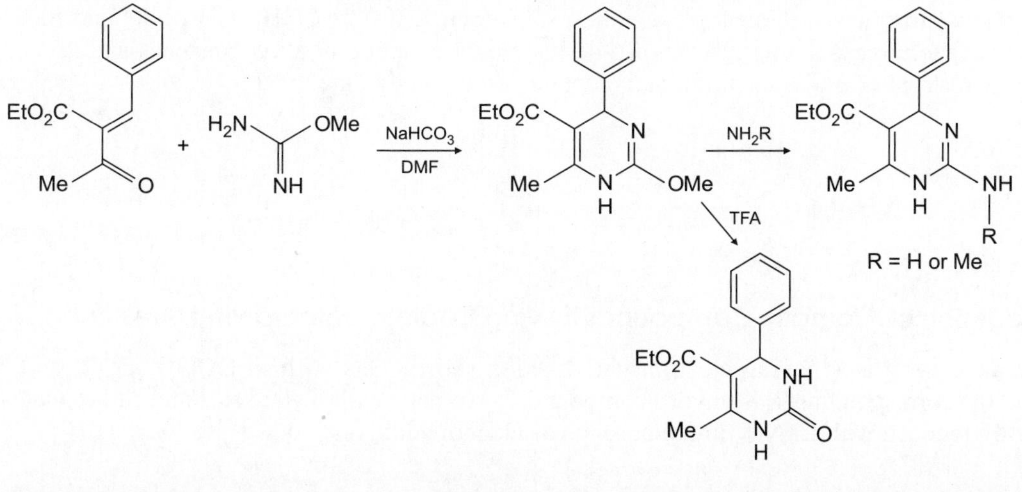

R = H or Me

Atul Kumar has reported first enzymatic synthesis for Biginelli reaction via yeast catalysed protocol in high yields.

42. BIRCH REDUCTION

The reduction of aromatic rings by solution of alkali metals in liquid ammonia was discovered by Wooster and Godfrey, who reacted toluene with sodium in ammonia followed by the addition of water. However, the real development of this reaction was worked out by Birch. This reaction is generally referred to as the Birch reduction; although in some cases it is simply called metal-ammonia reduction.

The Birch reduction offers access to substituted 1,4-cyclohexadienes (partial reduction of aromatic rings).

The reaction has been widely used for alicyclic synthesis from benzenoid compounds but it has a less profound effect on heterocyclic synthesis. The Birch reduction is, therefore, an organic reduction of aromatic rings with sodium and an alcohol in liquid ammonia to form 1,4-cyclohexadienes. Lithium and potassium can substitute for sodium, and alcohols are ethanol and tert-butanol.

Benzene $\xrightarrow[C_2H_5OH]{Na,\ NH_3}$ 1,4-dihydrobenzene

Naphthalene $\xrightarrow[C_2H_5OH,\ (C_2H_5)O]{Na,\ NH_3}$ 1,4,5,8-tetrahydronaphthalene

The catalytic hydrogenation of aromatics lead to the corresponding fully hydrogenated compound while only a partial and controlled reduction takes place by the Birch reduction.

Benzene $\xrightarrow[\substack{Catalytic \\ hydrogenation}]{3H_2}$ Cyclohexane

Reduction of Aromatic Compounds having Electron-Donor Substituents

Aromatic substrates with electron-donating substituents such as OCH_3, CH_3, NR_2 and $Si(CH_3)_3$ provides the corresponding 2,5-dihydro derivatives. Reduction of alkyl benzenes and aryl ethers require a stronger acid than ammonia; alcohols are typically employed.

$\xrightarrow[ROH]{M,\ NH_3}$ R = CH_3, OCH_3, NR_2, etc.

Reduction of Aromatic Compounds having Electron-Withdrawing Substituents

Aromatic substrates having electron-withdrawing substituents such as COOH, COR, and NO_2 yield the corresponding 1,4-dihydro compounds. Aromatic carboxylic acids and carboxylates are readily reduced with Li/NH_3 in the absence of alcohol additives.

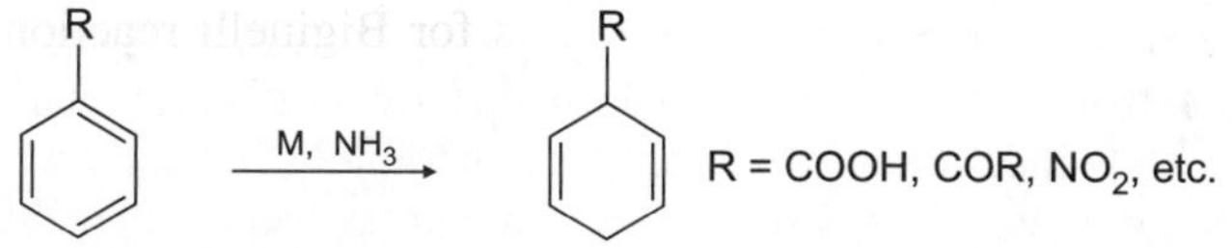

R = COOH, COR, NO$_2$, etc.

In Birch reduction, the alkali metals (e.g. Li, Na, K) dissolved in liquid ammonia work as solvated electrons (M$^+$ e$^-$ are in solution) which are responsible for the reduction. These solvated electrons are added to the aromatic ring and give a radical anion. The added alcohols (EtOH or t-BuOH) are used as protonating agents and supply a proton to the anions. Alcohol in the reaction also suppress the formation of amide, NH$_2^-$ ion (highly basic in nature), which may otherwise isomerise 1,4-diene to more stable 1,3-diene.

- Under the (relatively controlled and mild) reaction conditions, reduction stops at the dihydro stage.
- The rate of reduction is influenced by the substituents on the ring—as the intermediates are negatively charged, the rate is, not surprisingly, increased by electron-withdrawing substituents.
- Substituents also dictate the regiochemistry of protonation.

Mechanism:

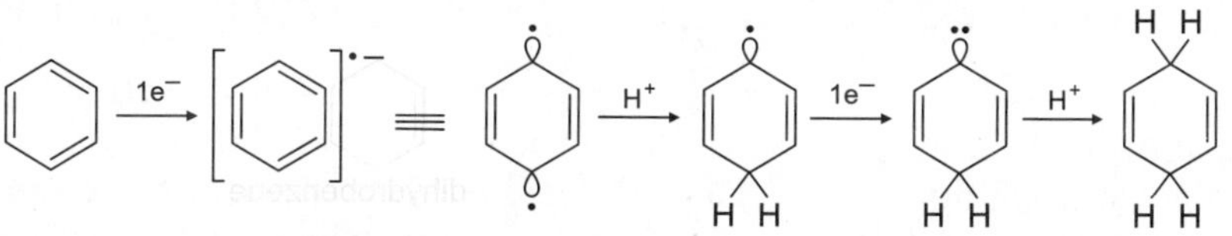

Why the 1,3-diene is not formed, even though it would be more stable through conjugation?

In the Birch reduction, 1,4-diene (less stable, non-conjugated diene) is formed as major product (about 80%) along with minor quantities of 1,3-diene (more stable, conjugated diene). The unusual formation of lesser stable 1,4-diene as major product is because the reaction is kinetically controlled rather than thermodynamically. In kinetically controlled reactions, the product with the lowest-energy transition state predominates while in thermodynamically controlled reactions, the lowest-energy product predominates. The valence bond terms also explain, electron-electron repulsions in the radical anion (the intermediates in the reaction) will preferentially have the nonbonding electrons separated as much as possible in a 1,4-relationship.

Alternatives to the Reaction

- Metals: Li, K, Na occasionally Ca or Mg.
- Co-solvents: Diethyl ether, THF, glymes.
- Proton sources (where appropriate): t-BuOH and EtOH are most common, also MeOH, NH$_4$Cl, and water.

Uses and Aplication

The most common applications for this reaction include reduction of alkynes, conjugated carbonyls and Birch-type reductions of aromatic rings.

- The Birch reduction is a very powerful tool for the hydrogenation of benzene and its derivatives to yield the cyclohexa-1,4-dien compounds.
- The Birch reduction is an alternative to hydrogenation that yields cyclohexadienes. It has the potential for widespread use in the synthesis of drugs and complex natural products.
- The Birch reaction is attractive because of the high yields, thus it is one of the most highly used synthetic reactions in organic chemistry.

- The reaction provides steric control.
- Catalytic hydrogenations of benzene and benzene derivatives usually yield completely hydrogenated products, since alkenes that are intermediately formed are more easily reduced than aromatic rings. However, a selective reduction can be accomplished by employing a one-electron transfer mechanism as with the **Birch reduction** that yields cyclohexadienes.
- **Reduction of alkynes** – The Birch reduction provides a useful route to (E)-alkenes. Alkynes are selectively converted into trans alkenes when they are reduced by a solution of sodium (or lithium) in liquid ammonia that contains stoichiometric amounts of an alcohol, such as ethanol. As in the case of Birch reduction of aromatic systems, the first step of this reaction is a one-electron transfer into an antibonding π orbital of the alkyne, which yields a radical anion. Subsequently, protonation of the radical anion, an additional one-electron transfer, and a concluding protonation yield a trans alkene.

43. BISCHLER-MÖHLAU INDOLE SYNTHESIS OR BISCHLER REACTION

The Bischler-Möhlau indole synthesis involves preparation of 2-substituted indoles by heating ω-haloketone, ω-hydroxyketone, or ω-anilinoketone with excess aniline. The reaction is based on an intramolecular electrophilic cyclization and proceeds via the cyclization of a 2-arylaminoketone intermediate. This reaction can also be applied for the synthesis of benzofuran and benzothiophene.

44. BISCHLER-NAPIERALSKI REACTION (BISCHLER-NAPIERALSKI CYCLIZATION)

This reaction is named after August Bischler and Bernard Napieralski in 1893. The Bischler-Napieralski synthesis involves the synthesis of 3,4-dihydroisoquinolines by the cyclodehydration of β-phenylethanamides using a dehydrating agent. This reaction is one of the important method for construction of an isoquinoline skeleton.

β-aryl ethylamine   Isoquinoline

The reaction is an intramolecular electrophilic aromatic substitution and is carried out in refluxing acidic conditions with a dehydrating agent. It allows the synthesis of 3,4-dihydroisoquinolines from the α-ethylamides of electron-rich arenes. The Bischler-Napieralski reaction involves an initial dehydration step of the amide followed by cyclization.

The reaction works well with electron-rich aromatics such as dimethoxy compounds. A monomethoxybenzene works to a much lesser extent. Phosphorus oxychloride ($POCl_3$) is widely used as the cyclization reagent, however, phosphorus pentoxide (P_2O_5), anhydrous zinc chloride ($ZnCl_2$), etc. are also used for the purpose.

The Bischler-Napieralski phosphoryl chloride ($POCl_3$) is widely used and cited for this purpose. Additionally, $SnCl_4$ and BF_3 etherate have been used with phenethylamides, while Tf_2O and polyphosphoric acid (PPA, a.k.a. Eaton's reagent) have been used with phenethylcarbamates. For reactants lacking electron-donating groups on the benzene ring, phosphorus pentoxide (P_2O_5) in refluxing $POCl_3$ is most effective. Depending on the dehydrating reagent used, the reaction temperature varies from room temperature to 100 °C. Dehydration reagents such as PCl_5, $POCl_3$, $SOCl_2$, $ZnCl_2$ can be used to promote loss of the carbonyl oxygen.

45. BLAISE KETONE SYNTHESIS; BLAISE-MAIRE REACTION

Formation of ketones by treatment of acid halides with organozinc compounds is known as the Blaise ketone synthesis.

The reaction also works with organocuprates.

In **Blaise-Maire reaction**, β-hydroxy acid chlorides are converted into β-hydroxyketones, which provide α,β-unsaturated ketones using boiling dilute sulfuric acid.

46. BLAISE REACTION

The Blaise reaction was discovered by Blaise in 1901. It closely resembles the Reformatsky reaction. It describes zinc-mediated chemical reaction of nitriles with α-haloesters. Either β-enamino esters or β-keto esters are obtained depending on the reaction conditions. Work up with 50% aq. K_2CO_3 delivers β-enamino esters in the organic phase. An additional acidification of the product phase with 1 M aq. HCl hydrolyzes the β-enamino esters to give β-keto esters.

The Blaise reaction can be truncated to produce β-amino-α,β-unsaturated esters, which are useful for the synthesis of heterocycles and β-amino acids.

Its use in synthetic chemistry is limited due to problems of low yield, narrow scope and undesired side reactions, such as self-condensation of the α-haloesters. However, these early problems are greatly improved by recent modifications, such as use of activated zinc, tetrahydrofuran as solvent and an excess of α-haloester, etc. Additional use of ultrasonic assistance and zinc oxide addition in the Blaise reaction are found more advantageous.

47. BLANC REACTION (CHLOROMETHYLATION) OR BLANC CHLOROMETHYLATION

The Blanc reaction, comparable to Friedel-Crafts alkylation, describes preparation of chloromethylated arenes by the introduction of chloromethyl group into aromatic rings when aromatic compounds are treated with formaldehyde and hydrogen chloride in the presence of zinc chloride.

The Blanc reaction closely resembles to Friedel-Crafts alkylation since the rate-determining step is the electrophilic aromatic substitution as in the Friedel-Crafts reaction and a Lewis acid catalyst is used in both the reactions.

Blanc reaction is normally performed in polar solvents, e.g. CCl_4, dimethyl formamide (DMF), N, N-dimethyl acetamide (DMAc) or N-methyl pyrrolidone (NMP), and these reactions are usually carried out in a heterogeneous medium in the presence of Lewis acid, e.g. $ZnCl_2$ or $SnCl_4$. The use of Lewis acid is important to activate formaldehyde. Formation of the electrophilic species can be formulated as follows:

$$CH_2O + HCl + ZnCl_2 \longrightarrow CH_2OH^+ ZnCl_3^-$$

However, stoichiometric amount of Lewis acid to substrate is required. These Lewis acid catalysts, in general, are not recommended for the inherent problems of corrosiveness, high susceptibility to water, difficulty in catalyst recovery, environmental hazards and waste control after the reaction.

This reaction has been widely applied in synthesis of a variety of pharmaceuticals, agrochemicals, dyes, flavors and fragrances, monomers, additives and modifier of polymer, in which chloromethyl group can easily be changed to CH_2OH, CHO, CH_2CN, CH_2NH_2, CH_3, CH_2R, etc.

48. BLANC REACTION-BLANC RULE

Blanc rule describes cyclization of dicarboxylic acids on heating with acetic anhydride. Either cyclic anhydrides or ketones are obtained in this reaction depending on the respective positions of the carboxyl groups, for example; 1,4- and 1,5-dicarboxylic acids on heating forms their respective anhydride, while diacids in which the carboxy groups are in 1,6 or further removed positions (i.e. 1,7-dicarboxylic acids) yield cyclic ketones. This generalization is called the Blanc rule.

For example, glutaric acid (1,5 diacid) and succinic acid (1,4 diacid) on treatment with acetic anhydride form an intramolecular anhydride while adipic acid (1,6 diacid) under the same treatment forms cyclopentanone.

Glutaric acid

Succinic acid

Adipic acid

49. BODROUX-CHICHIBABIN ALDEHYDE SYNTHESIS

The Bodroux-Chichibabin aldehyde synthesis involves synthesis of aliphatic or aromatic aldehydes from the reaction of a Grignard reagent and ethyl orthoformate. Other orthoformates have also been used for this reaction. In the reaction, aldehyde is obtained with one carbon longer than the corresponding Grignard reagent.

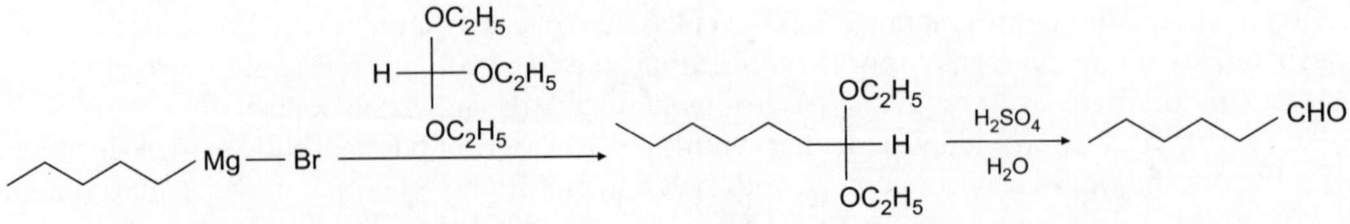

50. BODROUX REACTION OR BODROUX AMIDE SYNTHESIS

An aminomagnesium halide, formed when a primary or secondary amine reacts with a Grignard reagent at room temperature, reacts with a simple aliphatic or aromatic ester to yield a substituted amide.

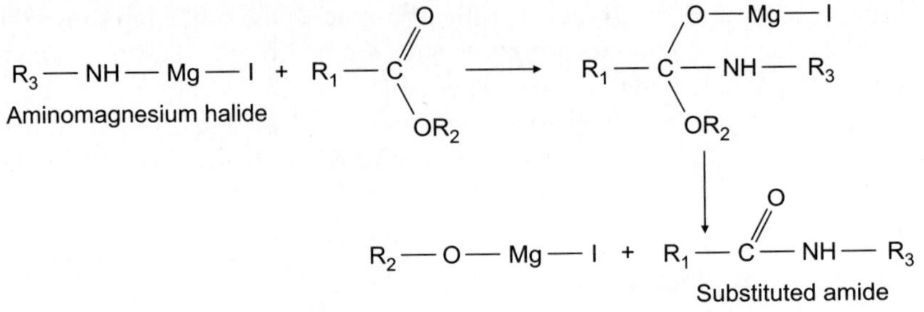

51. BOGERT-COOK SYNTHESIS

The Bogert-Cook synthesis (1933) involves condensation of β-phenylethylmagnesium bromide with cyclohexanones followed by cyclodehydration of the tertiary alcohol with concentrated sulfuric acid with formation of octahydrophenanthrene derivatives and accompanied by a small amount of spirane. When an alkyl or alkoxyl group is at the meta position to the side chain in the phenyl ring, cyclization can proceed to yield the 5- or the 7-substituted phananthrene derivatives. Reactivity of the aromatic nucleus at the cyclization position has been found to be important factor for the production of the intermediates in this reaction.

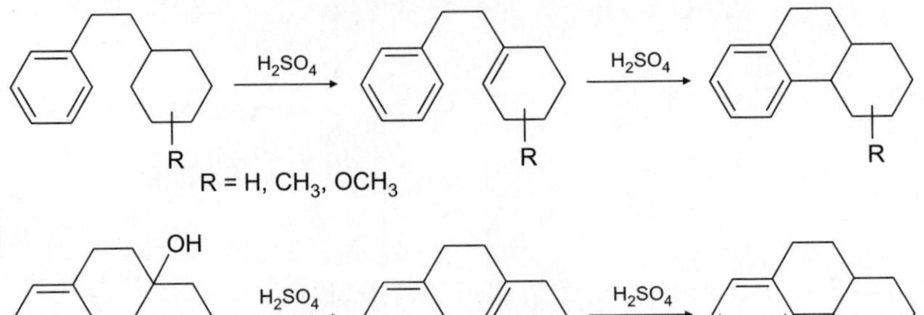

52. BOGER PYRIDINE SYNTHESIS

Boger pyridine synthesis is a [4+2] inverse electron demand aza-Diels–Alder reaction between enamines and 1,2,4-triazine. The 1,2,4-triazenes readily undergo inverse demand Diels-Alder reaction with electron-rich dienophiles with well-defined regioselectivity. This makes them attractive precursors to pyridines, as addition across C-3/C-5 is favored for all dieneophiles, with the exception of some ynamines. The most popular version of this reaction uses a pyrrolidine enamine or a ketone and pyrrolidine as the dienophile; this is called the Boger pyridine synthesis.

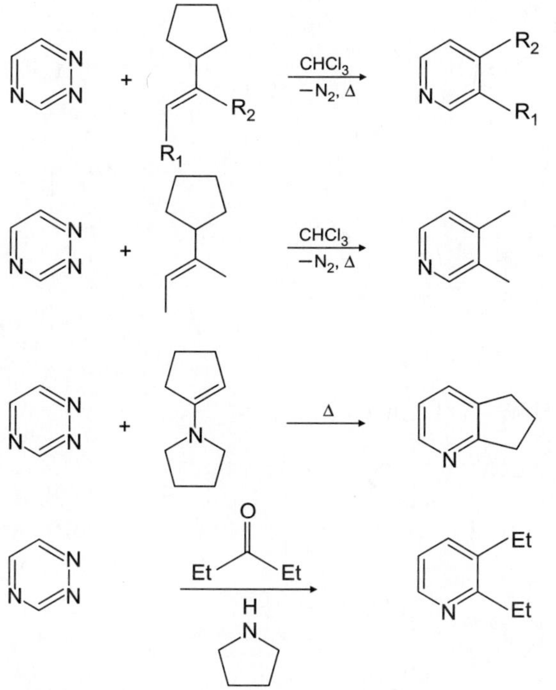

53. BOHN-SCHMIDT REACTION

The introduction of hydroxyl groups into anthraquinone (i.e. hydroxylation of anthraquinones) molecules containing at least one hydroxyl group by treatment with fuming sulfuric acid or sulphuric acid and boric acid by the oxidizing action of fuming sulfuric acid (i.e. oleum) is known as Bohn-Schmidt reaction. This reaction has a wide application in the preparation of dyes.

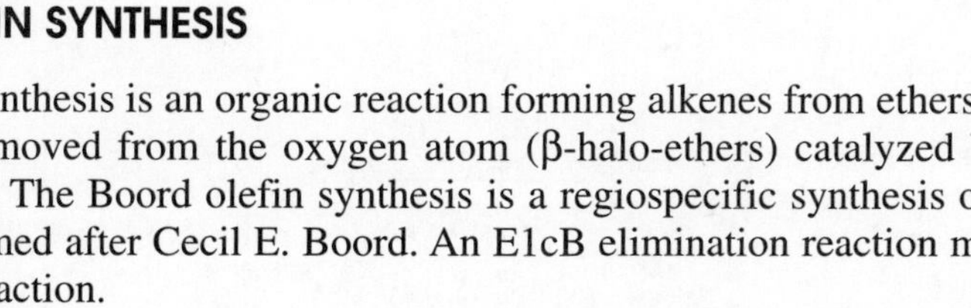

54. BOORD OLEFIN SYNTHESIS

The Boord olefin synthesis is an organic reaction forming alkenes from ethers carrying a halogen atom, 2 carbons removed from the oxygen atom (β-halo-ethers) catalyzed by a metal such as magnesium or zinc. The Boord olefin synthesis is a regiospecific synthesis of different types of olefins which is named after Cecil E. Boord. An E1cB elimination reaction mechanism has been proposed for this reaction.

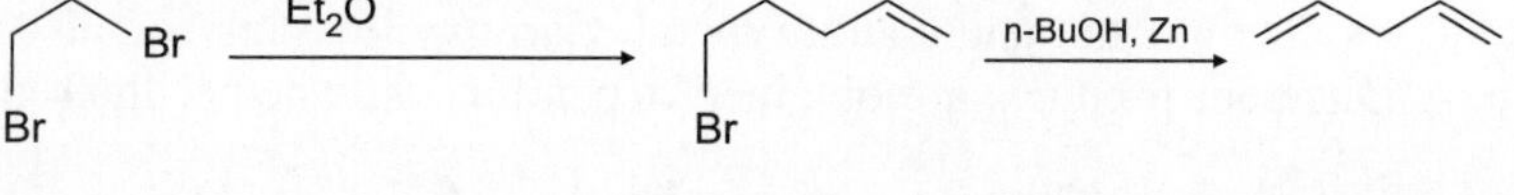

$$CH_3CHO + CH_3CH_2OH \xrightarrow{HCl}$$

55. BORSCHE-DRECHSEL CYCLIZATION

Borsche-Drechsel cyclization is a classic laboratory organic synthesis for carbazole which involves acid catalyzed rearrangement reaction and ring-closing of arylhydrazones of cyclohexanone to tetrahydrocarbazoles followed by oxidation of tetrahydrocarbazoles into carbazoles by red lead.

56. BOUVEAULT ALDEHYDE SYNTHESIS

Bouveault aldehyde synthesis involves reaction of Grignard or organic lithium reagents on N,N-disubstituted formamides, such as DMF in ether to yield the homologous aldehydes. The resulting aldehyde is one carbon longer than the primary alkyl halide taken in starting in the reaction.

In this reaction, an alkyl or aryl halide is first transformed to the corresponding organometallic reagent, then it is added with a N,N-disubstituted formamides such as DMF. The metals Li, Mg, Na, and K can be employed while forming the corresponding organometallic reagent of an alkyl or aryl halide.

57. BOUVEAULT-BLANC REDUCTION

Bouveault-Blanc reduction is an organic reaction in which the esters are reduced to primary alcohols in the presence of absolute alcohol and sodium metal. Carboxylic esters can also be reduced with dissolving sodium. Different products are obtained depending on whether the reduction is carried out in ethanol or xylene. The reaction of esters with sodium in ethanol is referred to as the Bouveault-Blanc reduction. This reaction is an inexpensive and large-scale alternative to lithium aluminium hydride reduction of esters. This dissolving metal reduction is also related to the Birch reduction.

Before the advent of widely available hydride reagents, reduction of esters to primary alcohols was generally performed with the Bouveault-Blanc reduction using alkali metals in ethanol. Now this reaction has been largely replaced by the use of metal hydrides, such as lithium aluminum hydride (LAH) or sodium borohydride because of hazards associated with alkali metal handling and the vigorous reaction conditions (typically refluxing toluene or xylene) in this process.

Classical Bouveault-Blanc reductions are typically performed using one of two procedures. In one method, the substrate to be reduced is dissolved in alcohol and sodium metal is added rapidly to the solution. The second method begins with sodium in an inert solvent such as toluene, to which the substrate is added rapidly as a solution in alcohol. In both cases, it is important to mix the sodium and the alcohol as fast as possible or the reaction fails to achieve complete conversion of the ester substrate. While the Bouveault-Blanc reduction can be successfully employed in large-scale continuous or batch processes, the reaction conditions may result in excessive foaming and even fires.

58. BOYLAND-SIMS OXIDATION

The Boyland-Sims oxidation involves oxidation of aromatic amines with alkaline persulfate to give predominantly the o-amino aryl sulfates and subsequent acid-catalyzed hydrolysis of o-amino aryl sulfates into o-hydroxy aryl amines with para-sulfate is formed in small amounts in certain cases depending on the type of aromatic amines.

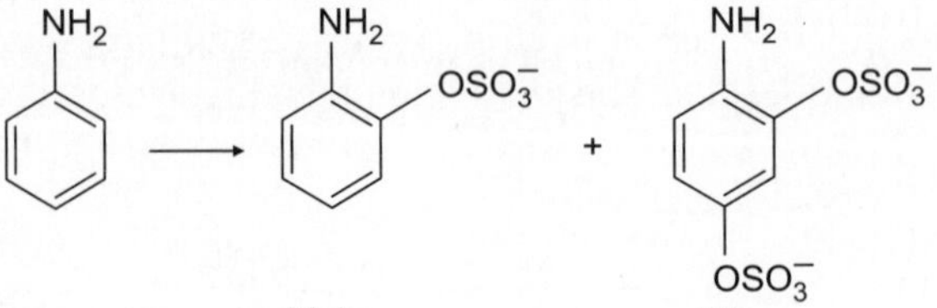

The yields of products are typically low to moderate, but the simplicity of the reaction frequently recommends its use. The products are aromatic sulfates whose orientation relative to the phenolic group is preferentially ortho. These sulfates are useful in synthesis themselves or may be hydrolyzed in acid to the dihydric phenols (or aminophenols).

The reaction involves nucleophilic displacements on peroxide oxygen of the peroxydisulfate ion. In Boyland-Sims oxidation, the nucleophile is a neutral aromatic amine. The mechanism of the reaction is not clear, but there is evidence that the $S_2O_8^{2-}$ ion attacks at the ipso position, and then a migration follows.

59. BRADSHER CYCLIZATION (BRADSHER CYCLOADDITION)

This reaction was first reported by C.K. Bradsher and T.W.G. Solomons in 1958. It involves synthesis of a variety of polycyclic aromatic compounds, especially the phenanthrene series through [4+2] addition of a common dienophile with cationic aromatic azadienes such as acridizinium or isoquinolinium.

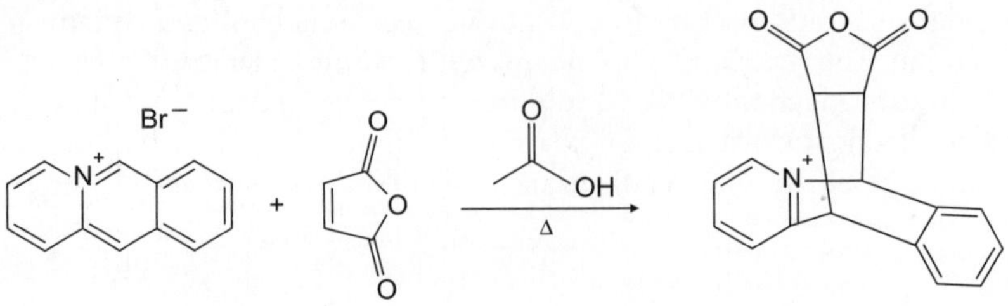

60. BRADSHER REACTION

The Bradsher reaction is an intramolecular cyclization reaction under the influence of an acid, also known as the acid-catalyzed aromatic cyclodehydration of ortho-acyl diarylmethanes to form anthracenes.

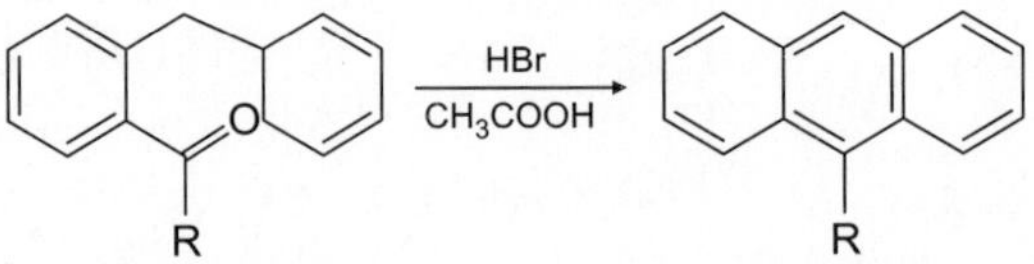

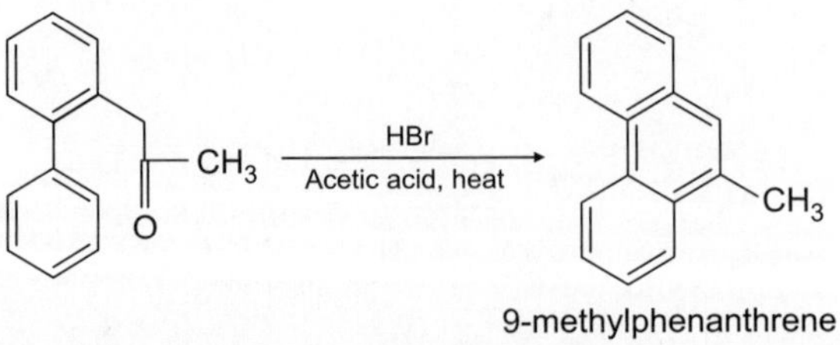

9-methylphenanthrene

61. BROOK REARRANGEMENT

The Brook rearrangement involves base-catalyzed intramolecular 1,2-anionic migration of a silyl group from a carbon atom to an oxygen atom in α-, β- and γ-silyl alcohols, yielding silyl ethers via a reversible process involving a pentacoordinate silicon intermediate. It is named after the Canadian chemist Adrian Gibbs Brook.

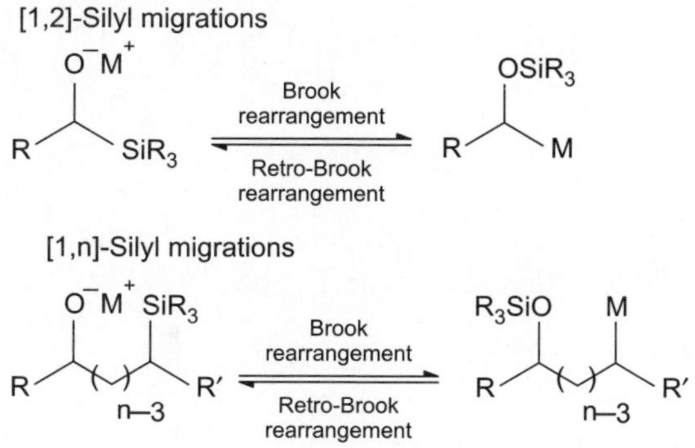

[1,2]-Silyl migrations

[1,n]-Silyl migrations

The base may be an amine, sodium hydroxide, an organolithium reagent or an alkali metal alloy such as Na/K. The silyl substituents can be aliphatic or aromatic. The alcohol in this reaction can be secondary or tertiary with aliphatic or aryl groups. The migratory aptitude of silyl groups in this rearrangement has been found general over a range of homologues, and [1,n]-carbon to oxygen migrations.

The reverse processes, intramolecular migration of a silyl group from oxygen to carbon, was first reported by Speier in 1952 are now called retro-Brook rearrangements.

62. BUCHERER-BERGS SYNTHESIS

The Bucherer-Bergs reaction is a multicomponent reaction between a carbonyl compound (aldehydes or ketones), potassium cyanide and ammonium carbonate, which leads to the formation of hydantoins.

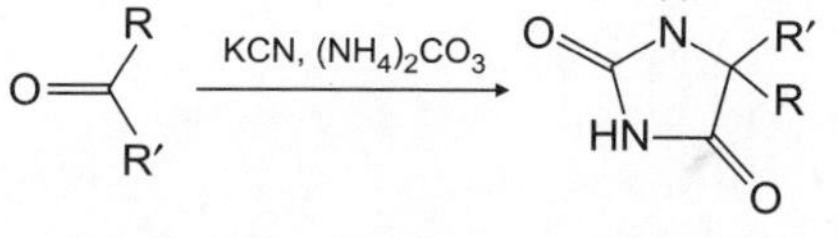

Same product, hydantoins can also be obtained when the preformed cyanohydrin react with ammonium carbonate:

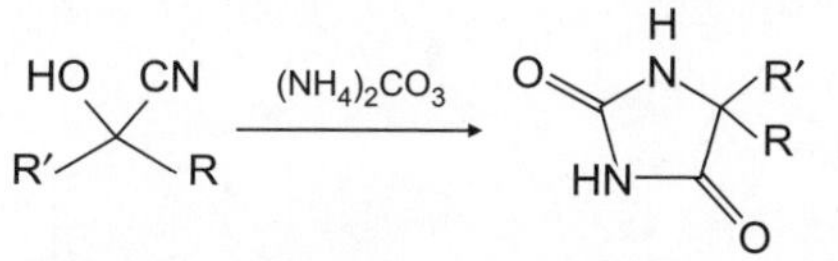

This multicomponent reaction is not stereoselective resulting in formation of racemic mixture of hydantoins, if the prochiral ketone was used as a substrate.

The Bucherer-Bergs reaction for the preparation of hydantoin suffers from the limitation of "one point of diversity", i.e. only changes in the structure of the starting ketone or aldehyde will lead to variations in the final hydantoin. Ketones react easier than aldehydes in the reaction.

Montagne and co-workers reported a modification of this multicomponent reaction that uses nitriles and organometallic reagents as the starting materials. A key advantage of this method is that it creates two points of chemical diversity by combining the R and R_1 groups together in the same vessel as hydantoin assembly. This reaction tolerates considerable variation in the structure of the organometallic reagent and the nitrile, and appears to have broad scope.

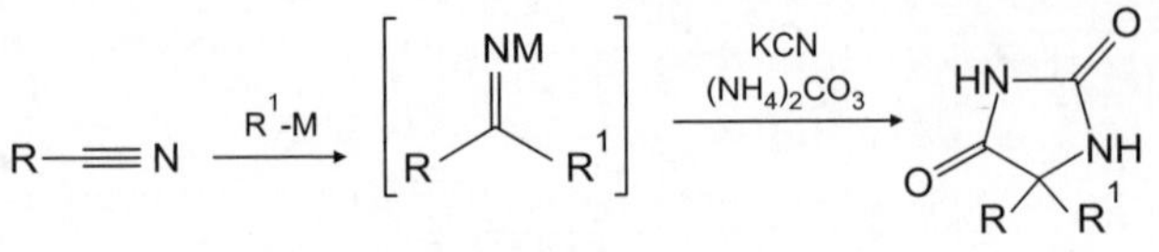

where M = Li or MgX

The Bucherer-Bergs reaction is equivalent to the Strecker synthesis with "additional CO_2," and considered often as a variant of the Strecker synthesis, gives better yield, and proceeds via formation of an intermediate hydantoin.

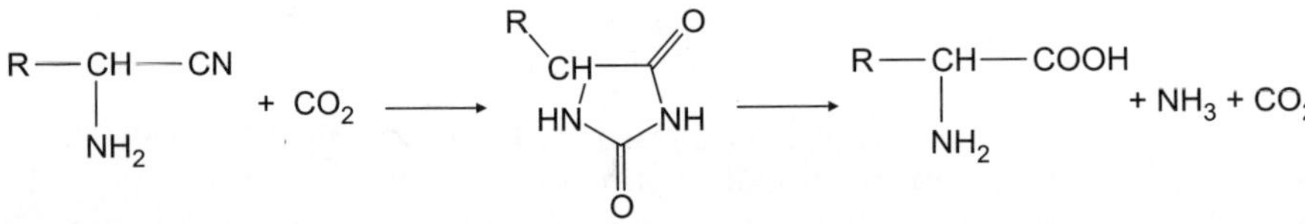

Hydantoins are important precursors in organic synthesis of natural and non-natural amino acids, via acid-, base- or enzyme-catalyzed hydrolysis and here the Bucherer-Bergs synthesis is most commonly employed reaction for the synthesis of hydantoins.

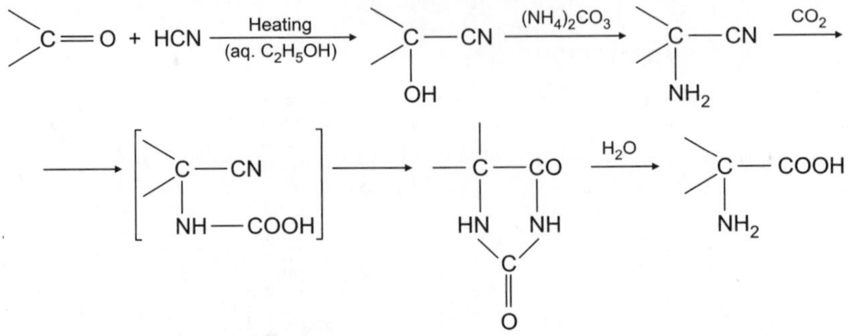

63. BUCHERER CARBAZOLE SYNTHESIS

This reaction is named after Hans Theodor Bucherer. It involves formation of carbazoles from 1- or 2-naphthols (or 1- or 2-naphthylamines), aryl hydrazines using sodium bisulfite. This reaction requires long-reaction time and high temperature.

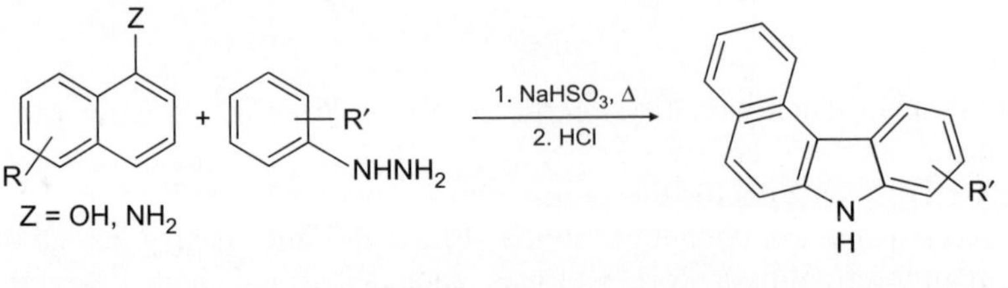

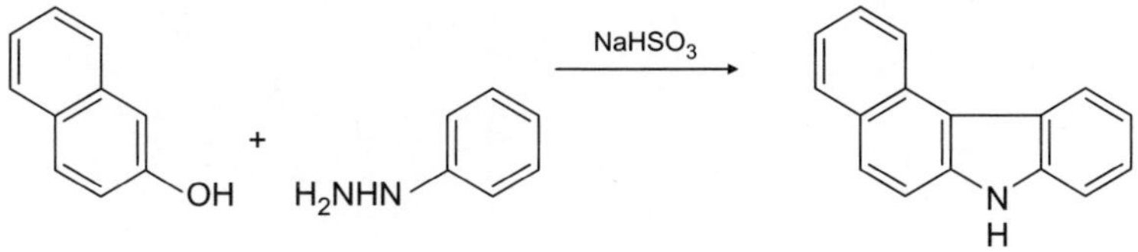

64. BUCHERER REACTION

This reaction was first discovered by a French chemist Robert Lepetit in 1898, but Theodor Bucherer a German chemist discovered the reversibility and applicability of the reaction. This reaction is also known by the name Bucherer-Lepetit reaction.

It involves a reversible conversion (or interconversion) of naphthylamines to naphthols in the presence of an aqueous sulfite or bisulfite and ammonia. The reaction proceeds via intermediate formation of tetralonesulfonic and tetraloneiminosulfonic acids.

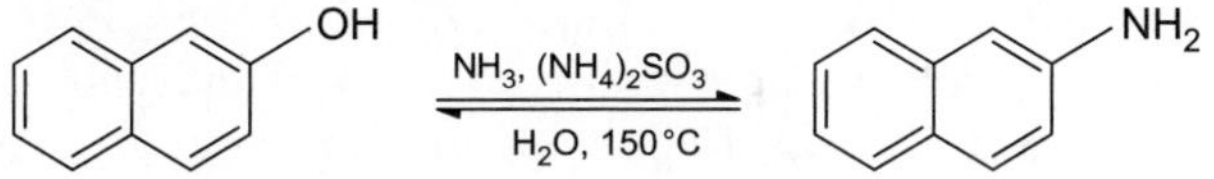

When the primary amines are used instead of ammonia, N-substituted naphthylamines are obtained. The Bucherer reaction provides a relatively simple method to synthesize aromatic amines from hydroxyaromatic compounds or to obtain hydroxyl compounds from aromatic amines.

Although, the classic Bucherer reaction requires high temperature and long reaction time, it may be carried out at room temperature with the aid of microwave.

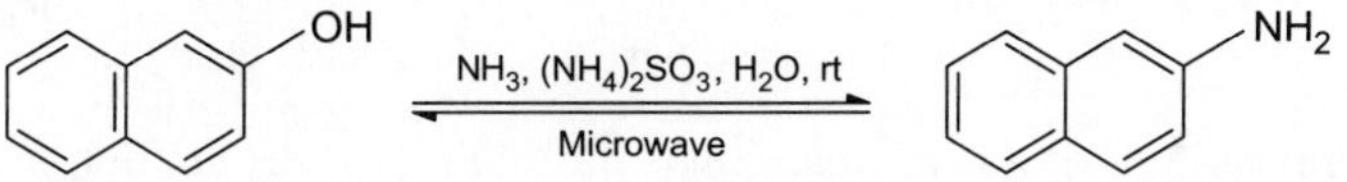

In reaction mechanism, naphthol is protonated in the first step, at a carbon center of high electron density (C-2 or C-4). This leads to resonance stabilized adducts. In the next step, a bisulfite anion adds at C-3. The addition product tautomerize to give the more stable tetralone sulfonate, the tetralone carbonyl group is then attacked by a nucleophilic amine (e.g. ammonia). Subsequent dehydration leads to the cation which again is stabilized by resonance. This compound is deprotonated to the imine or the enamine, which upon elimination of the bisulfite leads to the formation of naphthylamine.

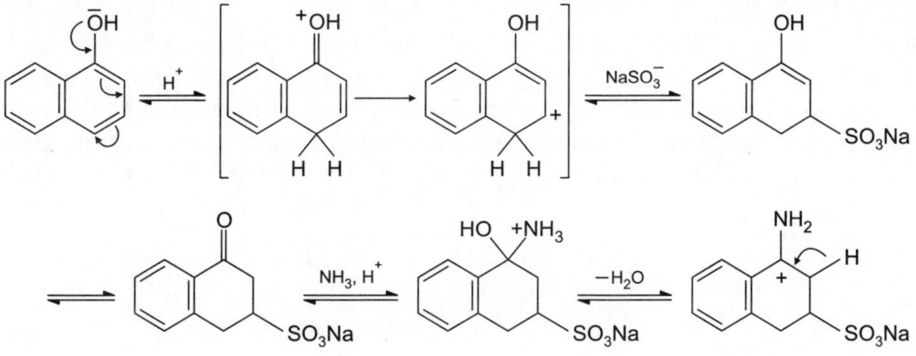

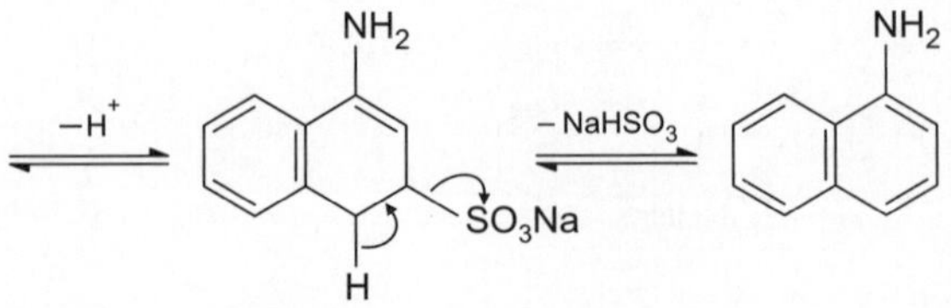

65. BUCHNER-CURTIUS-SCHLOTTERBECK REACTION

Reaction of carbonyl compounds with diazoalkanes to deliver homologated ketones is generally referred to as the Büchner-Curtius-Schlotterbeck reaction.

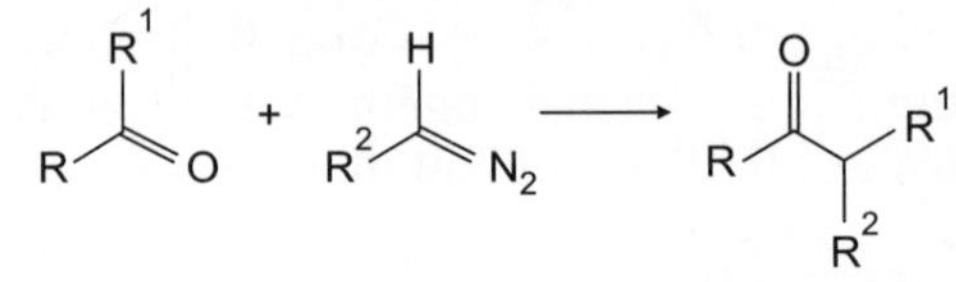

The reaction mechanism involves nucleophilic addition of diazoalkane to a carbonyl group followed by 1,2-alkyl migration. The alpha-epoxides and aldehydes are also produced in this type of reaction.

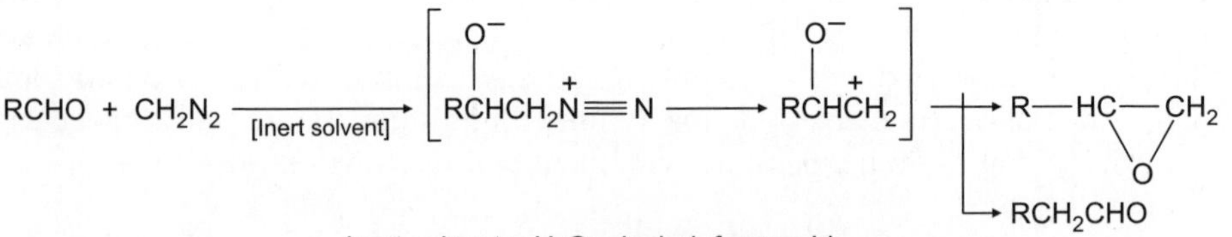

Inert solvent = H_2O, alcohol, formamide

The Buchner-Curtius-Schlotterbeck reaction is particularly useful in one-carbon ring expansion of cyclic ketones with diazoalkanes.

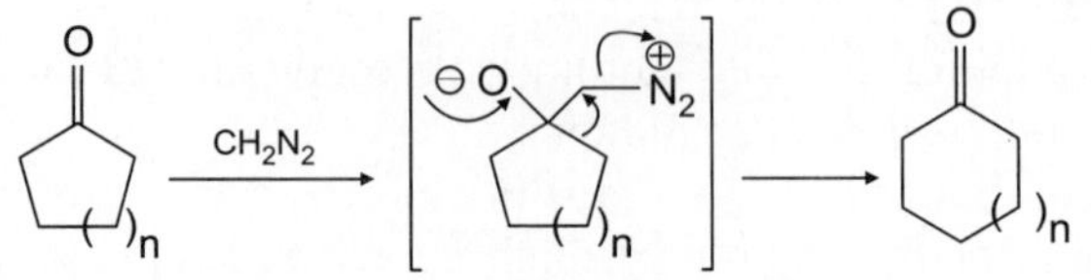

66. BUCHNER METHOD OF RING ENLARGEMENT

The Buchner ring expansion is a two-step process involving formation of a carbene from ethyl diazoacetate as the first step, which cyclopropanates an aromatic ring. The ring expansion occurs in the second step, with an electrocyclic reaction opening the cyclopropane ring at high temperatures. This reaction is generally used to access 7-membered rings, but often suffers from the side reactions of the carbene moiety. This reaction was named after E. Buchner and T. Curtius who first used it in 1885.

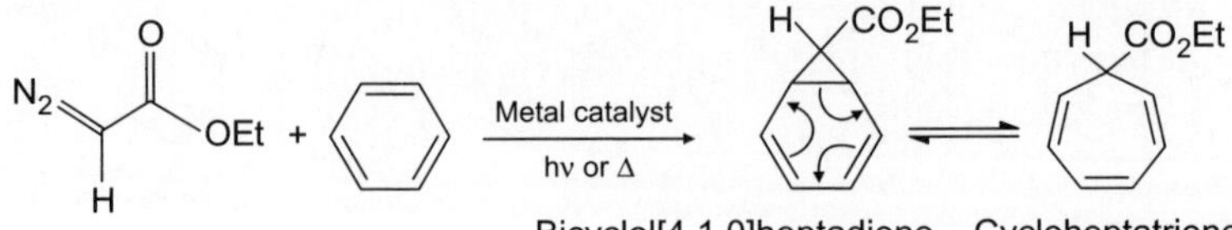

Bicyclol[4.1.0]heptadiene   Cycloheptatriene

The choices for metals include Cu, Rh and Ru with a variety of ligands.

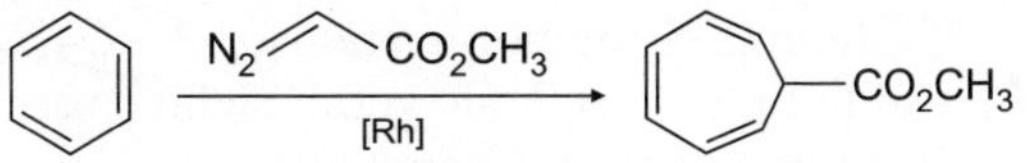

Synthetically useful intramolecular version of the Buchner reaction is also possible as given below:

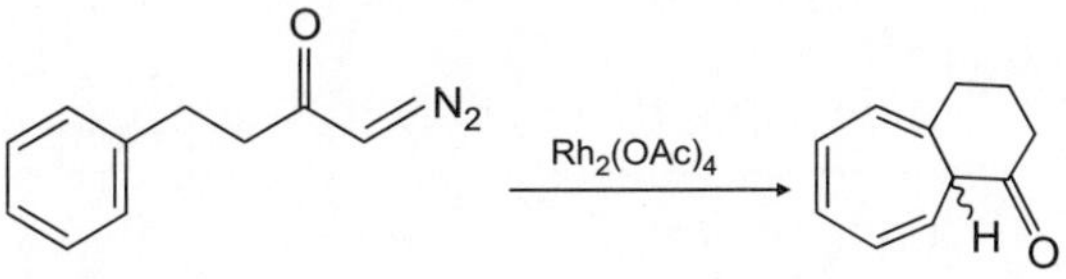

67. BUCHWALD-HARTWIG CROSS COUPLING REACTION

Buchwald-Hartwig cross coupling reaction involves palladium catalysed coupling of aryl halides with amine nucleophiles in the presence of stoichiometric amounts of base.

Usually $Pd(OAc)_2$ or Pd_2dba_3 tri(dibenzylideneacetone)dipalladium) are employed as palladium source for this reaction. Apart from the aryl halides, the pseudohalides (for example, triflates) and primary or secondary amines can be employed as the starting material for the reaction.

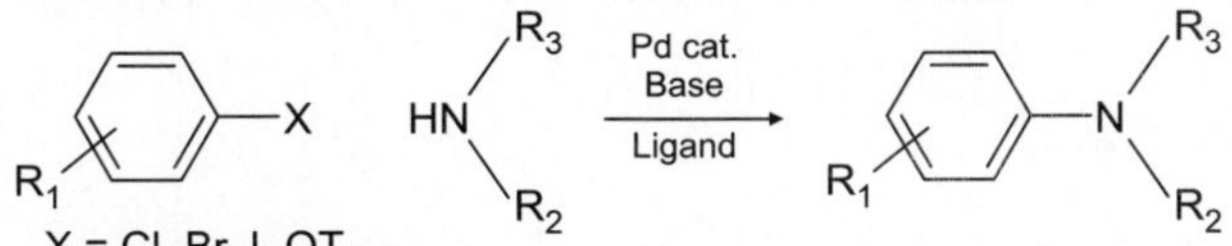

X = Cl, Br, I, OT_f
R_2 = Alkyl, aryl, H
R_3 = Alkyl, aryl

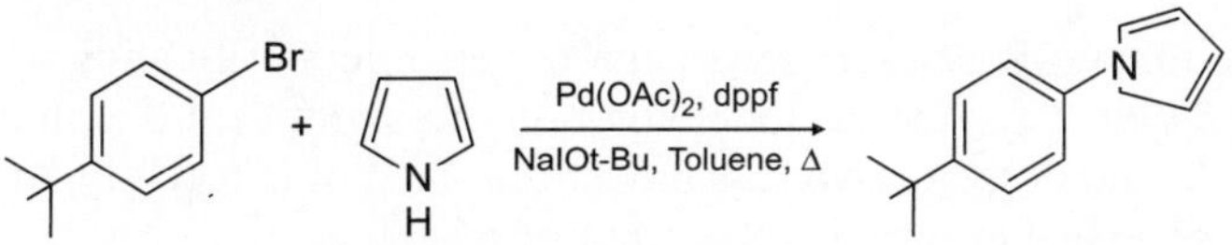

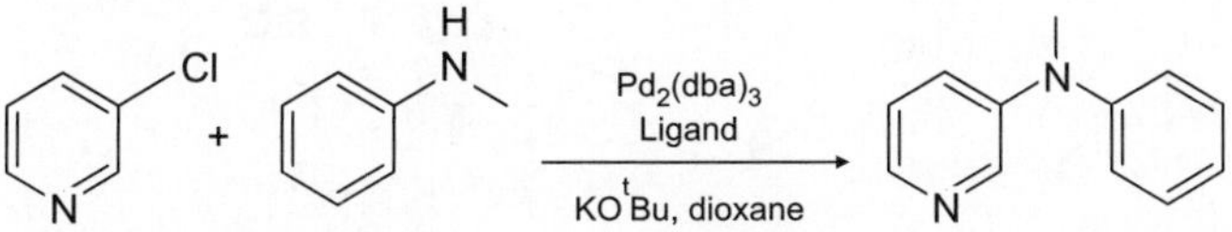

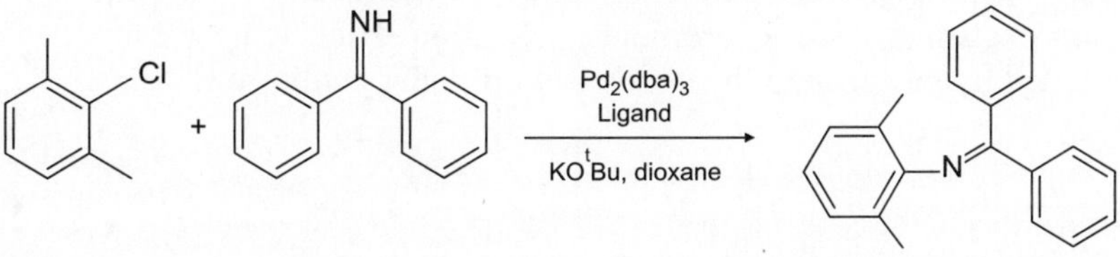

This reaction provides a mild access to aryl amine, found to tolerate a variety of functional groups and provide reproducible yields, therefore, replacing to harsher methods like Goldberg reaction, nucleophilic aromatic substitution, etc. For instances, traditional copper-mediated Ullmann couplings generally require harsh reaction conditions besides most Cu(I) salts are insoluble in organic solvents.

68. BOHLMANN-RAHTZ PYRIDINE SYNTHESIS

Bohlmann and Rahtz, in 1957, reported the preparation of 2,3,6-trisubstituted pyridines.Their method employs the Michael addition of acetylenic ketone with enamines. The aminoketones are typically isolated and subsequently heated at temperature greater than $120\,°C$ to facilitate the cyclohydration.

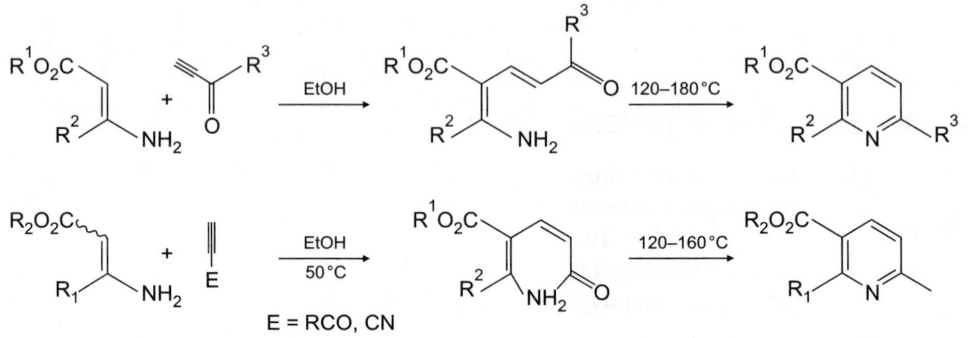

The Bohlmann-Rahtz pyridine Synthesis allows the generation of substituted pyridines in two steps. Condensation of enamines with ethynylketones leads to an aminodiene intermediate that after heat-induced *E/Z* isomerization, undergoes a cyclodehydration to yield 2,3,6-trisubstituted pyridines.

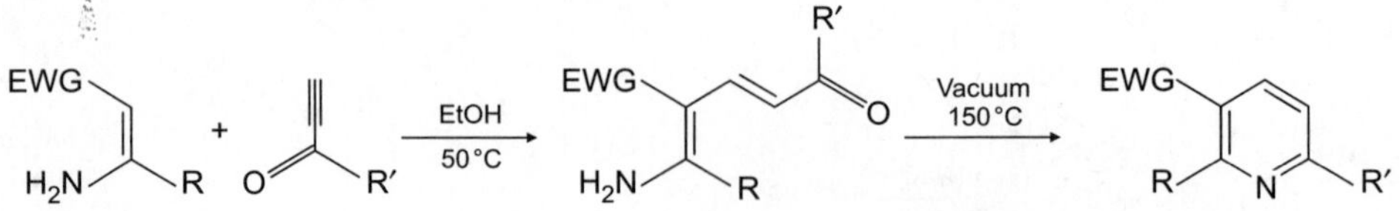

Although the Bohlmann-Rahtz synthesis is more versatile, purification of the intermediate and high temperatures required for the cyclodehydration are significant drawbacks that have limited its synthetic utility. Some of the drawbacks have been overcome recently, making the Bohlmann-Rahtz synthesis more valuable for the generation of pyridines.

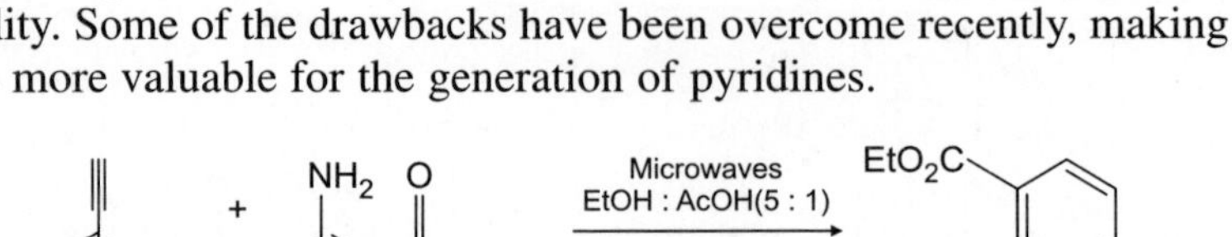

Bagley group reported a mild, single-step variant of the reaction that utilizes either acetic acid or amberlyst ion-exchange resin to promote cyclization. High temperatures in the dehydration step can be avoided by performing the condensation under acidic conditions.

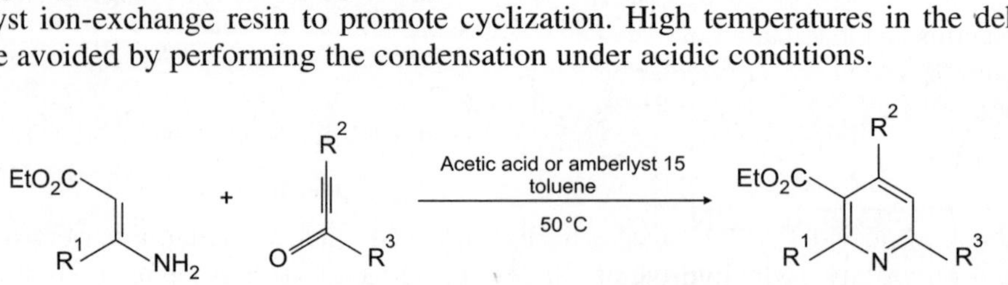

69. CAMPS QUINOLINE SYNTHESIS

Camps quinoline synthesis or the Camps cyclization involves the formation of hydroxyquinolines from the treatment of o-acylaminoacetophenones (N-acyl o-acylanilines) with a base in alcohol (e.g. alcoholic sodium hydroxide). Two isomers are produced; the relative proportions of these isomeric hydroxyquinolines produced are dependent upon the reaction conditions and structure of the starting material as well as mainly determined by the residue on the amino nitrogen.

This reaction found application for the synthesis of 3- and 4-substituted quinolines. The reaction found to introduce a variety of substituents into position 3 and 4 as well as o-amino derivatives of acetophenone, benzophenone, benzoylacetic ester, benzoylcarbinol, propiophenone, etc. all undergo this reaction.

70. CANNIZZARO REACTION

Aldehydes which lack an α-hydrogen when heated in presence of a strong base undergoes a disproportionation* reaction to give an alcohol and a carboxylic acid, this is called Cannizzaro reaction. One molecule of aldehyde is reduced to the corresponding alcohol, while a second one is oxidized to the carboxylic acid. In other words, Cannizzaro reaction is an example of self-reduction and oxidation. The reaction was named after its discoverer Stanislao Cannizzaro.

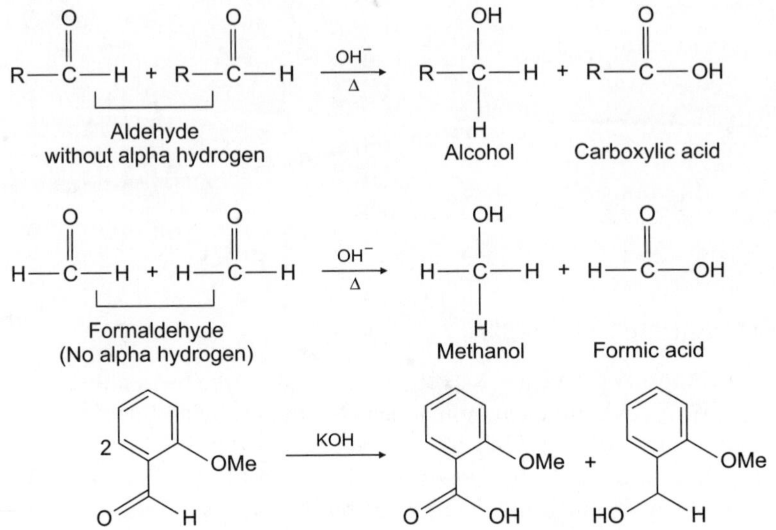

The Cannizzaro reaction is an example of a hydride transfer reaction. The reaction begins with nucleophilic attack of hydroxide on the carbonyl centre. In a strongly basic medium, the below given anion A and the dianionic species B may result. The reaction can proceed from both species A or B respectively. The strong electron-donating effect of one or even two O^--substituents allows for the transfer of a hydride ion H^- onto another aldehyde molecule.

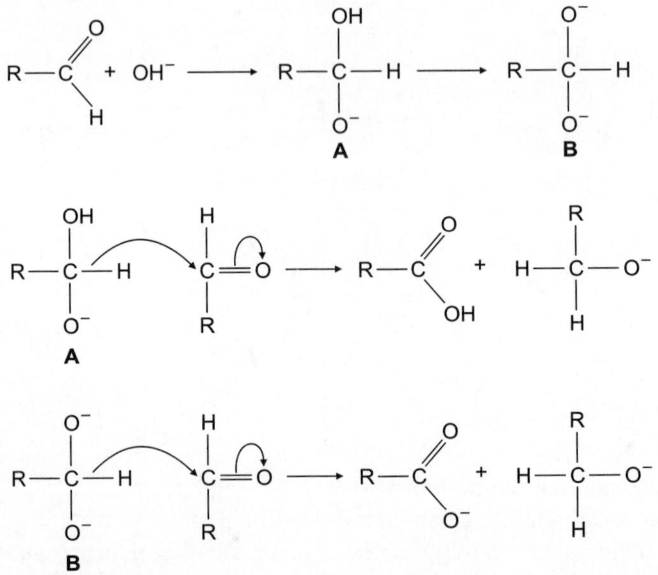

The reaction requires strong basic conditions and only works with aldehydes that are nonenolisable, i.e. does not have any α-protons ($C_6H_5 - CHO$, $CCl_3 CHO$, $(CH_3)_3C–CHO$, HCHO, etc.). Under ideal conditions, the reaction produces only 50% of the alcohol and the carboxylic acid. The latter is obtained only after acidification of the highly basic reaction mixture, typically 30% base. To avoid the low yields, it is more common to conduct the crossed Cannizzaro reaction with a sacrificial aldehyde. In this variation, the reductant is formaldehyde, which is oxidized to sodium formate and the corresponding alcohol is obtained in a high yield, although the atom economy is still low.

*A disproportionation reaction is a type of redox reaction in which a single substance is simultaneously oxidised and reduced.

Other Examples of Cannizzaro Reaction

The applicability of Cannizzaro reaction in organic synthesis is limited because even under ideal conditions the yield is not more than 50% for either acid or alcohol formed. Apart from this, the aldehydes with a hydrogen atom alpha to the carbonyl, i.e. R_2CHCHO, under the same reaction conditions preferred Aldol condensation reaction to give different type of products, therefore further restricts the scope of the Cannizzaro reaction.

Crossed Cannizzaro Reaction

The crossed Cannizzaro reaction was reported in 1985. It utilises formaldehyde as a reductant to yield the desired alcohol from the corresponding aldehyde. When a mixture of formaldehyde and a non enolizable aldehyde is treated with a strong base, the latter is preferentially reduced to alcohol while formaldehyde is oxidized to formic acid.

Two different aldehydes each having no α-hydrogen atom, exhibit crossed Cannizzaro reaction when heated in alkaline solution.

As previously mentioned, the Cannizzaro reaction is not ideal for synthesis as there are many other methodologies that are more economical. However, a solvent free Cannizzaro reaction is another modification of Cannizzaro reaction, which requires no solvent or heating, and the products can be separated by a simple water extraction.

It is important to consider the Cannizzaro reaction as a potential side reaction when treating an aldehyde under strong basic conditions. Therefore, the Cannizzaro reaction should be kept in mind as a source of potential side products.

71. CARROLL REARRANGEMENT

The Carroll rearrangement, a variant to the ester Claisen rearrangement, is a useful method for preparing γ, δ-unsaturated ketones from allylic acetoacetates. The reaction has found limited use in synthetic organic chemistry, probably because of the harsh thermal conditions (130–220 °C) needed to induce the [3,3] sigmatropic rearrangement. However, these thermal barriers are lowered through modifications to the starting β-ketoester.

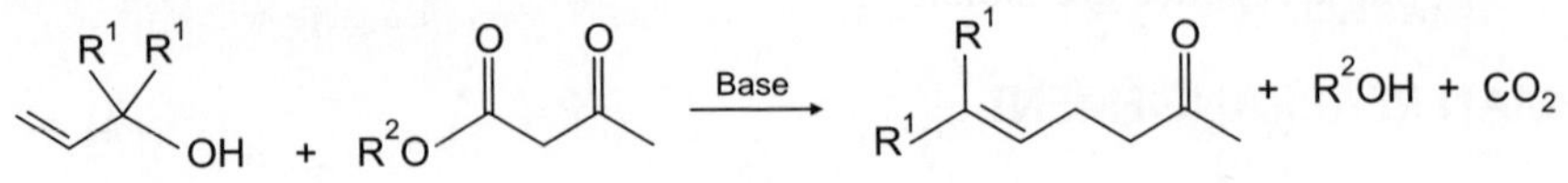

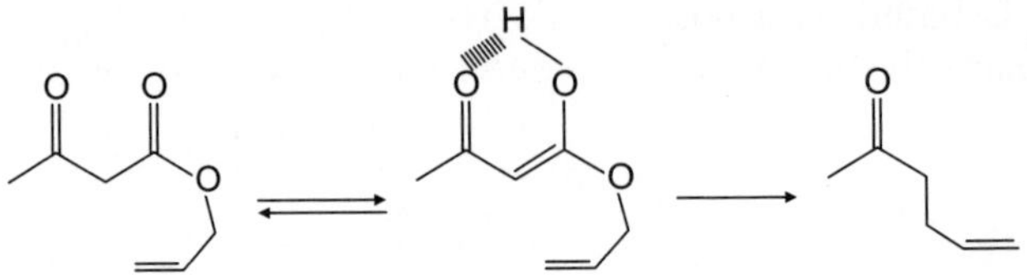

72. CASTRO-STEPHENS COUPLING (STEPHENS-CASTRO COUPLING, CASTRO REACTION)

The reaction was discovered in 1963 by Castro and Stephens. The Castro-Stephens coupling is a cross coupling reaction between a cuprous acetylide and an aryl halide in boiling pyridines forming a arylacetylene. In other words, the coupling of terminal alkynes and aryl halides using copper(I) salt is called Castro-Stephens coupling.

$$\mathrm{H{-}C{\equiv}C{-}R} \xrightarrow{\mathrm{Cu}} $$

$$\mathrm{ArX + Cu{-}C{\equiv}C{-}R} \xrightarrow[\text{Heat}]{\text{Pyridine}} \mathrm{Ar{-}C{\equiv}C{-}R}$$

$$\mathrm{RX + CuC{\equiv}C{-}R'} \xrightarrow[\text{Pyridine, reflux}]{} \mathrm{RC{\equiv}C{-}R'}$$

R = Aryl, alkenyl

$$\mathrm{ArX + CuC{\equiv}C{-}R'} \xrightarrow[\Delta]{\text{Pd/pyridine}} \mathrm{Ar{-}C{\equiv}C{-}R'}$$

X = I, Br, Cl
R = Alkyl, aryl, vinyl

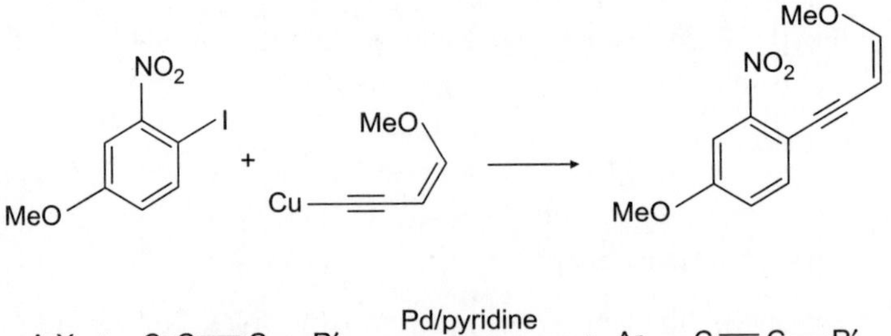

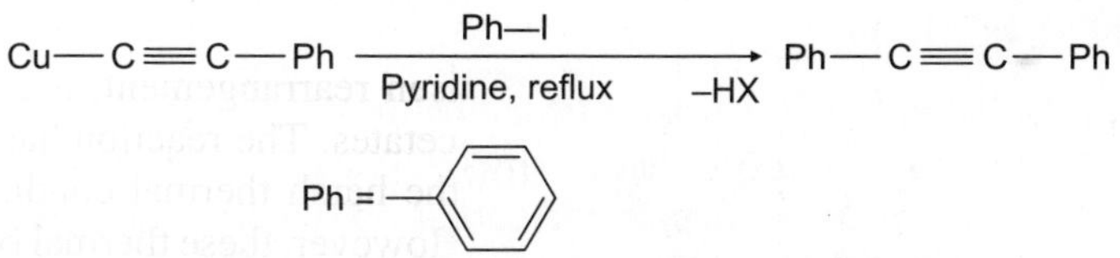

$$Ph = \langle \rangle$$

The Castro-Stephens coupling is a 2-step conversion and requires stoichiometric amount of Cu(I), that is, copper acetylides are isolated and they react with another reagent in the next step.

73. CHAPMAN REARRANGEMENT

Chapman rearrangement is a thermal rearrangement of aryl imidates to N,N-diaryl amides. The reaction mechanism is based on a nucleophilic aromatic substitution and found to proceed in either polar or nonpolar solvents or even in a solid state.

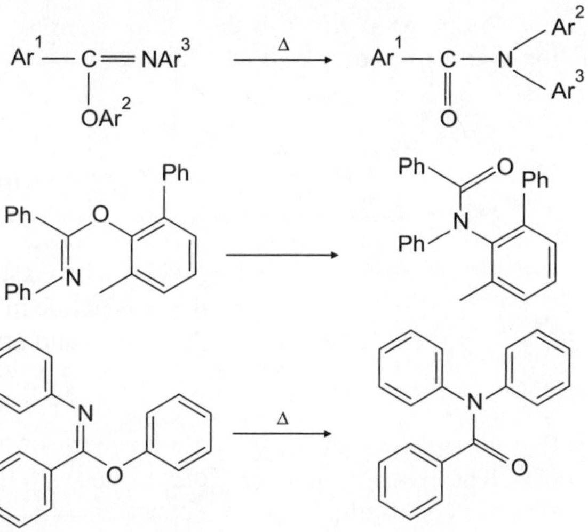

74. CHICHIBABIN PYRIDINE SYNTHESIS

The Chichibabin pyridine synthesis (Aleksei Chichibabin, 1924) affords substituted pyridines by the condensation reaction of aldehyde and ammonia over alumina catalyst.

This reaction is found useful for the synthesis of alkyl-substituted pyridines. However, low product yield and high prevalence of by-products makes it difficult to isolate pure pyridine product. Besides, aldehydes and ammonia gas also reacts with acetylene or acetonitrile to give pyridine derivatives. This reaction also takes place with aliphatic and aromatic ketones, α,β-unsaturated aldehyde, and keto acids.

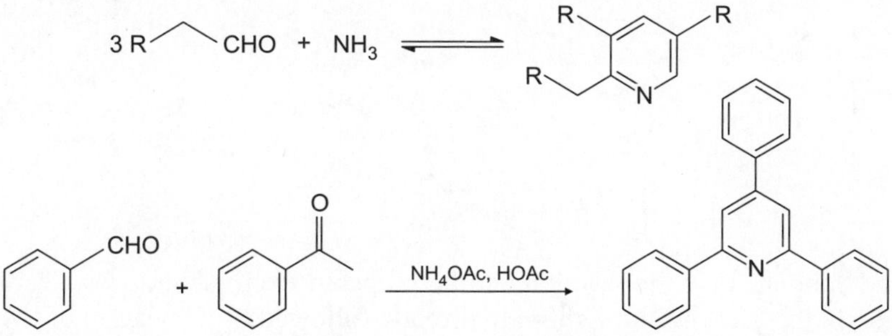

75. CHICHIBABIN REACTION

The reaction was discovered in 1914 by Aleksei Chichibabin. The Chichibabin reaction is a method for the production of 2-aminopyridine derivatives by the reaction of pyridine with sodium amide in an inert solvent, such as toluene or benzene. Apart from the amination of pyridines, reaction also works with other heterocyclic nitrogen compounds when they proceed with alkali-metal amides.

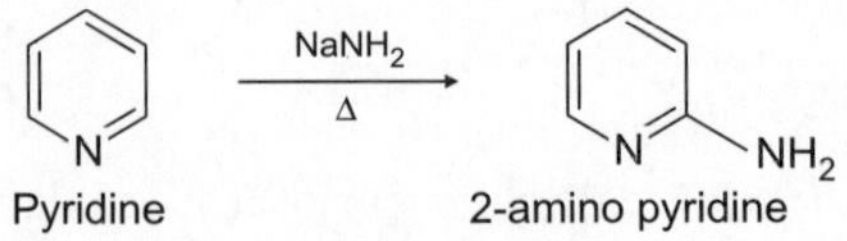

This is a nucleophilic aromatic substitution reaction, where an amide anion (NH_2^-) substitutes the hydride ion (H^-) from the C-2 or C-4 of the pyridine ring. Hence, hydride (H^-) is the leaving group and amide anion (NH_2^-) is nucleophile. The 2nd and 6th positions are favoured over the other positions. This reaction is useful in preparation of nitrogen-containing aromatic amines.

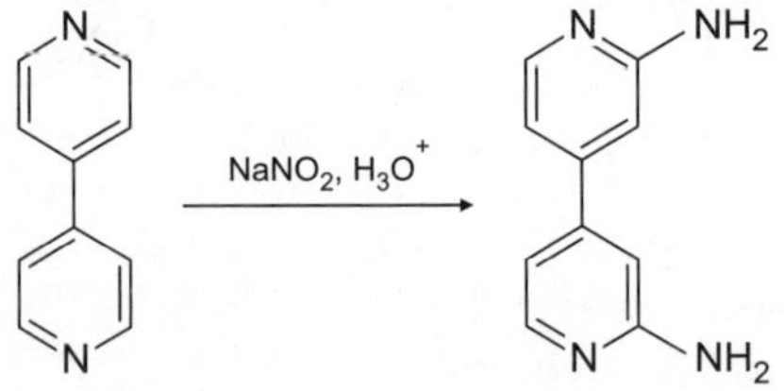

76. CHUGAEV REACTION

Chugaev reaction (xanthate ester pyrolysis) was named after its discoverer Lev Aleksandrovich Chugaev (a Russian chemist). The reaction involves formation of olefins from pyrolysis of the corresponding xanthate esters via cis-elimination. This pyrolysis which is usually carried out at temperatures ranging from 100–250 °C, proceeds via an intramolecular cis-elimination without the rearrangement of the carbon skeleton.

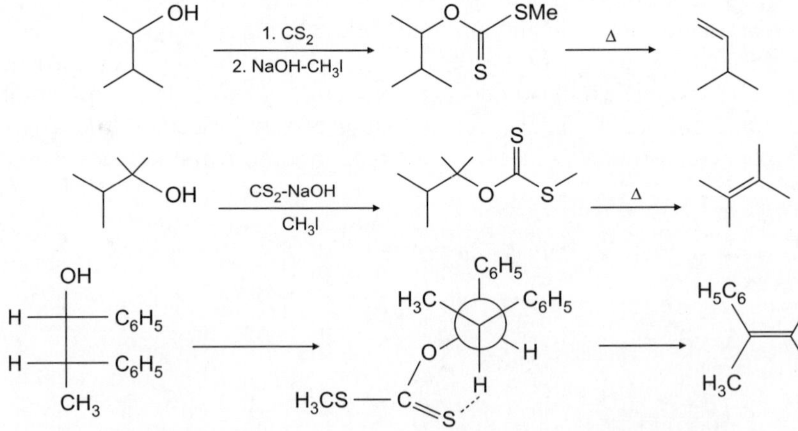

The required xanthates for the reaction can be prepared from alcohols by treating them with carbon disulfide in the presence of sodium hydroxide followed by alkylation of the intermediate sodium xanthate. Often, methyl iodide is used as the alkylating agent.

$$ROH + CS_2 + NaOH \longrightarrow RO\overset{\overset{\displaystyle S}{\|}}{C}S^- Na^+ \xrightarrow{CH_3I} RO\overset{\overset{\displaystyle S}{\|}}{C}SCH_3 + NaI$$

77. CIAMICIAN-DENNSTEDT REARRANGEMENT

The Ciamician-Dennstedt rearrangement entails the ring expansion of the pyrrole ring to 3-chloropyridine by heating with chloroform in alkaline solution. In this reaction, cyclopropanation of a pyrrole occurs with dichlorocarbene generated from $CHCl_3$ and NaOH. Subsequent rearrangement takes place to give 3-chloropyridine.

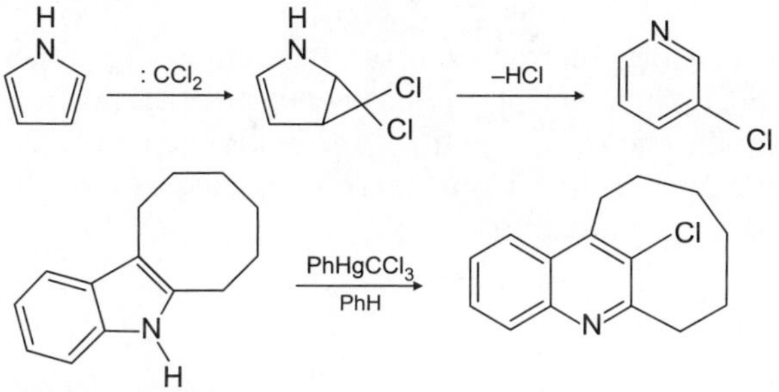

78. CLAISEN CONDENSATION

Claisen condensation was name after Rainer Ludwig Claisen (1881). The Claisen condensation occurs between esters containing α-hydrogens, promoted by a base such as sodium ethoxide, affords β-ketoesters. The driving force is the formation of the stabilized anion of the β-keto ester. If two different esters are used, an essentially statistical mixture of all four products is generally obtained, and the preparation does not have high synthetic utility.

However, if one of the ester partners has enolizable α-hydrogens and the other does not (e.g. aromatic esters or carbonates), the mixed reaction (or crossed Claisen) can be synthetically useful. If ketones or nitriles are used as the donor in this condensation reaction, a β-diketone or a β-ketonitrile is obtained respectively.

Crossed Claisen Condensation

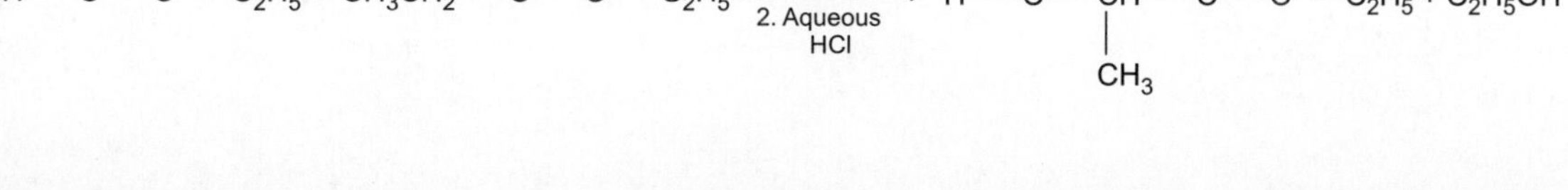

The use of stronger bases, e.g. sodium amide or sodium hydride instead of sodium ethoxide, often increases the yield.

When planning a Claisen condensation with an ester, it is important to use alkoxide ion that has the same alkyl group as the alkoxyl group of the ester. This is to avoid the possibility of trans esterification.

Like the aldol condensation, the Claisen condensation involves nucleophilic attack by a carbanion on an electron-deficient carbonyl compound. In the aldol condensation, nucleophilic attack leads to addition (the typical reaction of aldehydes and ketones). In the Claisen condensation, nucleophilic attack leads to substitution (the typical reaction of acyl compounds).

79. CLAISEN REARRANGEMENT

Allylic aryl ethers, when heated rearrange to o-allylphenols in a reaction called the Claisen rearrangement. The Claisen rearrangement was discovered in 1912 and its mechanism was proposed in 1960's, which was found to be similar to the Cope rearrangement.

The Claisen rearrangement is like Cope rearrangement, except that one of the aliphatic methylenes is replaced with a heteroatom.

The Claisen rearrangement is a highly stereoselective [3,3]-sigmatropic rearrangement, which involves sigmatropic conversion of allyl vinyl ethers into homoallyl carbonyl compounds or allyl aryl ethers to yield o-allyl substituted phenols respectively.

In the Claisen rearrangement, a strong C=O double bond is generated. Since C=O double bonds (172 kcal/mol) are stronger than C=C double bonds (148 kcal/mol), therefore formation of a stable carbonyl drives the reaction to the right and highly favoured at equilibrium.

The Claisen rearrangement is actually used more often than the Cope. The two examples below show the syntheses of a ketone and an aldehyde that are an important intermediates in the fragrance industry.

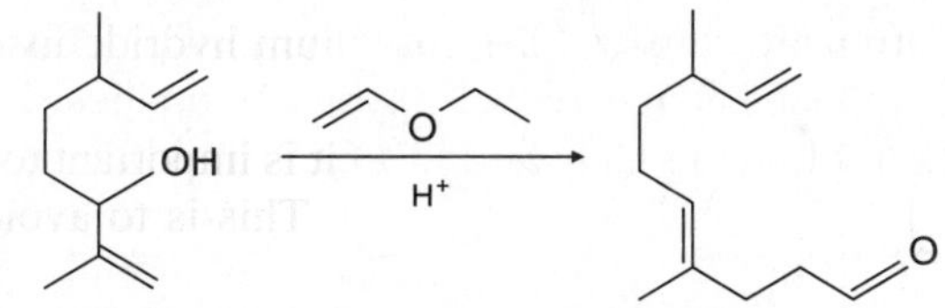

Some other rearrangements related to the Claisen rearrangement are given below:

Claisen Rearrangement

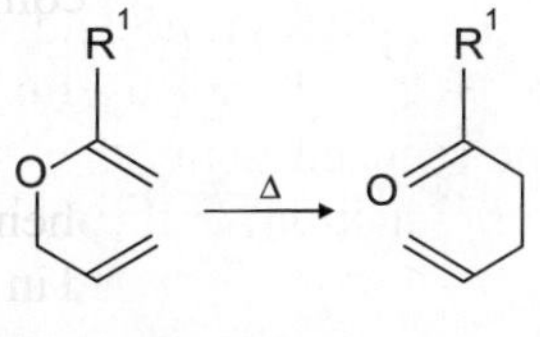

Johnson-Claisen Rearrangement (when R1 = OR)

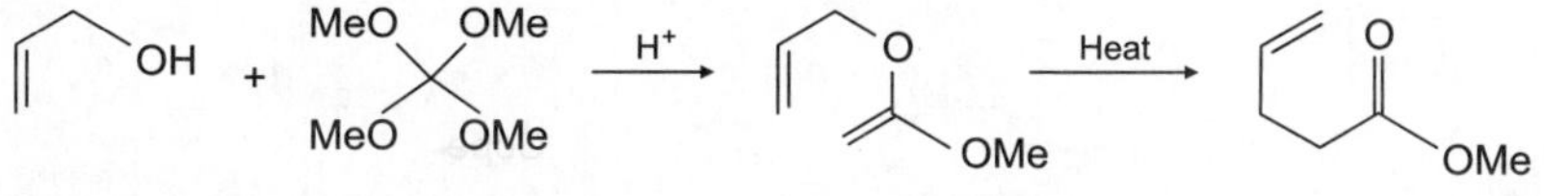

Eschenmoser-Claisen Rearrangement (when R1 = NR2)

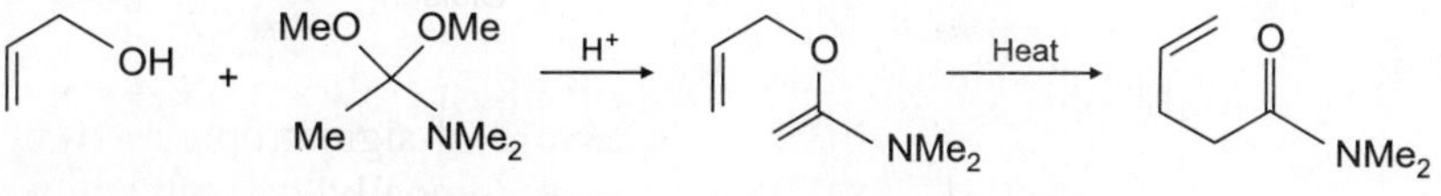

Ireland-Claisen Rearrangement (when R1 = OSiR3 or OLi)

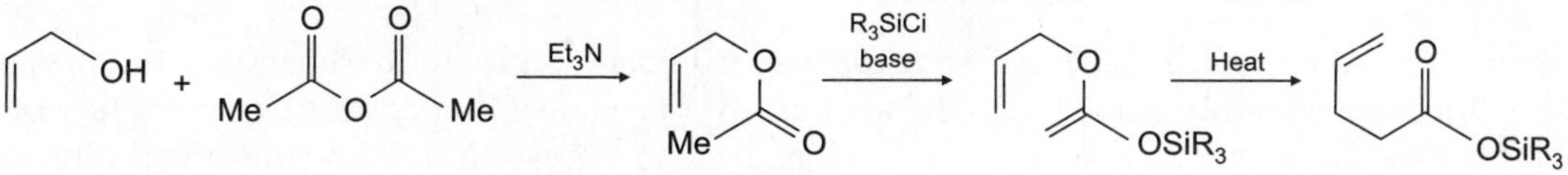

80. CLAISEN-SCHMIDT CONDENSATION

The Claisen-Schmidt condensation is a mixed or crossed aldol condensation involving an aromatic aldehyde. Therefore, this reaction can be described as the condensation of an aromatic aldehyde with an aliphatic aldehyde or ketone in the presence of a base or an acid to form an α,β-unsaturated aldehyde or ketone. The Claisen-Schmidt condensation always involves dehydration of the product of the mixed addition to yield a product in which the double bond (produced during dehydration)

$$C_6H_5CHO + H_3C-\underset{\underset{O}{\|}}{C}-R \xrightarrow{NaOH} C_6H_5HC=CH-COR$$

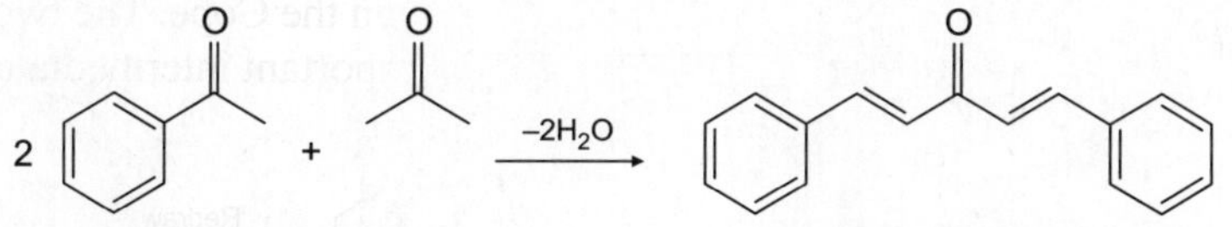

Benzaldehyde   Acetone   Dibenzalacetone
(1,5-diphenyl-1,4-pentadien-3-one)

is conjugated to both the aromatic ring and the carbonyl group. This reaction is useful for the preparation of chalcone, flavanone and macrocycles type compounds. The reaction is named after two of its investigators, Rainer Ludwig Claisen and J.G. Schmidt.

81. CLEMMENSEN REDUCTION

This reaction is named after Erik Christian Clemmensen. The Clemmensen reduction involves reduction of ketones (or aldehydes) to alkanes using zinc amalgam (Zn treated with mercury metal (Hg)) and aqueous hydrochloric acid. Therefore, the C=O group of ketones and aldehydes can be converted into a methylene group (–CH$_2$ group), i.e. corresponding methylene compounds. Because this reaction uses aqueous HCl, it is not useful for compounds that are sensitive to acid. Acid sensitive substrates can be reduced with other methods such as in the Wolff-Kishner reduction (utilizes basic conditions for reduction) or through a milder method such as the Mozingo reduction.

$$RCOR^1 \xrightarrow[\text{HCl}]{\text{Zn(Hg)}} RCH_2R^1$$

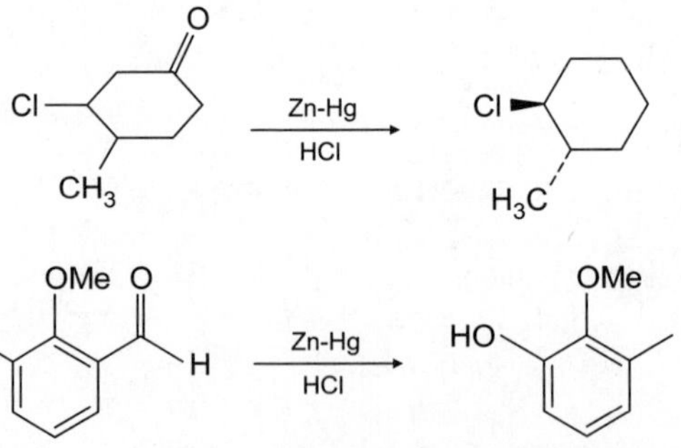

82. COMBES QUINOLINE SYNTHESIS

Combes quinoline synthesis describes acid-catalyzed condensation of unsubstituted anilines and β-diketones to assemble quinolines. The acid-catalyst promotes ring closure or cyclization of an intermediate Schiff base. This method provides a rapid access to the 2,4-substituted quinoline skeleton.

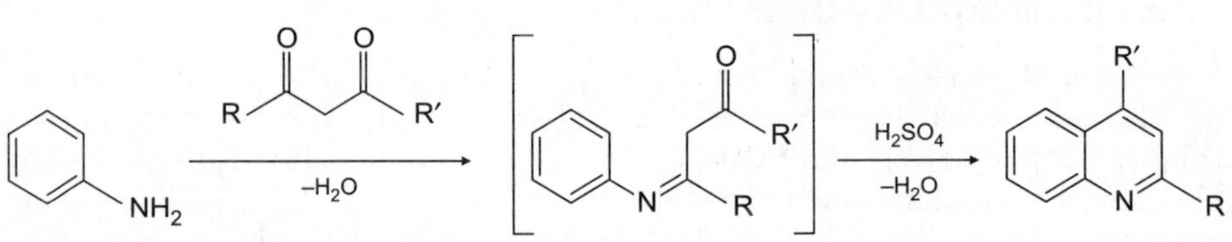

The polyphosphoric acid is reported to be a better catalyst than sulfuric acid for this cyclization.

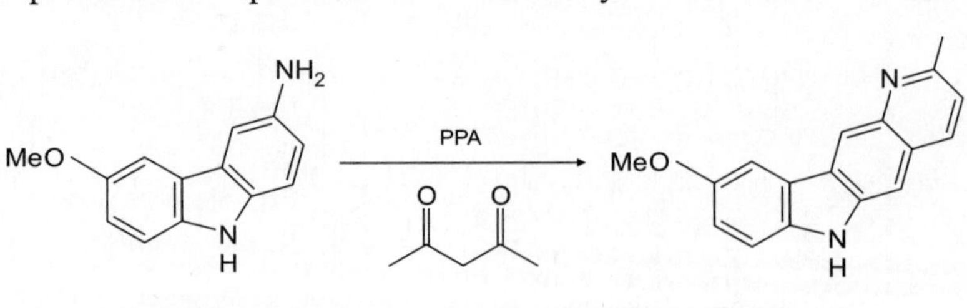

83. CONRAD-LIMPACH CYCLIZATION

Conrad-Limpach cyclization involves thermal condensation of arylamines with the carbonyl group of β-keto esters followed by cyclization of the intermediate Schiff base to 4-hydroxyquinolines.

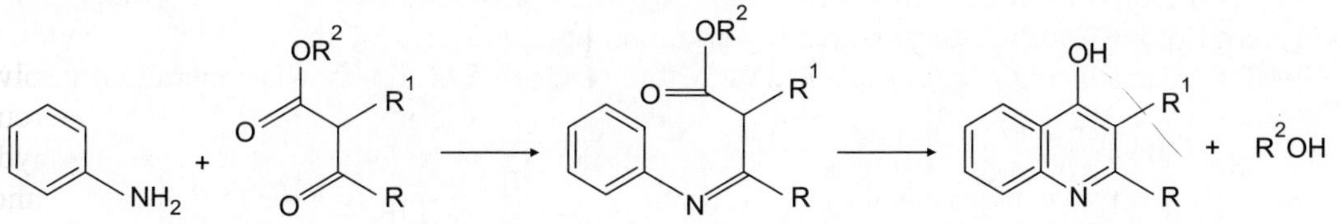

84. COPE ELIMINATION REACTION

The Cope reaction or Cope elimination was developed by Arthur C. Cope. It involves synthesis of olefins and a hydroxylamine from thermal decomposition (i.e. pyrolysis) of N-oxides of tertiary amines (can easily be prepared *in situ* from tertiary amines with an oxidant such as peracid/ mCPBA). Amine oxides undergo elimination to form the least substituted alkene. The reaction involves an intramolecular 5-membered cyclic transition state leading to a syn elimination product.

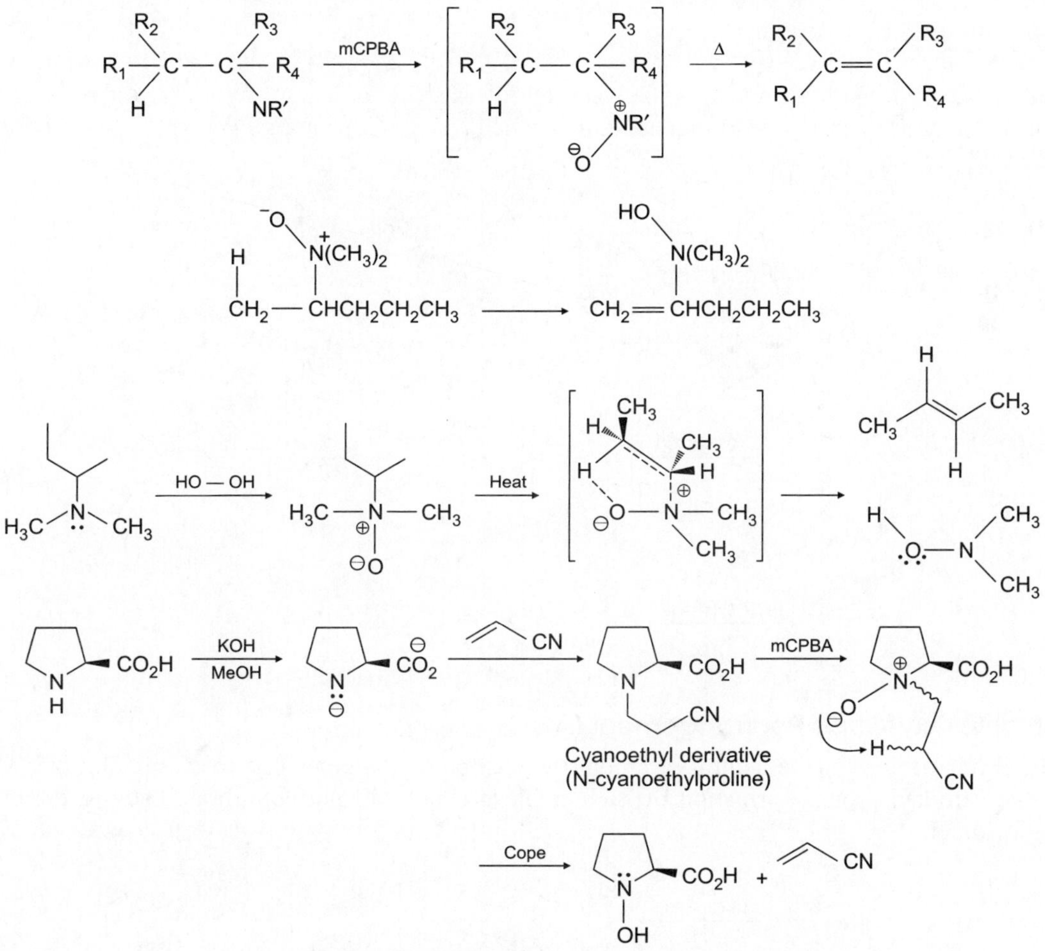

Cyanoethyl derivative
(N-cyanoethylproline)

85. COPE REARRANGEMENT; OXY-COPE REARRANGEMENT

The Cope rearrangement is a highly stereoselective [3,3]-sigmatropic rearrangement of 1,5-dienes which involves thermal isomerization of a 1,5-diene to a regioisomeric 1,5-diene. The main product is the thermodynamically more stable regioisomer. This rearrangement is comparable to its aza and oxa (Claisen rearrangement) variations both in terms of mechanism and synthetic utility. This rearrangement was developed by Arthur C. Cope. This reaction is generally reversible.

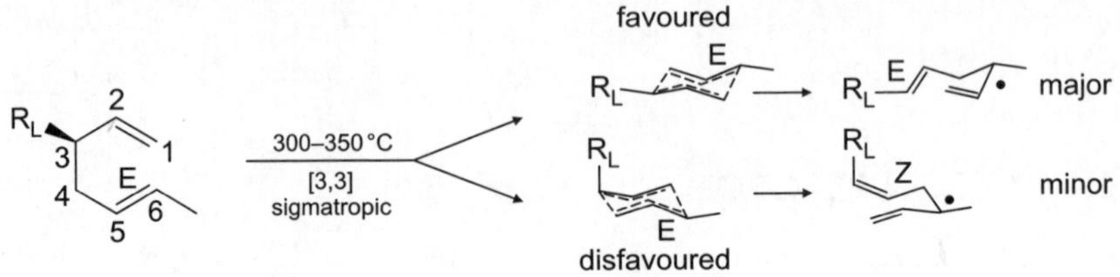

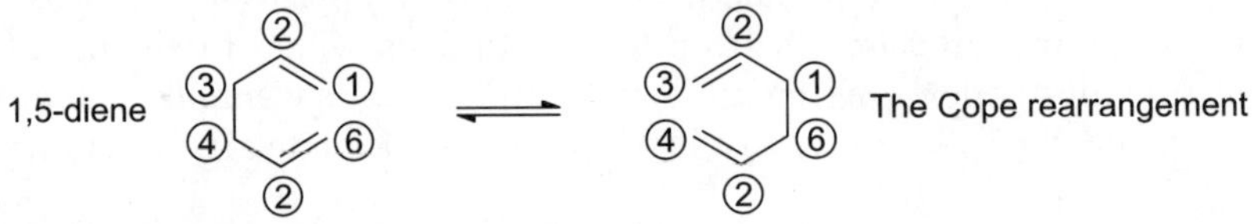

Oxy-Cope Rearrangement

If a hydroxy group is a substituent between the olefins (i.e. on an sp^3-hybridized carbon of the starting isomer) in a Cope rearrangement, it is called an oxy-Cope rearrangement. This reaction is generally irreversible.

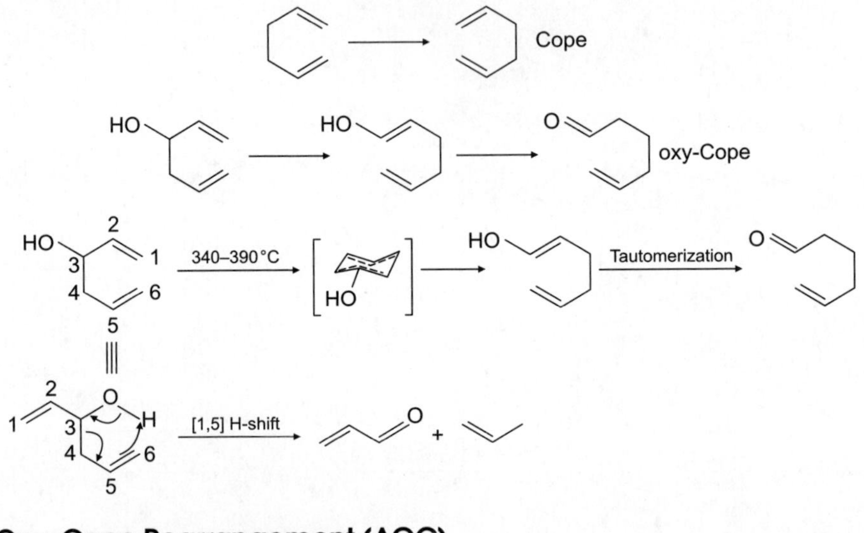

Anionic Oxy-Cope Rearrangement (AOC)

The driving force for the neutral or anionic oxy-Cope rearrangement is the formation of a very stable carbonyl group. Formation of such stable product will not equilibrate back to the other regioisomer.

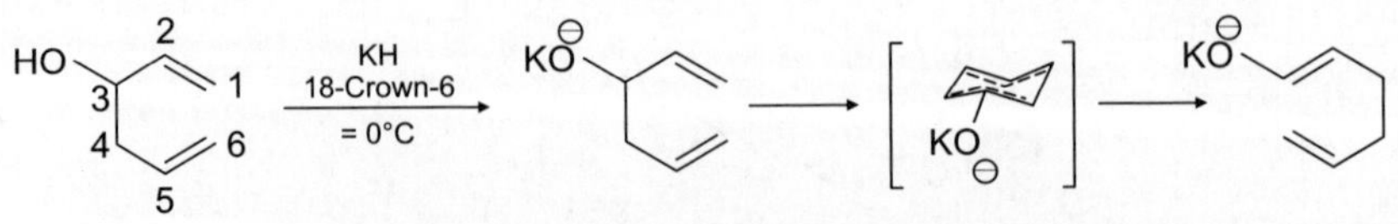

It has been found that if the starting alcohol is deprotonated first, e.g. with KH, the subsequent oxy-Cope is much faster, even the reaction may be conducted at room temperature.

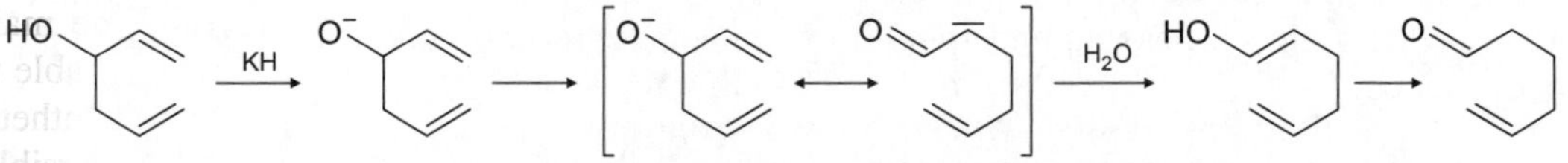

86. COREY-BAKSHI-SHIBATA REDUCTION (CBS)

The Corey-Bakshi-Shibata reduction or simply the CBS reduction describes enantioselective reduction of ketones by borane in the presence of a chiral oxazaborolidine catalyst. The chiral oxazaborolidine is known as the Corey-Bakshi-Shibata catalyst (CBS catalyst).

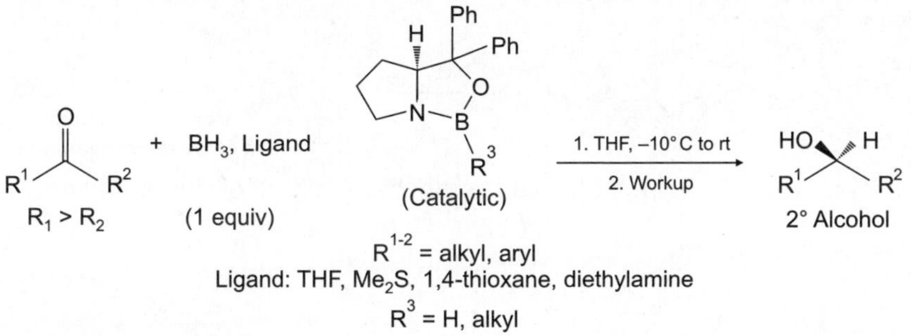

R^{1-2} = alkyl, aryl
Ligand: THF, Me$_2$S, 1,4-thioxane, diethylamine
R^3 = H, alkyl

Advantages of CBS Catalyst

- Ease of preparation
- Air and moisture stability
- Short reaction time
- High enantioselectivity
- Typically high yields
- Recovery of catalyst precursor by precipitation as the HCl salt
- Prediction of the absolute configuration from the relative steric bulk of the two substituents attached to the carbonyl group.

87. COREY-KIM OXIDATION

The Corey-Kim oxidation allows the synthesis of aldehydes and ketones from primary alcohols and secondary alcohols respectively using N-chlorosuccinimide/dimethylsulfide, followed by treatment with a base (triethylamine). Oxidation of alcohols occurs via their alkoxysulfonium salts which upon addition of base rearranges intramolecularly to aldehydes or ketones. Use of dimethylsulfide, a poisonous and volatile liquid with a very bad odour downgrades this reaction for the synthetic applications. The reaction is named after Elias James Corey and Choung Un Kim.

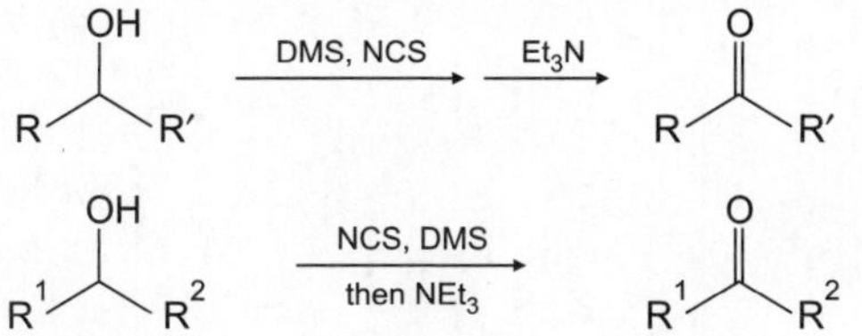

NCS = N-chlorosuccinimide; DMS = Dimethylsulfide

Reaction DMS with NCS afford S,S-dimethylsuccinimidosulfonium chloride (Corey-Kim reagent). This electrophilic reagent is then attacked by the nucleophilic oxygen of the alcohol to form a S-O bond. The addition of TEA results in deprotonation of one of the methyl groups to form a zwitterionic species that undergoes a rearrangement reaction to release DMS gas and the final product.

88. COREY-WINTER OLEFIN SYNTHESIS

Corey-Winter olefin synthesis involves transformation of diols to the corresponding olefins in a series of reactions with 1,1'-thiocarbonyldiimidazole and trimethylphosphite. This transformation takes place via the cis- (or syn-) elimination of cyclic thionocarbonate in the presence of trimethyl- or triethylphosphite. The reaction was named after Elias James Corey and Roland Arthur Edwin Winter.

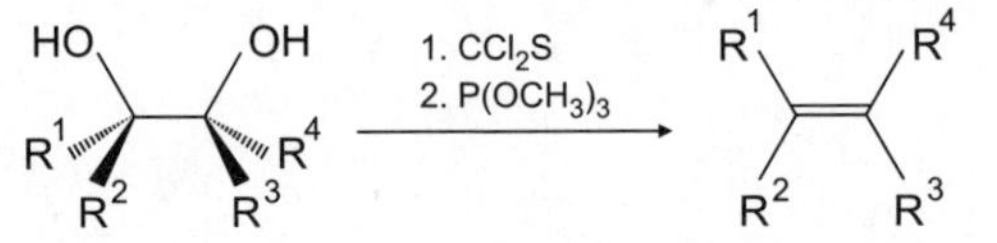

In this reaction, thiocarbonyldiimidazole is often used instead of thiophosgene.

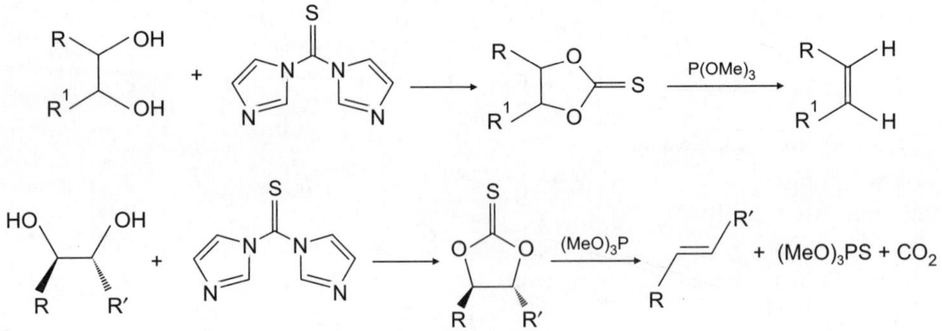

89. CORNFORTH REARRANGEMENT

The Cornforth rearrangement of 4-acyloxazoles is a thermal rearrangement reaction of 4-carbonyl substituted oxazoles to their isomeric oxazoles through a ring-opening process to form a zwitterionic dicarbonyl nitrile ylides followed by a ring closure (i.e. 1,5-dipolar cyclization). This rearrangement of 4-acyloxazoles occurs with the organic acyl residue and the C_5 substituent changing positions via the postulated dicarbonyl nitrile ylides.

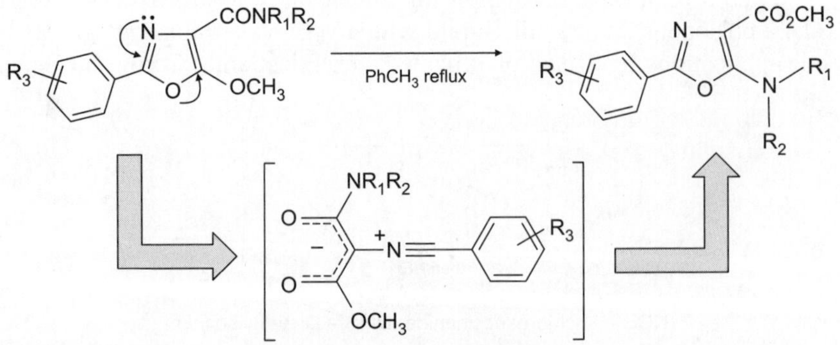

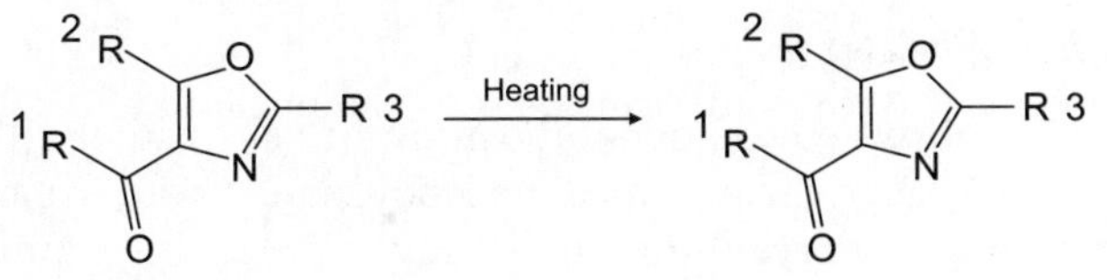

90. CRAIG METHOD

Craig method of **dediazoniation** (releases nitrogen N_2) describes introduction of a halogen into the α-position of aminopyridines by treatment with sodium nitrite in hydroalcohalic acid followed by warming.

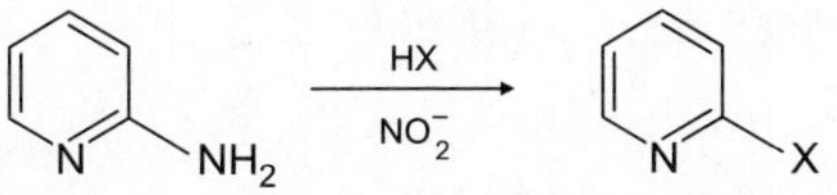

For this, 2-aminopyridine reacts with sodium nitrite, hydrobromic acid and excess bromine to 2-bromopyridine or in the Craig diazotization-bromination method 2-aminopyridine is first reacted to form perbromide; then diazotized with sodium nitrite; followed by a reaction with sodium hydroxide to form the desired 2-bromopyridine.

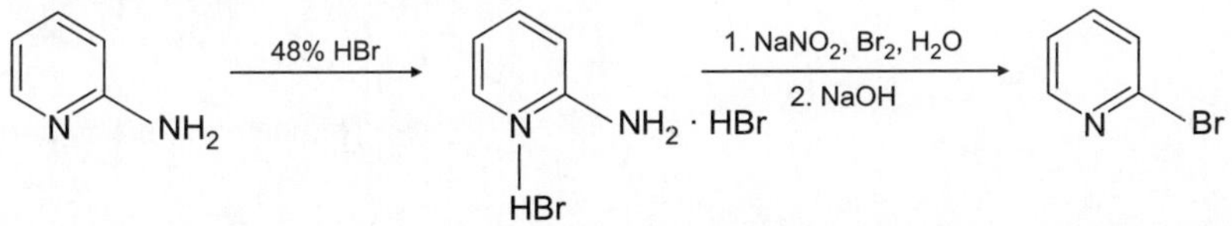

91. CRIEGEE GLYCOL CLEAVAGE OR CRIEGEE REACTION OR GLYCOL CLEAVAGE

In glycol cleavage, the carbon-carbon bond in a vicinal diol (glycol) is oxidatively cleaved and replaced with two carbon-oxygen double bonds. Reaction products are either ketones or aldehydes depending on the substitution pattern in the diol.

The Criegee glycol cleavage refers to the lead tetraacetate ($Pb(OAc)_4$) mediated oxidative cleavage of vicinal glycols (or 1,2-diols) to the two corresponding carbonyl compounds. The reaction mechanism is similar to the Baeyer-Villiger oxidation.

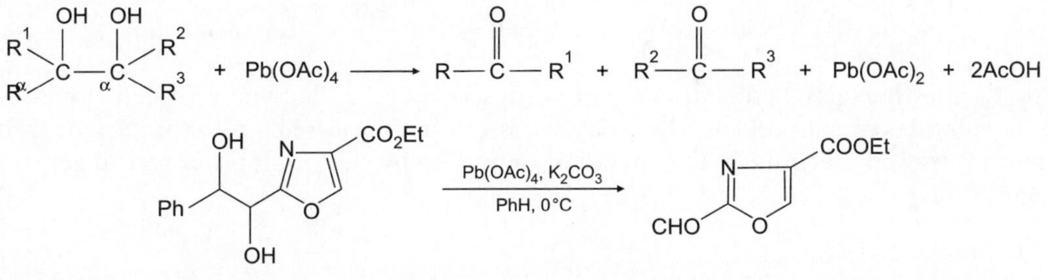

When the glycol cleavage is mediated by periodic acid it is called malaprade periodic acid oxidation (first reported by the French chemist Léon Malaprade in 1934). With the application of periodic acid water soluble diols can be cleaved (the water sensitive lead tetra-acetate is used in organic solvents).

92. CURTIUS REARRANGEMENT

In 1890, Theodor Curtius observed the decomposition of acyl azide with loss of nitrogen to generate aniline after hydrolysis of the intermediate isocyanate. Curtius rearrangement describes a thermal or photochemical rearrangement of acyl azides into amines via isocyanate intermediates. These intermediates may be isolated, or their corresponding reaction or hydrolysis products may be obtained. The reaction goes through an acyl-nitrene species with retention of configuration and by loss of N_2. This rearrangement is closely related to the Lossen reaction as well as the Hofmann rearrangement. The Curtius reaction can thus be applied to convert carboxylic acids into primary amines. The reaction can also be considered as the nitrated analogue of Wolff rearrangement.

The thermal rearrangement:

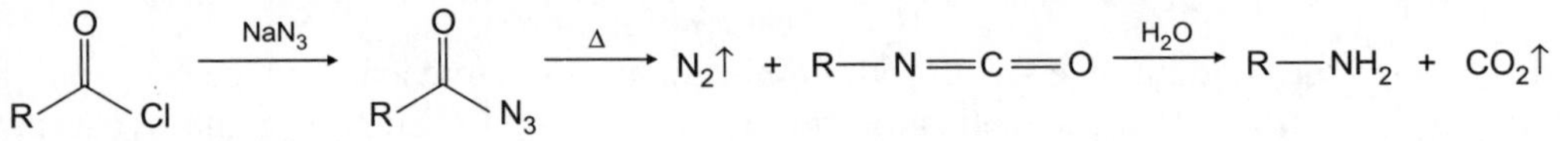

The photochemical rearrangement:

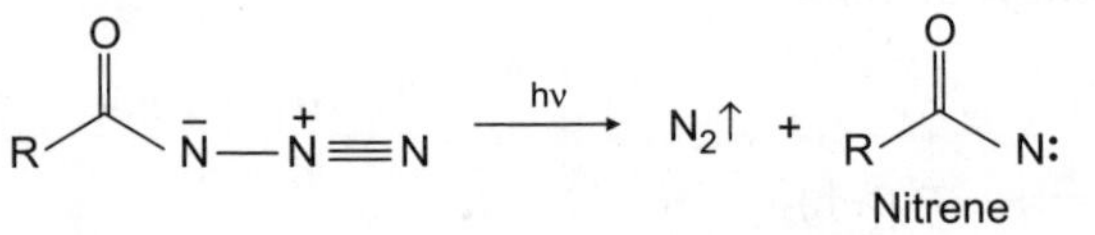

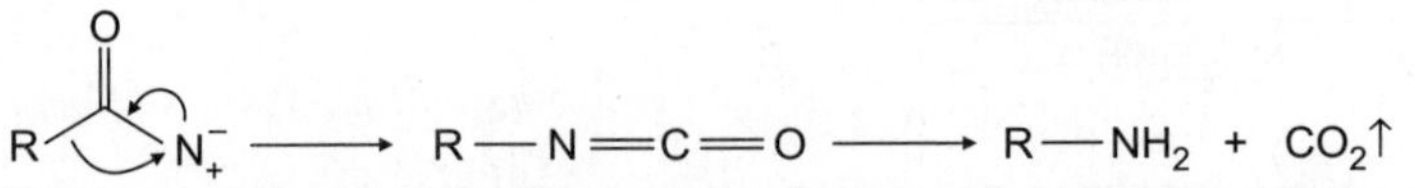

This rearrangement is valuable for synthesizing amines from the corresponding acid. This stereospecific rearrangement provides carbamates or amines in good overall yields and selectivity.

Curtius rearrangement allows the formation of a new C-N bond from carbonyl containing compounds. The reaction sequence including subsequent reaction with water which leads to amines is named the Curtius reaction.

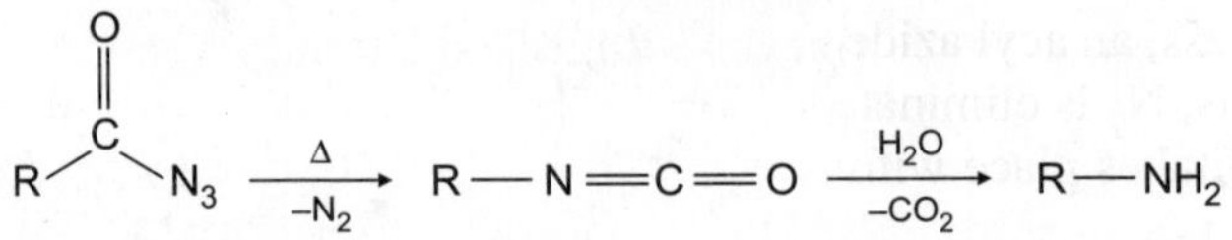

Acyl azides for the reaction are prepared from activated carboxylic acid derivatives such as acyl chlorides or anhydrides. The isocyanates can be transformed into (protected) amines by hydrolysis and alcoholysis after workup or *in situ* by performing the rearrangement in alcoholic solvents.

The isocyanates can be transformed into boc-protected amines with tert-butanol and cbz-protected amines with benzyl alcohol.

Although, the Curtius rearrangement can be catalyzed by Lewis acids or protic acids, but good yields can also be obtained without catalyst.

In the general process, an acyl azide is converted by heating into the corresponding isocyanate. During the thermolysis, N_2 is eliminated and at the same time a [1,2]-shift of the substituent next to the carbonyl group takes place with retention of configuration.

93. DAKIN REACTION OR DAKIN OXIDATION

Dakin reaction involves preparation of phenols from aryl aldehydes or aryl ketones by their oxidation with basic hydrogen peroxide followed by hydrolysis of the resulting aryl formate or alkanoate intermediates. Overall, formyl or acetyl groups in phenolic aldehydes or ketones are replaced by a hydroxyl group or in other words the carbonyl group is oxidized, and the hydrogen peroxide is reduced. The Dakin reaction proceeds by a mechanism analogous to that of the Baeyer-Villiger reaction.

Aryl aldehydes or aryl ketones in Dakin reaction usually need to have one electron-donating group like hydroxyl, alkoxyl or amino groups in the ortho- or para- positions. Such groups found to facilitate this reaction, i.e. they help the migration of the aryl group. Absence of such groups results in substituted benzoic acid as reaction product.

This reaction should not be confused with the Dakin-West reaction, though both are named after Henry Drysdale Dakin. The Dakin oxidation is used for the synthesis of benzenediols and alkoxyphenols, e.g. catechol from o-hydroxy and o-alkoxy phenyl aldehydes and ketones, guaiacol, indolequinones, etc.

94. DAKIN-WEST REACTION

The Dakin-West reaction involves preparation of α-acetamido alkyl methyl ketones from the reaction of α-amino acids with acetic anhydride in the presence of base such as pyridine, via oxazoline (azalactone) intermediates. Overall, the Dakin-West reaction transforms an amino-acid into a keto-amide using an acid anhydride and a base. This reaction was named after Henry Drysdale Dakin and Randolph West.

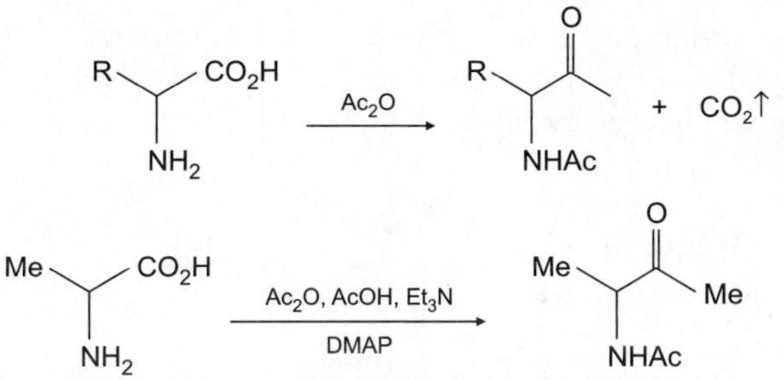

It has been found that with pyridine as a base and solvent refluxing conditions or high temperature is usually required for the reaction; however, with the addition of 4-dimethylamino-pyridine (DMAP) the reaction can even take place at ambient temperatures. This reaction should not be confused with the Dakin reaction.

95. DARZENS CONDENSATION (DARZENS-CLAISEN REACTION, DARZENS GLYCIDIC ESTER CONDENSATION)

The Darzens reaction (also known as the Darzens condensation or glycidic ester condensation) was discovered by the organic chemist Auguste George Darzens in 1904. It involves formation of α,β-epoxy esters also called a "glycidic ester" from base-catalyzed condensation of α-haloesters with carbonyl compounds (aldehydes or ketones).

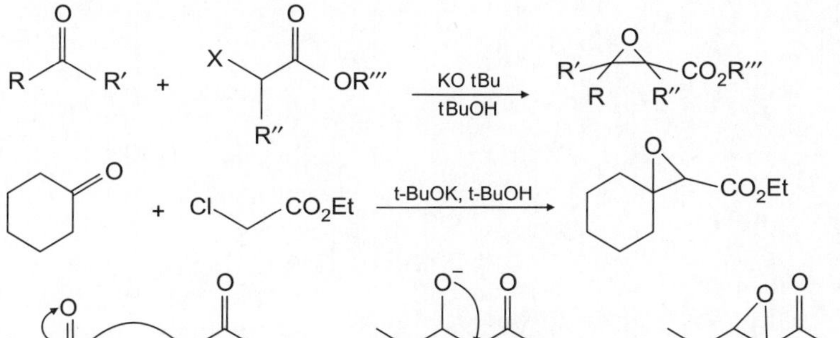

Mechanistically, this reaction proceed with an initial Knoevenagel-type reaction followed by an internal S_N2 reaction. The negatively charged oxygen attacks the carbon with the halogen, forming the epoxide. Chlorine is used as the halogen because it is an excellent leaving group.

This reaction enables the construction of highly functionalized oxiranes. The Darzens condensation plays an important role in catalytic asymmetric C-C bond forming reactions, and is one of the most powerful methodologies for the synthesis of α,β-epoxy carbonyl compounds and derivatives thereof. Although dramatic progress has been achieved in the past two decades, the classical Darzens condensation is still performed in the presence of a strong base (RONa, ROK, NaNH$_2$) in organic solvents (THF, toluene, CH$_2$Cl$_2$) under anhydrous conditions and at low temperatures (−78–0 °C). Aromatic aldehydes and ketones give good yields but aliphatic aldehydes react poorly for the obvious reason that they undergo aldol condensation.

96. DARZENS-NENITZESCU SYNTHESIS OF KETONES OR NENITZESCU SYNTHESIS

This reaction involves formation of an α,β-unsaturated ketone by lewis acid (e.g. aluminum chloride) catalyzed acylation of olefins with acid chlorides or anhydrides. When the reaction is carried out with a saturated hydrocarbon instead of olefins the product is the saturated ketone.

97. DARZENS SYNTHESIS OF TETRALIN DERIVATIVES

This reaction was discovered by Auguste George Darzens in 1926. It describes intramolecular ring-closing reaction of α-benzyl-α-allylacetic acid type compounds by moderate heating in concentrated sulfuric acid to yield tetralin derivatives.

98. DELEPINE REACTION (DELEPINE AMINE SYNTHESIS)

The Delépine reaction involves the synthesis of primary amines from alkyl halides by the reaction with hexamethylenetetramine (urotropine) followed by acidic hydrolysis (e.g. ethanolic hydrochloric acid solution) of the resulting quartenary ammonium salt. This reaction works well for active halides such as benzyl, allyl halides and α-halo-ketones.

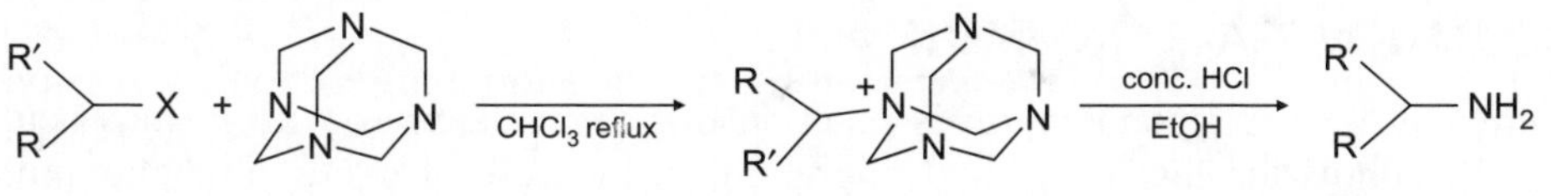

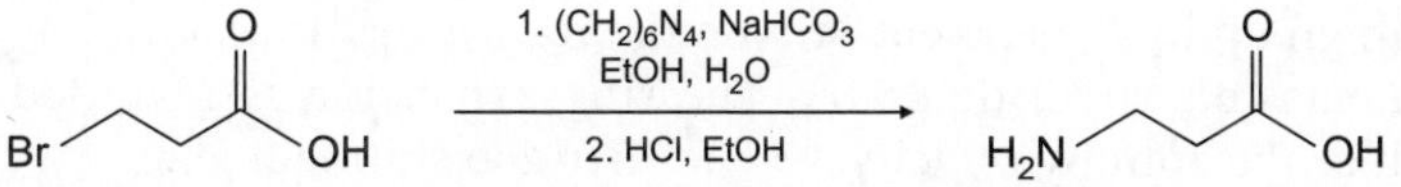

An S_N2 reaction leads to the hexamethylenetetramine salt which when refluxed in concentrated ethanolic hydrochloric acid solution is converted to the primary amine. During acidic hydrolysis, semiaminals are formed first; these further decompose to yield formaldehyde or the diethylacetal, ammonium salt and the amine hydrochloride. This reaction was named after Stéphane Marcel Delépine and found importance in the transformation of alkyl halides into primary amines. The reaction also allows selective access to the primary amines under mild reaction conditions within short reaction times and with no side reactions.

99. DE MAYO REACTION

The de Mayo reaction is a [2 + 2] photochemical cyclization in which the enol of a 1,3-dicarbonyl compound reacts with an alkene (or another species with a C=C bond) to form a cyclobutanol intermediate which then undergoes a retro-aldol reaction to yield a 1,5-diketone.

The enol originates in most cases from the tautomerization of a 1,3-dicarbonyl compound. Overall, the two carbon atoms in the C=C double bond are added between the two carbonyl groups of the diketone in the reaction.

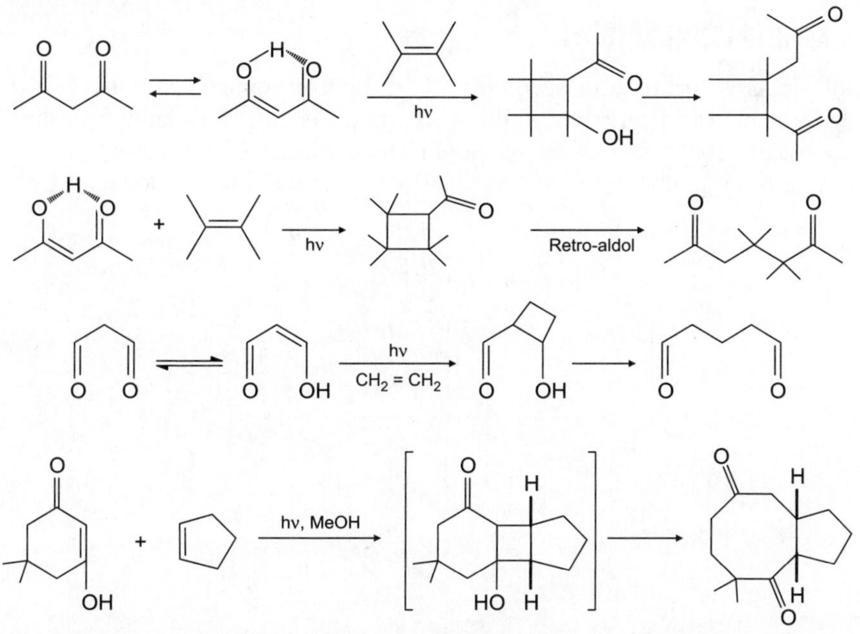

[2+2] cycloaddition being the first part of reaction results in formation of a cyclobutane ring. The subsequent retro-aldol cleavage is favoured by the relative instability of the cyclobutane ring.

100. DEMJANOV REARRANGEMENT

The Demjanov rearrangement involves reaction of primary amines with nitrous acid to give rearranged alcohols through C-C bond migration. In the case of cyclic aliphatic amines, ring enlargement or contraction occurs. The rearrangement is competitive with the substitution of the diazonium leaving group by the solvent or with the formation of carbocations that may undergo other rearrangements (e.g. hydride shift). The ring expansion is favoured in the Demjanov rearrangement, since the entropy of activation for hydride shift is higher.

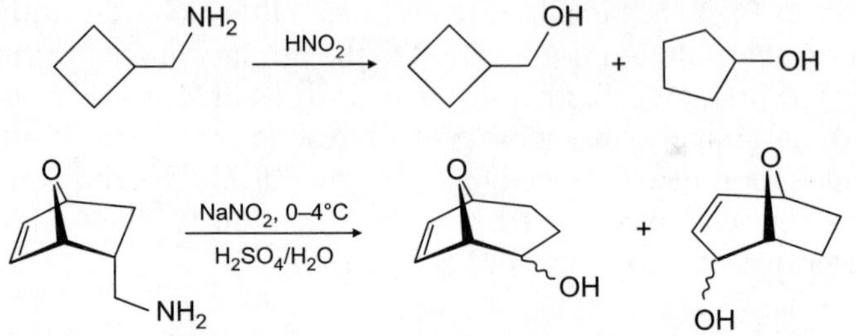

Reaction starts with the formation of nitrosonium ion (an electrophile), which is attacked by the primary amino group and in a series of proton transfer, the diazonium ion is formed. Finally, the diazonium ion undergoes a [1,2]-alkyl shift accompanied by the loss of nitrogen. Mechanistically, the reaction is said to be a carbocation rearrangement of primary amines via diazotization to give alcohols through C-C bond migration. In the reaction, HNO_2 forms a diazo compound with the NH_2 group, which therefore converts NH_2 into a good leaving group. The reaction is named after the Russian chemist Nikolai Jakovlevich Demjanov.

101. DESS-MARTIN OXIDATION

Oxidation of alcohols (primary or secondary) to the corresponding carbonyl compounds using triacetoxyperiodinane (a hypervalent iodine compound) is generally known as the Dess-Martin periodinane oxidation or Dess-Martin oxidation. The oxidation is carried out in dichloromethane or chloroform at room temperature, and is usually completed within 2 hours.

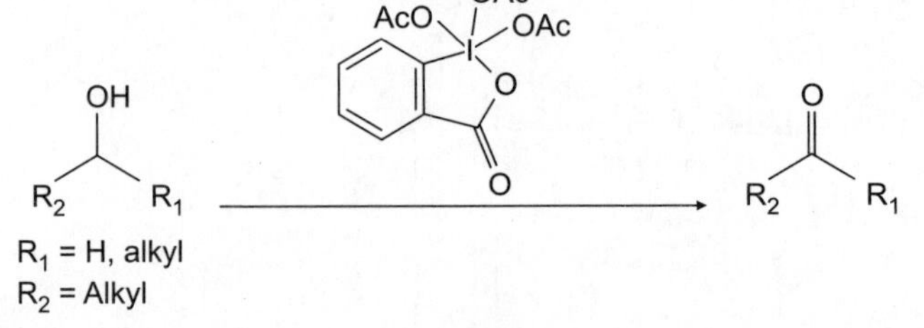

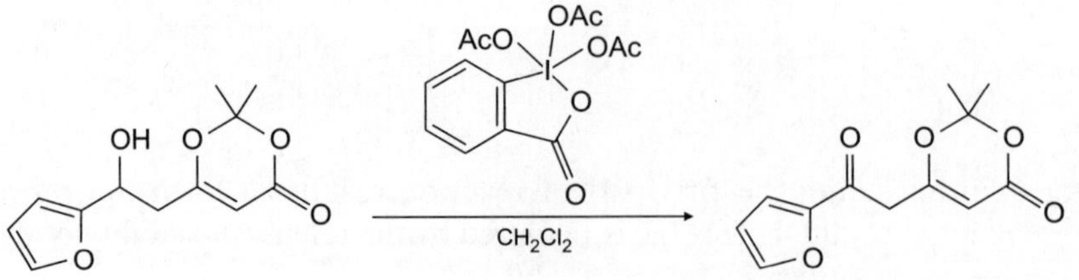

Dess-Martin Periodinane

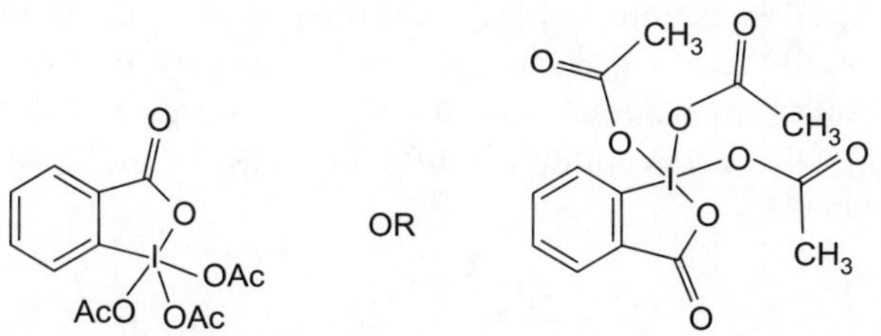

There are wide number of hypervalent iodine reagents (iodine in +3 and +5 oxidation states). Dess-Martin periodinane (DMP) [1,1,1-triacetoxy-1,1-dihydro-1,2-benziodoxol-3(1H)-one] is among the most widely used reagents for the mild oxidation of alcohols. It is named after its discoverers Daniel Benjamin Dess and James Cullen Martin who developed the reagent in 1983.

The neutral and milder conditions (room temperature and neutral pH) of the oxidation reaction makes it a suitable choice in synthesis over chromium- and DMSO-based oxidants of sensitive functionality. Other advantages such as high chemoselectivity, avoiding the use of toxic (i.e. chromium-based) chemicals, using stoichiometric amounts of reagent, a long shelf life and ease of workup make the Dess-Martin periodinane a reliable and easy-to-use oxidizer in organic synthesis. The reagent can be readily prepared from 2-iodobenzoic acid.

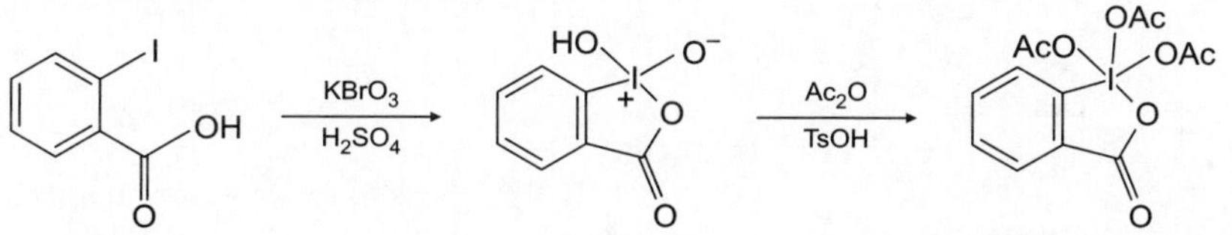

Dess-Martin periodinane

102. D-HOMO REARRANGEMENT OF STEROIDS

The D-homo rearrangement occurs at the D-ring of steroids and involves an acid (Lewis acid) or base-catalyzed acyloin rearrangement of a cyclopentane ring to yield a 6-membered D-ring (cyclohexane ring). Therefore, a cyclopentane ring expands to a cyclohexane ring. It has been found that the products resulting from the base treatment differ greatly from those caused by the acid treatment. This rearrangement serves application for the preparation of a series of steroid derivatives.

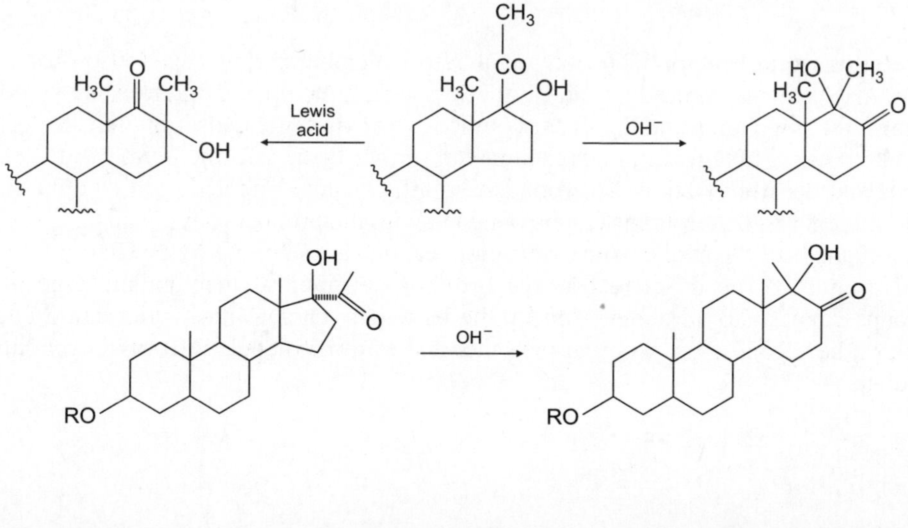

103. DIECKMANN REACTION

This reaction was named after the German chemist Walter Dieckmann. The Dieckmann condensation (occasionally known as Dieckmann ring closure) involves a base-catalyzed intramolecular condensation of α,ω-diesters to form cyclic β-keto esters. This reaction is the intramolecular version of the Claisen condensation and is particularly useful in the formation of five- and six-membered rings.

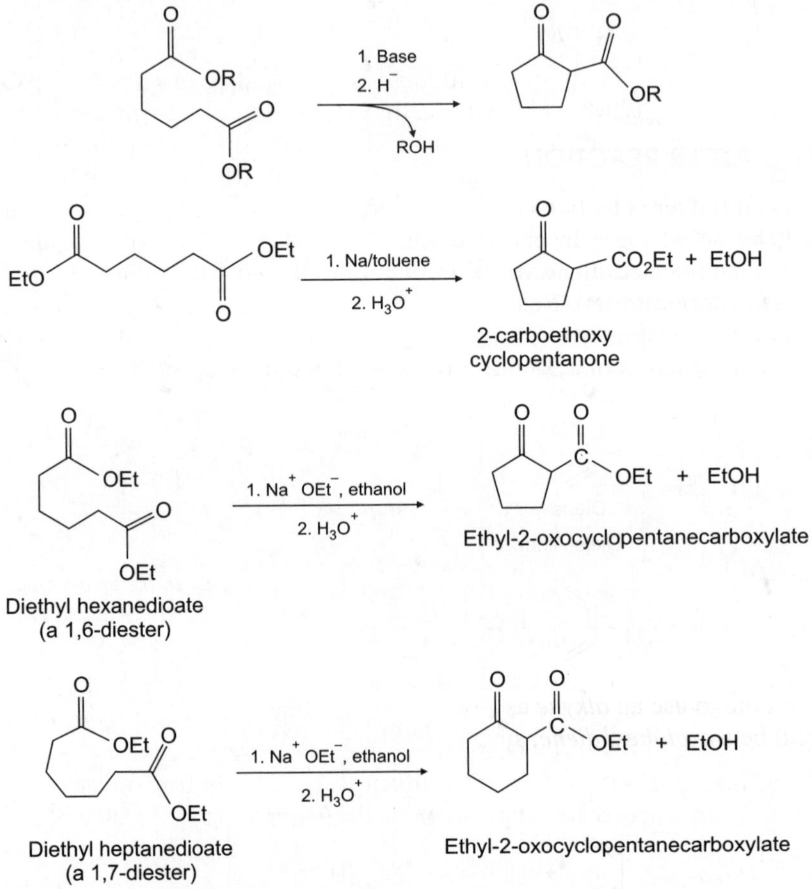

2-carboethoxy
cyclopentanone

Ethyl-2-oxocyclopentanecarboxylate

Diethyl hexanedioate
(a 1,6-diester)

Diethyl heptanedioate
(a 1,7-diester)

Ethyl-2-oxocyclopentanecarboxylate

The reaction works best for the formation of 5- to 7-membered rings, where 1,6-diesters provide 5-membered β-keto ester as product and 1,7-diesters yield 6-membered β-keto ester as product. With larger rings, the low yields and the faster competitive acyloin condensation may occur.

The β-keto esters obtained in this reaction can further be transformed into cyclic ketone upon hydrolysis and decarboxylation. This condensation is usually effected with sodium alkoxide in alcoholic solvent and better yields can be obtained with the products having an enolizable proton.

The mechanism of the Dieckmann cyclization reaction is analogous to the Claisen condensation mechanism and can be described as the ordinary tetrahedral mechanism. One of the two ester groups converts to an enolate ion by the base. The enolate ion works like a nucleophile, then attacks the second ester group at the other end of the molecule to form a cyclic tetrahedral intermediate.

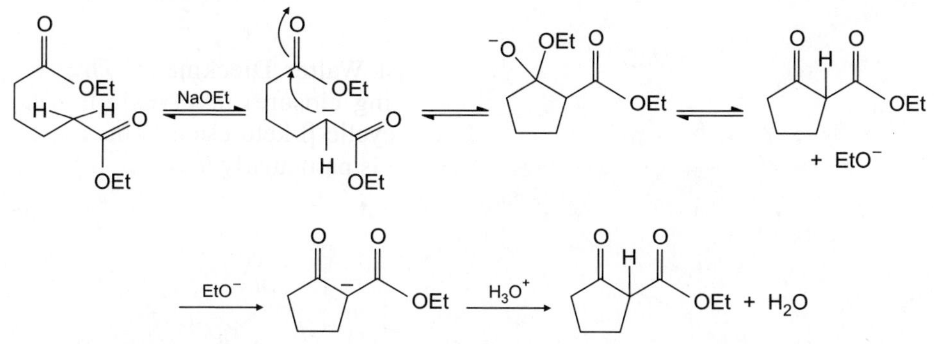

104. THE DIELS-ALDER REACTION

The reaction was discovered by two German chemists, Otto Diels and Kurt Alder, for which they were awarded the nobel prize in chemistry in 1950. The Diels-Alder reaction involves the cycloaddition of a conjugated **diene** with an alkene **(dienophile)** to form a compound containing a six-member ring **(cycloadduct)**. Reaction takes place in a single step (a *concerted reaction*). In conventional terminology, this is a 1,4-addition of a **diene** and a **dienophile**. The simplest Diels-Alder reaction is the reaction of 1,3-butadiene and ethylene to yield cyclohexene.

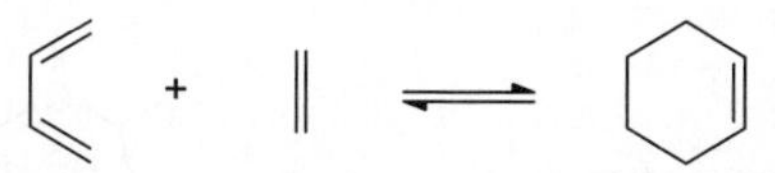

Diene   Dienophile   Cyclohexene

Diels-Alder addition of ethylene and butadiene

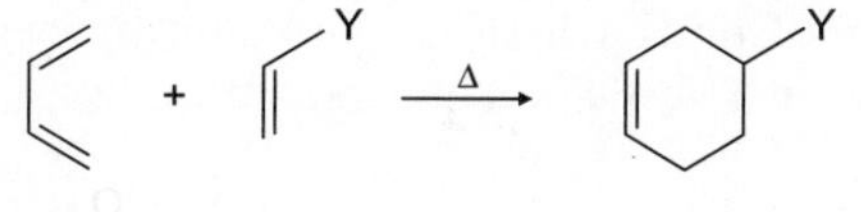

It is also possible to use an *alkyne* as the dienophile in a Diels-Alder reaction. In this situation, the product will be a *cyclohexadiene*, since only one pi bond of the alkyne will be involved in the cycloaddition.

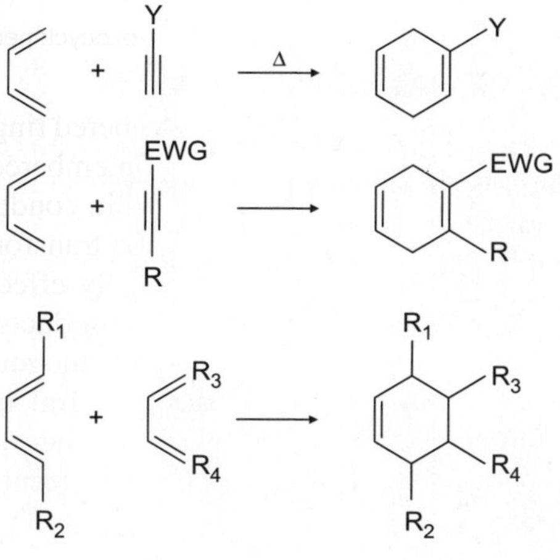

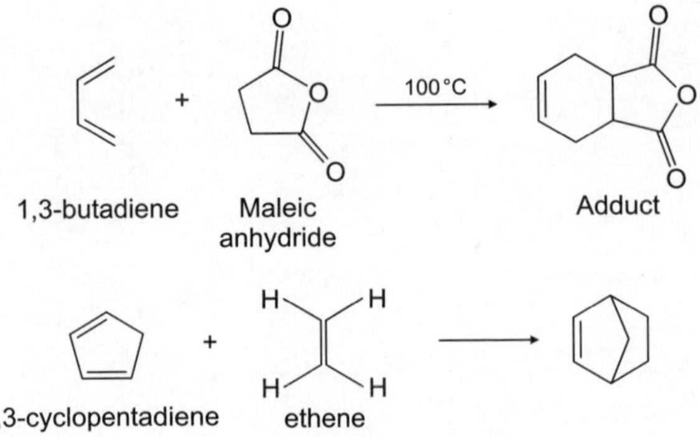

1,3-butadiene   Maleic   Adduct
anhydride

1,3-cyclopentadiene   ethene

In all Diels-Alder reactions, three π bonds, two in a diene and one in a dienophile, reorganize to give a six-membered ring containing one π bond and two new sigma (σ) bonds. Since the reaction forms a cyclic product, via a cyclic transition state, it can also be described as a "cycloaddition".

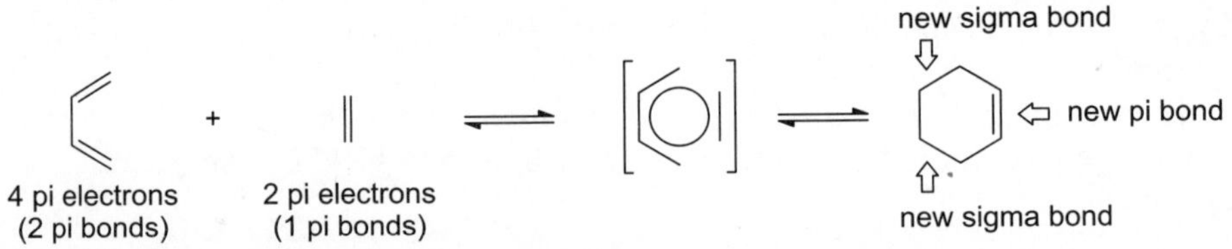

4 pi electrons   2 pi electrons
(2 pi bonds)   (1 pi bonds)

Diels-Alder reaction normally makes a cyclohexene product, but it makes a bicyclic ring structure if the diene is part of a cyclic system. An example is provided by the reaction of cyclopentadiene and maleic anhydride. The Diels-Alder reaction is usually thermodynamically favourable because the π-bonds are converted into σ-bonds, which are energetically more stable than π-bonds. Diels-Alder reactions are thermal reactions which are initiated by heat and usually require higher temperatures because of the higher activation energy.

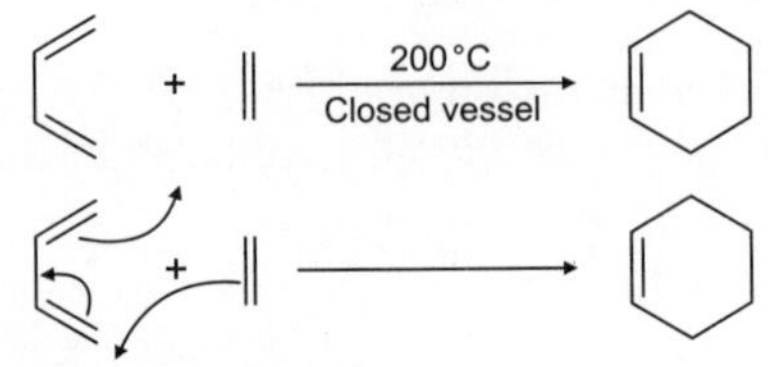

105. DIENONE-PHENOL REARRANGEMENT

Dienone-phenol rearrangement is an acid-promoted rearrangement of 4,4-disubstituted cyclohexadienones to 3,4-disubstituted phenols. The rearrangement involves the creation of a very stable aromatic system on acid treatment. Moreover, this aromatization works as driving force for these rearrangements. It is an exothermic reaction due to the formation of a very stable aromatic product. The rearrangement goes through a 1,2-alkyl shift, i.e. one of the alkyl groups in the 4-position undergoes 1,2-shift.

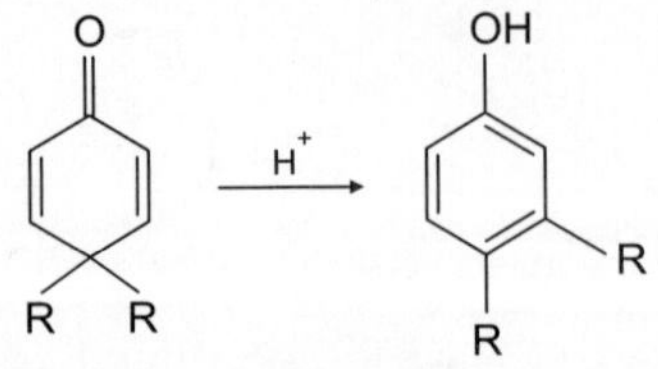

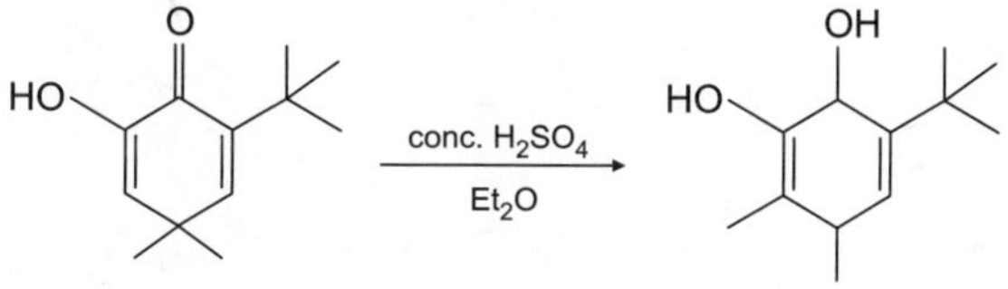

106. DIMROTH REARRANGEMENT

The Dimroth rearrangement involves an isomerization of certain 1,2,3-triazoles that consists of a translocation of endo- or exocyclic nitrogen atoms. Rearrangement occurs through a ring-opening-ring-closure sequence and can be catalyzed by acids, bases, heat or light. The reaction was discovered in 1909 by Otto Dimroth.

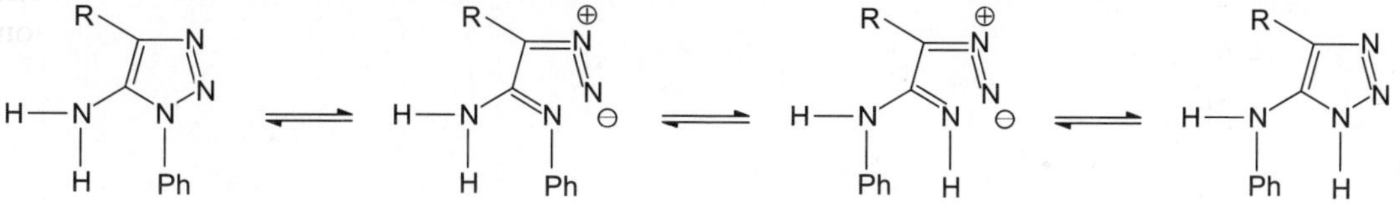

107. DOEBNER-MILLER REACTION; BEYER METHOD FOR QUINOLINES

Lewis acids catalyzed synthesis of 2,3-disubstituted quinolines from the condensation between primary aromatic amines (e.g. anilines) and α,β-unsaturated carbonyl compounds (mostly α,β-unsaturated aldehydes) to give 2,3-disubstituted quinolines is known as the Doebner-Miller reaction.

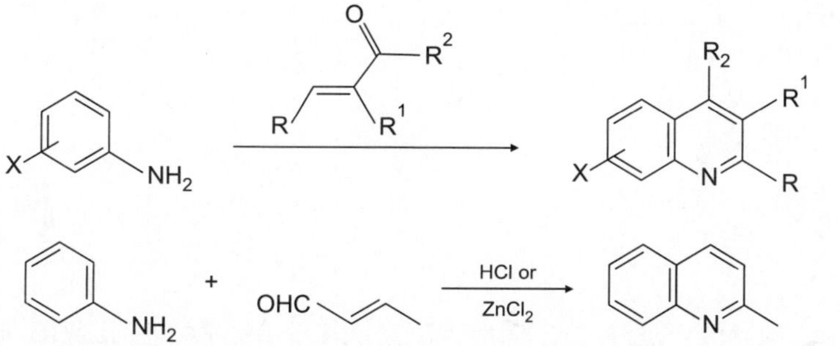

Various acids like Lewis acids (such as tin tetrachloride and scandium(III) triflate) and Bronsted acids (such as p-toluenesulfonic acid, perchloric acid, amberlite and iodine) can be used to catalyze the reaction. Doebner-von Miller reaction is a variant of the Skraup quinoline synthesis.

When the α,β-unsaturated carbonyl compound is prepared *in situ* from two carbonyl compounds (via an aldol condensation of two molecules of aldehyde or an aldehyde and methyl ketone), the reaction is referred to as the Beyer method for quinolines.

108. DOEBNER REACTION

The Doebner reaction involves formation of substituted cinchoninic acids (or quinolinic acid) in a three-component coupling reaction between aromatic amine such as aniline, pyruvic acid, and an aldehyde.

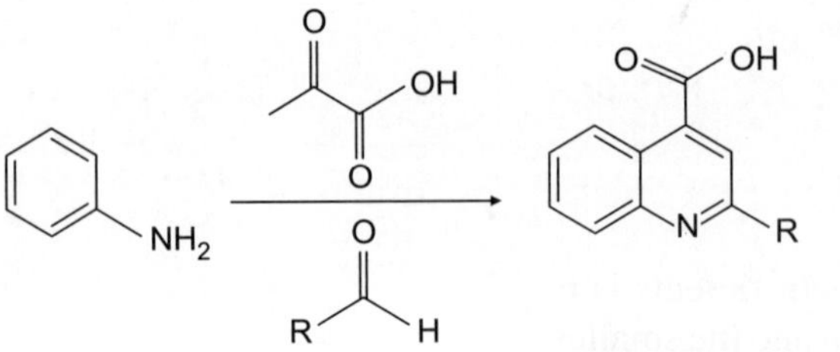

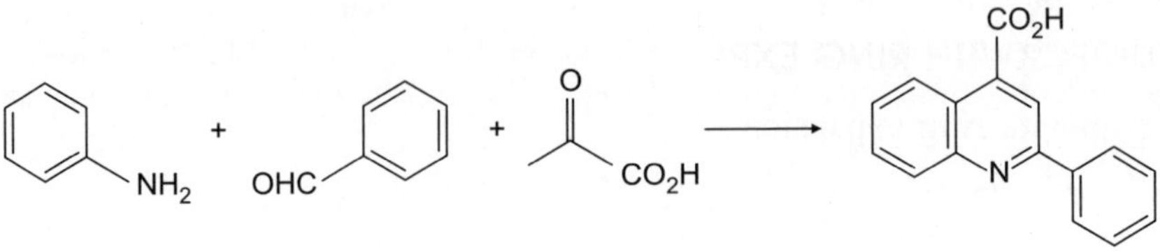

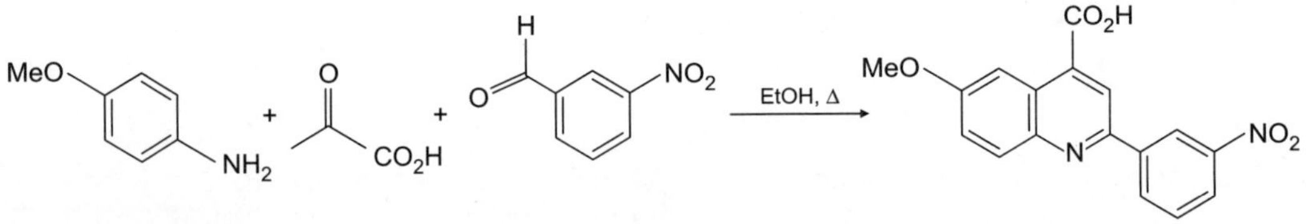

109. DOERING-LAFLAMME ALLENE SYNTHESIS

In Doering-LaFlamme allene synthesis, an olefin reacts with bromoform and an alkoxide to yield the 1,1-dibromocyclopropane. Subsequent reaction of 1,1-dibromocyclopropane with an active metal (magnesium or sodium metal) produce an allene.

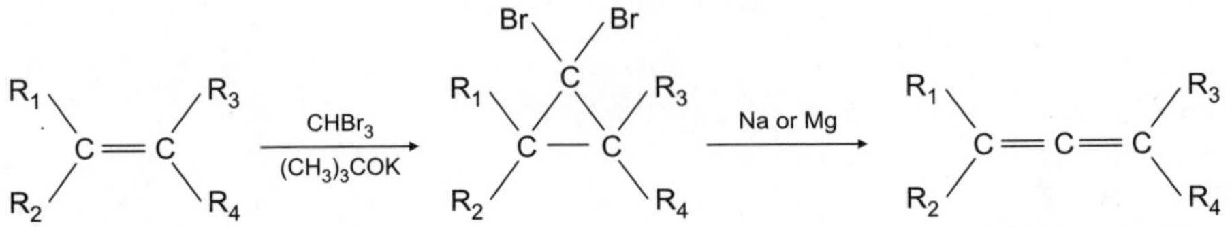

110. DÖTZ REACTION (WULFF–DÖTZ REACTION OR BENZANNULATION REACTION OF THE FISCHER CARBENE COMPLEXES)

The Dötz reaction or the Wulff-Dötz reaction or the benzannulation reaction of the Fischer carbene complexes involves the reaction of chromium carbenes with alkynes. This process, a formal [3 + 2 + 1] cycloaddition, generates substituted phenols by sequential coupling of the alkyne, the carbene and one carbonyl ligand at a $[Cr(CO)_3]$ template.

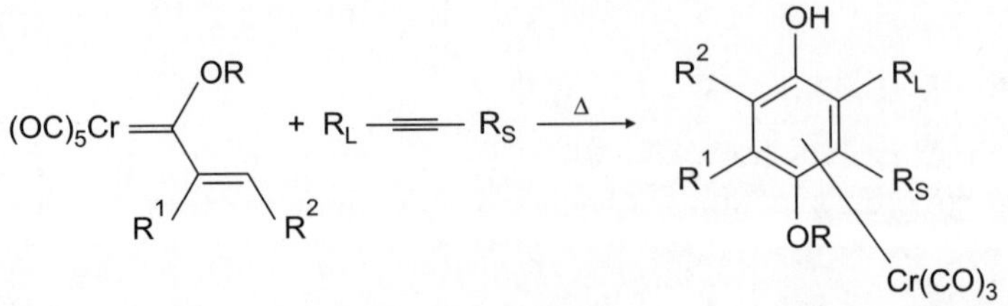

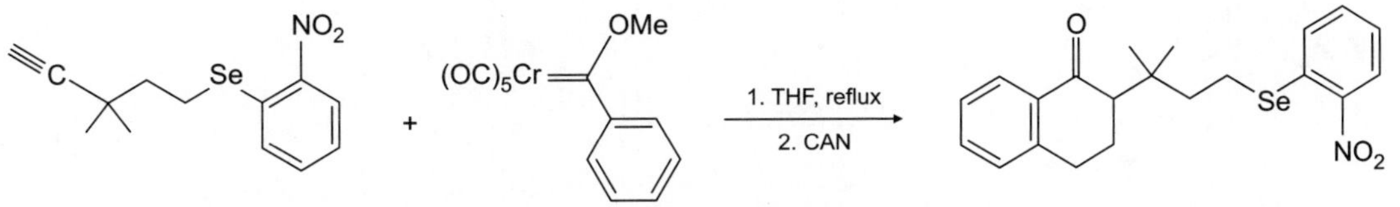

The position of the substituents is highly predictable with the largest alkyne substituent (R_L) neighbouring the phenol and the smallest alkyne substituent (R_S) neighbouring the methoxy group. Hence, this reaction is more useful for terminal alkynes than internal alkynes.

111. DOWD-BECKWITH RING EXPANSION REACTION

The Dowd-Beckwith ring expansion reaction describes radical-mediated ring expansion of a cyclic β-keto ester.

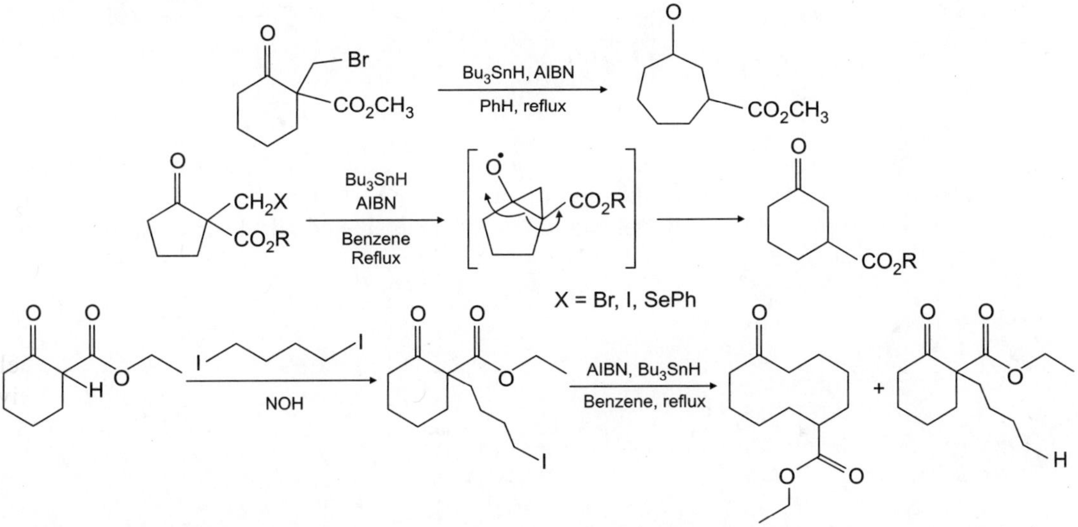

X = Br, I, SePh

The cyclic β-keto ester for the reaction can be obtained through Dieckmann condensation. This reaction has been applied for the preparation of cyclic ketones.

112. DUFF REACTION

This reaction was named after James Cooper Duff. It involves formylation of phenols or aromatic amines with hexamethylenetetramine (as the formyl carbon source) in the presence of an acidic catalyst.

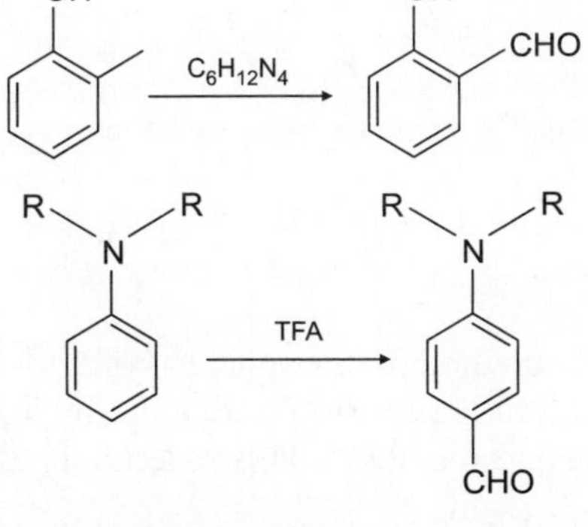

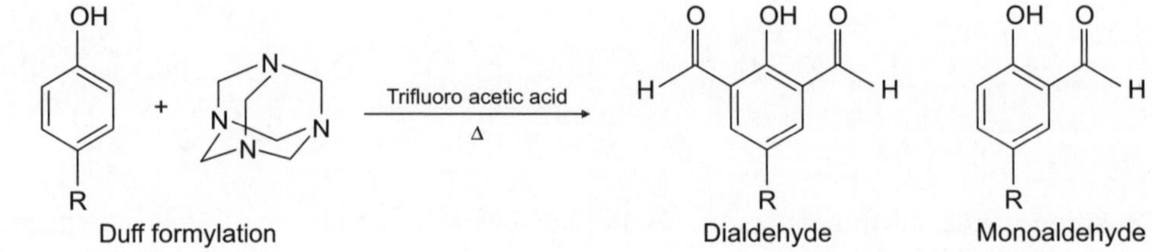

Duff formylation   Dialdehyde   Monoaldehyde

Ortho substitution is generally observed in this reaction, although the formylation may occur at para position if the ortho positions are blocked. This reaction is an useful route to aldehydes for compounds bearing sensitive amide functions.

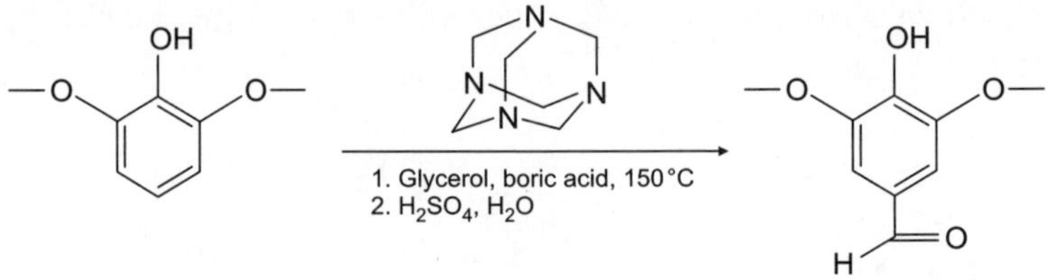

1. Glycerol, boric acid, 150 °C
2. H_2SO_4, H_2O

A mechanism has been proposed that involves an iminium ion $CH_2^+NR_2$ as the electrophilic species in this electrophilic aromatic substitution reaction. Reaction starts with aminoalkylation to give $ArCH_2NH_2$, followed by dehydrogenation to $ArCH=NH$ and hydrolysis of this to yield the aldehyde product.

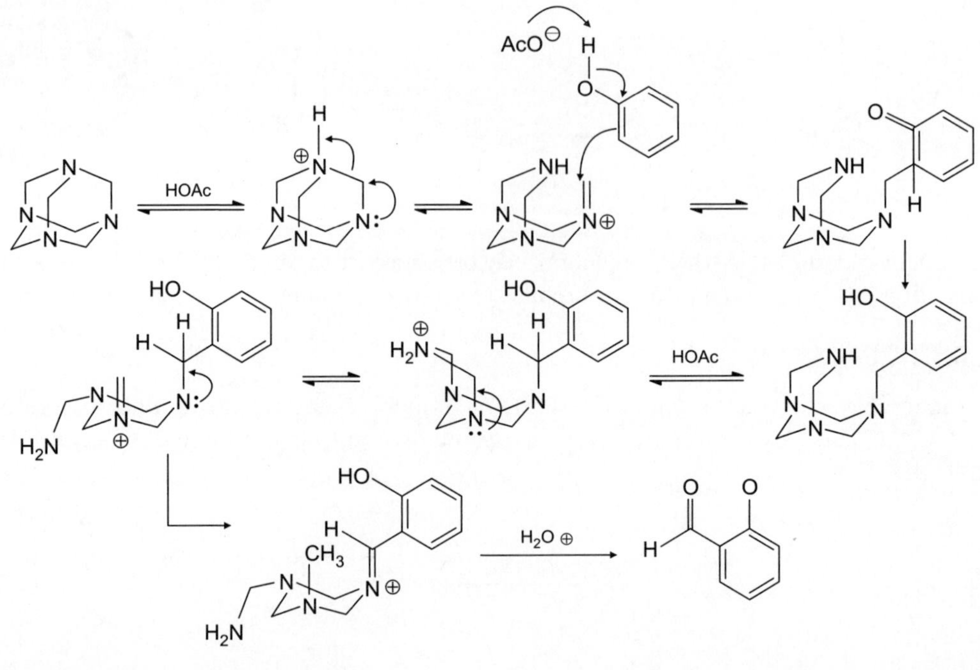

113. DUTT-WORMALL REACTION

Dutt-Wormall reaction is a classic method for the synthesis of azides involving reaction of diazonium salts with aryl- or alkylsulfonamides followed by alkaline hydrolysis to yield the corresponding sulfinic acid of the sulfonamide and the azide. This reaction is especially useful for the synthesis of aromatic azides.

$$\text{ArN}_2\text{Cl} \xrightarrow{\text{H}_2\text{NSO}_2\text{R}} \text{ArN} = \text{NNHSO}_2\text{R} \xrightarrow{\text{HO}^-} \text{ArN}_3 + \text{HO}_2\text{SR}$$

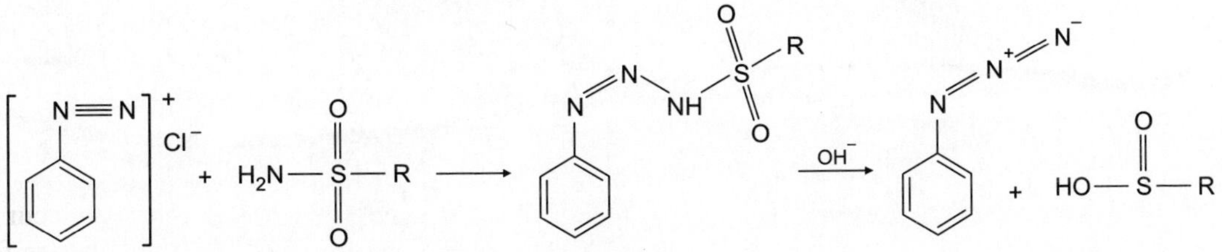

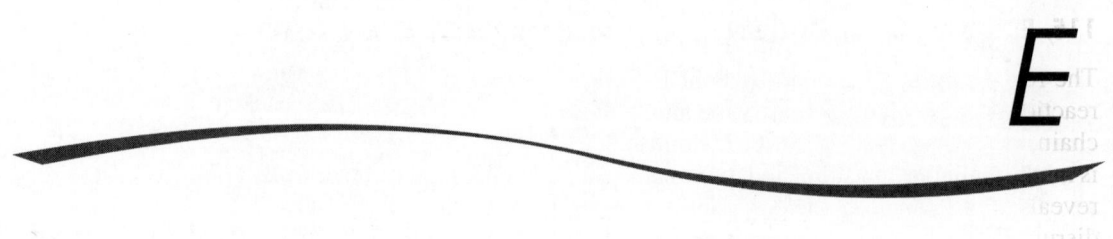

114. EASTWOOD REACTION (EASTWOOD DEOXYGENATION)

This reaction was reported by Grank and Eastwood in 1964. In Eastwood deoxygenation, a vicinal diol is treated with ethyl orthoformate at high temperature (140–180 °C), followed by thermal decomposition of the resulting cyclic orthoformate (160–220 °C) in the presence of a carboxylic acid (typically acetic acid). In other words, this reaction describes stereospecific conversion of vicinal diols into olefins:

115. EDMAN DEGRADATION

The reaction was developed by Pehr Edman in 1950. The method employs a series of chemical reactions to remove and identify the amino acid residue that is at the N-terminus of the polypeptide chain, i.e. the residue with a free α-amino group. At the same time, the next residue in the sequence is made available and subjected to the same round of chemical reactions. Reiteration of this process reveals the sequence of the polypeptide. Peptide bonds between other amino acid residues are not disrupted during the process. Amino acids are removed one at a time and identified as their phenylthiohydantoin derivatives.

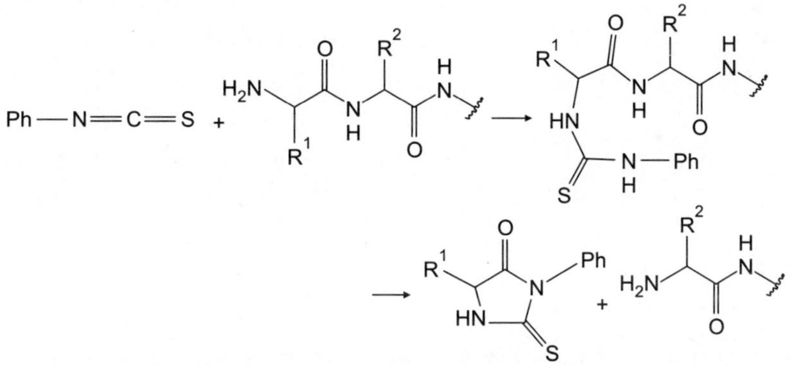

In Edman degradation, **phenylisothiocyanate (PITC)** reacts with the amino acid residue at the amino terminus under basic conditions to form a **phenylthiocarbamyl derivative (PTC-protein)**. First amino acid then cleaves off as its **anilinothialinone derivative (ATZ-amino acid)** and leaves the new amino terminus for the next degradation cycle. Edman method is used in sequencing amino acids in a peptide. This method allowed determination of extended sequences of peptides or whole proteins.

The Edman degradation method is described below in stepwise manner:

Formation of a phenylthiocarbamyl derivative (PTC-peptide) by using phenyl isothiocyanate (PITC), the Edman reagent, under mild alkaline conditions (N-methylpiperidine/water/methanol).

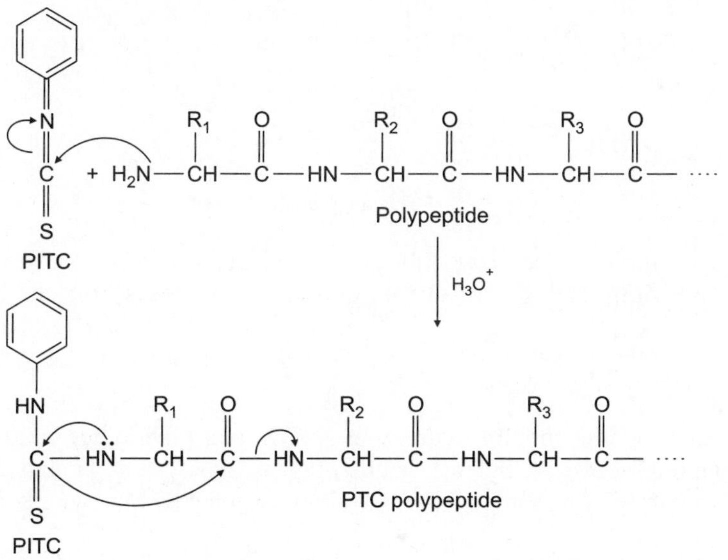

Then the terminal amino acid is cleaved (e.g. with anhydrous trifluoroacetic acid (TFA)) in the form of a thiozolinone derivative leaving the other peptide bonds intact and carries a free amino terminus. The thiozolinone (TZ) derivative is extracted in an organic solvent (e.g. N-butyl chloride).

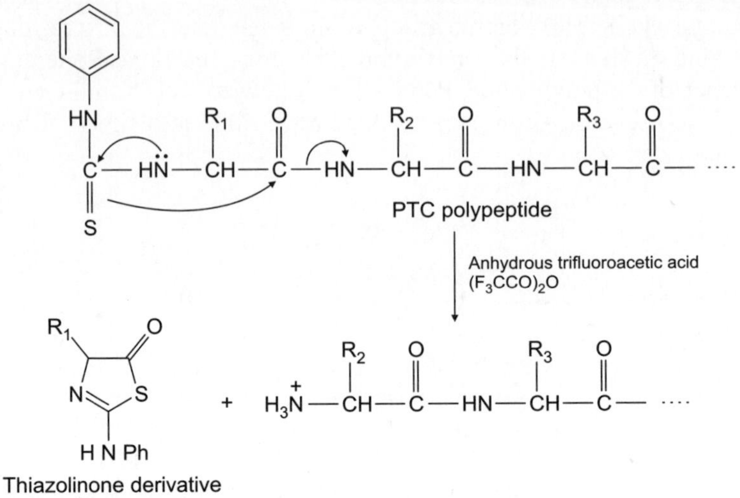

PTC polypeptide

Anhydrous trifluoroacetic acid
(F$_3$CCO)$_2$O

Thiazolinone derivative

The extracted TZ is treated with an acid (e.g. with 25 % TFA/water) to form phenylthiohydantoin (PTH) derivative and is detected from UV absorption. This sequence can be repeated to identify all amino acids.

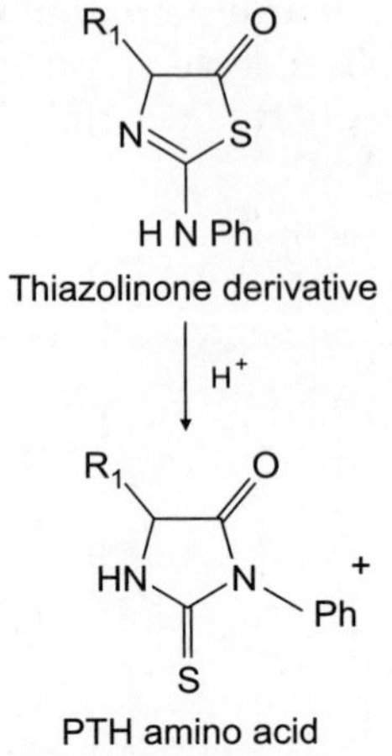

Thiazolinone derivative

H$^+$

PTH amino acid

The Edman degradation does not work if the N-terminal amino acid has been chemically modified or if it is concealed within the body of the protein since it proceeds from the N-terminus of the protein.

116. EHRLICH-SACHS REACTION

The base-catalyzed (such as sodium hydroxide, K$_2$CO$_3$ and piperidine) condensation between compounds containing active methylene groups (e.g. malonic esters, benzyl cyanide, etc.) and aromatic nitroso molecules to form N-phenylimines is generally known as the Ehrlich-Sachs reaction.

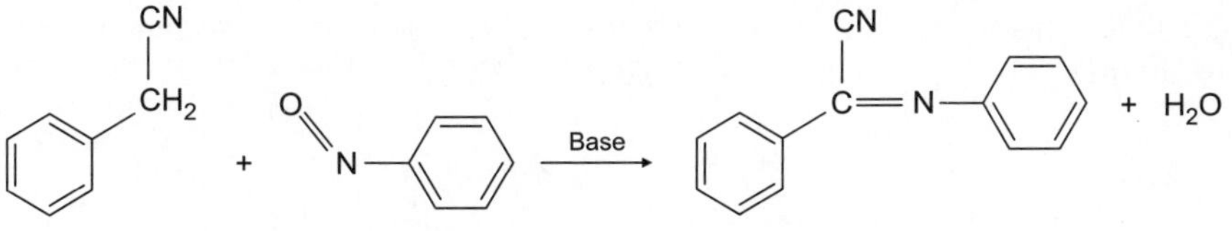

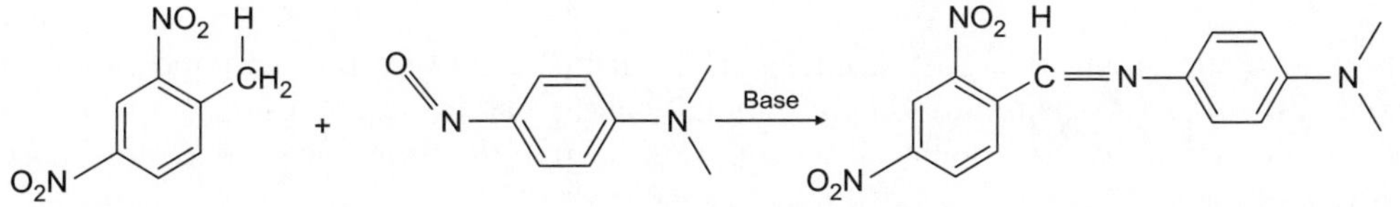

This reaction is used in the preparation of azomethine derivatives, which are used in the dye and drug industry.

117. EINHORN-BRUNNER REACTION

The Einhorn-Brunner reaction is the acid-catalyzed condensation of imides with hydrazines or semicarbazides to form a mixture of isomeric 1,2,4-triazoles. This cyclization reaction is catalyzed by an organic acid, such as acetic acid.

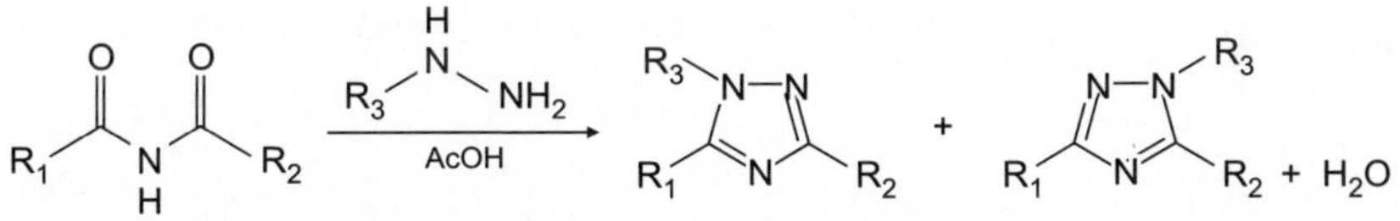

1,2,4-triazole derivatives find use in a wide variety of applications, most notably as antifungal such as fluconazole and traconazole.

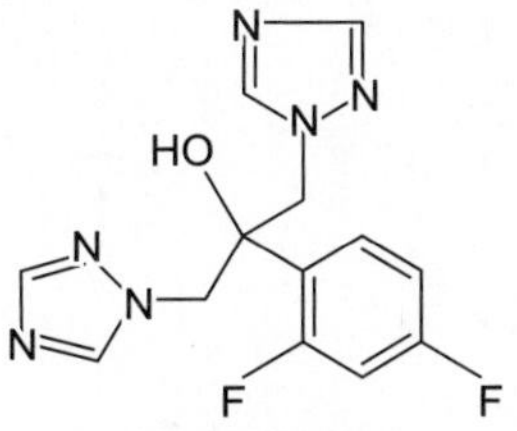

Fluconazole

118. ELBS PERSULFATE OXIDATION

The Elbs peroxydisulfate oxidation is the reaction of phenolate anions with peroxydisulfate ions to form a mixture of the ortho- and para-sulfates of the parent phenol. The para product predominates in the reaction. The reaction is named after the German inventor Karl Elbs.

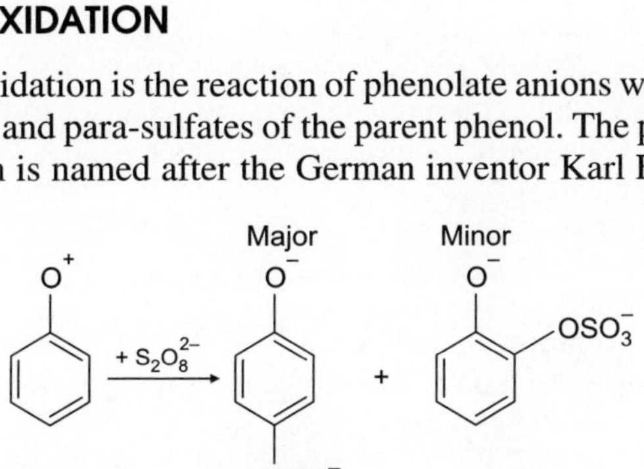

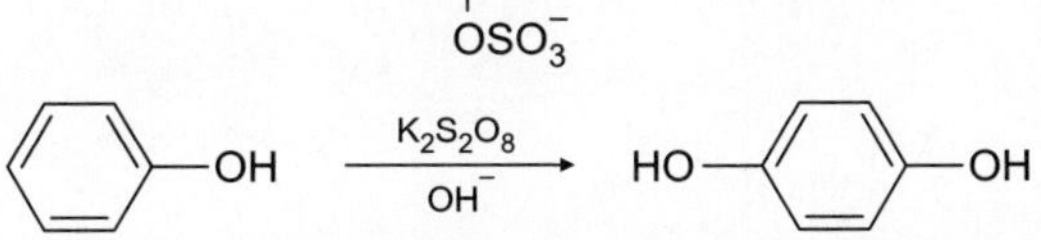

In the reaction mechanism (described for a phenol), first a resonance stabilized phenolate anion is obtained via deprotonation by a base:

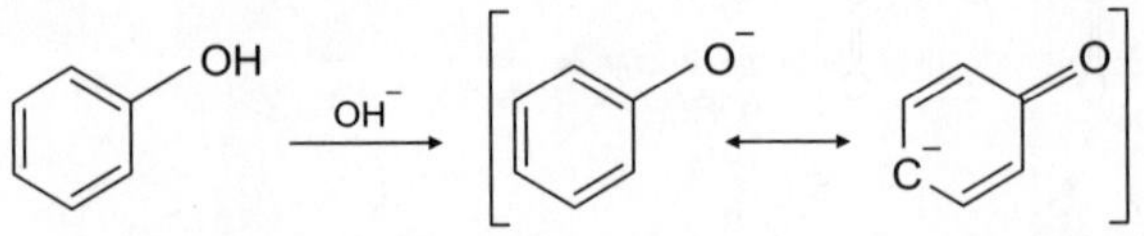

The phenolate anion undergoes a nucleophilic displacement on the peroxide oxygen of the peroxodisulfate (peroxydisulfate) ion preferentially at the para position, leading to formation of a cyclohexadienone derivative, which loses a proton to give the aromatic compound. Subsequent hydrolysis of the sulphate yields 1,4-dihydroxybenzene:

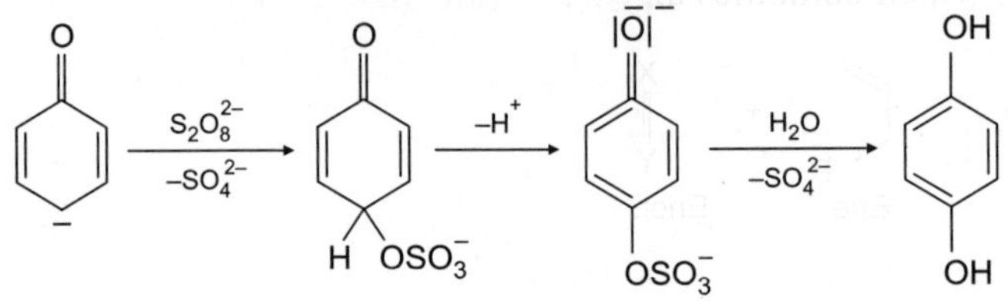

A similar reaction with amines is called the Boyland-Sims oxidation where aniline react with alkaline potassium persulfate, which after hydrolysis forms ortho-hydroxyl anilines.

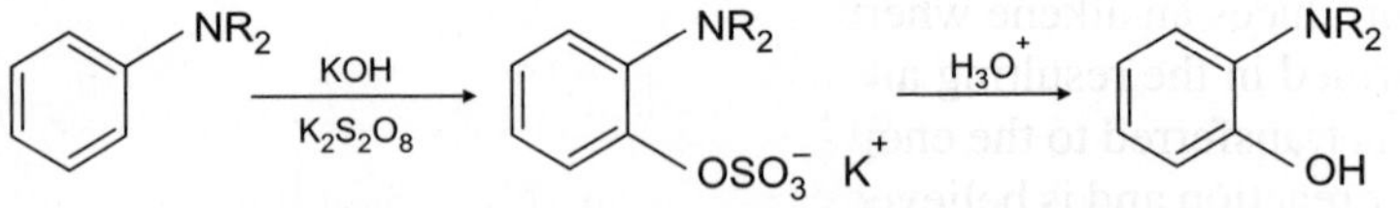

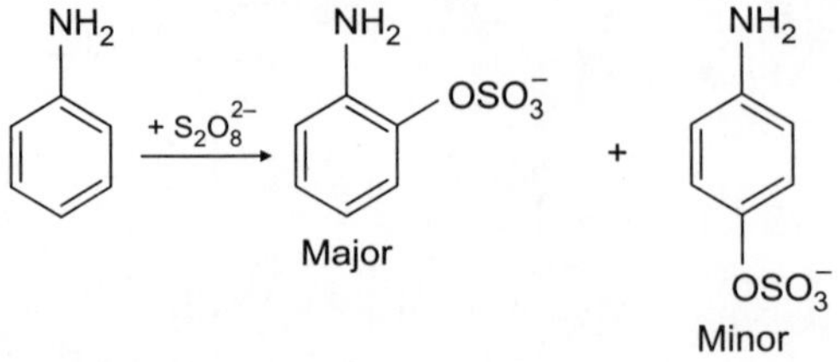

Both of these reactions are nucleophilic displacements on a peroxide oxygen of the peroxydisulfate ion. In the Elbs oxidation, the nucleophile is a phenolate anion (or a tautomer) and in the Boyland-Sims oxidation, it is a neutral aromatic amine. There is no radical involvement in either case. The products are aromatic sulfates whose orientation relative to the phenolic group is preferentially para in the Elbs oxidation and ortho in the Boyland-Sims case.

The yields of products are typically low to moderate, but the simplicity of the reactions frequently recommends their use.

119. ELBS REACTION

Elbs reaction describes formation of polyaromatics (especially anthracene) by intramolecular condensation of diaryl ketones containing a methyl or methylene substituent adjacent to the carbonyl group.

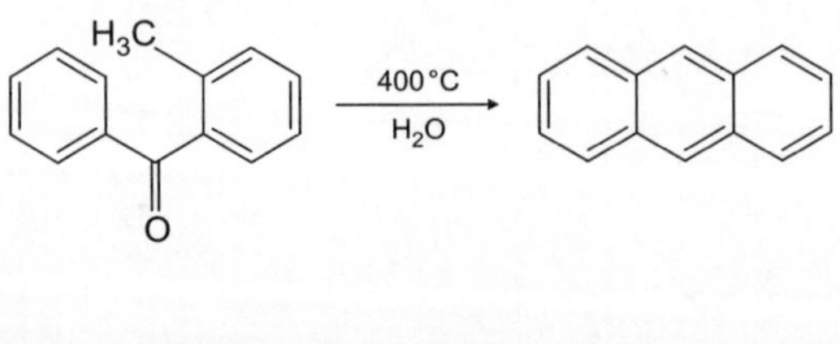

120. ALDER-ENE OR ENE REACTION

The Alder-ene or ene reaction or ene cyclization or ene functionalization or Alder-ene synthesis is a pericyclic reaction which combines an ene and enophile.

The ene is an alkene (or allene, arene, carbon-heteroatom bond, etc.) with an allylic hydrogen and the enophiles are normally electron deficient species comprising another unsaturated compound (carbonyls, thiocarbonyls, imines, alkenes, alkynes, etc.).

The reaction produces an alkene where the double bond is shifted to the allylic position. A new sigma bond is formed in the resulting alkene which connects the two unsaturated termini, and the allylic hydrogen is transferred to the enophile. The mechanism of the ene reaction is related to that of the Diels-Alder reaction and is believed to proceed via a six-membered aromatic transition state.

The Alder-ene reaction requires higher temperatures (450–500 °C) under forcing conditions because of the higher activation energy and stereoelectronic requirement of breaking the allylic C-H σ-bond. This has limited the synthetic use of the ene reaction. This reaction has very wide applications in organic synthesis. Frontier orbital considerations suggest that the major interaction between reacting components involves the HOMO* of the enes and the LUMO* of the enophiles.

Thermal ene reactions have several drawbacks, such as the need for very high temperatures and the possibility of side reactions, like proton-catalyzed olefin polymerization or isomerization reactions. The ene reactions can be catalyzed by Lewis acids [(BF$_3$O(CH$_2$CH$_3$) or Al(CH$_3$)$_2$Cl or Al(CH$_2$CH$_3$)Cl$_2$)], however, in such cases ene reactions are not necessarily concerted.

121. ERLENMEYER-PLÖCHL AZLACTONE AND AMINO ACID SYNTHESIS

Erlenmeyer-Plöchl Azlactone synthesis is the preparation of 5-oxazolones (or 'azlactones') by intramolecular condensation of acylglycines in the presence of acetic anhydride.

The azlactones on treatment with carbonyl compounds followed by hydrolysis to the unsaturated α-acylamino acid and by reduction yields the amino acid; drastic hydrolysis gives the α-oxo acid. The Erlenmeyer-Plöchl azlactone and amino acid synthesis are named after Friedrich Gustav Carl Emil Erlenmeyer. The azlactones are useful for the synthesis of α-ketos, α-amino acids and peptides.

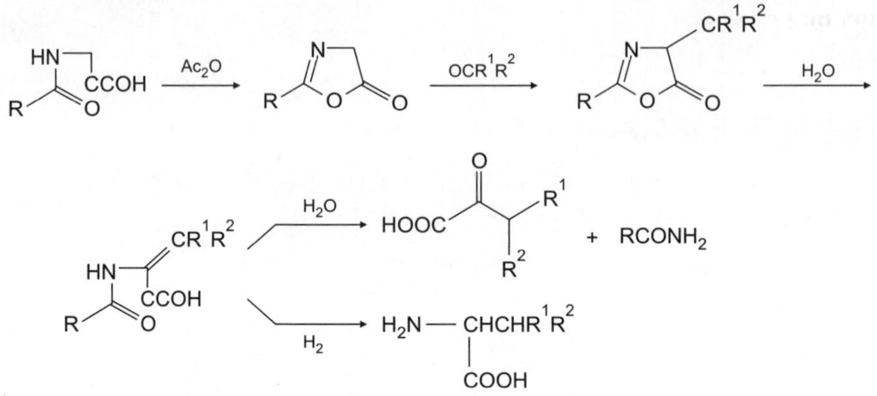

122. ESCHENMOSER COUPLING REACTION (SULFIDE CONTRACTION)

Eschenmoser coupling reaction describes preparation of vinylogous amides and urethanes by alkylation of secondary or tertiary thioamides with an electrophilic component (normally an α-bromocarbonyl system) followed by the sulfur extrusion to give the alkenic bond. The sulfide contraction as a synthetic tool was developed by Albert Eschenmoser.

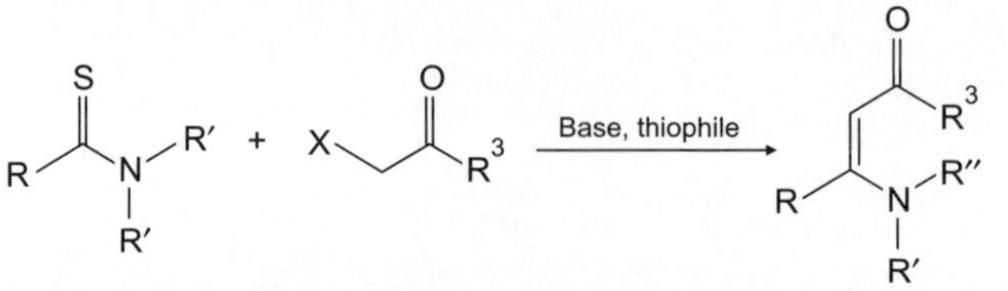

The reaction mechanism consists of two steps:

(i) Reversible *S*-alkylation of thioamides with electrophiles to form α-thioiminium salts or α-thioimine.

(ii) Deprotonation of the α-proton by a base, followed by sulfur extrusion to give the alkenic bond. The second step is believed to go through the formation of an episulfide intermediate.

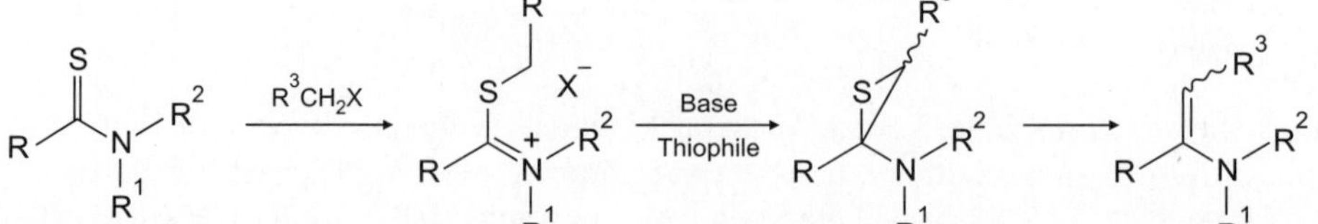

The reaction conditions are mild and it is tolerant of various functional groups. However, the reversibility of the first step of the mechanism is a notable limitation which does not allow the reaction to proceed forward in certain cases. Among the thioamides, the tertiary thioamides are preferred over the secondary substrates due to the failure of the sulfide-contraction of many secondary thiolactams instead the tertiary thioamides form a more reactive α-thioiminium intermediate.

The Eschenmoser coupling reaction is similar to the *aza Claisen-Schmidt condensation* and has been used for the preparation of vinylogous amides, especially for vitamin B_{12}.

123. ÉTARD REACTION

The Étard reaction involves the direct oxidation of an aromatic or heterocyclic bound methyl group to an aldehyde using hexavalent chromium compounds (chromyl chloride, CrO_2Cl_2). Chromyl chloride oxidizes methyl group to a chromium complex, which on hydrolysis gives corresponding benzaldehyde.

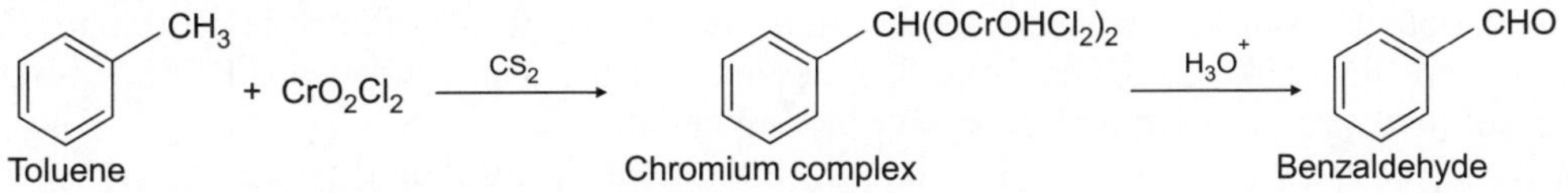

Toluene   Chromium complex   Benzaldehyde

The Étard reaction is most commonly used as a relatively easy method of converting toluene into benzaldehyde but synthesis of aldehyde products from reagents other than toluene is found to be difficult due to rearrangements. Apart from this lack of specificity, there is low yield of desired product because number of othe products are also formed such as a mixture of alcohol, carbonyl compounds (aldehyde or ketones), chloro-ketones, etc. Hence, this reaction is not a suitable tool for organic synthesis apart from converting toluene into benzaldehyde. It is named after the French chemist Alexandre Léon Étard.

124. EVANS ALDOL REACTION

Asymmetric enantioselective aldol condensation of the chiral N-acyl-oxazolidone via its dibutylboryl enolate with the appropriate aldehyde is generally known as the Evans aldol reaction. This reaction provides the most reliable means for controlling the vicinal syn stereogenic centers.

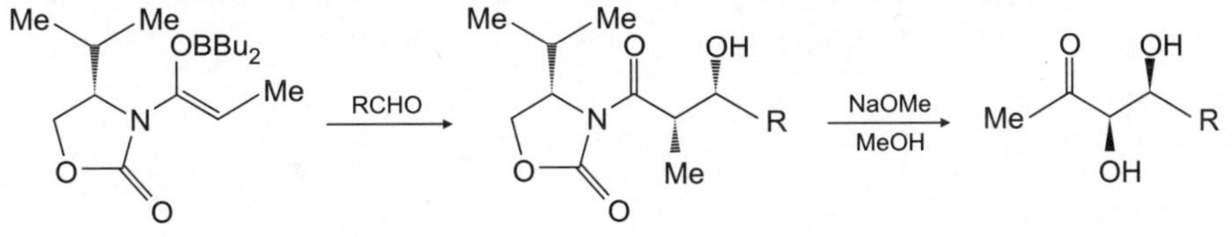

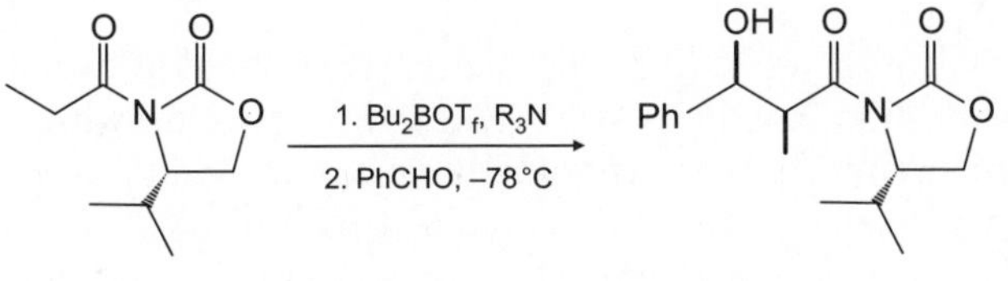

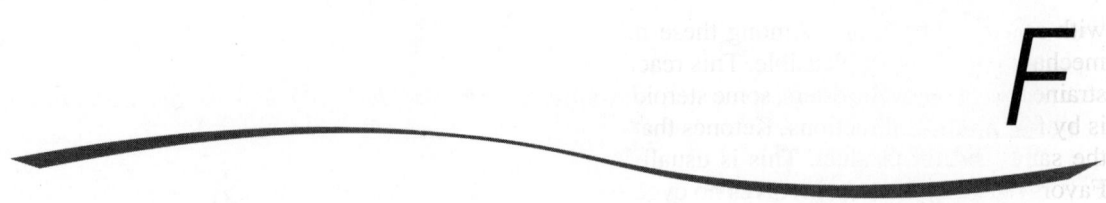

125. FAVORSKII-BABAYAN SYNTHESIS

Favorskii-Babayan synthesis or Favrskii reaction involves synthesis of acetylenic alcohols base-promoted, or catalyzed ethynylation of aldehydes and ketones using anhydrous KOH or NaOH. It should not be confused with the Favorskii rearrangement.

The reaction was discovered in the early 1900s by the Russian chemist Alexei Yevgrafovich Favorskii. In this reaction, nucleophilic attack of a terminal alkyne with acidic protons on a carbonyl group takes place. The reaction is catalyzed with a strong base (a hydroxide or alkoxide), typically the potassium hydroxide is used.

126. FAVORSKII REARRANGEMENT

The Favorskii rearrangement refers to the base-induced (alkoxide ions, hydroxide ions or amines) rearrangement of enolizable α-halo ketones (chloro, bromo, or iodo) to the corresponding carboxylic acid derivatives (acids, esters, and amides) with the same number of carbon atoms in the skeletons. Cyclic α-halo ketones undergo ring contraction by one carbon atom in the reaction.

In 1914, A. Favorskii observed a ring contraction when 2-chlorocyclohexanone reacted with alcoholic alkali. This reaction is akin to Wallach reaction (also known as Wallach degradation). Nature of product in **Favorskii rearrangement** varies with the base taken in the reaction. The free carboxylic acid (salt) or amide are obtained with hydroxide ions or amines as bases, respectively. Similarly, an ester is obtained when alkoxide ions are employed as base.

About six different mechanisms are given for this rearrangement. The detailed mechanism of a Favorskii rearrangement and the involvement of the postulated intermediates is expected to vary with reaction conditions. Among these mechanisms, the semibenzilic acid and cyclopropanone mechanisms are most plausible. This reaction is found to be applicable for the synthesis of highly strained esters, bicyclic esters, some steroids, etc. The most extensive use of Favorskii rearrangement is by far in ring contractions. Ketones that do not have enolizable hydrogen also rearrange to give the same type of product. This is usually called the quasi-Favorskii rearrangement. The quasi-Favorskii rearrangement involves no cyclopropanone intermediate and a semibenzilic mechanism is generally accepted.

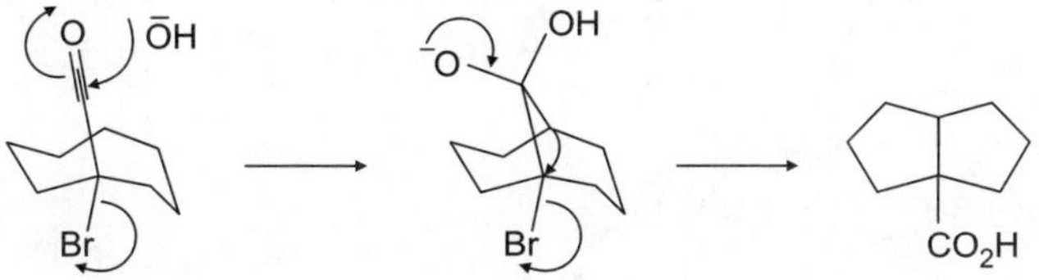

The synthesis of Pethidine (also known as Demerol or Meperidine) makes use of this rearrangement.

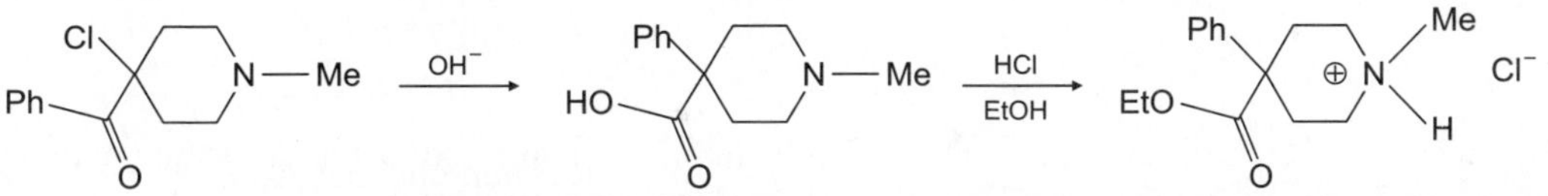

The Favorskii rearrangement akin to Wallach reaction (also known as Wallach degradation), which involves rearrangement of α,α′-dibromocyclohexanones to 1-hydroxycyclopentanecarboxylic acids, followed by oxidative decarboxylation to the ketones.

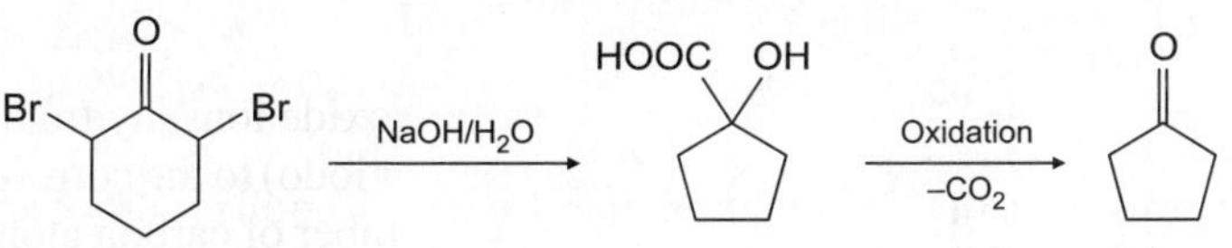

127. FEIST-BENARY SYNTHESIS

The Feist-Benary (FB) reaction is a very useful reaction for the construction of highly substituted furan derivatives via condensation of β-dicarbonyl (1,3-dicarbonyl compounds) compounds with α-haloketones in the presence of a base.

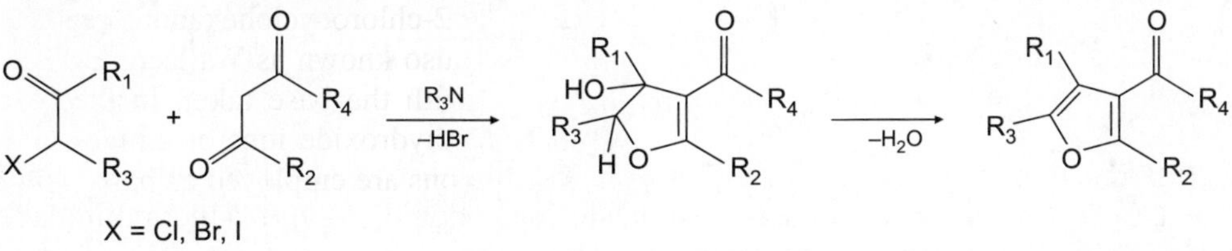

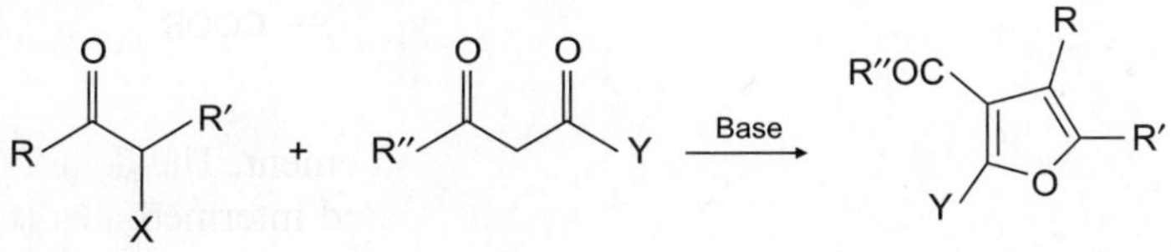

X = Cl, Br, I

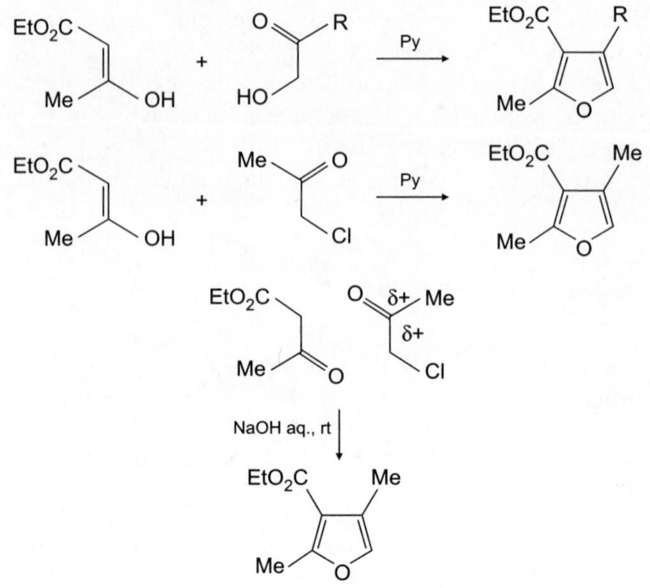

The bases used are normally ammonia and pyridine. When ammonia is used as the condensing agent, pyrrole derivatives are also formed as secondary products.

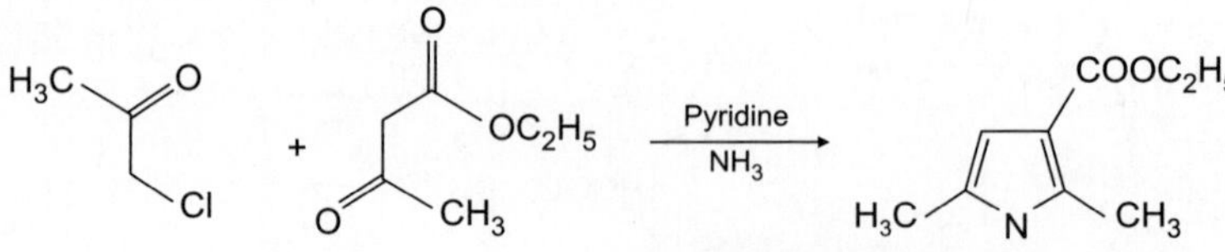

The overall mechanism of the FB reaction starts with an enolization of the beta dicarbonyl molecule and subsequent addition to the halocarbonyl compound, which then undergoes a new enolization followed by ring closure and expulsion of the halogen atom.

When the above stated Feist-Benary reaction is stopped at the hydroxydihydrofuran stage using a milder base, it is called the 'interrupted Feist-Benary' (IFB) reaction.

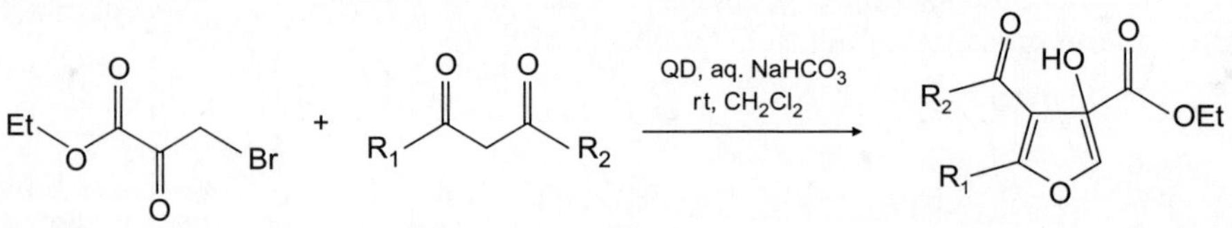

The IFB reaction is important for an easy access to substituted dihydrofurans which are constituents of many natural products. The basic ionic liquid, 1-butyl-3-methylimidazolium hydroxide is found to promote the interrupted Feist-Benary reaction at room temperature under organic solvent-free conditions to produce a variety of substituted hydroxydihydrofurans.

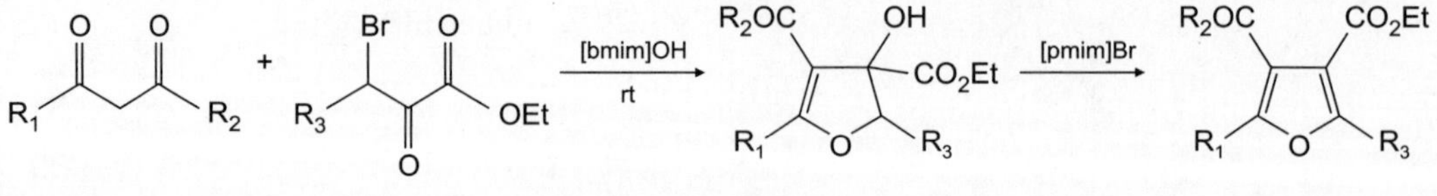

An enantioselective interrupted Feist-Benary reaction is also reported as given below:

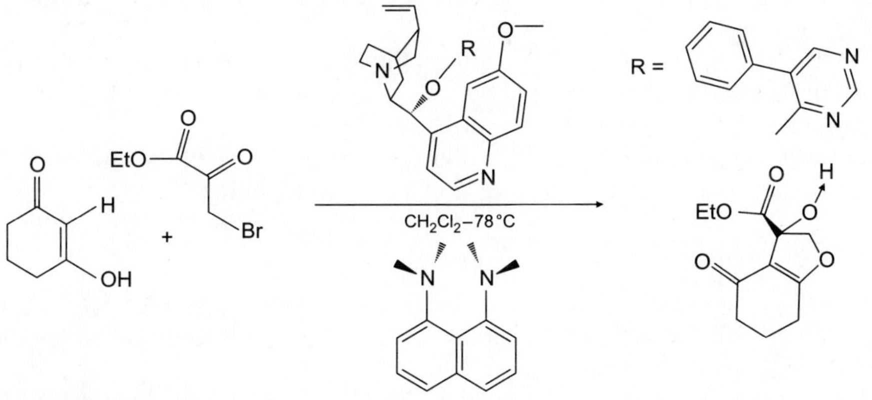

128. FENTON REACTION

The oxidation of organic substrates by iron (II) and hydrogen peroxide is called the *Fenton chemistry*, as it was first described by Henry John Horstman Fenton who first observed the oxidation of tartaric acid by H_2O_2 in the presence of ferrous iron ions.

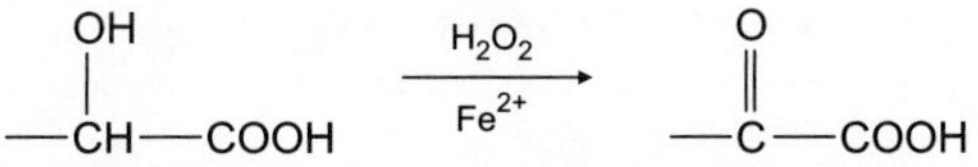

A mixture of hydrogen peroxide and ferrous sulfate is called Fenton's reagent. This reagent can be used to hydroxylate aromatic rings, though yields are usually low.

$$C_6H_6 + FeSO_4 + H_2O_2 \longrightarrow C_6H_5OH$$

The Fenton reagent is one of the most effective tools for the oxidation of organic pollutants. In reaction mechanism, it has been found that that free aryl radicals (formed by a process, e.g. HO. + ArH $\longrightarrow$ AR. + H_2O) are not intermediates. However, the HO. are the attacking species formed by

$$Fe^{2+} + H_2O_2 \longrightarrow Fe^{3+} + OH^- + HO\bullet$$

The rate-determining step is found to be the formation of HO. and not its reaction with the aromatic substrate. The efficiency of the Fenton reaction depends mainly on H_2O_2 concentration, Fe^{2+}/H_2O_2 ratio, pH and reaction time. The initial concentrations of the substrate being oxidize and its character as well as temperature, also have a substantial influence on the final efficiency. Nowadays, the Fenton's reaction is used to treat a large variety of water pollution such as phenols, formaldehyde, BTEX, pesticides, rubber chemicals and so on.

Factors Affecting the Fenton Reaction

pH: A pH in the range of 3 to 5.5 is found to be optimum for the course of the reaction. Usually, the reaction is not favored at pH values below or higher than this range because of the decomposition of hydrogen peroxide or Fe(III) precipitation. Reactions at higher pH slow down, which in turn affect the degradation process efficiency. The reason is the loss of Fe^{2+} ions in solution caused by their coagulation in the form of $Fe(OH)_3$. On the other hand, low pH causes an excessive decomposition of H_2O_2 to oxygen and water.

Temperature: It has been found that the efficiency of the Fenton reaction is higher at the higher temperatures. At temperatures over 40°C, the decomposition of hydrogen peroxide to oxygen and water occurs.

Ratio Fe²⁺:H₂O₂: The ratio of ***Fe²⁺:H₂O₂*** affects the speed of production of OH radicals and their degradation. The higher the quantity of Fe^{2+}, the slower the process. Similarly, the higher the peroxide concentration, the higher the speed of Fenton reaction.

Concentration of inorganic anions: Some anions (e.g. SO_4^{2-}, Cl^-, HPO_4^{2-}, and HCO_3^-) degrade hydroxyl radicals, or form complex compounds with iron (incapable of further reaction).

129. FERRIER REARRANGEMENT

Ferrier rearrangement is an important acid catalyzed glycosidation reaction essentially involving acetylated glycals, to afford 2,3-unsaturated glycosides through nucleophilic allylic displacement. It was discovered by Robert J. Ferrier.

The well-known Ferrier rearrangement involves the reaction of a suitably protected 1,2-glycal with an alcohol under Lewis acid catalysis to form the corresponding 2,3-unsaturated 1-o-glycosides. Such 1-o-glycosides have been transformed into a variety of useful intermediates both in organic synthesis as well as specifically in carbohydrate chemistry. Likewise, there have been reports on aza-1h and thia-1i Ferrier rearrangements.

The requirement of an acid catalyst to bring about the Ferrier rearrangement precludes its applicability to substrates that are sensitive to acidic conditions.

130. FINKELSTEIN REACTION

The Finkelstein reaction refers to the synthesis of alkyl halides from alkyl halides thus involving the exchange of one halogen for another, especially in primary alkyl halides.

The reaction is named after the German chemist Hans Finkelstein and works via an S_N2 mechanism. This means the reaction is most successful when using primary halides. The secondary, tertiary, vinyl, and aryl halides are less reactive in Finkelstein reaction.

This reaction is a good way to prepare 1° iodides, which are often useful, because they are much more reactive partners in S_N2 reactions than the corresponding chloride or bromide. Alkyl iodides are prepared from alkyl bromide or chloride with potassium or sodium iodide in acetone. The halide exchange is a reversible reaction. The reaction is driven to completion by taking the advantage of differential solubility of metal halide salts in acetone solvent. When an alkyl chloride or bromide is treated with sodium iodide in acetone, the equilibrium for the reaction is shifted towards the right (products) by the precipitation of sodium chloride or bromide; these two salts are insoluble in acetone, but sodium iodide is soluble.

Since the mechanism is S_N2, the reaction is much more successful for primary halides than for secondary or tertiary halides which are much less reactive for the S_N2 reaction conditions. However, for the Finkelstein reaction on and the secondary and tertiary substrates Lewis acids may be used, e.g. $ZnCl_2$, $FeCl_3$ or Me_3Al.

The Finkelstein reaction is also used to prepare alkyl fluoride by treatment of other alkyl halides with any one of fluorinating agents, such as anhydrous HF, AgF, KF, HgF_2, Et_3N, HF, etc. The (F^-) is a poor leaving group and forms stable C-F bond. The reverse reaction thus does not take place easily, and the equilibrium lies far to the right. Phase-transfer catalysis of the exchange reaction is found to be effective while preparing both fluorides and iodides.

By modifying the Finkelstein reaction, an alcohol can also be converted to an alkyl iodide via forming a tosylate with tosyl chloride in presence of Et_3N in dichloromethane.

131. FISCHER-HEPP REARRANGEMENT (NITROSAMINE REARRANGEMENT)

The reaction was first described by the German chemist Otto Philipp Fischer and Eduard Hepp in 1886, and is of importance because para-NO secondary anilines cannot be prepared in a direct reaction.

The conversion of N-nitroso or N-nitrosamine aromatic amines into carbon nitroso compound (p-nitroso aromatic amines) in the presence of HCl or HBr is generally referred to as the Fischer Hepp rearrangement. In other words, the Fischer-Hepp rearrangement involves conversion of secondary aromatic nitrosamines to p-nitrosoarylamines. This reaction is found to be intermolecular rearrangement.

132. FISCHER INDOLE SYNTHESIS

The Fischer indole synthesis involves heating of aryl hydrazone of an aldehyde or ketone with a catalyst leading to elimination of ammonia to form an indole.

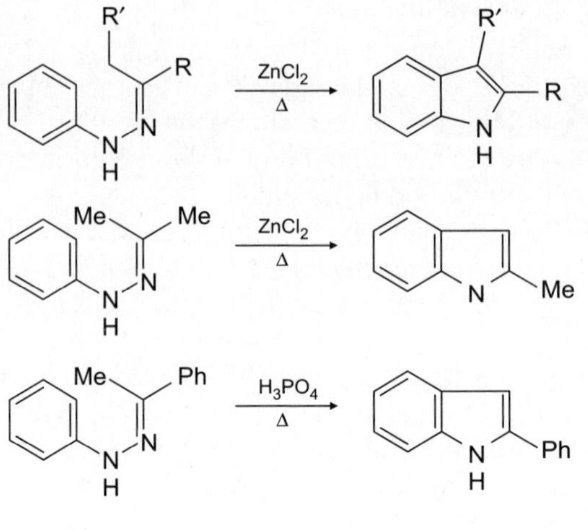

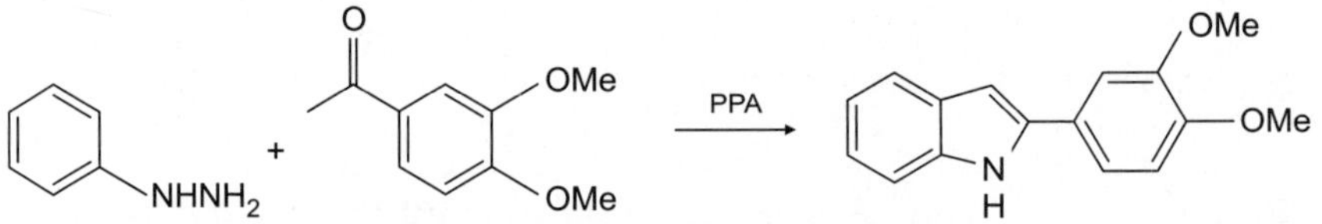

A number of catalysts are found to catalyze the reaction, but zinc chloride is most frequently employed. Other catalysts like metal halides, Bronsted (e.g. HCl, H_2SO_4, PPA) and Lewis acids (e.g. BF_3/AcOH, $ZnCl_2$, $FeCl_3$, $AlCl_3$) and certain transition metals (e.g. $CoCl_2$, $NiCl_2$, TsOH) have found application also. However, the reaction can be performed without a catalyst. Recently, this reaction is found to be facilitated by microwave irradiation and solid-phase Fischer indole syntheses are also known. Arylhydrazones used as starting material can be easily prepared by the treatment of aldehydes or ketones with phenylhydrazine or by aliphatic diazonium coupling.

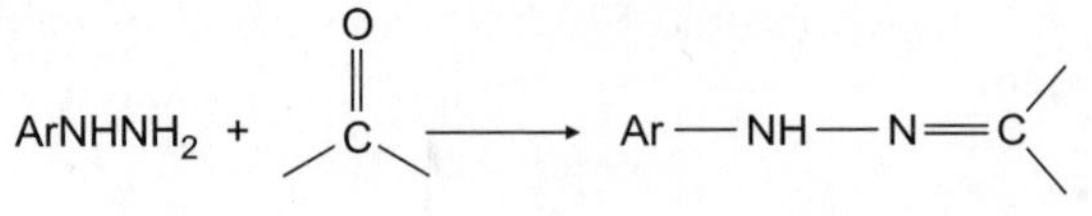

Unsymmetrical phenyl hydrazones give a mixture of two differently substituted indoles, the ratio of which depends on steric factors and reaction conditions.

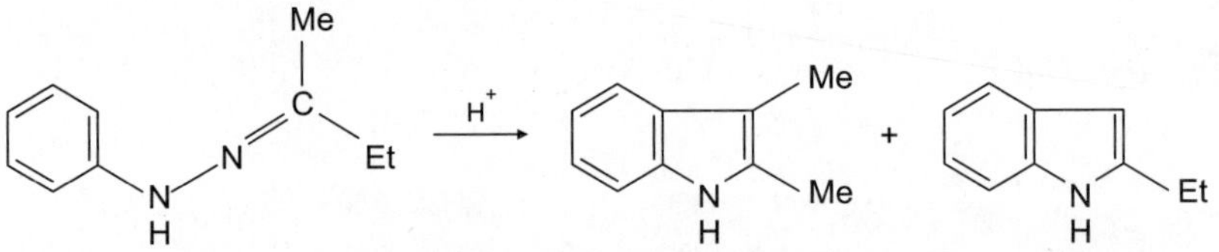

Mechanistically, the Fischer-indole synthesis consists of three steps including a [3,3]-sigmatropic rearrangement as the key step. Initially, the phenyl hydrazone undergoes a reversible rearrangement to give the reactive ene-hydrazine, which subsequently undergoes a [3.3]-sigmatropic rearrangement to form a new carbon-carbon bond and the cationic species. Finally, a cyclization takes place with subsequent elimination of ammonia to yield the indole. The main function of the catalyst seems to be to speed up the formation of the new carbon-carbon bond in second step.

133. FISCHER OXAZOLE SYNTHESIS

The synthesis of oxazoles from the condensation of equimolar amounts of aldehyde cyanohydrins and aromatic aldehydes in dry ether in the presence of dry hydrochloric acid is known as the Fischer oxazole synthesis or Fischer synthesis. This method was discovered by Hermann Emil Fischer in 1896. The Fischer synthesis is found suitable for the preparation of 2,5-diaryloxazoles.

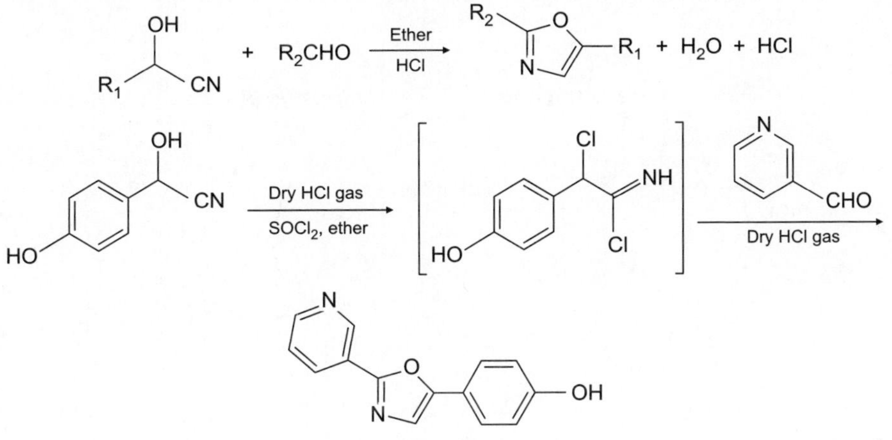

Halfordinal

134. FISCHER PEPTIDE SYNTHESIS

Fischer synthesized peptides by acid chloride method. In the Fischer peptide synthesis, an α-chloro (or α-bromo) acyl chloride is condensed with an amino acid ester. The ester is then hydrolyzed to the acid and then the acid is converted to the acid chloride enabling the extension of the peptide chain by another unit and so on. The terminal chloride is finally converted to an amino group with ammonia or an amino group completing the peptide synthesis.

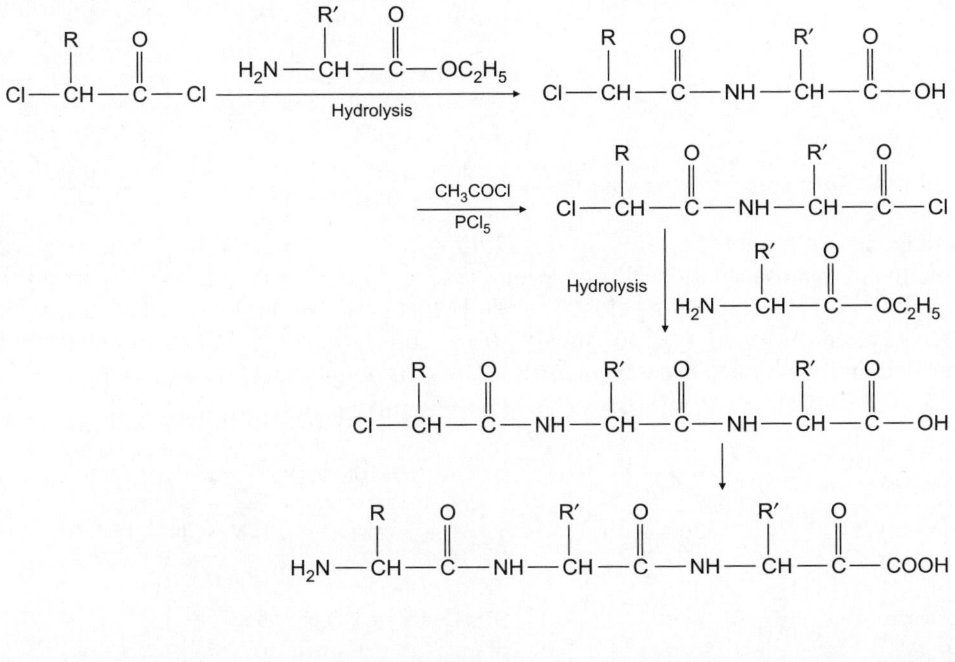

135. FISCHER PHENYLHYDRAZINE SYNTHESIS

The synthesis involves formation of arylhydrazines, e.g. phenylhydrazine, by diazotization of aniline followed by the reduction of the resulting diazonium salt with an excess amount of sodium sulfite, and then hydrolysis of substituted hydrazine sulfonic acid salt with hydrochloric acid.

Phenylhydrazine was the first hydrazine derivative characterized, reported by Emil Fischer in 1875. The process is used in industries as well as on small scale for production of arylhydrazines.

$$ArN_2^+ \xrightarrow{NaSO_3^-} ArN=NSO_3Na \xrightarrow{NaHSO_3} ArNNHSO_3Na \xrightarrow{HCl} ArNHNH_2$$

with SO_3Na over $ArNNHSO_3Na$

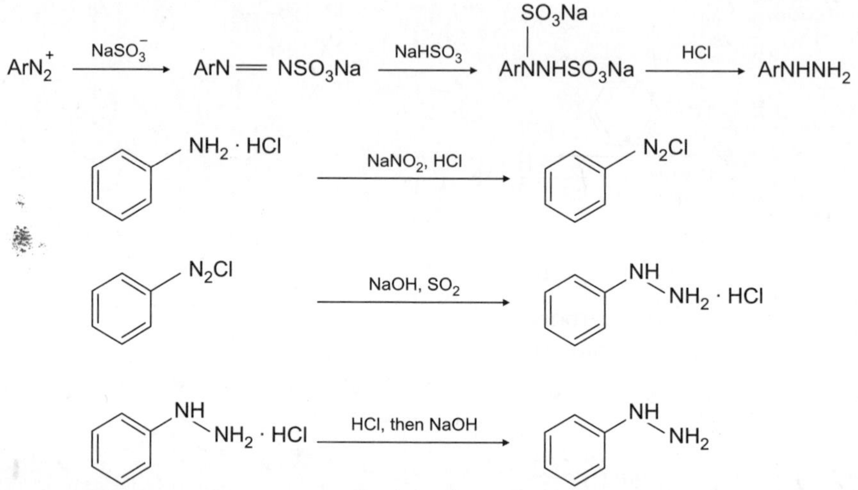

136. FISCHER PHENYLHYDRAZONE AND OSAZONE REACTION

When one equivalent of an aldehyde or ketone (i.e. compounds that contain a carbonyl group) reacts with phenylhydrazine, the product is called a phenylhydrazone. For example, the reaction between benzaldehyde and phenylhydrazine yields the phenylhydrazone of benzaldehyde as shown below:

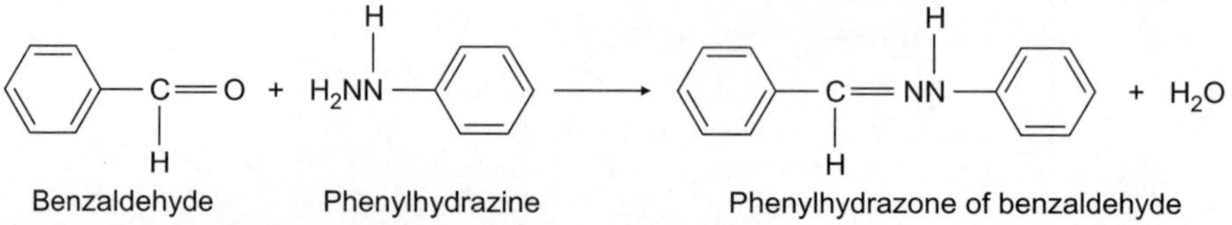

Benzaldehyde   Phenylhydrazine   Phenylhydrazone of benzaldehyde

The tendency of monosaccharides to form syrups, that do not crystallize, made the purification and isolation of monosaccharides difficult. Emil Fischer found that when phenylhydrazine is added to an aldose or a ketose, a yellow crystalline solid that is insoluble in water is formed. He called this derivative an osazone ("ose" for sugar; "azone" for hydrazone). Osazones are easily isolated and purified and were once used extensively to identify monosaccharides.

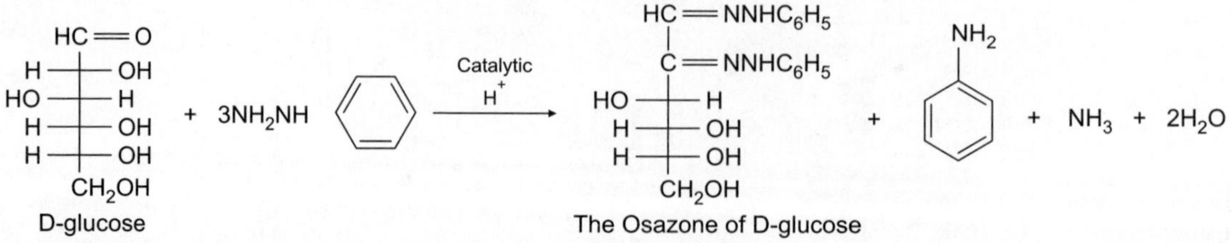

D-glucose   + 3NH₂NH ... Catalytic H⁺ ... The Osazone of D-glucose   +   + NH₃ + 2H₂O

An osazone contains two phenylhydrazone groupings. Aldehydes and ketones react with one equivalent of phenylhydrazine forming phenylhydrazones. Aldoses and ketoses, in contrast, react with three equivalents of phenylhydrazine, forming osazones. One equivalent functions as an oxidizing agent and is reduced to aniline and ammonia. Two equivalents form imines with carbonyl groups. The reaction stops at this point, regardless of how much phenylhydrazine is present.

Because the configuration of the number-2 carbon is lost during osazone formation, C-2 epimers form identical osazones. For example, D-idose and D-glucose, which are C-2 epimers, both form the same osazone.

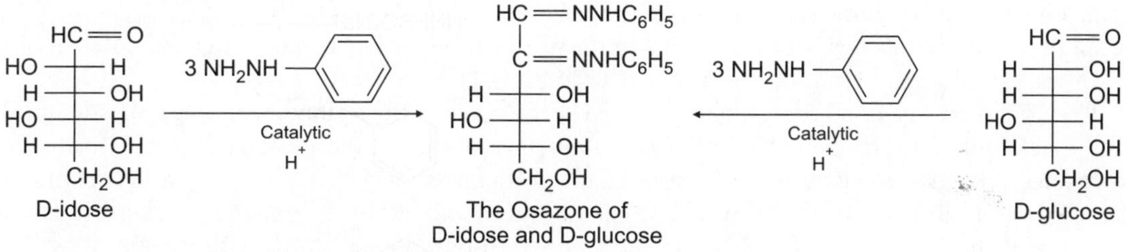

D-idose   The Osazone of D-idose and D-glucose   D-glucose

The number-1 and number-2 carbons of ketoses react with phenylhydrazine, too. Consequently, D-fructose, D-glucose, and D-mannose all form the same osazone.

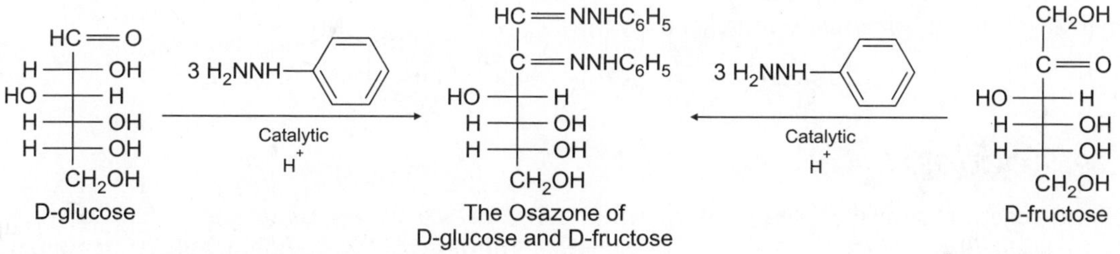

D-glucose   The Osazone of D-glucose and D-fructose   D-fructose

137. FISCHER-SPEIER ESTERIFICATION

The synthesis of ester by refluxing carboxylic acid with an excess amount of alcohol in the presence of Lewis or Brønstedt acid is known as the Fischer-Speier esterification.

$$CH_3COOH + HOC_2H_5 \longrightarrow CH_3COOC_2H_5 + H_2O$$
Acetic acid   Ethyl alcohol   Ethyl acetate

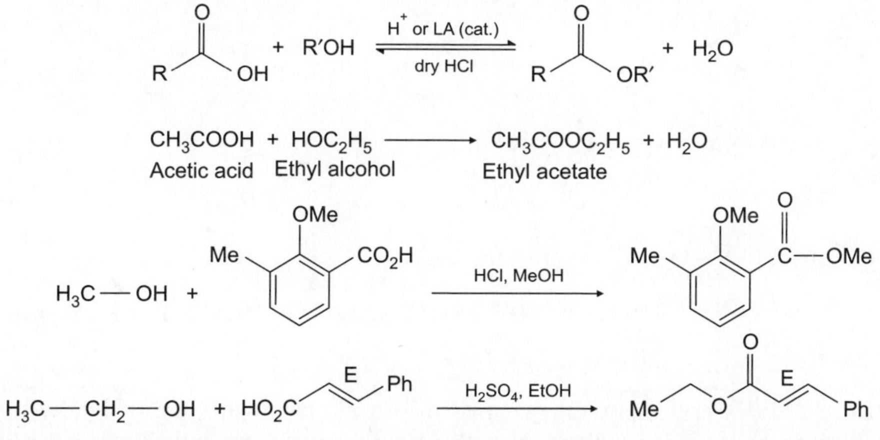

Most carboxylic acids are suitable for the reaction, but the alcohol should generally be a primary or secondary alkyl. Tertiary alcohols are prone to elimination, and phenols are usually too unreactive to give useful yields.

Commonly used catalysts for a Fischer esterification includes sulfuric acid, tosic acid, and Lewis acids such as scandium(III) triflate. Fischer esterification is an example of nucleophilic acyl substitution.

138. FISCHER-TROPSCH (FT) SYNTHESIS

Fischer-Tropsch synthesis is the process that converts synthesis gas, i.e. a mixture of carbon monoxide and hydrogen, into a wide range of long chain hydrocarbons and oxygenates. As such, the Fischer-Tropsch synthesis constitutes a practical way for the chemical liquefaction of solid (coal) or gaseous (natural gas) carbon resources.

In the catalytic Fischer-Tropsch (FT) synthesis, one mole of CO reacts with two moles of H_2 to form mainly aliphatic straight-chain hydrocarbons (C_xH_y).

The common Fischer-Tropsch catalysts are based on iron (Fe), ruthenium (Ru), or cobalt (Co) as the active metal. The costs for Fe-based catalysts are low, but these catalysts suffer from a low wax selectivity, deactivation, and inhibition of the productivity by water at large syngas conversions. Despite the high activity of Ru-based catalysts, their utilization is limited to scientific studies because of the high price of ruthenium. However, Co-based catalysts are stable and allow high syngas conversions, promoting the formation of heavy wax.

The Fischer-Tropsch reaction is the chemical heart in the gas-to-liquid technology. The highly exothermic Fischer-Tropsch reaction converts synthesis gas into a large range of linear hydrocarbons, schematically represented as:

$$nCO + 2nH_2 \longrightarrow -(CH_2)_n^- + nH_2O \qquad \Delta H = -165 \text{ kJ mole}^{-1}$$

139. FLOOD REACTION

Flood reaction describes synthesis of trialkylsilyl halide by the treatment of the mixture of hexaalkyldisiloxanes ($R_3SiOSiR_3$) and concentrated sulfuric acid with sodium halide or ammonium halide or by treatment of the intermediate silane sulfates with hydrogen chloride in the presence of ammonium sulphate.

Organosilyl halides are important reagents in organic chemistry, e.g. trimethylsilyl chloride Me_3SiCl, dichloromethylphenylsilane, dimethyldichlorosilane, methyltrichlorosilane, trimethylchlorosilane, etc.

140. FORSTER DIAZOKETONE SYNTHESIS

The synthesis of diazoketones, especially the cyclic diazoketones from α-keto oximes by the reaction with chloramines is known as the Forster reaction.

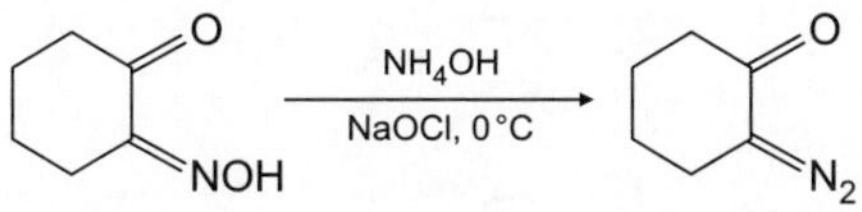

In the Forster reaction, the α-methylene position of a ketone, as in following compound reacts with a nitrite to form an oxime followed by the reaction with a chloramine providing the α-diazo ketone.

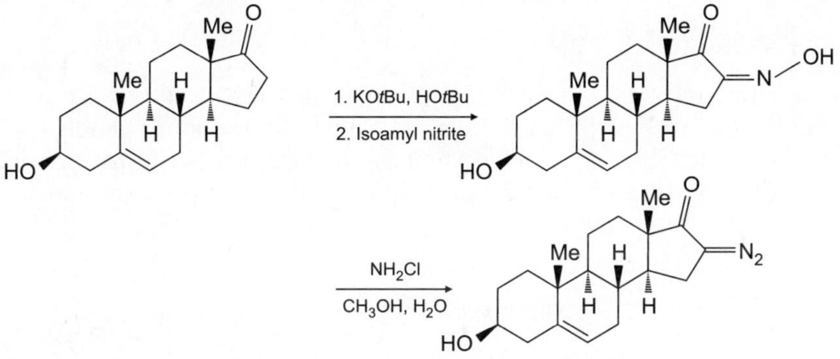

This reaction has been modified to generate chloramine *in situ* via the reaction of ammonia with sodium hypochlorite.

An α-keto oxime is treated with chloroamine to give the intermediate, which suffers dehydration to give the diazo ketone.

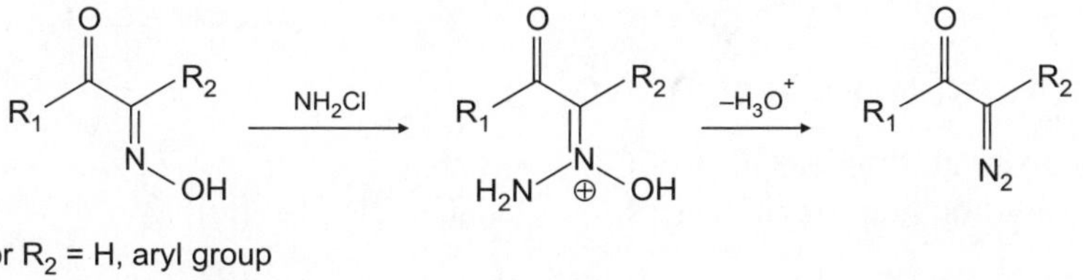

R_1 or R_2 = H, aryl group

An alternative to this transformation is the condensation of an oxime directly with phenyl-hydrazine.

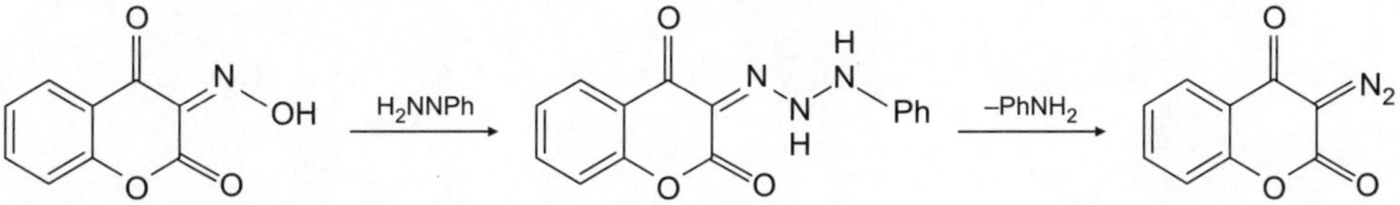

141. FORSTER REACTION (OR FORSTER-DECKER METHOD OR DECKER-BECKER SECONDARY AMINE SYNTHESIS)

In Forster reaction, secondary amines are formed by the condensation of a primary amine with an aldehyde to form a Schiff base, followed by alkylation with alkyl halide and subsequent hydrolysis.

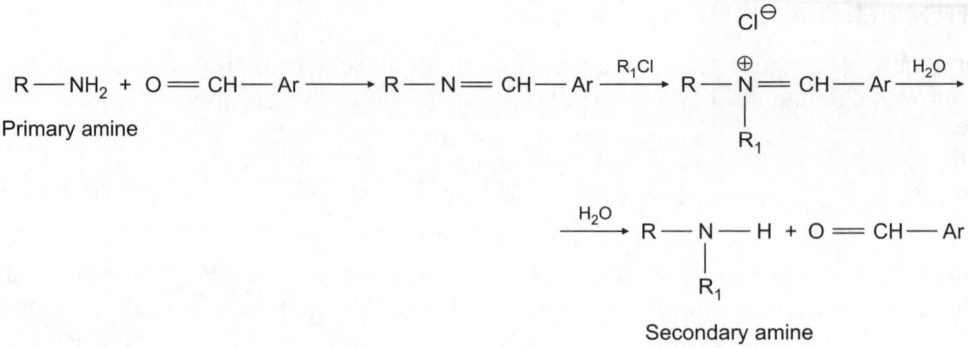

$$R—NH_2 + O = CH—Ar \longrightarrow R—N = CH—Ar \xrightarrow{R_1Cl} \overset{Cl^{\ominus}}{\underset{R_1}{R—\overset{\oplus}{N} = CH—Ar}} \xrightarrow{H_2O}$$

Primary amine

$$\xrightarrow{H_2O} \underset{R_1}{R—N—H} + O = CH—Ar$$

Secondary amine

142. FRANCHIMONT REACTION

The reaction was first reported by A.P.N. Franchimont in 1872. It describes the dimerization of carboxylic acid to 1,2-dicarboxylic acids. In the so-called Franchimont reaction, two molecules of α-bromocarboxylic acids undergo condensation in absolute alcohol in the presence of sodium cyanide or potassium cyanide to give 1,2-dicarboxylic acids after hydrolysis and decarboxylation.

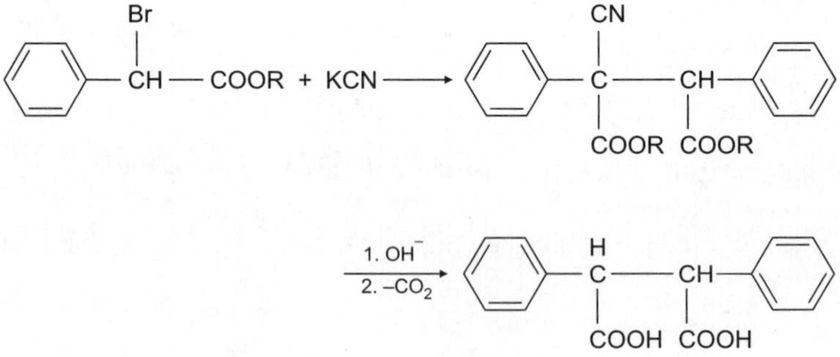

143. FRANKLAND-DUPPA REACTION

Reaction of an oxalate ester (ROCOCOOR) with an alkyl halide R′X, zinc and hydrochloric acid to form the α-hydroxycarboxylic esters RR′COHCOOR is referred to as the Frankland-Duppa reaction.

In 1865, Frankland and Duppa reported this reaction between diethylzinc and diethyl oxalate to give the ethyl ester of α-ethyl-α-hydroxybutyric acid.

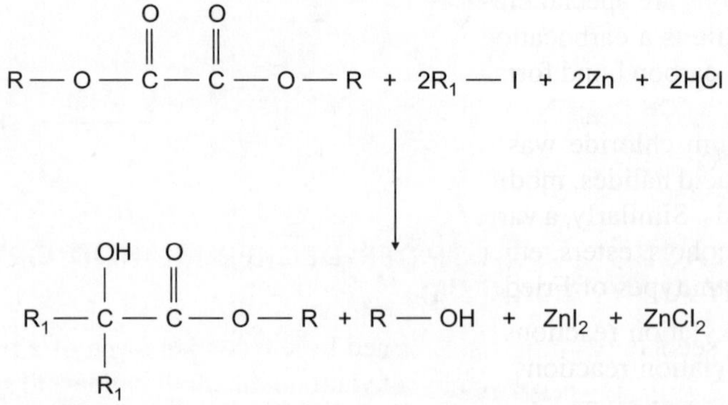

$$R—O—\overset{O}{\overset{\|}{C}}—\overset{O}{\overset{\|}{C}}—O—R + 2R_1—I + 2Zn + 2HCl$$

$$\underset{R_1}{R_1—\overset{OH}{\overset{|}{C}}—\overset{O}{\overset{\|}{C}}—O—R} + R—OH + ZnI_2 + ZnCl_2$$

144. FRANKLAND SYNTHESIS

The original diethylzinc synthesis by Frankland was an oxidative addition of iodoethane to zinc metal with hydrogen gas as a "protective" blanket (this reaction is called the Frankland synthesis). The reactivity of zinc metal is increased in so-called Rieke zinc obtained by reduction of zinc chloride by potassium metal.

The preparation of dialkyl zinc from zinc and alkyl iodide is called the Frankland reaction.

$$2R — X + Zn \longrightarrow R — R + ZnX_2$$
$$2C_2H_5 — I + Zn \longrightarrow C_2H_5 — C_2H_5 + ZnI_2$$

With the synthesis of diethylzinc in 1849, E. Frankland laid the foundation stone for modern organometallic chemistry. Since Frankland's synthesis of EtZnI, a multitude of applications have been found for such simple organometallic compounds. For instance, alkylzinc iodides are used for Simmons-Smith cyclopropanation, alkylzinc bromides are used for nickel-catalyzed Negishi reactions, and EtZnCl generated *in situ* has been shown to be a good chain transfer agent in the polymerization of olefins.

145. FREUND REACTION; GUSTAVSON REACTION; HASS CYCLOPROPANE PROCESS

The Freund reaction (1881) describes preparation of cyclopropane derivatives by the reaction of 1,3-dihalopropanes with sodium. Cyclopropane was discovered in 1881 by August Freund.

$$BrCH_2CH_2CH_2Br + 2\ Na \longrightarrow (CH_2)_3 + 2\ NaBr$$

The Gustavson reaction (1887) (modification to Freund reaction) involves preparation of cyclopropane derivatives by an intramolecular coupling reaction of 1,3-dihalopropanes with zinc dust.

$$Cl \diagup \diagdown \diagup Cl + Zn \longrightarrow \triangle + ZnCl_2$$

The Hass cyclopropane process (another modification to the Freund reaction) involves preparation of cyclopropane derivatives by the reaction of 1,3-dihalopropanes with zinc dust in aqueous alcohol in the presence of a catalytic amount of sodium iodide.

146. FRIEDEL-CRAFTS REACTION

The reaction was named after its development in 1887 by Charles Friedel and James Mason Crafts. These reactions allow attaching substituents to an aromatic ring. Friedel-Crafts alkylation and acylation reactions are special class of electrophilic aromatic substitution (EAS) reactions in which the electrophile is a carbocation or acylium ion. These reactions are highly useful in that they involve carbon-carbon bond formation and allow alkyl and acyl groups to be substituted onto aromatic rings.

Initially, aluminium chloride was used as catalyst but now a variety of acidic catalysts are available including acid halides, modified zeolites, cation exchange resins, superacids, solid super acids and proton acids. Similarly, a variety of alkylating agents are also available, e.g. alkyl halides, alkenes, alkynes, alcohols, esters, ethers, carbonyl compounds, mercaptain and thiocyanates, etc.

There are two main types of Friedel-Crafts reactions:

1. Friedel-Crafts alkylation reactions
2. Friedel-Crafts acylation reactions

1. **Friedel-Crafts acylation:** The Friedel-Crafts acylation involves the electrophilic substitution of acyl group on aromatic ring when arenes are treated with acyl chlorides or anhydrides in

presence of Lewis acids. The reaction allows the synthesis of monoacylated products. It is the most important method for the preparation of aryl ketones.

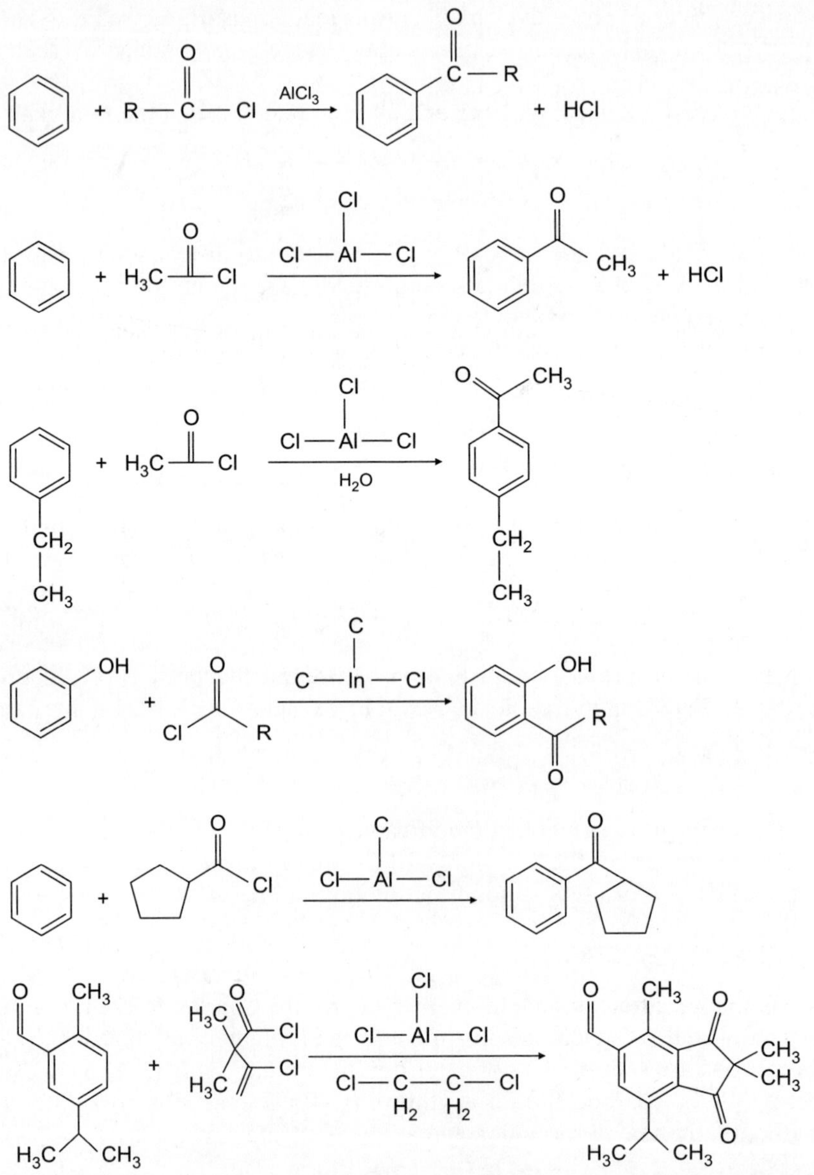

In acylation, the electrophilic species is the acyl cation or **acylium ion** (i.e. RCO$^+$) formed by the "removal" of the halide by the Lewis acid catalyst which is stabilized by resonance as shown above. This resonance stability of **acylium ion** prevents the problems associated with the rearrangement of simple carbocations. Due to the electron-withdrawing effect of the carbonyl group, the ketone product (acylated product) is always less reactive than the original molecule, so multiple acylations do not occur. Also, there are no carbocation rearrangements, as the carbonium ion is stabilized by a resonance structure in which the positive charge is on the oxygen. In other words, drawbacks like polyalkylation, and rearrangement of the intermediate carbocation as known from the *Friedel-Crafts alkylation* are not found for the Friedel-Crafts acylation.

Compounds containing ortho-para directing groups (e.g. alkyl, hydroxy, alkoxy, halogen, and acetamido groups) are acylated and give mainly the para products but the Friedel-Crafts acylation is usually prevented by meta directing groups. Many aromatic heterocyclic systems, including furans, thiophenes, pyrans, and pyrroles can also be acylated in good yield; however, pyridine as well as quinoline are unreactive.

The acylated products in the Friedel-Crafts acylation can be easily converted to the corresponding alkanes via Clemmensen reduction or Wolff-Kishner reduction.

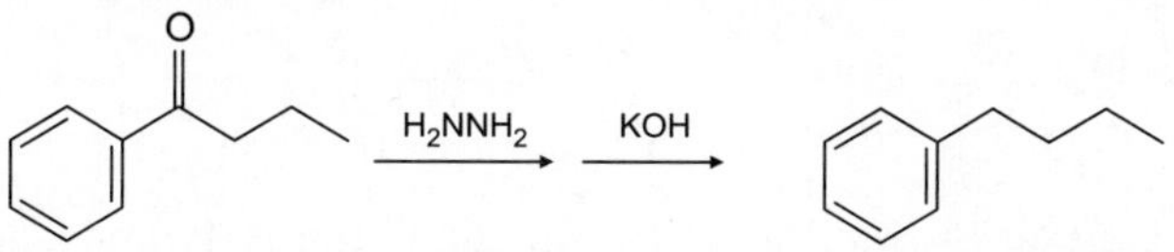

Another important use of the Friedel-Crafts acylation is the intramolecular reaction to effect the ring closure. This variant reaction is used mostly to close six-membered rings, but has also been done for five- and seven-membered rings, as well as for heterocycles, which close less readily.

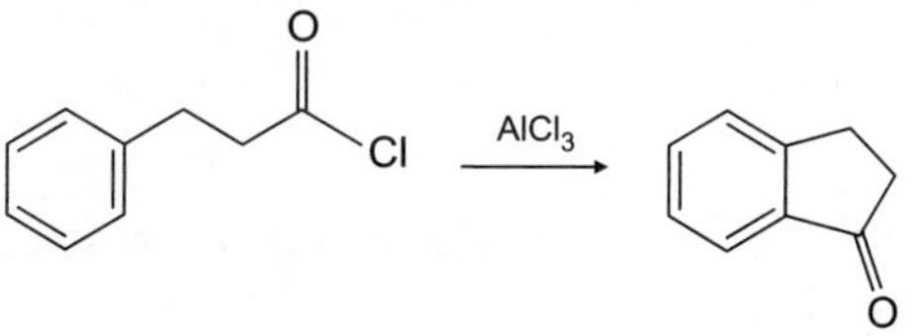

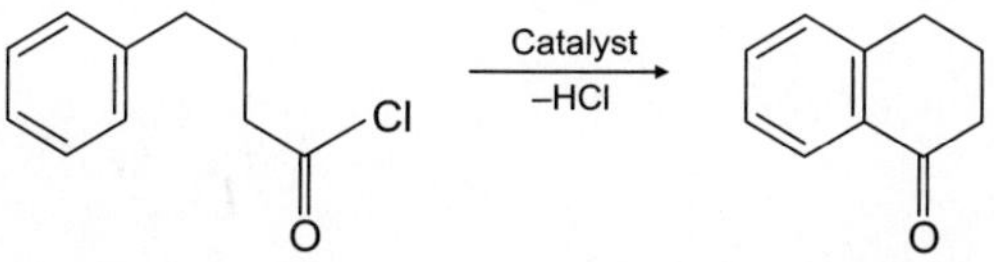

2. **Friedel-Crafts alkylation:** The alkylation of aromatic compounds catalyzed by aluminum chloride or other Lewis acids is known as the Friedel-Crafts alkylation. This reaction involves the electrophilic substitution of alkyl group on aromatic ring when arenes are treated with alkyl halides in presence of Lewis acids (BF_3, $FeCl_3$ or $ZnCl_2$, etc.).

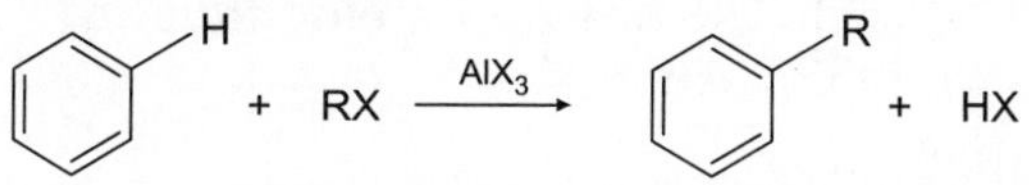

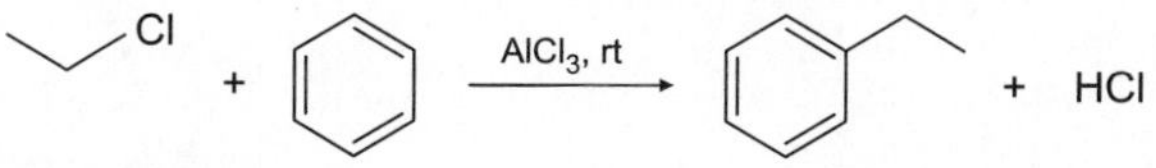

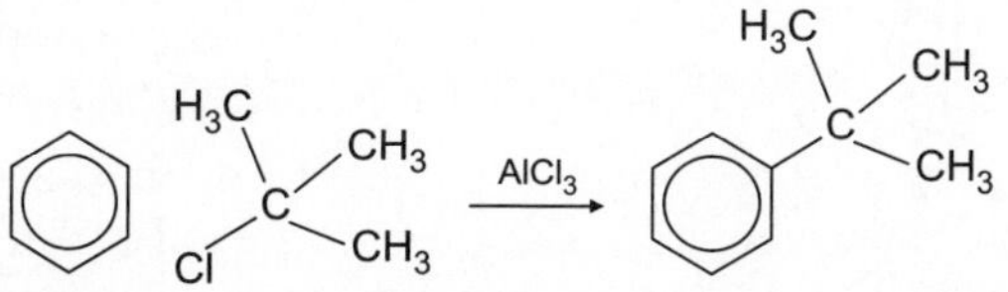

Friedel-Crafts alkylation which was first discovered by Charles Friedel and his co-worker, James Crafts, in 1877 is a reaction of very broad scope. However, the reaction suffers with the disadvantages of polyalkylation since introduction of alkyl group on arene makes it more reactive towards electrophilic substitution due to electron donating effect of alkyl substituent and hence the reaction does not stop to monoalkylation but further hydrogens on arenes get substituted by the alkyl group results in polyalkylated products. If the objective is to introduce an alkyl group onto an aromatic ring, acylation followed by Clemmensen reduction or Wolff-Kishner reduction are usually the preferred routes.

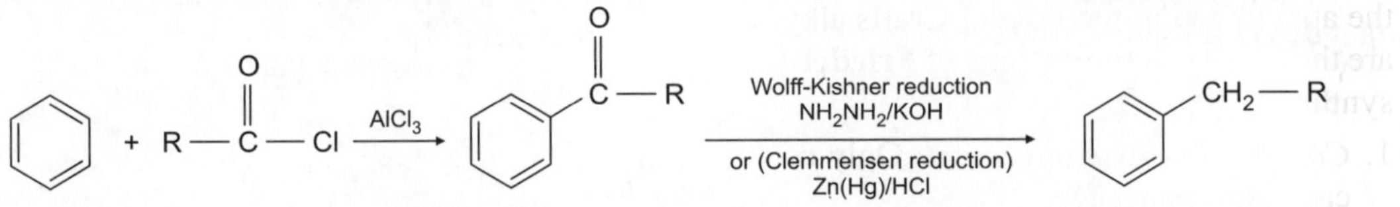

The intramolecular variant of the Friedel-Crafts alkylation is synthetically important for the closure of rings especially for six-membered rings, but five- and seven-membered ring products are also accessible. Formation of tetralin in following reaction involves intramolecular Friedel-Crafts reaction.

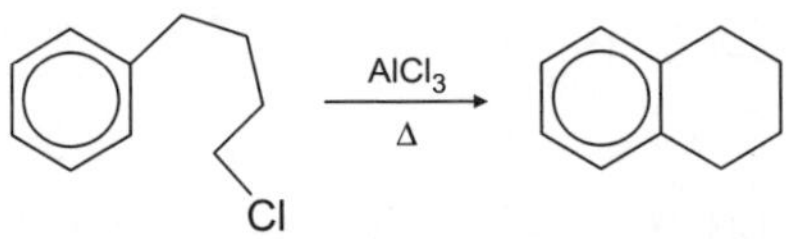

Using a reagent containing two groups, such as given below is another way of effecting ring closure through Friedel-Crafts alkylation.

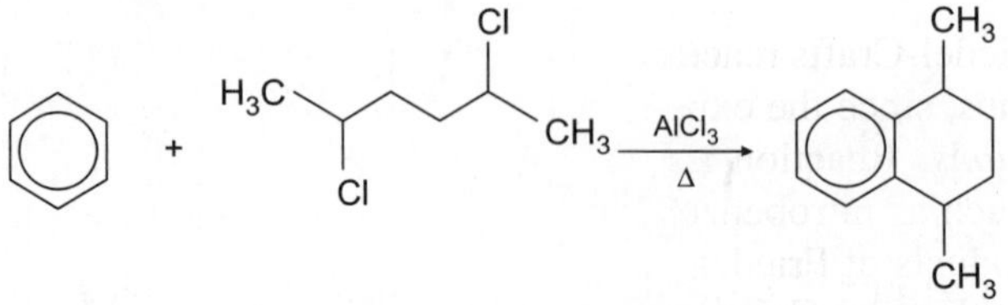

Mechanistically, the Friedel-Crafts reaction begins with coordination of the Lewis acid to the halogen of the alkyl halide (step 1, alkyl halide activation). This coordination places a partial positive charge on the carbon bearing the halogen, making it more electrophilic for the electrophilic attack (step 2, electrophilic attack). The loss of a proton from the ring-system re-establishes aromaticity, the driving force for the reaction (step 3).

Step 1: Alkyl halide activation

Step 2: Electrophilic attack

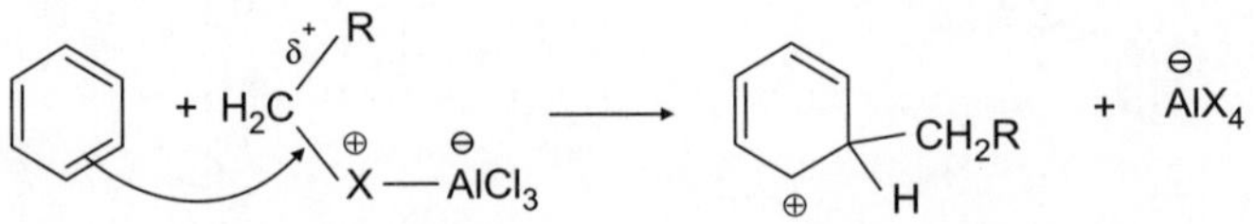

Step 3: Proton loss

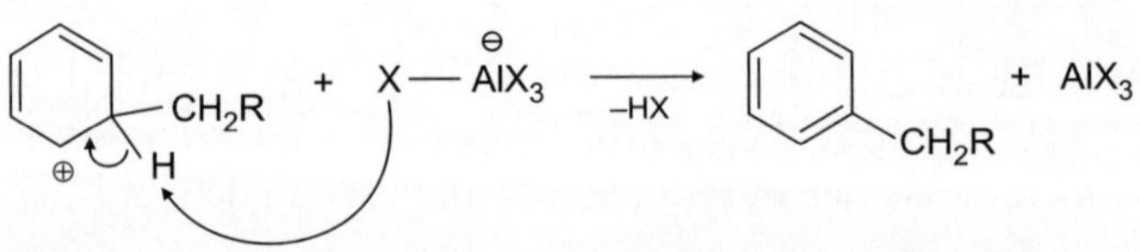

The alkylation of benzene or substituted benzene is accompanied by two important side reactions:

(a) Polyalkylation of the benzene system

(b) Carbocation rearrangement

Both side reactions lead to drop in the yield of the desired products significantly as well as result in a mixture of products that is difficult to separate. Because of these undesired side reactions, the application of the Friedel-Crafts alkylation reaction in organic synthesis is limited. Following are the important **limitations of Friedel-Crafts alkylation** that reduced its applicability in organic synthesis.

1. *Carbocation rearrangement:* Only certain alkylbenzenes can be made due to the tendency of cations to rearrange.

sec-butylbenzene
(unexpected)
65%

Butylbenzene
(expected)
35%

That is why the Friedel-Crafts reaction is almost useless for the preparation of derivatives with primary alkyl groups, since the expected product is either minor or not present at all.

2. *Compound limitations:* Reaction is not successful on deactivated rings and fails when used with compounds such as nitrobenzene and other strong deactivating systems.

3. *Polyalkylation:* Products of Friedel-Crafts are even more reactive than starting material. Alkyl groups produced in Friedel-Crafts alkylation are electron-donating substituents meaning that the products are more susceptible to electrophilic attack than what we began with. For synthetic purposes, this is a big problem. This effect favors the formation of di- or even poly-substituted products.

1,4-bis(*t*-butyl) benzene
Major product

t-butylbenzene
Minor product

Unreacted
benzene

This limitation can be avoided in particular cases, if benzene (or the benzene derivative) is used in very large excess. Such approach is, of course, not always practical, due to availability of the particular benzene derivative.

4. The scope of the reaction is also limited by the reactivity of certain starting materials. The reaction works only with benzene or activated benzene derivatives (e.g. substituted with alkyl group, benzene ring, OH, OR, NH_2, NHR, NR_2, F, Cl, Br, I). Meta directors such as NO_2, CN, NR_3^+, C=O, CF_3, SO_3H slow down the reaction, e.g. nitrobenzene cannot be alkylated. Similarly, naphthalenes and other fused ring aromatic substrates are so reactive that they may undergo side reactions towards the catalyst, and give low yields of monoalkylated product.

Heterocyclic rings are also tend to be poor substrates for a Friedel-Crafts alkylation. Functional groups like $-OH$, $-NH_2$, $-OR$, that coordinate to the Lewis acid also should not be present on the aromatic ring.

Alkyl halides, alkenes, and alcohols are the most important reagents for the reaction. However, other types of reagent have also been employed. Among the alkyl halides, the tertiary halides are particularly good substrates since they form relatively stable tertiary carbocations. When alkyl halides are used, the reactivity order is F > Cl > Br > I. Alkenes are especially good alkylating agents, generally proceeding by formation of an intermediate carbocation that reacts with the electron-rich aromatic ring, and the final product incorporates a H and Ar from ArH to a C-C double bond. Alcohols are more active than alkyl halides, but if a Lewis acid catalyst is used, more catalyst is required, since the catalyst complexes with the OH group. However, proton acids, such as H_2SO_4, are often used to catalyze alkylation with alcohols. Other alkylating agents are ethers, thiols, sulfates, sulfonates, alkyl nitro compounds, and even alkanes and cycloalkanes, under conditions where these are converted to carbocations.

From alkyl halides: $RCl + AlCl_3 \longrightarrow R^+ + AlCl_4^-$

From alcohols and Lewis acids: $ROH + AlCl_3 \longrightarrow ROAlCl_2 \longrightarrow R^+ + {}^-OAlCl_2$

From alcohols and proton acids: $ROH + H^+ ROH_2^+ \longrightarrow R^+ + H_2O$

From alkenes (a supply of protons is usually required):

Regardless of the reagent used a catalyst is nearly always required. As a catalysts for the Friedel-Crafts reaction, the Lewis acids such as $AlCl_3$, $TiCl_4$, SbF_5, BF_3, $ZnCl_2$ or $FeCl_3$ are used. Protic acids such as H_2SO_4 or HF are also used, especially for reaction with alkenes or alcohols. Recent developments include the use of acidic polymer resins, e.g. Nafion-H, as catalysts for Friedel-Crafts alkylations and the use of asymmetric catalysts. Aluminum chloride and boron trifluoride are the most common, but many other Lewis acids and also proton acids, such as HF and H_2SO_4 can be taken. For active halides, a trace of a less active catalyst, such as $ZnCl_2$, may be enough. Unreactive halides, such as chloromethane, require more powerful catalyst, such as $AlCl_3$ even in larger amounts.

147. FRIEDLÄNDER SYNTHESIS

Base-catalyzed condensation of ortho-aminobenzaldehydes or ortho-aminoarylketones with another aldehyde or ketone with at least one methylene α-adjacent to the carbonyl to form quinoline derivatives is known as the Friedländer synthesis. The reaction is named after German chemist Paul Friedländer.

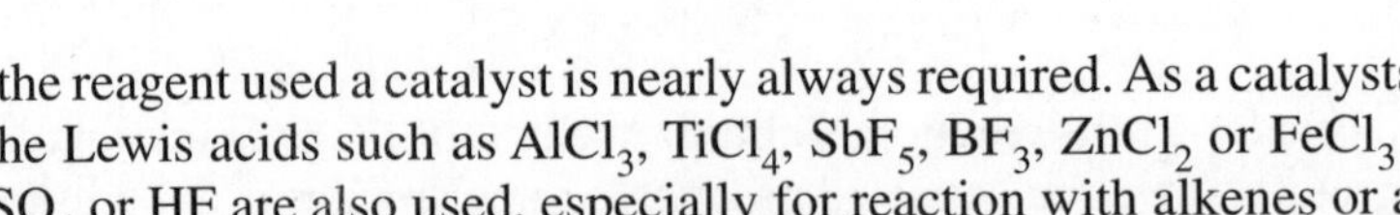

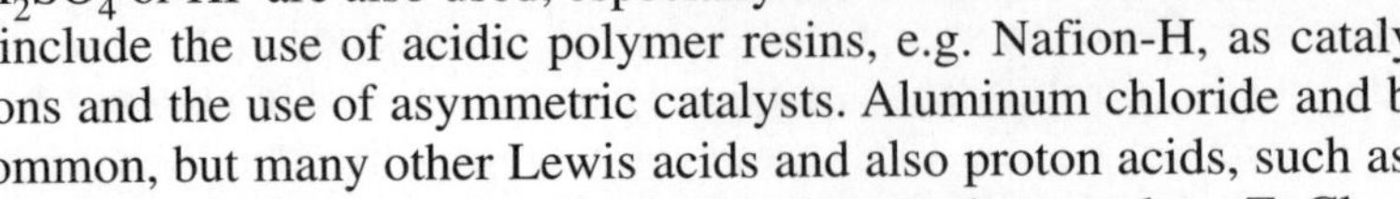

Usually the reaction is carried out in the presence of a basic catalyst, but can also be promoted by acid or heat. Among the basic catalysts, KOH, NaOH or piperidine are used while trifluoroacetic acid, HCl, H_2SO_4, polyphosphoric acid or p-toluenesulfonic acid are used as acidic catalysts. Friedländer reaction if uncatalyzed requires more drastic conditions of temperatures but often results in better yields.

After an initial amino-ketone condensation two different types of products are possible depending on whether the reaction is catalyzed by acid or base. An imine (Schiff base) is obtained when the reaction is mediated with an acid while an α,β-unsaturated carbonyl compound is there with base catalysts. The actual mechanistic pathway varies with substrate and reaction conditions. The next step in both cases involves a cyclocondensation to produce a quinoline derivative.

The Friedländer quinoline synthesis is especially useful for the preparation of 3-substituted quinolines, which are less accessible by other methods.

148. FRIES REARRANGEMENT

Rearrangement of phenolic esters to o- and/or p-phenolic ketones by heating with Lewis acid catalysts. The Fries rearrangement, named after the German chemist Karl Theophil Fries, enables the preparation of acyl phenols.

Both o- and p-acylphenols are produced in the reaction. One of the two products can be favoured by changing reaction conditions, such as the temperature, solvent, and amount of catalyst used. At high temperatures, the ortho-product is formed, while the para-product is formed at low temperatures (below $100\,°C$).

The Fries rearrangement is known to be catalyzed by aluminum halides, zinc chloride, titanium tetrachloride, boron trifluoride and trifluoromethanesulfonic acid. Apart from these, transition-metal-catalyzed Fries rearrangements have also been reported.

In the reaction mechanism, the Lewis acid catalyst forms complexes with the substrate at either one of the oxygen centers, or even both when used in excess. The complex then dissociate into an ion-pair that is held together by the solvent cage, and that consists of an acylium ion and a Lewis acid-bound phenolate. A π-complex is then formed, which further reacts via electrophilic aromatic substitution in the ortho- or para-position. After hydrolysis, the product is liberated.

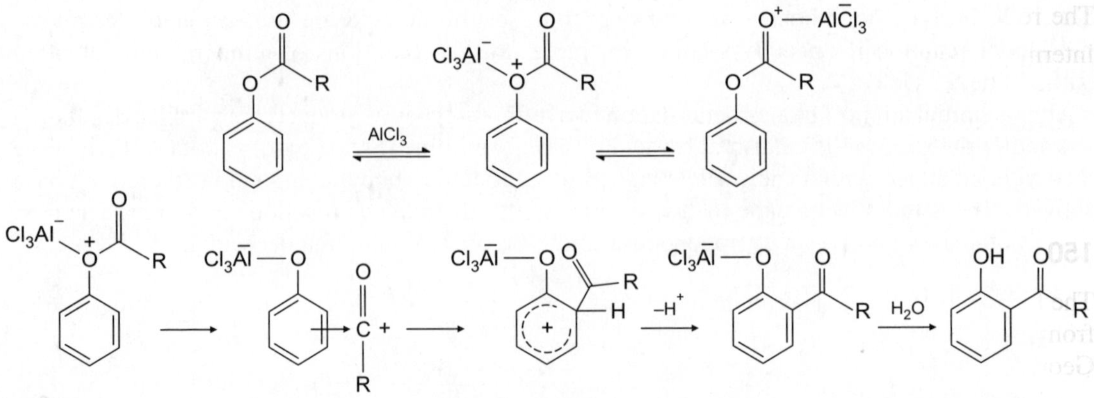

The Fries rearrangement is also found to conduct with UV light, in the absence of a catalyst (the photo-Fries rearrangement). Similar to the Fries rearrangement, both ortho- and para-products are observed in the photo-Fries rearrangement. This photochemical variant of *Fries rearrangement*, proceeds via radical species as intermediate where the excited state ester dissociated to a radical pair. The photo-Fries rearrangement is found predominantly intramolecular in nature.

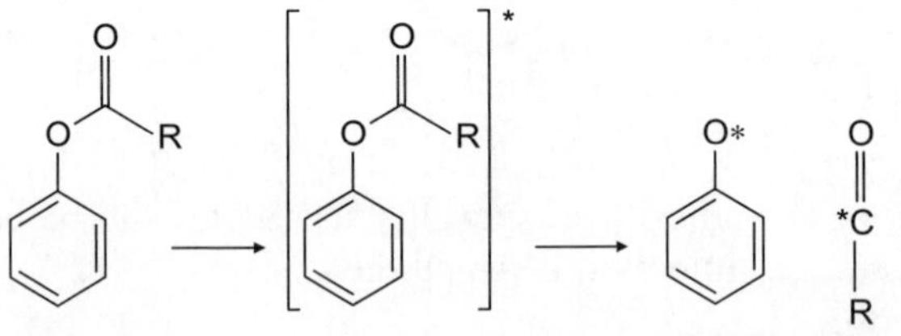

*free radical

149. FRITSCH-BUTTENBERG-WIECHELL (FBW) REARRANGEMENT

The reaction was discovered in 1894 by the three scientists—Paul Ernst Moritz Fritsch, Wilhelm Paul Buttenberg, Heinrich G. Wiechell. In Fritsch-Buttenberg-Wiechell rearrangement, a 1,1-diaryl-2-bromoalkene rearranges to a 1,2-diaryl alkyne promoted by the strong base (e.g. NaOEt).

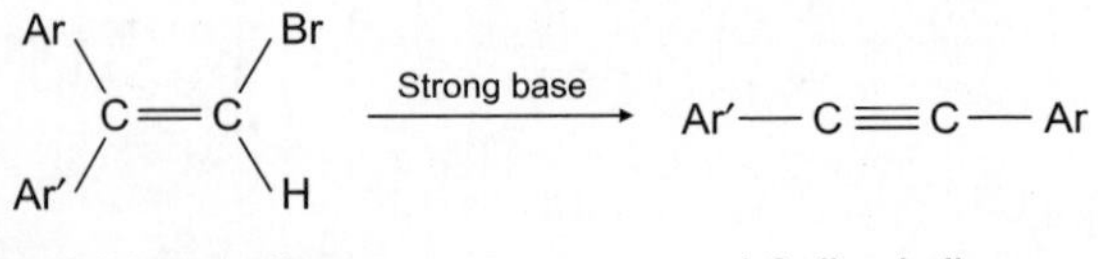

1,1-diaryl-2-bromoalkene   1,2-diaryl alkyene

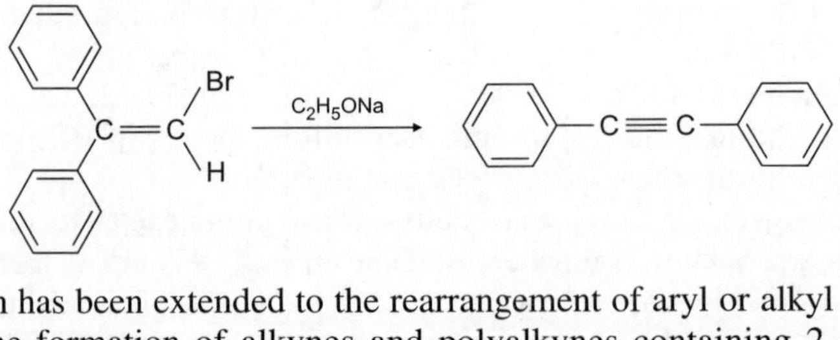

Now the reaction has been extended to the rearrangement of aryl or alkyl olefinic halides and found useful for the formation of alkynes and polyalkynes containing 2–10 acetylene units.

The reaction mechanism comprises α-elimination of hydrogen halide which leads to a carbine intermediate and subsequent migration of an aryl group to give diarylacetylenes.

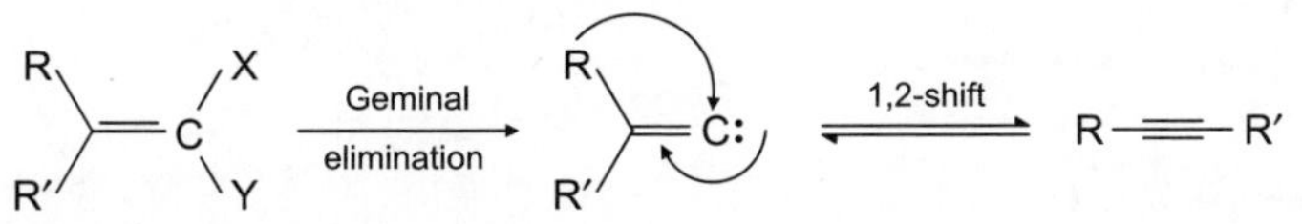

150. FUJIMOTO-BELLEAU REACTION

The Fujimoto-Belleau reaction involves synthesis of cyclic α-substituted α,β-unsaturated ketones from enol lactones and Grignard reagents. The reaction is named after the Bernard Belleau and George I. Fujimoto. Grignard reagents for the reaction are prepared from primary halides.

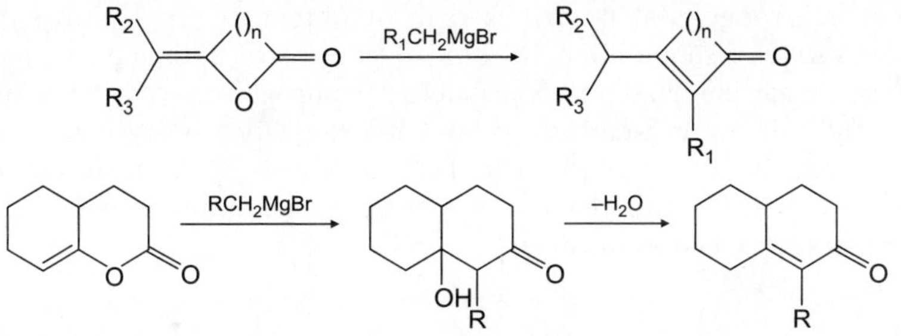

The Fujimoto-Belleau reaction begins with a Grignard reaction, followed by a hydrogen-shift tautomerisation (an enol-keto tautomerisation), an adol reaction then loss of water through a reverse aldol resulting in the final product.

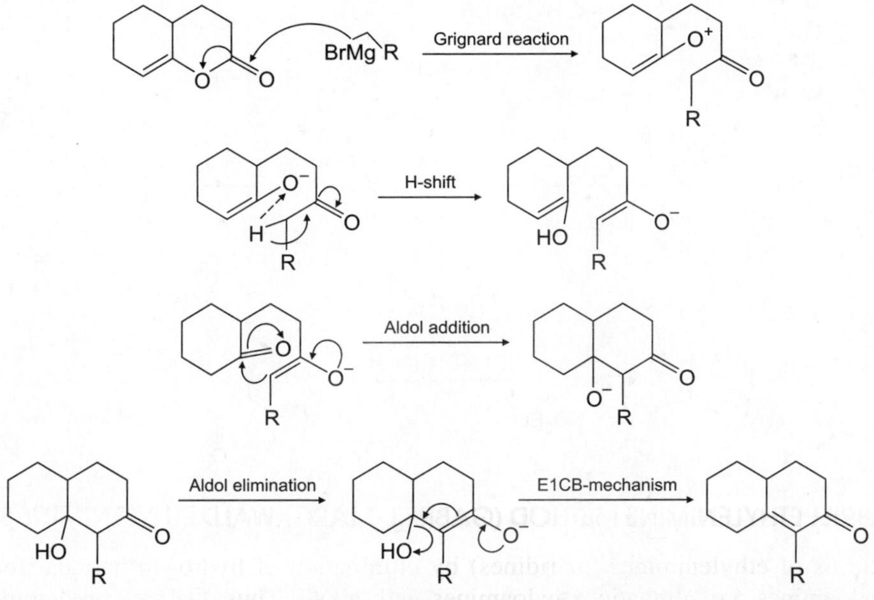

This reaction has been found to be especially useful for introducing a radioactive carbon into the benzo ring, such as in benzazepine and into the steroid scaffold. Among its many applications, the Fujimoto-Belleau reaction provided the final step completing the steroid nucleus in Woodward's nobel prize-winning total synthesis of cortisone.

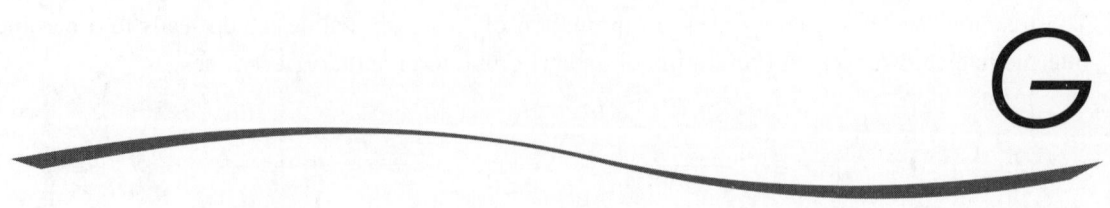

151. GABRIEL-COLMAN REARRANGEMENT (PHTHALIMIDOACETIC ESTER → ISOQUINOLINE REARRANGEMENT, GABRIEL ISOQUINOLINE SYNTHESIS)

Gabriel-Colman rearrangement describes rearrangement of alkyl phthalimidoacetate into isoquinoline derivatives (isoquinoline 1,4-diol) by means of alkoxide treatment. Phthalimido derivatives containing an enolizable carbon attached to nitrogen can only be taken in this reaction. Under acidic conditions, the 3-carboethoxyl-4-hydroxyisocarbostyril has been observed to predominate, whereas in neutral conditions, the tetrahydroisoquinolinedione is the dominating structure of the product. For this reaction to occur, the phthalimido derivatives must contain an enolizable carbon attached to nitrogen.

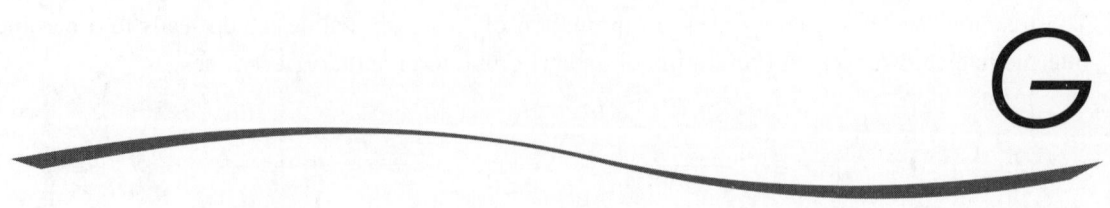

$$Y = CO$$
$$Ar = C_6H_4, C_6H_3R$$
$$X = Halogen$$

OR

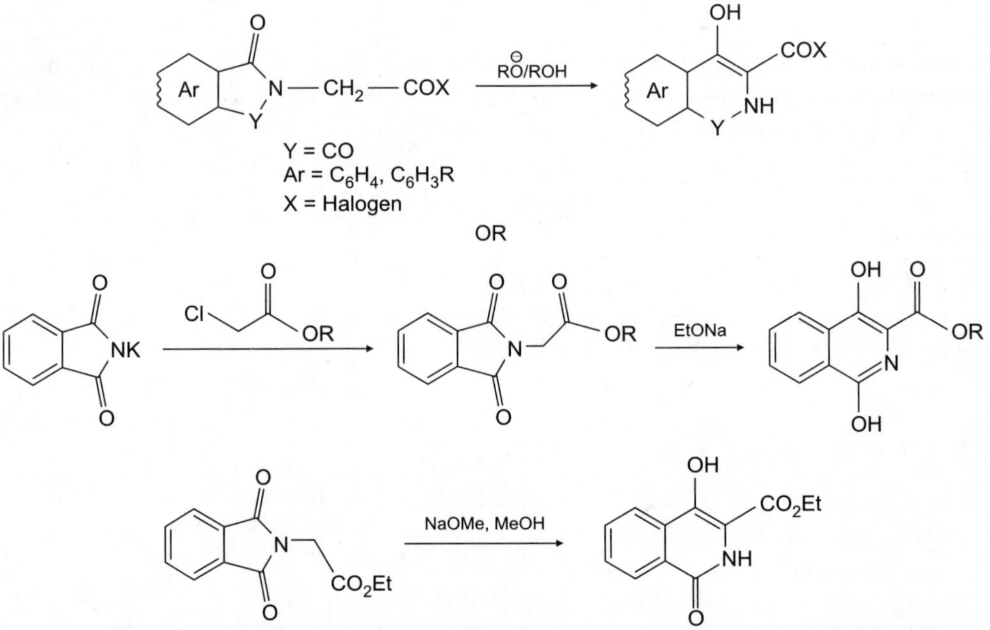

152. GABRIEL ETHYLENIMINE METHOD (GABRIEL-MARCKWALD ETHYLENIMINE SYNTHESIS)

The synthesis of ethylenimines (aziridines) by elimination of hydrogen halides from aliphatic vicinal haloamines, i.e. aliphatic α-haloamines with alkali. Thus, Gabriel ethylenimine method involves cyclization of haloamines and now found to be useful for the preparation of four-(azetidines), five-, and six-membered cyclic amines.

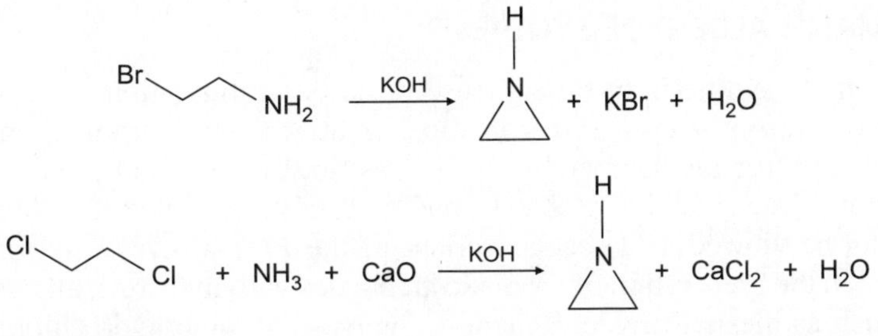

153. GABRIEL SYNTHESIS

Gabriel synthesis involves conversion of alkyl halides to primary amines by treatment with potassium phthalimide and subsequent hydrolysis. This synthesis utilizes the anion of phthalimide as a nucleophile in S_N2 reaction with alkyl halides. Alkaline hydrolysis of the N-alkylated phthalimide release the 1° amine free of 2° and 3° amines. Therefore, primary amines can be made cleanly by the Gabriel synthesis. It has been observed that using hydrazine, NH_2NH_2, to release the 1° amine often gives superior results. The reaction is generally used for the synthesis of aliphatic primary amines. Aromatic primary amines cannot be prepared by this method because aryl halides do not undergo nucleophilic substitution with the anion formed by phthalimide. The Gabriel synthesis is named for the German chemist Siegmund Gabriel.

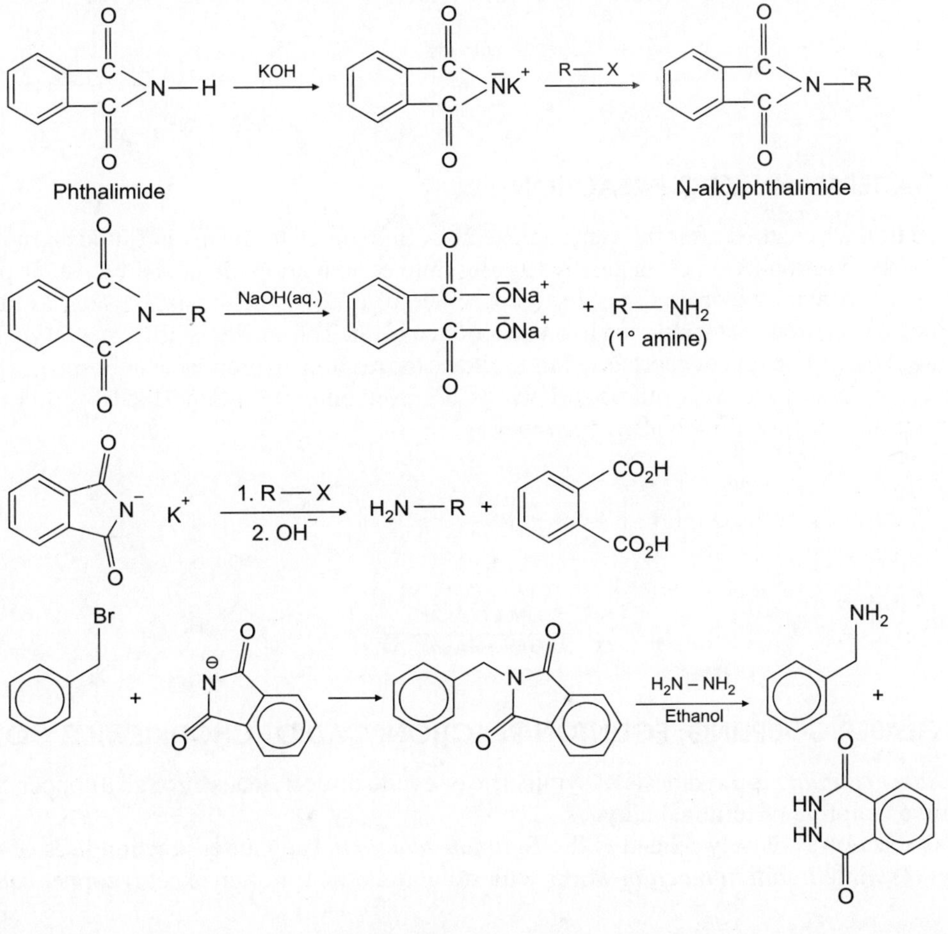

154. GATTERMANN ALDEHYDE SYNTHESIS

Gattermann aldehyde synthesis in broad sense describes formylation of aromatic compounds. The preparation of a formyl-substituted aromatic derivative from an aromatic substrate by reaction with hydrogen cyanide (or another metallic cyanide as zinc cyanide) and gaseous hydrogen chloride in the presence of a Lewis acid (e.g. $ZnCl_2$, $AlCl_3$) catalyst is called the *Gattermann synthesis*. This reaction can be viewed as a special variant of the *Friedel-Crafts acylation* reaction. This reaction is useful in the preparation of aromatic aldehydes with hydroxyl, alkoxyl, and even mutli-alkyl groups, such as mesitaldehyde. Reaction involves aldimine hydrochloride functions as an intermediate.

$$ArH + Zn(CN)_2 \xrightarrow{HCl} ArCH = \overset{+}{N}H_2Cl^- \xrightarrow{H_2O} ArCHO$$

$$HCl + HCN \xrightarrow{AlCl_3} \underset{\substack{\text{Imidoformyl} \\ \text{chloride}}}{NH = CHCl} \xrightarrow{ArH} \underset{\text{Aryl imine}}{HCl + ArCH = NH}$$

$$\downarrow H_2O$$

$$ArCHO + NH_3$$

155. GATTERMANN-KOCH REACTION

The reaction was named after the German chemists Ludwig Gattermann and Julius Arnold Koch. It involves the introduction of an aldehyde group into certain aromatic nuclei by means of carbon monoxide, hydrogen chloride (gas), in the presence of a Lewis acid catalyst such as aluminum chloride. The cuprous chloride used as a carrier of catalyst. The carrier is not necessary when high pressure is used. Addition of cuprous chloride allows the reaction to proceed at atmospheric pressure. Benzaldehyde and many aromatic aldehydes are conveniently synthesized by this reaction. This method works well with alkyl benzenes.

156. GLASER COUPLING; EGLINTON REACTION; CADIOT-CHODKIEWICZ COUPLING

The *Glaser coupling* is a synthesis of symmetric or cyclic bisacetylenes through a copper catalyzed oxidative coupling of terminal alkynes.

This reaction is closely related is the *Eglinton reaction*. The Glaser reaction uses of catalytic copper (I) while *Eglinton reaction* works with stoichiometric (or often excess) copper catalysts as oxidizing agent. Both the reactions proceed by very similar mechanisms. This *Glaser coupling* reaction was first reported by Carl Andreas Glaser in 1869.

$$2\,R\!-\!\!\equiv\!\!-H \xrightarrow[\text{Base, O}_2]{\text{Cu(I)-cat.}} R\!-\!\!\equiv\!\!\equiv\!\!-R$$

The Eglinton reaction is an oxidative homo coupling of terminal alkynes, and allows the synthesis of symmetric or cyclic bisacetylenes via reaction of the terminal alkyne with a stoichiometric (or often excess) amount of a copper(II) salt (copper (II) acetate [Cu(OAc)$_2$]) in pyridine. The Eglinton reaction is named after Geoffrey Eglinton. This reaction is particularly applicable to cyclizations. For the synthetic application viewpoint in organic chemistry, the Eglinton reaction is more convenient than the Glaser method but suffers with a drawback of stoichiometric amounts of copper salt.

$$2\,R\!-\!\!\equiv\!\!-H \xrightarrow[\text{pyridine}]{\text{Cu(OAc)}_2} R\!-\!\!\equiv\!\!\equiv\!\!-R$$

Heterocoupling or preparation of unsymmetrical bis-acetylenes can be achieved by using the Cadiot-Chodkiewicz reaction. It involves reaction of terminal alkyne with a haloalkynes in the presence of a copper catalyst such as copper(I) bromide and an aliphatic amine base to yield an unsymmetrical coupling product. The reaction product is a diacetylene or dialkyne.

$$R\!-\!C\!\equiv\!C\!-\!H \;+\; Br\!-\!C\!\equiv\!C\!-\!R_1 \xrightarrow[\text{NEt}_3]{\text{CuBr}} R\!-\!C\!\equiv\!C\!-\!C\!\equiv\!C\!-\!R_1$$

157. GOMBERG-BACHMANN REACTION

The Gomberg-Bachmann reaction which is named for the Moses Gomberg and the Werner Emmanuel Bachmann describes as base-promoted aryl-aryl coupling reaction via a diazonium salt.

The reaction involves alkali dependent formation of diaryls (mostly unsymmetrical diaryls) by condensation of aryl diazonium salts with aromatic or heterocyclic compounds. It has been found that this reaction works well for the aromatic diazonium salts containing an electron-withdrawing group. The arene compound in this reaction is coupled through an intermediate aryl radical.

The intramolecular variant of this reaction is called the *Pschorr reaction* which is carried out in strongly acidic solution, and in the presence of copper powder. The Gomberg-Bachmann reaction is usually conducted in a two-phase system, an aqueous alkaline solution, that also contains the arenediazonium salt, and an organic layer containing the other aromatic reactant. Yields can be improved by the use of a phase transfer catalyst. Otherwise, yields often are below 40%, due to various side reactions taking place. The Pschorr reaction generally gives better yields.

158. GOMBERG FREE RADICAL REACTION

Generation of free radicals by abstraction of the halogen from triarylmethyl halides with metals is referred to as the Gomberg reaction.

159. GOULD-JACOBS REACTION

The Gould-Jacobs reaction involves a sequence of reactions for preparing 4-hydroxyquinoline derivatives. Reaction covers condensation of an aniline or an aniline derivative 1 with either alkoxy methylenemalonic ester or acyl malonic ester to yield anilidomethylenemalonic ester 3 followed by cyclization to the 4-hydroxy-3-carboalkoxyquinoline, then subsequent saponification to form acid 5, and finally decarboxylation to give the 4-hydroxyquinoline.

R = Alkyl
R' = Alkyl, aryl, or H
R'' = Alkyl or H

In other words, alkylidene β-diesters or analogous substructures derived from condensation with Meldrum acid or electron-withdrawing group containing esters (e.g. malonic ester or cyanoacetic acid esters) readily undergo pericyclic annulation reactions at high temperatures, if they are located in a favourable position relative to aromatic or unsaturated systems.

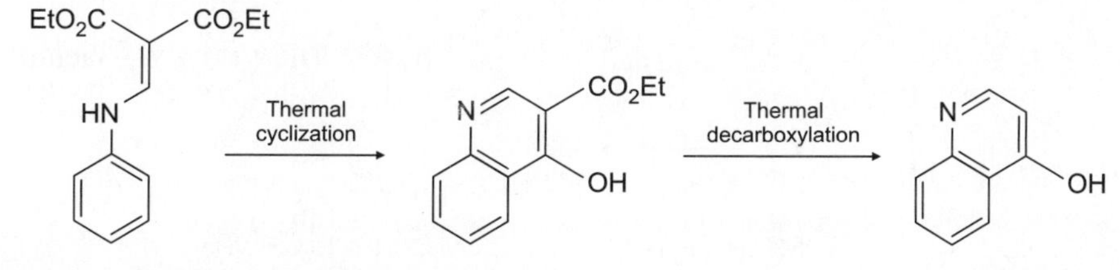

160. GRAEBE-ULLMANN SYNTHESIS

The preparation of carbazoles involving the pyrolysis of benzotriazoles prepared from ortho-amino-diphenylamine (2-aminodiphenylamines) and nitrous acid is known as the Graebe-Ullmann reaction. The importance of this reaction is in preparation of carbazole derivatives.

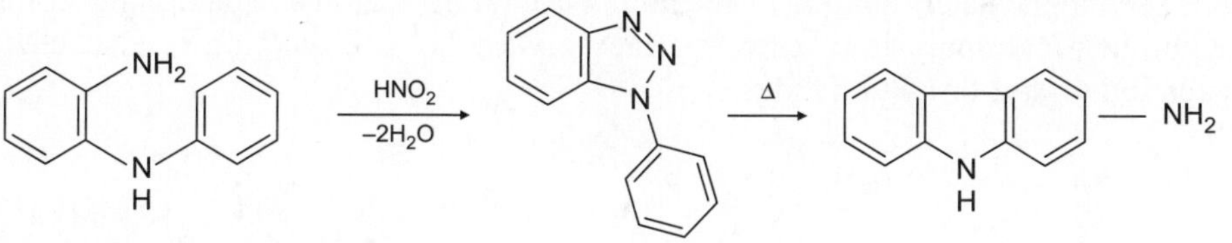

161. GRIESS DIAZO REACTION

Primary aromatic amines react with nitrous acid or other nitrosating agents to yield diazonium salts via a number of intermediate species. This reaction is called Diazo reaction. Since it was discovered by Johan Peter Griess (1858), it is referred to as Griess Diazo reaction.

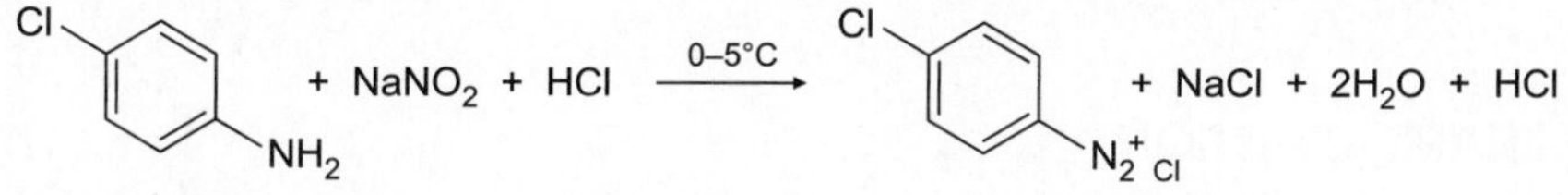

The various steps involved in the Diazo reaction are as under:

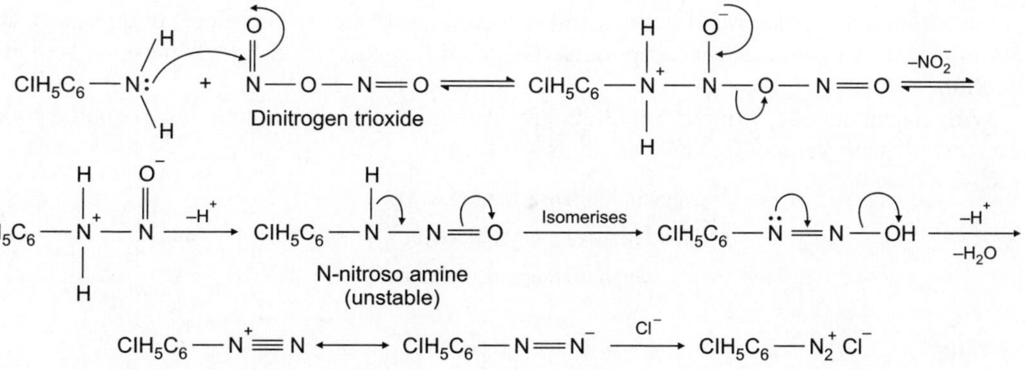

Diazotization coupling reaction (Griess reagent)

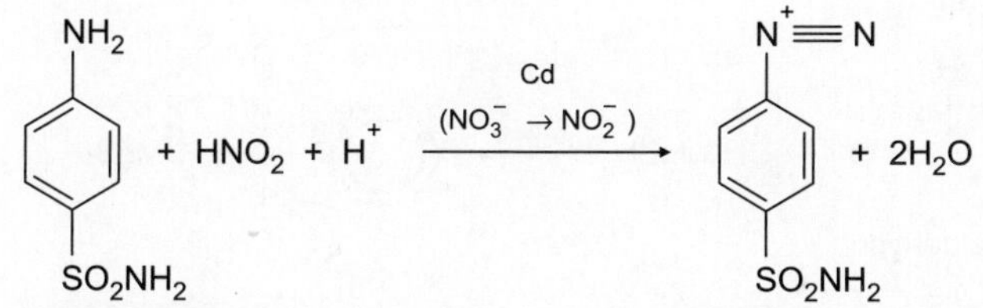

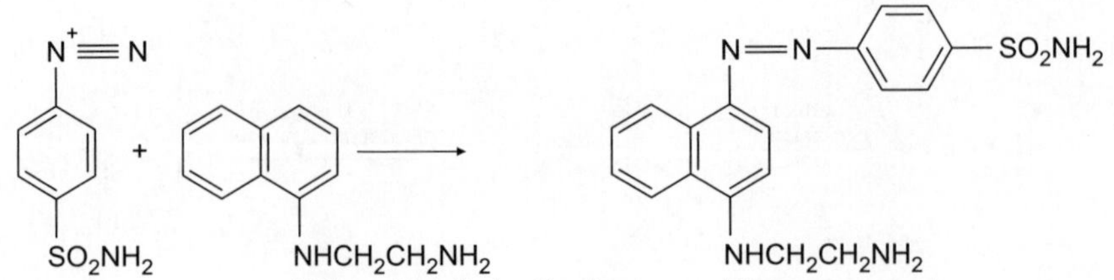

N-1-napthylethylendiamine

Diazo compound

162. GRIGNARD DEGRADATION

Grignard degradation is a stepwise dehalogenation of a polyhalo compound through its Grignard reagent, which on treatment with water yields a product containing one halogen atom less. This reaction is used in the determination of the number of halogen atoms in an organic compound and also in the chemical analysis of certain triacylglycerols. The reaction is also useful for the dehalogenation of aryl or alkyl halides.

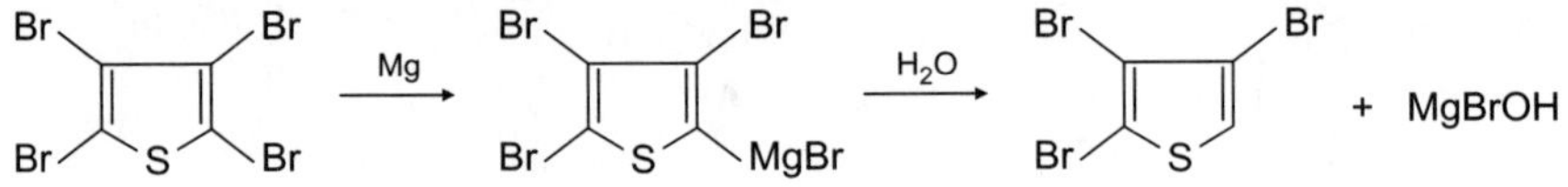

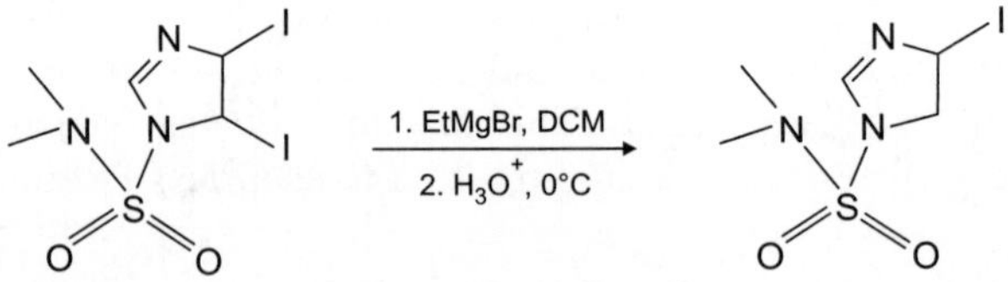

163. GRIGNARD REACTION

In 1912, Victor Grignard received the Nobel Prize in chemistry for his work on the reaction that bears his name, a carbon-carbon bond-forming reaction by which almost any alcohol may be formed from appropriate alkyl halides and carbonyl compounds. Therefore, traditionally, it is the addition of organomagnesium compounds (Grignard reagents) to carbonyl compounds to generate alcohols.

With formaldehyde, primary alcohols are formed; with other aldehydes, secondary alcohols are formed; with ketones, tertiary alcohols are formed.

$$\text{Grignard reagent + formaldehyde} \longrightarrow 1^\circ \text{ ROH}$$
$$\text{Grignard reagent + other aldehydes} \longrightarrow 2^\circ \text{ ROH}$$
$$\text{Grignard reagent + ketones} \longrightarrow 3^\circ \text{ ROH}$$

$$H_2C{=}O + CH_3MgBr \xrightarrow{\text{Ether}} H_2\overset{\overset{\large CH_3}{|}}{C}{-}OMgBr \xrightarrow{H_2O} CH_3CH_2OH + Mg(OH)Br$$

Formaldehyde

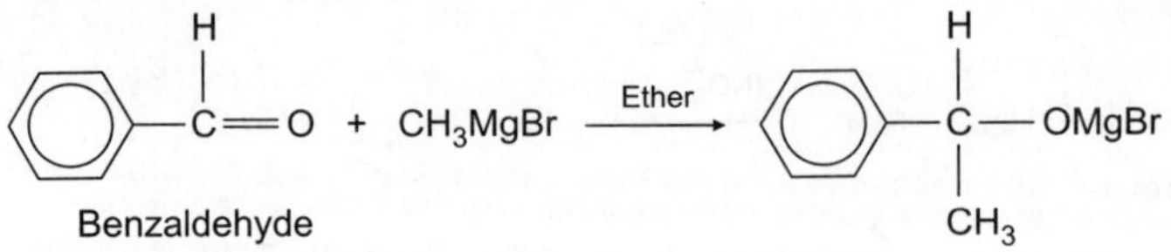

Benzaldehyde

$$\text{H}_2\text{O} \downarrow$$

$$\text{C}_6\text{H}_5\text{—CHOH} \quad + \quad \text{Mg(OH)Br}$$
$$\text{CH}_3$$

Acetone

$$\underset{\parallel}{\underset{O}{\text{CH}_3\text{CCH}_3}} + \text{CH}_3\text{CH}_2\text{MgBr} \xrightarrow{\text{Ether}} \underset{\text{OMgBr}}{\overset{\text{CH}_2\text{CH}_3}{\text{CH}_3\text{CCH}_3}} \xrightarrow{\text{H}_2\text{O}} \underset{\text{OH}}{\overset{\text{CH}_2\text{CH}_3}{\text{CH}_3\text{CCH}_3}} + \text{Mg(OH)Br}$$

A more modern interpretation extends the scope of the reaction to include the addition of Grignard reagents to a wide variety of electrophilic substrates.

$$\underset{O}{\overset{|}{\text{—C—}}} + \text{RMgX} \longrightarrow \underset{\text{OMgX}}{\text{R—C—}} \xrightarrow{\text{H}_2\text{O}} \underset{\text{OH}}{\overset{R}{\text{—C—}}}$$

$$\text{RC} \equiv \text{N} + \text{R}^1\text{MgX} \longrightarrow \underset{\text{R}^1}{\text{R—C}=\text{NMgX}} \xrightarrow{\text{H}_2\text{O}} \text{R}\overset{O}{\underset{\parallel}{—\text{C}—}}\text{R}^1$$

$$\underset{\text{R}}{\overset{O}{\underset{}{\text{R—C—R}}}} \xrightarrow[\text{2. H}_2\text{O}]{\text{1. RMgX}} \underset{\text{R}}{\overset{\text{OH}}{\text{R}\diagup\diagdown\text{R}}}$$
1. addition
2. protonation yields a 3° alcohol

$$\text{R—X} \xrightarrow[\text{ether}]{\text{Mg}} \begin{bmatrix} \text{R—MgX} \\ \updownarrow \\ \overset{\ominus}{\text{R}} \overset{\oplus}{\text{MgX}} \end{bmatrix}$$

$$\underset{\text{R}}{\overset{O}{\underset{}{\text{R—C—H}}}} \xrightarrow[\text{2. H}_2\text{O}]{\text{1. RMgX}} \underset{\text{R}}{\overset{\text{OH}}{\text{R}\diagup\diagdown\text{H}}}$$
1. addition
2. protonation yields a 2° alcohol

$$\text{Ph—Br} \xrightarrow{\text{Mg}°} \text{Ph—MgBr}$$

$$\underset{\text{Ph}\diagdown\diagup\text{OEt}}{\overset{O}{\parallel}} \xrightarrow{\text{Ph—MgBr}} \underset{\text{Ph}\diagdown\diagup\text{Ph}}{\overset{O}{\parallel}} \xrightarrow{\text{Ph—MgBr}} \underset{\text{Ph}}{\overset{\text{OH}}{\text{Ph}\diagup\diagdown\text{Ph}}}$$

$$\text{Mg} + \text{C}_6\text{H}_5\text{—Br} \xrightarrow{\substack{\text{Anhydrous diethyl} \\ \text{ether (ether)}}} \text{C}_6\text{H}_5\text{—MgBr}$$

$$\text{C}_6\text{H}_5\text{—MgBr} + \underset{}{\overset{O}{\parallel}} \xrightarrow[\text{2. Dilute HCl}]{\text{1. Ether}} \text{Ph}_3\text{C—OH}$$

164. GROB FRAGMENTATION

Grob fragmentation is an elimination reaction involving cleavage of an amine of a suitable β-leaving group (an elimination reaction) in the presence of a base, where the leaving group is such (as a halogen or tosyl group). The elimination taking place when an electrofuge and nucleofuge are situated in positions 1 and 3 on an aliphatic chain. This reaction is an useful skeletal transformation, involving specific carbon-carbon bond cleavage with accompanying conversion of certain sigma-bonds to pi-bonds.

The reaction was named after the British chemist Cyril A. Grob. The reaction product is an electrofugal fragment (carbonium ion, acylium ion), an unsaturated fragment (alkene, alkyne, imine) and a nucleofugal fragment (leaving group such as tosyl, hydroxyl).

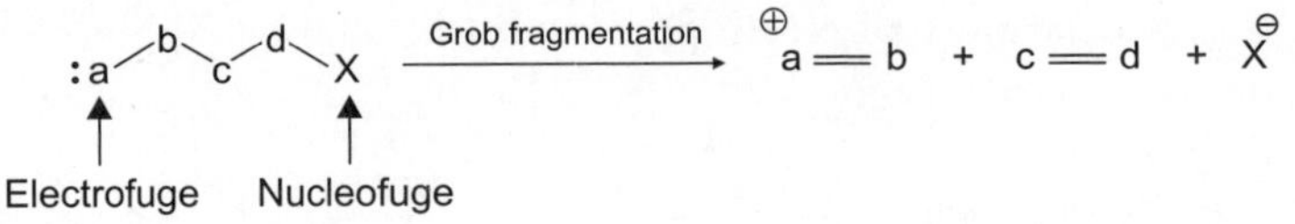

An **electrofuge** is a leaving group, which does not retain the bonding pair of electrons from its previous bond with another species.

A **nucleofuge** is a leaving group, which retains the lone pair from its previous bond with another species

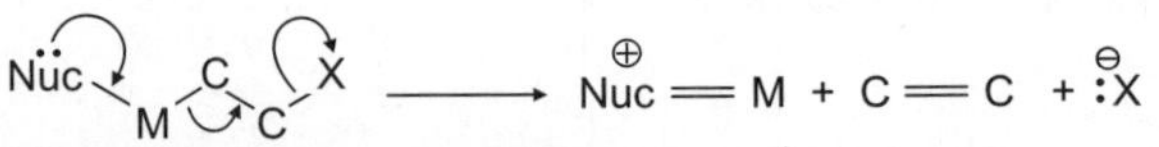

Three potential mechanisms for the Grob:

There are three potential mechanisms as the bond break sequence may occur as a one or two stepped process:

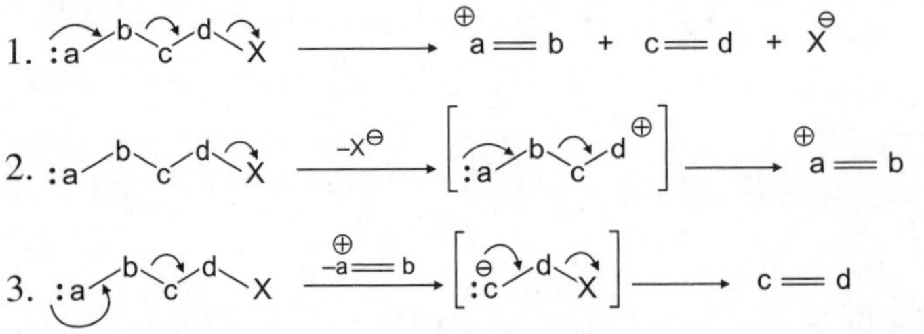

Certain structural and stereoelectronic requirements determine which mechanism is used. These are similar mechanisms to the known 1,2 elimination to give olefins.

1. One stepped process, simultaneous loss of the electrofugal group and nucleofugal group

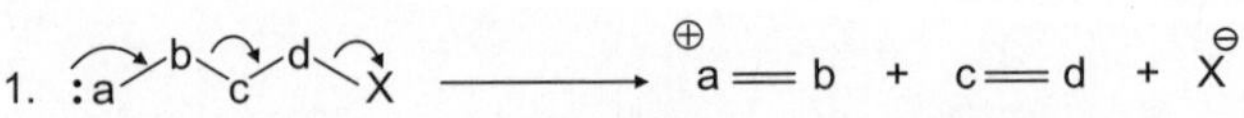

2. Two stepped process, firstly loss of X generating a carbocation then break down to the two olefinic species if following the fragmentation route (similar to E1 or S_N1)

   However the carbonium ion can further react via elimination, substitution, or ring-closure.

   The rate-determining step is the ionization to the carbonium ion. The tendency to ionize is greater when a tertiary and thus stable carbonium ion is formed.

   The leaving ability of X- is important as it can lead to an increased ionization rate (e.g. Cl < Br < I)

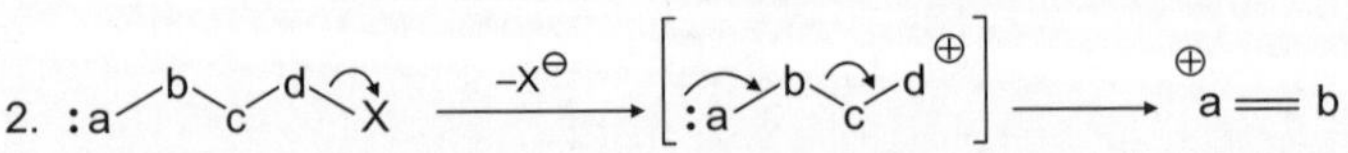

3. Two stepped process, firstly loss of electrofugal group generating a carbanionic species then break down to give olefin and release of X (rarer)

This can only occur if the carbanion is stabilized by electron-withdrawing substituents and if X is a poorer leaving group.

165. GRUNDMANN ALDEHYDE SYNTHESIS

The Grundmann aldehyde synthesis involves conversion of an acyl halide into an aldehyde of the same chain length. The overall conversion takes place via the diazo ketone, to the acetoxy ketone, reduction with aluminum isopropoxide and hydrolysis to the glycol, and cleavage with lead tetra-acetete. Because of the Rosenmund reduction and DIBAL-H accomplish similar transformations; this reaction sequence is not practiced much currently.

166. GUARESCHI-THORPE CONDENSATION

The Guareschi-Thorpe condensation involves formation of pyridine derivatives (2-pyridone) from the condensation of cyanoacetic ester with a 1,3-diketone in the presence of ammonia. The reaction is named after Icilio Guareschi and Jocelyn Field Thorpe.

167. GUERBET REACTION

The Guerbet reaction can be summarized as a dimerization of alcohols (I) with liberation of one equivalent of water to give branched alcohols (II).These 'typical' Guerbet alcohols have an even number of carbon atoms, with a minimum of six. The reaction sequence, which bears his name (named after Marcel Guerbet), is related to the Aldol reaction and occurs at high temperatures under catalytic conditions. The product is an alcohol with twice the molecular weight of the reactant alcohol minus a mole of water. The reaction proceeds by a number of sequential steps:

A. Oxidation of alcohol to aldehyde
B. Aldol condensation after proton extraction

C. Dehydration of the aldol product
D. Hydrogenation of the allylic aldehyde

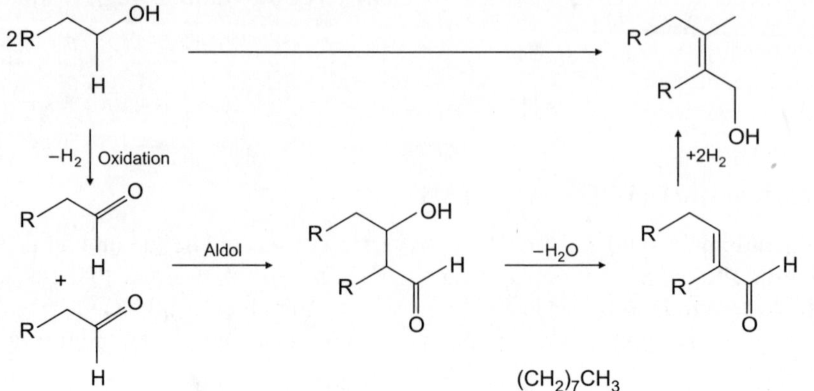

$$2 CH_3(CH_2)_9OH \xrightarrow{\text{Catalyst}} CH_3(CH_2)_9\overset{\displaystyle (CH_2)_7CH_3}{\underset{|}{C}}HCH_2OH$$

Many catalysts have been described in the literature such as nickel, oxides of copper, lead, zinc, chromium, molybdenum, tungsten, etc. There are advantages and disadvantages for each type. The Cannizzaro reaction is a major side reaction.

168. GUTKNECHT PYRAZINE SYNTHESIS

It involves the self-condensation of two α-aminocarbonyl compounds to give 3,6-dihydropyrazine. These are dehydrogenated with mercury(I) oxide or copper(II) sulfate, or sometimes with atmospheric oxygen. The aromatization can be accomplished under very mild conditions This synthesis is regio-selective but not substrate specific. As the α-aminoketones are decomposed in air quickly, thus the general strategy is to prepare them *in situ* (produced by reduction of isonitroso ketones).

169. HALLER-BAUER REACTION

The base-induced (e.g. by sodium amide) cleavage of non-enolisable ketones leading to a carboxylic acid derivative and a neutral fragment in which the carbonyl group is replaced by hydrogen, is referred to as the Haller-Bauer (HB) reaction. This reaction has utility for the preparation of amides of the types ArCONH, and tert-RCONH through hydrolysis, the corresponding carboxylic acids. The reaction is frequently applied to formation of trisubstituted acetic acid.

170. HAMMICK REACTION

Hammick reaction or Hammick coupling involves thermal decarboxylation of α-picolinic or related acids in the presence of excess amount of carbonyl compounds to form 2-pyridyl-carbinols.

171. HANTZSCH DIHYDROPYRIDINE SYNTHESIS (PYRIDINE SYNTHESIS)

Arthur Hantzsch described preparation of 1,4-dihydropyridine more than a century ago. Original synthesis involved a three components (acetoacetic ester, benzaldehyde and ammonia or ammonium salts) coupling reaction in refluxing ethanol.

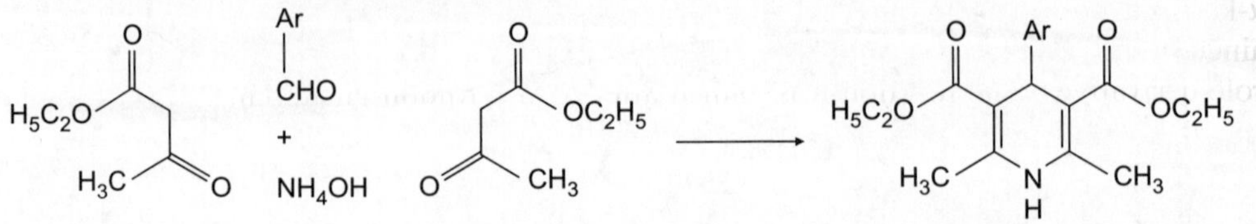

The Hantzsch pyridine synthesis or Hantzsch dihydropyridine synthesis is a multi-component organic reaction which reaction allows synthesis of dihydropyridine derivatives by condensation of one mole of an aldehyde with two mols of a β-keto ester in the presence of ammonia. Subsequent oxidation (or dehydrogenation) with an oxidizing agent, gives pyridine-3,5-dicarboxylates, which may also be decarboxylated to yield the corresponding pyridines.

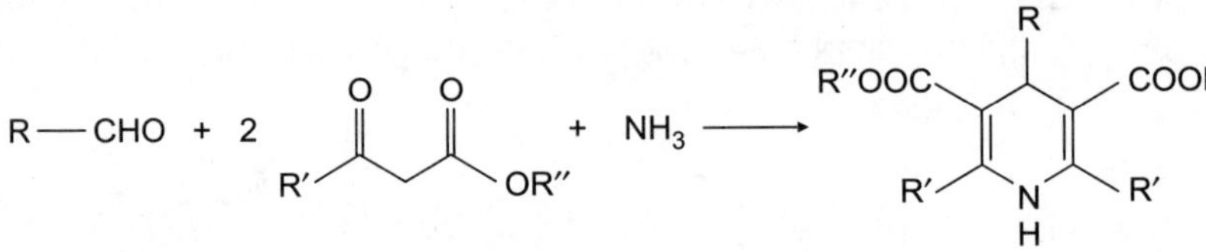

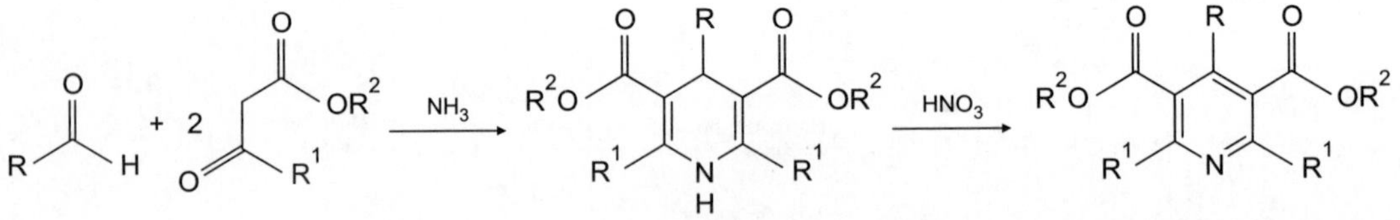

In this reaction, ammonium acetate or ammonia serve as nitrogen donor and the driving force for this second reaction step is aromatization. This reaction was reported in 1881 by Arthur Rudolf Hantzsch. A 1,4- dihydropyridine dicarboxylate is also called a 1,4-DHP compound or a Hantzsch compound.

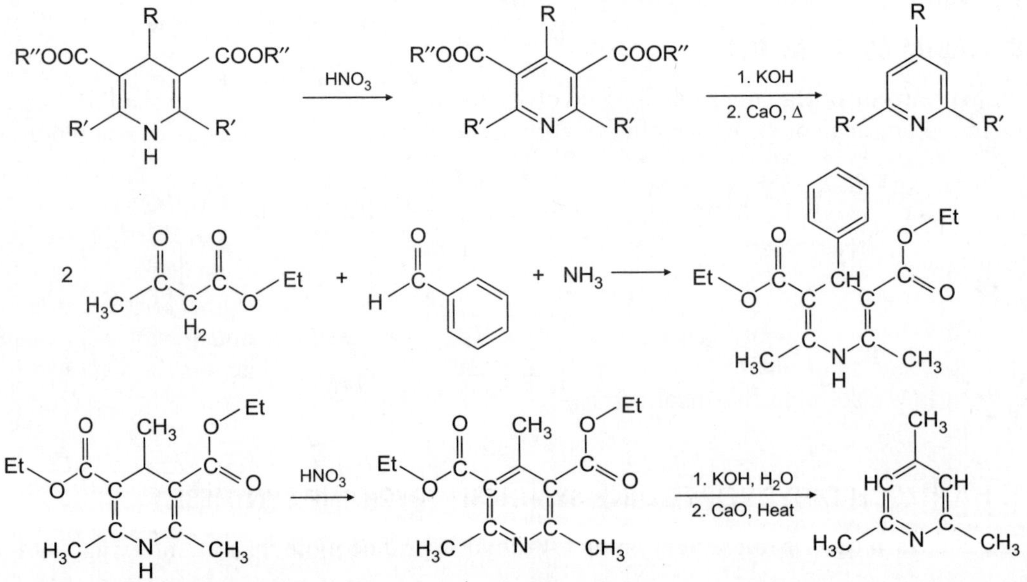

172. HANTZSCH PYRROLE SYNTHESIS

Hantzsch pyrrole synthesis involves formation of pyrrole derivatives from the chemical reaction of α-keto esters with ammonia (or primary amines) and α-haloketones. Substituted pyrroles are obtained in the reaction. It is applied in the preparation of 2,5-dialkyl- or 2,4,5-trialkylsubstituted pyrrole derivatives. The reaction was named after Arthur Rudolf Hantzsch.

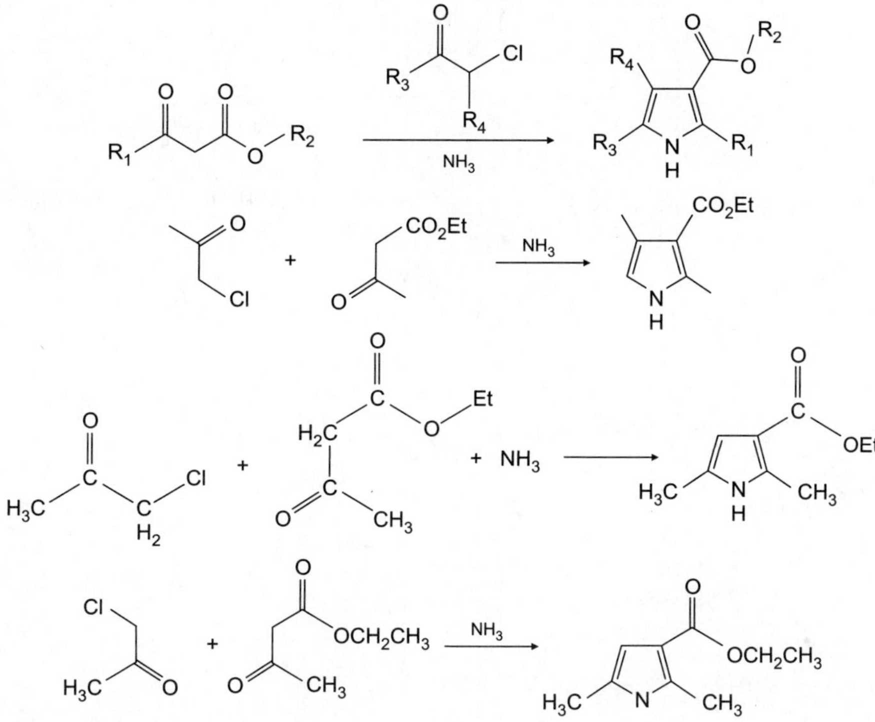

173. HARRIES OZONIDE REACTION (OZONOLYSIS)

Ozone readily attacks ethylenic linkages and from the products, carbonyl compounds can be obtained. Ozonolysis is the cleavage of an alkene or alkyne with ozone to form organic compounds in which the multiple carbon-carbon bond has been replaced by a double bond to oxygen. The outcome of the reaction depends on the type of multiple bond being oxidized and the workup conditions. Depending on the workup, different products may be isolated, for example: reductive workup (e.g. with NaBH$_4$) gives either alcohols or carbonyl compounds, while oxidative work-up leads to carboxylic acids or ketones.

The process results in separation of the carbon atoms originally joined by the double bond. The identities and yields of carbonyl products provide information on the positions of the double bonds in the alkene. Hence, ozonolysis is frequently used in structure determination as well as for synthetic purposes.Treatment of olefins with ozone is a method of cleaving olefinic linkages. On hydrolysis or catalytic hydrogenation, the initially formed ozonide yields two molecules of carbonyl compounds.

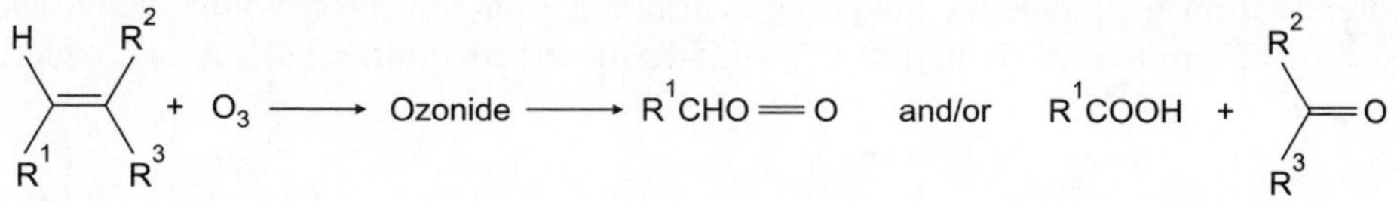

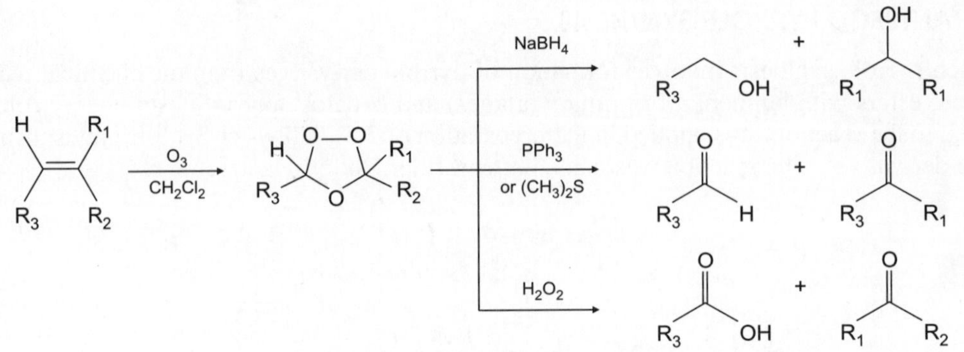

CH$_3$(CH$_2$)$_7$CH = CH(CH$_2$)$_7$COOH $\xrightarrow[\text{2. H}_2\text{O}]{\text{1. O}_3}$ OHC(CH$_2$)$_7$COOH + HOOC(CH$_2$)$_7$COOH + CH$_3$(CH$_2$)$_7$COOH + CH$_3$(CH$_2$)$_7$CHO

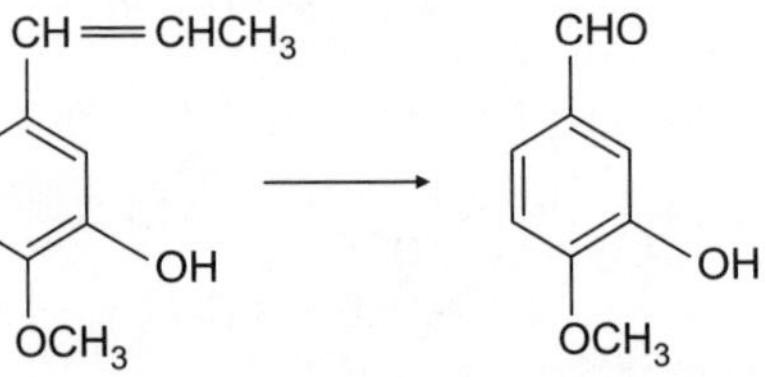

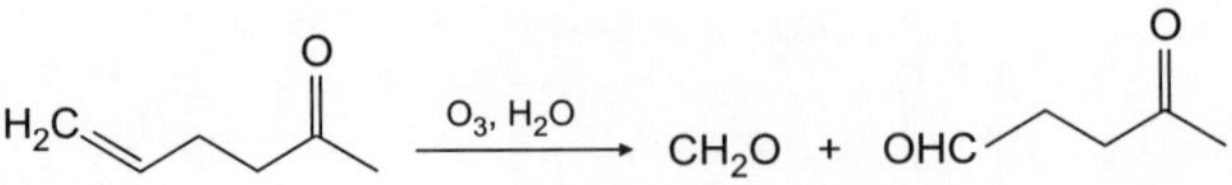

174. HAWORTH METHYLATION

Haworth methylation describes formation of methylated methyl glycosides from monosaccharides with dimethyl sulfate and sodium hydroxide. The glycosidic methyl group is hydrolyzed with acid to yield the free methylated sugar. It is used to elucidate the sugar structures, including the linkages between monosaccharide units in disaccharides or oligosaccharides and the ring structures. It has wide application in carbohydrate chemistry, for the structural analysis of carbohydrates.

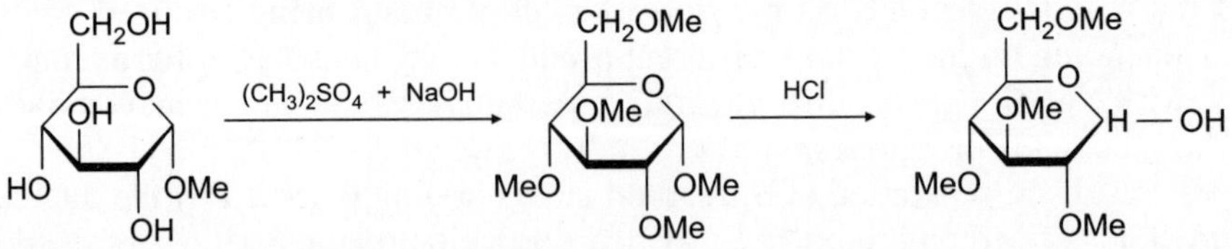

175. HAWORTH PHENANTHRENE SYNTHESIS

Haworth synthesis or Haworth phenanthrene synthesis involves multistep preparation of phenanthrenes from naphthalenes including a Friedel-Crafts acylation with succinic anhydride followed by a Clemmensen reduction or Wolff-Kishner reduction, cyclization, reduction, and dehydrogenation.

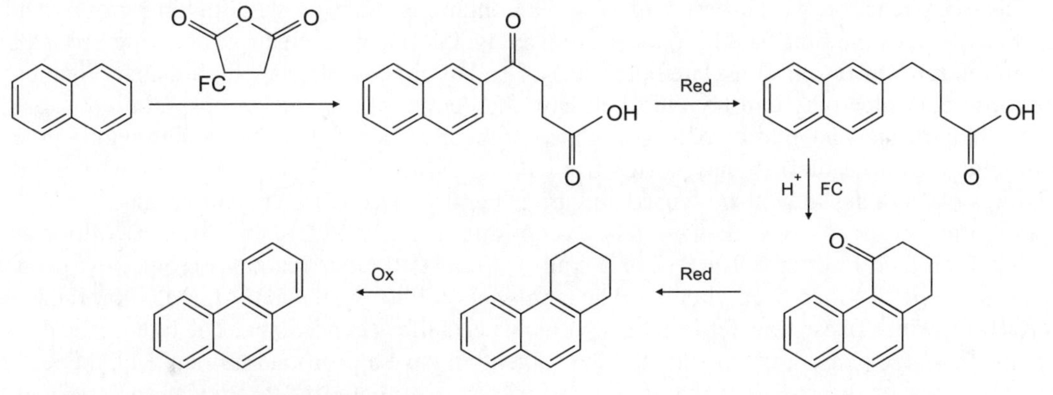

FC = Friedel-Crafts acylation
RED = Reduction
OX = Oxidation

176. HAYASHI REARRANGEMENT

This reaction involves sulfuric acid or phosphorus pentoxide catalyzed rearrangement of ortho-benzoylbenzoic acids. This rearrangement has been found limited applications.

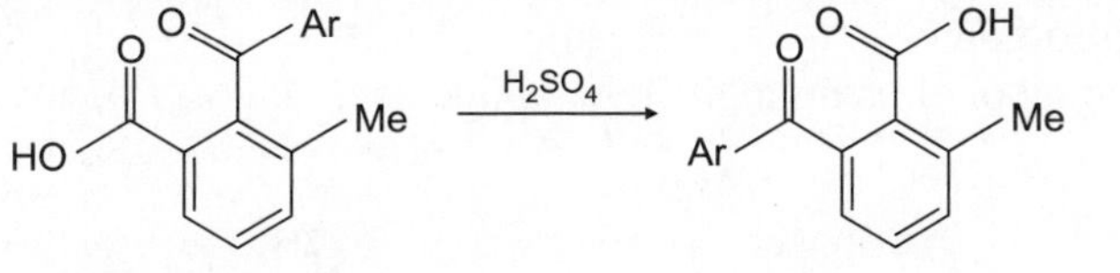

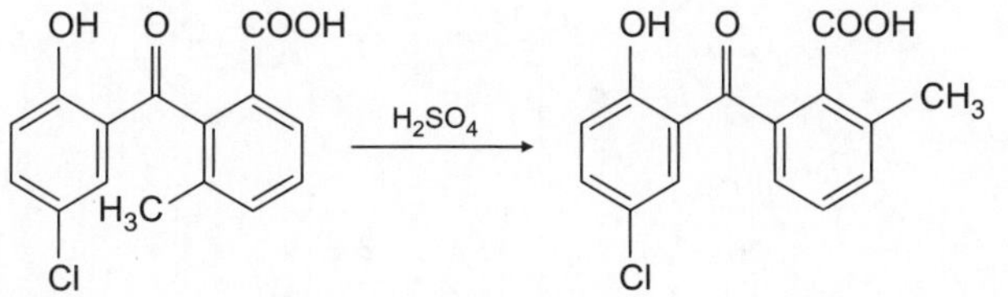

177. HECK REACTION

The Heck arylation reaction was invented independently by Mizoroki and Heck in 1970. It involves stereo specific palladium-catalyzed coupling reaction of an unsaturated halide (or triflate) with an alkene and a base. This reaction allows establishing a bond between olefins and aromatic rings, thus a versatile method for the coupling sp^2 hybridised centres.

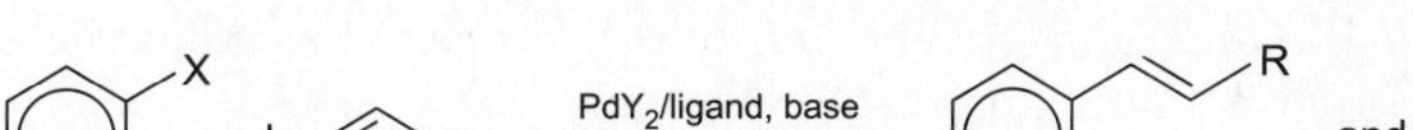

R = Alkenyl, aryl, alyl, alkynyl, benzyl   X = Halide, triflate   R′ = Alkyl, alkenyl, aryl, CO_2R, OR, SiR_3

X = I, Br, Cl, COCl, OSO_2R, SO_2Cl, N_2^+ X, C(O)OC(O)Ar, SiR_2OH
Y = Cl, OAc

The Heck reaction works best with alkenes containing electron-withdrawing groups and in most cases gives the β-arylated products exclusively. Olefins with electron-donating groups give rise to mixtures of α- and β-arylated products. If palladium complexes with bidentate ligands are used the regioselectivity can be determined by the choice of leaving groups. Non-coordinating anions like triflate lead mainly to the α-arylated products, whereas halides predominantly give the β-products. Simple olefins may suffer from Pd-catalyzed isomerization reactions leading to mixtures. Acetylenes may also be used and are generally more reactive than olefins.

Pd-catalysts are used with very few exceptions, usually $PdCl_2$ or $Pd(OAc)_2$ alone or in combination with Ph_3P or o-Tol_3P (2 or 3 equiv). Almost all Heck reactions require the presence of a base, which is often triethylamine. The combination $Pd(OAc)_2$, $MHCO_3$ (M = Na, K), KOAc or K_2HPO_4 with a phase transfer salt is also often used (Jeffery conditions). The ligand, the counter ion, the base, the phase transfer salt, and the solvent all have a profound influence on the rate and the selectivity of the reaction; many of these effects are related to the oxidation state and the coordination chemistry of the catalyst.

Typical solvents for the Heck reaction are dipolar nonprotic solvents like DMF and NMP.

178. HELFERICH METHOD

It refers to glycosidation of an acetylated sugar by heating with a phenol in the presence of a metal halide or p-toluenesulfonic acid as catalyst, or the Helferich method may refer to:

1. Glycosylation of an alcohol using a glycosyl acetate as glycosyl donor and a Lewis acid (e.g. a metal halide) as promoter
2. Glycosylation of an alcohol using a glycosyl halide as a glycosyl donor and a mercury salt as promoter

179. HELL-VOLHARD-ZELINSKY REACTION

The Hell-Volhard-Zelinsky reaction refers to α-bromination of carboxylic acids using Br_2/PBr_3.

Reaction of carboxylic acids (aliphatic acid containing α-hydrogen) with bromine in the presence of a catalytic amount of phosphorus or phosphorus tribromide replaces the hydrogen of α carbon atom. This reaction is evidenced to involve the enol form of the intermediate acyl halide.

By application of the *Hell-Volhard-Zelinsky reaction*, an α-hydrogen of a carboxylic acid can also be replaced by chlorine to give an α-chlorocarboxylic acid.

Further, the α-bromo or α-chloro carboxylic acids are versatile intermediates for further synthetic transformations.

$$CH_3CHCH_2CHCO_2H \xleftarrow[\text{2. H}^+]{\text{1. OH}^-\text{(aq)}}$$

with CH_3 and OH substituents

$$CH_3CHCH_2CHCO_2H \xrightarrow{NH_3\text{(aq)}} CH_3CHCH_2CHCO_2H$$

with Br substituent on left, NH_2 (leucine) on right, CH_3 substituents

$$\xrightarrow[\text{2. H}^+]{\text{1. EtO}^-\text{/EtOH}} CH_3CHCH=CHCO_2H$$

with CH_3 substituent

The reaction is named after three chemists, Carl Magnus von Hell, Jacob Volhard and Nikolay Zelinsky.This reaction has been used to prepare α-halo aliphatic acids.

180. HENKEL REACTION (RAECKE PROCESS, HENKEL PROCESS)

Henkel reaction involves the thermal rearrangement or disproportionation of alkaline salts of aromatic acids to symmetrical diacids in the presence of a metallic salt and carbon dioxide. It is also known as Raecke process or Henkel process. This reaction has been applied for the production of terephthalic acid at an industrial scale.

181. HENRY REACTION (NITROALDOL REACTION)

The Henry or nitroaldol reaction is a base-catalyzed aldol-type condensation it involves essentially a coupling reaction between a carbonyl compound and an alkylnitro compound bearing a hydrogens, the overall transformation enables the formation of a carbon-carbon bond with the concomitant generation of a new difunctional group, namely the β-nitroalcohol function. Henry or reaction facilitate the joining of two molecular fragments, under mild conditions, with the formation of two asymmetric centers at the new carbon-carbon juncture.

Typically, further transformations involving the newly formed β-nitroalkanol functionality, such as oxidation, reduction, and dehydration, will follow thereby depending on the requirements.

The base catalysed 1,2-addition of nitromethane to n-butanal yields the nitroalcohol.

182. HERON REARRANGEMENT (HETEROATOM REARRANGEMENTS ON NITROGEN)

HERON is an acronym for heteroatom rearrangements on nitrogen, where the rearrangement of amides, with two heteroatoms (e.g. oxygen) substituted at nitrogen (RC(O)NXY), to esters and 1,1-diazenes via the migration of oxygen from the nitrogen to the carbonyl carbon. It has been reported that the anomeric interactions at nitrogen are enhanced when the second heteroatom (*Y*) is strongly electronegative. This reaction has an importance for the preparation of ester with a bulky alcohol moiety.

183. HERZ REACTION

The reaction was named after Richard Herz which involves the chemical conversion of an aniline-derivative to a 1,3,2-benzothiazathiolium chlorides, the so-called Herz salt. In other words, the term "Herz reaction" describes the condensation of aromatic amines and disulfur dichloride to give the corresponding 1,2,3-benzodithiazolium chlorides (Herz salts) and their subsequent hydrolysis to afford 2-aminobenzenethiols. This reaction is found useful in preparation of o-amino thiophenols and benzothiozoles.

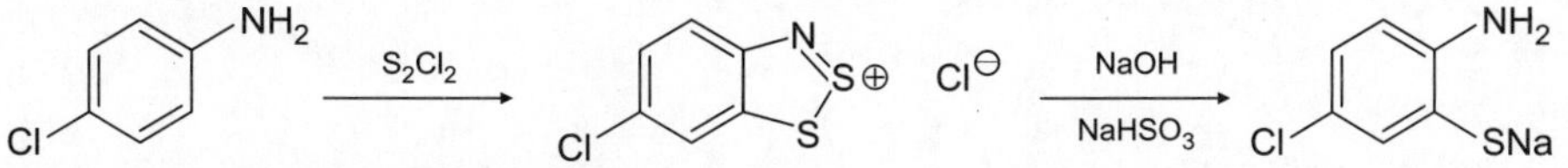

184. HILBERT-JOHNSON REACTION

The synthesis of pyrimidine nucleosides from the reaction of 2,4-dialkoxypyrimidines with halogenated sugar (glycosyl halides). This reaction is found to be highly efficient and proceeds with high regioselectivity and stereoselectivity. This reaction has been used for the synthesis of a variety of 2-oxopyridimine nucleosides and analogues. The Hilbert-Johnson reaction also works for nonaromatic aglycons.

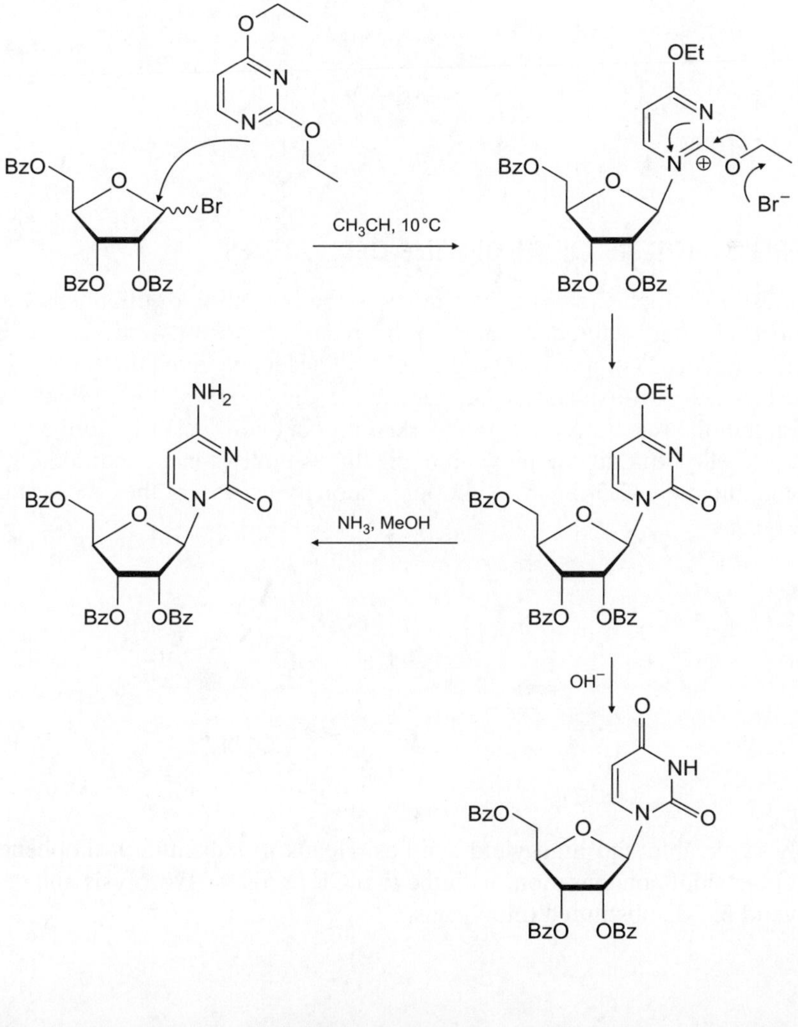

185. HINSBERG OXINDOLE AND OXIQUINOLINE SYNTHESIS

The Hinsberg synthesis starts with a secondary aromatic amine which reacts with the adduct of glyoxal sodium bisulfite. In a primary arylamide, one obtains a glycine or glycinamide derivative.

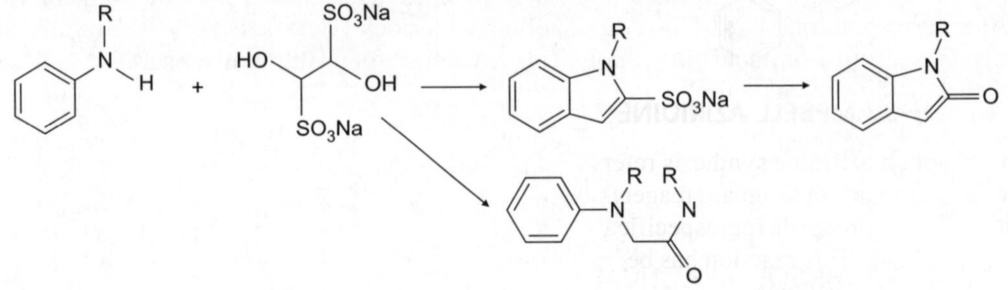

186. HINSBERG SULFONE SYNTHESIS

Hinsberg sulfone synthesis refers to the formation of sulfonylquinol derivatives by addition of quinones to cold dilute aqueous solutions of sulfinic acids. This synthesis can also be extended to nonaqueous aprotic solvents (e.g. THF).

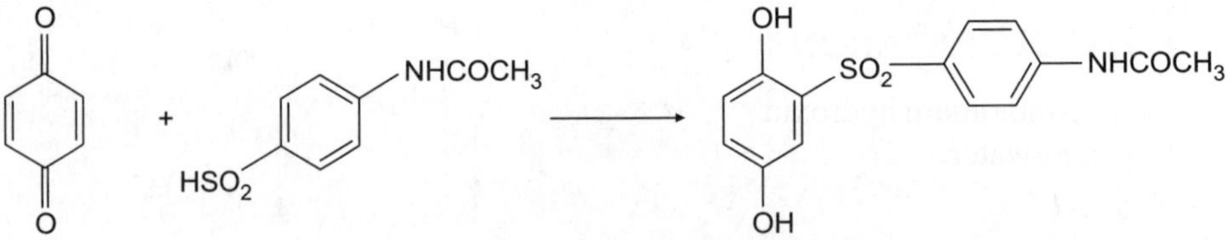

187. HINSBERG SYNTHESIS OF THIOPHENE DERIVATIVES

Hinsberg synthesis of thiophene derivatives refers to the formation of thiophene carboxylic acids from α-diketones and dialkyl thiodiacetates. In this reaction two consecutive aldol condensations between a 1, 2-dicarbonyl compound and dialkyl thiodiacetates gives thiophene. The immediate product is an ester-acid (3,4-disubstituted thiophene-2,5-dicarboxylic acids) produced by a Stobbe-type mechanism but the reactions are often worked up via hydrolysis to afford an isolated diacid. Condensation of diethyl thiodiglycolate and α-diketones under basic conditions, which provides 3,4-disubstituted thiophene-2,5-dicarboxylic acids upon hydrolysis of the crude ester product with aqueous acid.

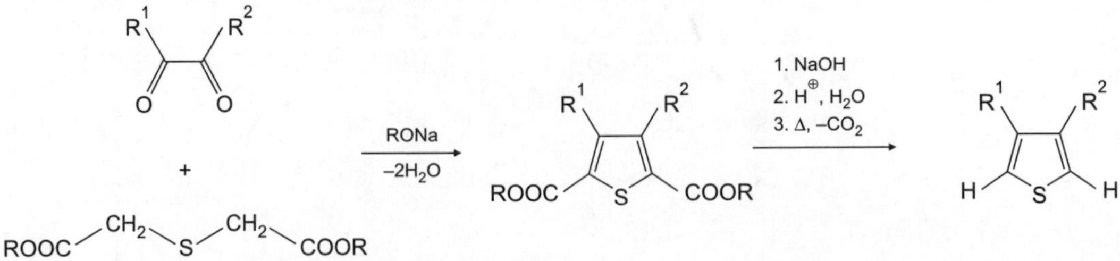

This widely applicable and high-yield synthesis leads to substituted thiophene dicarboxylic esters via a double aldol condensation, with the two CH_2 groups. Hydrolysis and decarboxylation of the esters yield 3,4-disubstituted thiophenes.

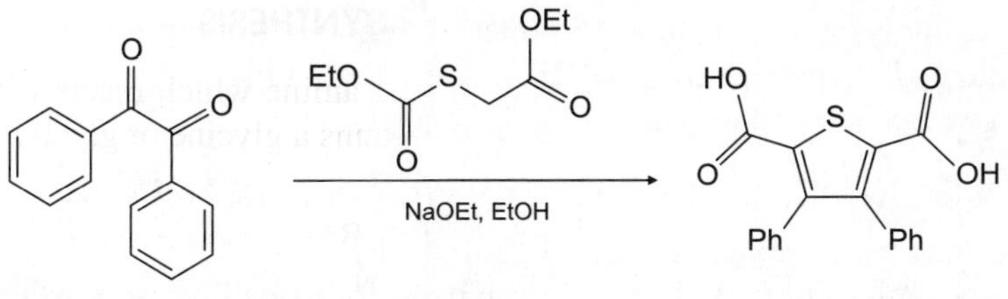

188. HOCH-CAMPBELL AZIRIDINE SYNTHESIS

Hoch-Campbell aziridine synthesis refers to the formation of aziridines by treatment of ketoximes with excess amounts of Grignard reagents and subsequent hydrolysis of the organometallic complex.

This reaction proceeds regiospecifically through a vinyl nitrene intermediate and solvents play an important role. This reaction has been found useful in the synthesis of aziridines, especially the fully substituted aziridines.

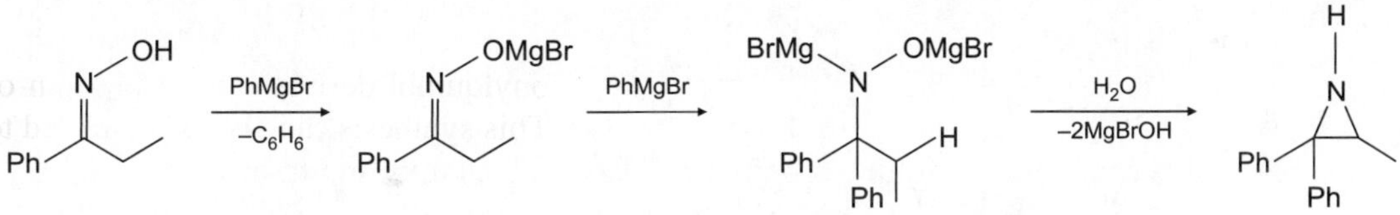

189. HOFMANN DEGRADATION (EXHAUSTIVE METHYLATION)

In this, quaternary ammonium hydroxide undergoes beta elimination reactions, producing an alkene, a tertiary amine and water.

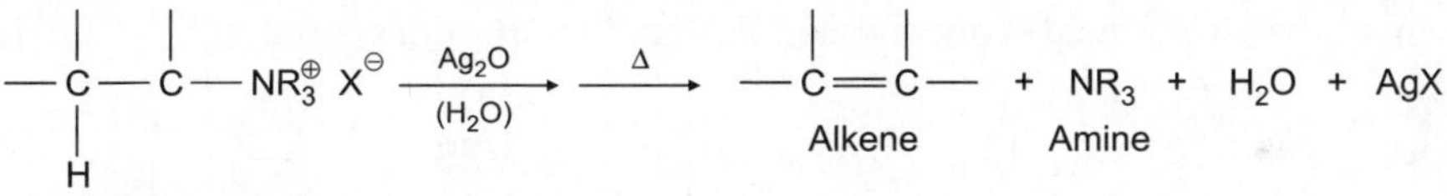

4° ammonium salt
R = CH$_3$ (mostly), ...
X = Cl, Br, I
Ag$_2$O/H$_2$O = Wet silver oxide, converts halide to hydroxide salt:

$$2R_4N^{\oplus}X^{\ominus} + Ag_2O + H_2O \longrightarrow 2R_4N^{\oplus} OH^{\ominus} + 2AgX\downarrow$$

4° ammonium halide   4° ammonium hydroxide

This method, known as the Hofmann exhaustive methylation, involves converting the amine to the quaternary ammonium salt with methyl iodide, carrying out the Hofmann elimination, and repeating these two steps in order until the nitrogen atom is lost as trimethylamine.

Its use is illustrated by applying it to the tropine nucleus.

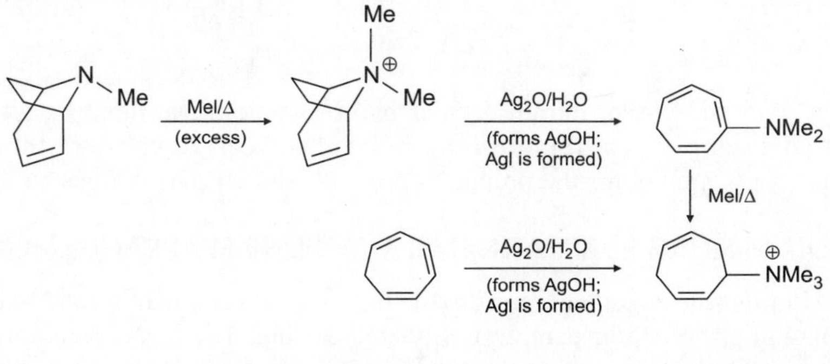

Coupled with ozonolysis as a method for determining the structure of the alkenes produced, the Hofmann exhaustive methylation (or **Hofmann degradation**) was a powerful tool for determining the structures of many amines.

190. HOFMANN ISONITRILE SYNTHESIS (CARBYLAMINE REACTION)

Aliphatic and aromatic primary amines on heating with chloroform and ethanolic potassium hydroxide form isocyanides or carbylamines, which are foul smelling substances. Secondary and tertiary amines do not show this reaction. This reaction is known as carbylamine reaction or isocyanide test and is used as a test for primary amines.

$$RNH_2 + CHCl_3 + 3KOH \longrightarrow R-NC + 3KCl + 3H_2O$$

Alkyl isocyanide
(unpleasant smell)

This reaction is found to be useful to convert primary amines into isonitriles.

191. HOFMANN-LÖFFLER-FREYTAG REACTION

The Hofmann-Löffler reaction (also referred to as Hofmann-Löffler-Freytag reaction, Löffler-Freytag reaction, Löffler-Hofmann reaction, as well as Löffler's method) involves formation of pyrrolidines or piperidines by thermal or photochemical decomposition of protonated N-haloamines in the presence of a strong acid (concentrated sulfuric acid or concentrated CF_3CO_2H).

n = 1,2
X = Cl, Br, I
R^1 = Alkyl, aryl, H
R^2 = Alkyl, acyl, H

N-halogenated amine   Cyclic amine

$$RCH_2(CH_2)_nNCIR^1 \xrightarrow[\Delta\ or\ h\nu]{H^+} RCHCl(CH_2)_nNH_2R^{1+} \longrightarrow$$

n = 3 or 4

This reaction proceeds via an intramolecular hydrogen atom transfer to a nitrogen-centered radical and is an example of a remote intramolecular free radical C-H functionalization. This reaction has been applied for the preparation of substituted pyrrolidines and piperidines.

192. HOFMANN-MARTIUS REARRANGEMENT (ANILINE REARRANGEMENT)

The Hofmann-Martius rearrangement refers to the thermal conversion of an an N-alkylated aniline to the corresponding ortho and/or para aryl-alkylated aniline. The reaction requires heat and the catalyst is an acid like hydrochloric acid. This rearrangement has proven valuable in the preparation of substituted aromatic amines.

193. HOFMANN REACTION

Hofmann reaction refers to the conversion of primary carboxylic amides to primary amines with one fewer carbon atom upon treatment with hypohalites or hydroxide *via* the intermediate isocyanate. Primary amine synthesized in the reaction with one less carbon than the starting amide.

194. HOFMANN-SAND REACTIONS

Hofmann-Sand reactions refers to olefin mercuration by the reaction between mercuric salt (acetates, halides, nitrates, or sulfates) and compounds containing a carbon-carbon double bond in aqueous or alcoholic solutions to form either addition products or coordination complexes or both. In alcoholic solutions the accelerated reaction produces alkoxyalkyl compounds. Overall reaction outcome depends on the various factors. The trans-olefin normally reacts more sluggishly than the cis-isomers and the reaction occurs most readily in methanol solution. The mercurate salt-olefin adducts further react with varity of compounds. This reaction is very useful for the determination of olefinic configuration and the conversion of cis-olefins into trans-olefins.

195. HOOKER REACTION

Hooker reaction refers to the oxidation of 2-hydroxy-3-alkyl-1,4-quinones with dilute alkaline permanganate with shortening of the alkyl side chain by a methylene group and simultaneous exchange of hydroxyl and alkyl or alkenyl group positions. In the Hooker reaction an alkyl chain in a certain naphthoquinone is reduced by one methylene unit as carbon dioxide in each potassium permanganate oxidation.

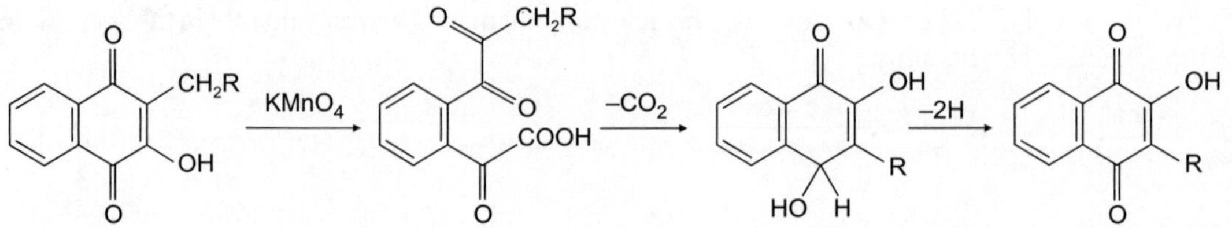

Mechanistically oxidation causes ring-cleavage at the alkene group, extrusion of carbon dioxide in decarboxylation with subsequent ring-closure.

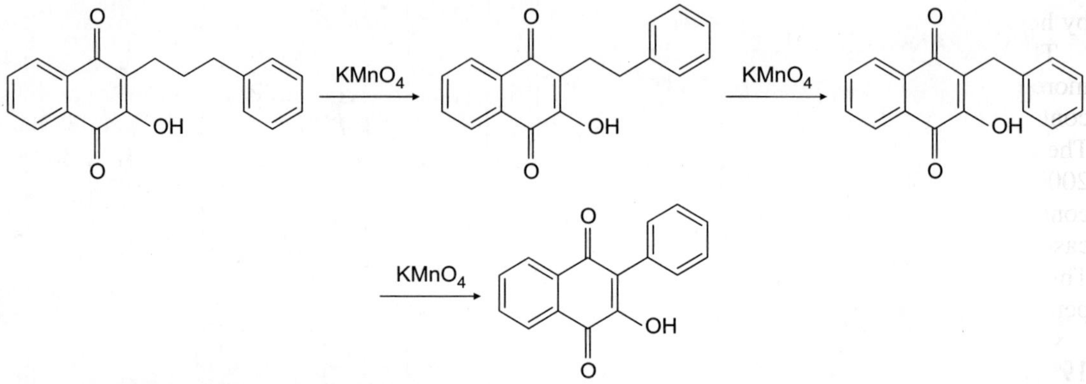

196. HOUBEN-FISCHER SYNTHESIS

In the Houben-Fischer synthesis aromatic nitriles are prepared by the basic hydrolysis of trichloromethyl aryl ketimines. Acidic hydrolysis, on the other hand, yields ketones.

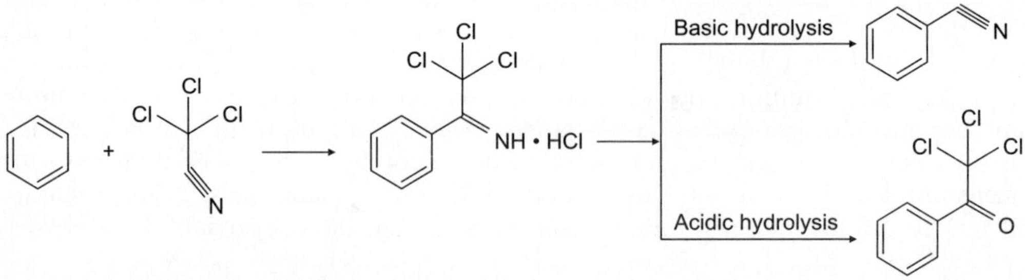

197. HOUBEN-HOESCH REACTION (THE HOESCH REACTION)

The Hoesch reaction or Houben-Hoesch reaction involves reaction of a nitrile with an arene compound to form an aryl ketone in the presence of hydrochloric acid and aluminum chloride as catalyst.

The reaction is a type of Friedel-Crafts acylation with hydrochloric acid and a Lewis acid catalyst and also said to be Friedel-Crafts acylation with nitriles and HCl.

$$\text{ArH} \quad + \quad \text{RCN} \quad \xrightarrow[\text{ZnCl}_2]{\text{HCl}} \quad \text{ArCOR}$$

In most cases, a Lewis acid is necessary; zinc chloride is the most common. The reaction is generally useful only with phenols, phenolic ethers, and some reactive heterocyclic compounds such as pyrrole, but it can be extended to aromatic amines by the use of BCl_3.

198. HOUDRY CRACKING PROCESS

Decomposition of petroleum or heavy petroleum fractions into more useful lower boiling materials by heating at 500°C and 30 psi, over a silica-alumina-magnanese oxide catalyst.

This reaction is an industrial process to degrade petroleum or heavy petroleum fractions into more useful lower-boiling materials by heating the oil at 500°C and 30 psi, over a clay catalyst containing silica, alumina, and manganese oxide and is known as the Houdry cracking process. The first commercial fixed-bed, three-case cracker was constructed (in 1936), which could crack 2000 barrels of oil daily. It has been found that during the thermolytical degradation of oil, continuous coking occurs which deactivates the catalyst. The typical cycle duration of the three-case cracker is found to be ~10 min and few factors affect the effectiveness of a catalyst. This process has wide application in oil refinery manufacturing plants to produce gas and other petrochemicals.

199. HUNSDIECKER REACTION (BORODINE REACTION)

The degradation of a carboxylic acid salt (silver(I) salts) in anhydrous medium by means of halogen to a halide of one less carbon atom than the original acid is known as the Hunsdiecker reaction (also called the Borodin reaction after Alexander Borodin). The silver(I) salts of carboxylic acids react with halogens to give unstable intermediates which readily decarboxylate thermally to yield alkyl halides. The reaction is believed to involve homolysis of the C-C bond and a radical chain mechanism.

Overall it involves conversion of silver carboxylate to halide by treatment with halogen. It is an example of a halogenation reaction. The reaction is named after Heinz Hunsdiecker and Cläre Hunsdiecker.It has been reported that the Hg(II), Ti(I), or Pb(IV) salt of the carboxylic acids also undergoes a similar reaction. This reaction has been applied in the preparation of aliphatic halides.

200. HYDROBORATION REACTION

Addition of boron hydrides to alkenes, allenes, and alkynes to form organoboranes, such that boron adds to the less substituted carbon. The reaction was first reported by Herbert C. Brown in the late 1950s.

Attack usually takes place on the less hindered side in a cis fashion.

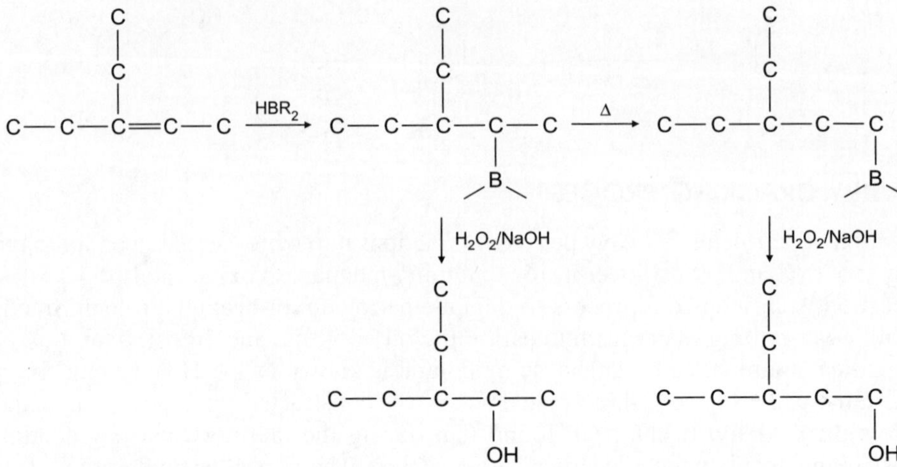

Hydroboration of alkenes/alkynes is one of the most versatile reactions available. Most commonly, the resulting alkyl borane intermediates are not isolated, but are used in subsequent reactions for forming a wide range of functional groups.

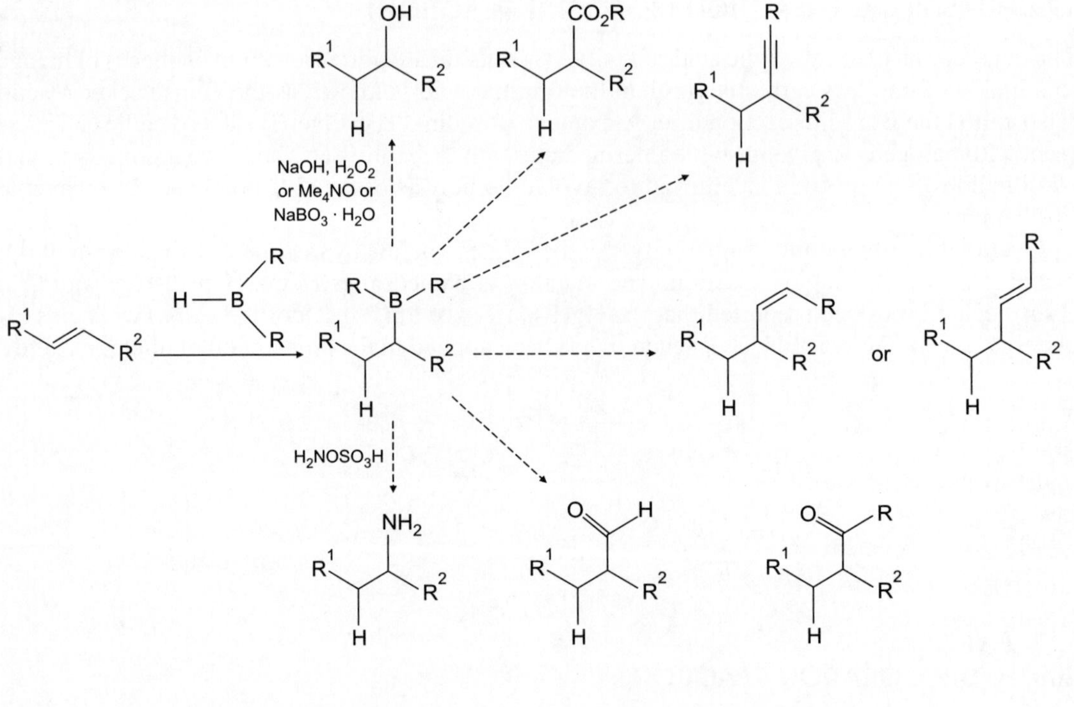

201. IVANOV REACTION

The addition of carboxylate enolates (enediolates or carboxylic acid dianions) of aryl acetic acids or substituted analogs (Ivanov reagents*) to electrophiles, particularly carbonyl compounds to give β-hydroxy ac ids is generally referred to as the **Ivanov reaction. The electrophiles may include aldehydes, ketones, isocyanates, and alkyl halides.**

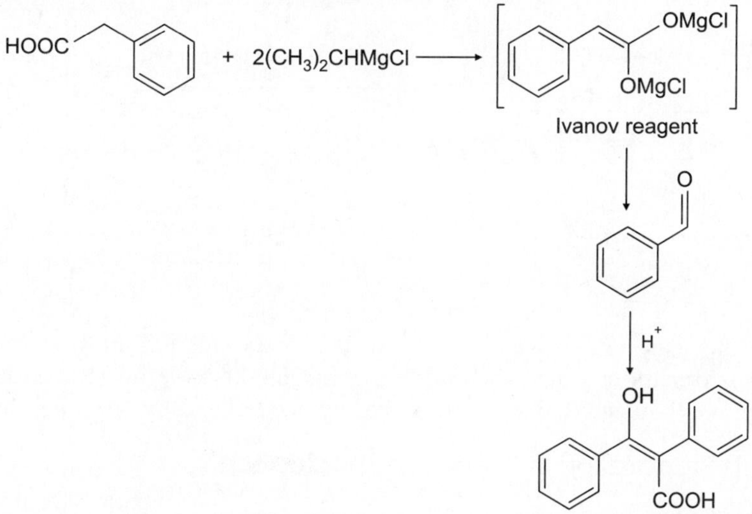

HOOC ... + 2(CH$_3$)$_2$CHMgCl ⟶ [... OMgCl / OMgCl]

Ivanov reagent

↓

O

↓ H$^+$

OH

Ph   Ph

COOH

The stereoselectivity of the Ivanov reaction known to proceed through the Zimmerman-Traxler model transition state which predicts a six-membered cyclic transition state leading to excellent stereoselectivity for anti-substituted products.

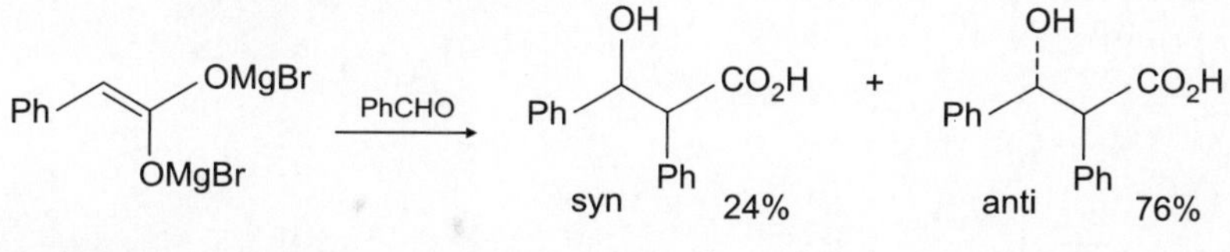

Ph ⁀ OMgBr / OMgBr   —PhCHO→   Ph ⁀ CO$_2$H (OH, Ph) syn 24%   +   Ph ⁀ CO$_2$H (OH, Ph) anti 76%

***Ivanov reagents** are reaction product of aryl acetic acids and excess Grignard reagent.

202. JACOBSEN EPOXIDATION

Jacobsen epoxidation is the chiral (salen) manganese (III)-catalyzed asymmetric epoxidation of alkenes. The enantio- and diastereo- selectivity in the reaction depends strongly on the nature of the substrate. The reaction is very easy to set up and not air-sensitive, often requires no solvent, the only reagents are acetic acid and water.

Compared to the sharpless epoxidation, the Jacobsen epoxidation allows a broader substrate scope for the transformation: good substrates are conjugated cis-olefins (R: Ar, alkenyl, alkynyl; R′: Me, alkyl) or alkyl-substituted cis-olefins bearing one bulky alkyl group.

Catalyst and its Preparation in Jacobsen Epoxidation

Other Examples

203. JACOBSEN REARRANGEMENT

Herzig described this type of rearrangement for the first time in 1881 using polyhalogenated benzenesulfonic acids but Oscar Jacobsen described the rearrangement of polyalkylbenzene derivatives in 1886 and the reaction took his name.

The Jacobsen rearrangement is sulfuric acid mediated conversion of polysubstituted (polyalkyl- or polyhalo-) aromatics into corresponding sulfonated aromatics with the migration of an alkyl group. The halogenated polyalkylaromatics undergo disproportionation of halogen atoms under the same reaction conditions. The reaction is limited to benzene rings with at least four substituents (alkyl and/or halogen groups), i.e. among the polyalkyl benzenes, only the tetraalkyl or pentaalkyl benzenes undergo this reaction. This rearrangement is also worked for halogenated aromatics and the fluorinated aromatics are the exceptions. The Jacobsen rearrangement can also be considered as a rearrangement of polyalkylbenzenes because the sulfo group is easily removed.

The rearrangement is believed to occur intermolecularly and that the migrating group is transferred to a polyalkylbenzene, not to the sulfonic acid (sulfonation only takes place after migration).

204. JANOVSKY REACTION

Janovsky found that aromatic nitro compounds treated with alkali, acetone and alcohol show a characteristic colour. The Janovsky reaction of aldehydes and ketones containing activated α-proton, e.g. α-methylene groups with aromatics containing multinitro groups, e.g. *m*-dinitro-benzenes to give a coloring molecules (usually intense purple coloration). The product formed is known as Janovsky adduct. The reaction is used for the detection of carbonyl compounds.

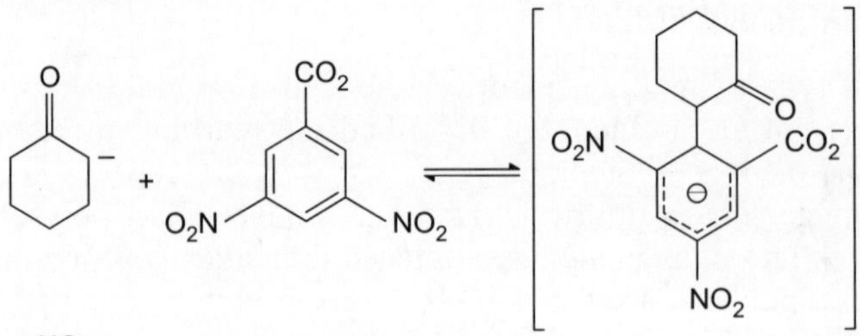

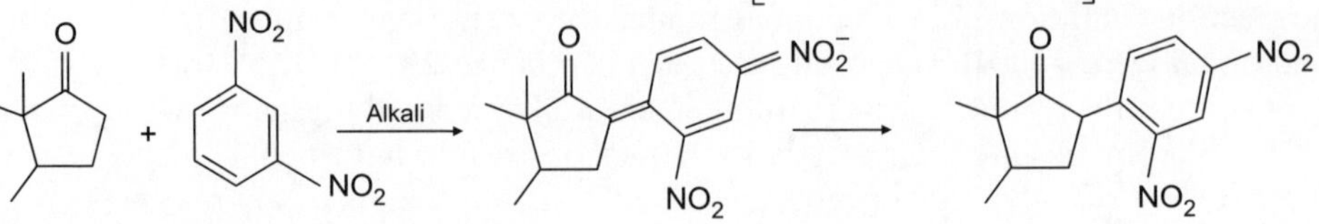

205. JAPP-KLINGEMANN REACTION

The Japp-Klingemann reaction, which is named after the chemists Francis Robert Japp and Felix Klingemann is used for the preparation of the hydrazones from β-keto acids (or β-keto esters or the carbonyl compounds containing an adjacent methylene group) and aryl diazonium salts.

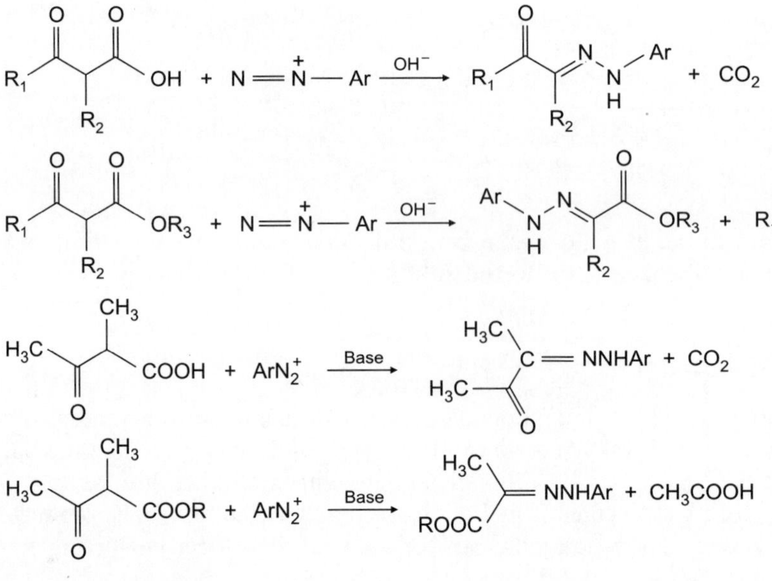

206. JONES OXIDATION

In presence of chromium trioxide and dilute sulphuric acid (Jones reagent*, i.e. $CrO_3 + H_2SO_4 + H_2O$), secondary alcohols oxidizes to ketones and primary alcohols are oxidised to carboxylic acids.

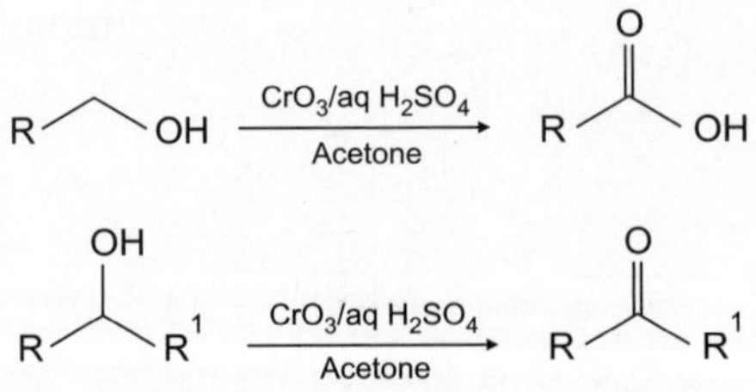

Jones reagent provides relatively mild conditions for oxidation it oxidizes alcohols efficiently and is relatively unreactive toward olefins. However, workup on large scale often require a large excess of the Chromium reagent which are toxic and mutagenic. Apart from this acidic reaction conditions may be a problem with acid-labile groups.

*Jones reagent is a mixture of chromium trioxide and dilute sulphuric acid ($CrO_3 + H_2SO_4 + H_2O$). As a reagent it has a bad reputation for causing fire and chromium trioxide in the reagent is highly toxic.

207. JOURDAN-ULLMANN-GOLDBERG SYNTHESIS

The Jourdan-Ullmann-Goldberg reaction involves formation of a nitrogen-carbon bond to an aromatic halide in the presence of copper and the Jourdan-Ullmann condensation or Jourdan-Ullmann reaction is stated as the copper-promoted synthesis of substituted diphenylamines, from amination of aryl halides with aniline in the presence of a base. Hence the reaction is useful in their further conversion into acridone and acridine derivatives.

It has been found that the electron-withdrawing groups on aryl halides facilitates the reaction compared with electron-donating substituents. Copper acetate and cuprous iodide are used as the catalysts for this reaction.

208. JULIA OLEFINATION (JULIA-LYTHGOE OLEFINATION)

Julia olefination (also known as the Julia-Lythgoe olefination) is named after the French chemist Marc Julia is a multistep synthesis enables the preparation of (E)-alkenes (trans-olefins). The reaction involves the addition of phenyl sulfones to aldehydes or ketones followed by alcohol functionalization and subsequent reductive elimination with sodium amalgam or SmI_2.

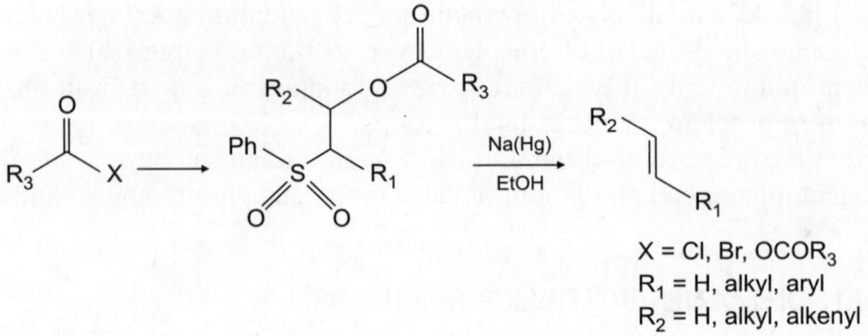

X = Cl, Br, OCOR$_3$
R$_1$ = H, alkyl, aryl
R$_2$ = H, alkyl, alkenyl

The Julia-Kociensky Olefination is an alternative procedure, which leads to the olefin in one step.

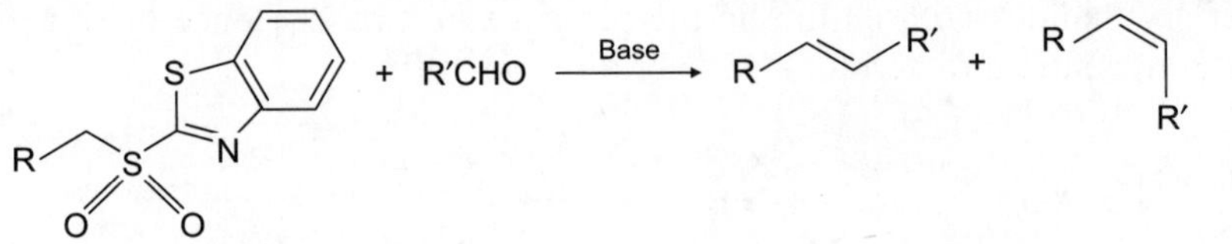

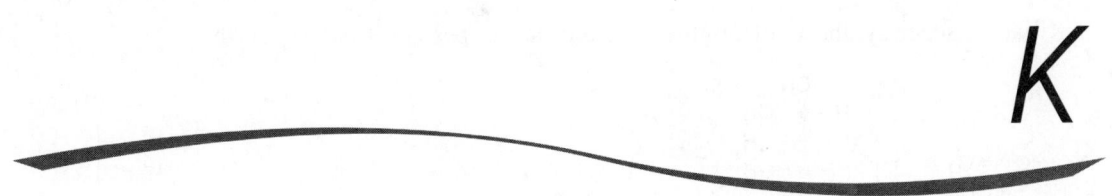

209. KENDALL-MATTOX REACTION

The Kendall-Mattox reaction involves synthesis of α,β-unsaturated ketone from an α-bromoketone. The first step of this reaction is the formation of a phenylhydrazone. Subsequently deprotonation of the phenylhydrazone eliminate the bromide ion followed by a prototropic shift reforms the phenylhydrazone, with the addition of the α,β -unsaturated double bond. The phenylhydrazone is then reacted with 2-oxopropanoic acid to produce the α,β-unsaturated ketone.

210. KILIANI-FISCHER SYNTHESIS

The Kiliani-Fischer synthesis, named after German chemists Heinrich Kiliani and Hermann Emil Fischer, is a method for extension of the carbon atom chain of aldoses. In other words, tetroses can be converted into pentoses, and pentoses can be converted into hexoses.

Kiliani-Fischer synthesis involves conversion of aldose into two one-carbon-higher epimeric homologs, which involves a nucleophilic addition of cyanide to the terminal carbonyl group of an aldose, hydrolysis of the cyanohydrins followed by reduction of the lactone yields the homologous aldose. The configurations of the other asymmetric carbons do not change, because no bond to any of the asymmetric carbons is broken during the course of the reaction. Addition of cyanide ion to the carbonyl group creates a new asymmetric carbon. Consequently, two cyanohydrins that differ only in configuration at C-2 are formed so the product of a Kiliani-Fischer synthesis is a mixture of two diastereomeric sugars, called epimers. For example, D-arabinose is converted to a mixture of D-glucose and D-mannose. The net result of all this is that the chain gets lengthened by one carbon, and the product is a mixture of two epimers that are different at the second carbon of the chain. Every other part of the chain stays the same.

Kiliani-Fischer synthesis of D-glucose from the proper D-tetrose (D-erythrose).

```
                        CN
                   H ──┼── OH
                   H ──┼── OH                              CN
                   H ──┼── OH                         HO ──┼── H
       HC = O          CH2OH                          HO ──┼── H
  H ──┼── OH    HCN ┌──                                H ──┼── OH
  H ──┼── OH         │      CN                          H ──┼── OH
       CH2OH         │  HO ──┼── H     COOH                 CH2OH
  D-erythrose        │   H ──┼── OH  HO ──┼── H   HC = O          CN
                     └── H ──┼── OH   H ──┼── OH  HO ──┼── H  HCN ┌──
                             CH2OH    H ──┼── OH   H ──┼── OH      │  H ──┼── OH
                              H2O      CH2OH  Na(Hg) H ──┼── OH    │  HO ──┼── H
                              ──→                    CH2OH        └── H ──┼── OH
                               H+                                     H ──┼── OH
                                                                        CH2OH
```

```
         CN                      COOH                    HC = O
    H ──┼── OH              H ──┼── OH              H ──┼── OH
   HO ──┼── H     H2O      HO ──┼── H    Na(Hg)   HO ──┼── H
    H ──┼── OH    ──→       H ──┼── OH    ──→       H ──┼── OH
    H ──┼── OH     H+       H ──┼── OH              H ──┼── OH
       CH2OH                  CH2OH                   CH2OH
```

Kiliani-Fischer synthesis of D-ribose starts with the proper D-triose, D-glyceraldehyde.

```
                         CN
                    HO ──┼── H
     HC = O          H ──┼── OH
  H ──┼── OH  HCN     CH2OH                              CN
     CH2OH      ┌──                                 HO ──┼── H
 D-glyceraldehyde│                                   H ──┼── OH
                 │      CN                            H ──┼── OH
                 │  H ──┼── OH     COOH                  CH2OH
                 └── H ──┼── OH H ──┼── OH   HC = O           CN
                        CH2OH   H ──┼── OH  H ──┼── OH HCN ┌──
                         H2O      CH2OH     H ──┼── OH      │  H ──┼── OH
                         ──→   Na(Hg)          CH2OH       │  H ──┼── OH
                          H+                              └── H ──┼── OH
                                                               CH2OH
```

```
         CN                   CN                    HC = O
    H ──┼── OH           H ──┼── OH           H ──┼── OH
    H ──┼── OH    H2O    H ──┼── OH   Na(Hg)  H ──┼── OH
    H ──┼── OH    ──→    H ──┼── OH    ──→    H ──┼── OH
       CH2OH      H+        CH2OH               CH2OH
                                             D-ribose
```

211. KISHNER CYCLOPROPANE SYNTHESIS

The Kishner's method involves formation of cyclopropane derivatives, by thermal decomposition of 2-pyrazolines which formed by reacting of α, β-unsaturated ketones or aldehydes with hydrazine. Extrusion of nitrogen takes place in pyrazolines in this reaction which requires. The reaction needs higher temperatures for removing the nitrogen or for cyclopropanation.

212. KNOEVENAGEL CONDENSATION; DOEBNER MODIFICATION

The Knoevenagel Condensation reaction is named after Emil Knoevenagel, who in 1896 identified the reactions of aldehydes and ketones that undergo carbon-carbon bond forming reactions with an activated methylene compound ($X\text{-}CH_2\text{-}Y$, where X and Y are electron withdrawing groups (EWG)) to produce an α,β-unsaturated compound.

R_1 and R_2 can be H or any other substituent group; X and Y are electron withdrawing groups.

Active methylene

Among the EWG used as X and Y are NO_2, quaternary pyridinium or similar heterocycles, CN, COR (and COAr), CONHR (and CONHAr), CO_2R, CO_2H, SO_2, S, Ar (and ortho, meta, para EWG), pyridine and similar electron deficient heterocycles. With respect to positions R_1 and R_2, virtually every aldehyde has been shown to undergo the reaction in the literature, with ketones being less reactive. The catalyst in this reaction is usually are weakly basic amine since the use of strong bases may induce self-condensation of the aldehyde or ketone. **Knoevenagel condensation** is considered as a modification of the aldol condensation since is the mechanism of the aldol reaction is similar to the Knoevenagel condensation in terms of the steps involved (e.g. enolate ion formation).

The Knoevenagel condensation is an important and widely employed method for carbon-carbon bond formation in organic synthesis with numerous applications in the synthesis of fine chemicals, hetero Diels-Alder reactions and in synthesis of carbocyclic as well as heterocyclic compounds of biological significance.

Doebner modification of Knoevenagel condensation involves aromatic aldehyde, malonic acid and pyridine. For example, the reaction product of acrolein and malonic acid in pyridine is trans-2,4-Pentadienoic acid with one carboxylic acid group and not two.

213. KNOOP-OESTERLIN AMINO ACID SYNTHESIS

Knoop-Oesterlin amino acid synthesis involves platinum, palladium or Raney nickel catalyzed hydrogenation of α-oxo acids in aqueous ammonia to form α-amino acids.

$$R—CO—COOH \;+\; NH_3 \,(aq.) \longrightarrow R—\underset{\underset{NH_2}{|}}{CH}—COOH$$

Reagents = Pt or Raney nickel catalyst + H_2

214. KNORR PYRAZOLE SYNTHESIS

The term Pyrazole was given by Ludwig Knorr in 1883. Formation of pyrazole or pyrazolone ring system from hydrazines or substituted hydrazine, hydrazides, semicarbazides, and aminoguanidines by condensation with 1,3-dicarbonyl compounds; substituted hydrazines yield two structurally isomeric pyrazoles:

R = H, alkyl, aryl, het-aryl, acyl, etc.

215. KNORR PYRROLE SYNTHESIS

The Knorr pyrrole synthesis involves formation of pyrrole derivatives by condensation of α-amino ketones as such or generated *in situ* from isonitrosoketones with a compound containing a methylene group α- to (bonded to the next carbon to) a carbonyl group. It is a widely used chemical reaction that synthesizes substituted pyrroles. The mechanism requires zinc and acetic acid as catalysts. It will proceed at room temperature. Because α-amino-ketones self-condense very easily, they must be prepared in situ. The usual way of doing this is from the relevant oxime. This reaction is highly regioselective and always gives high yields of pyrrole derivatives.

216. KNORR QUINOLINE SYNTHESIS

The Knorr quinoline synthesis is an intramolecular organic reaction which involves formation of α-hydroxyquinolines from β-keto esters and arylamines above 100°C. The intermediate anilide undergoes cyclization by dehydration with concentrated sulfuric acid. This reaction was first described by Ludwig Knorr in 1886.

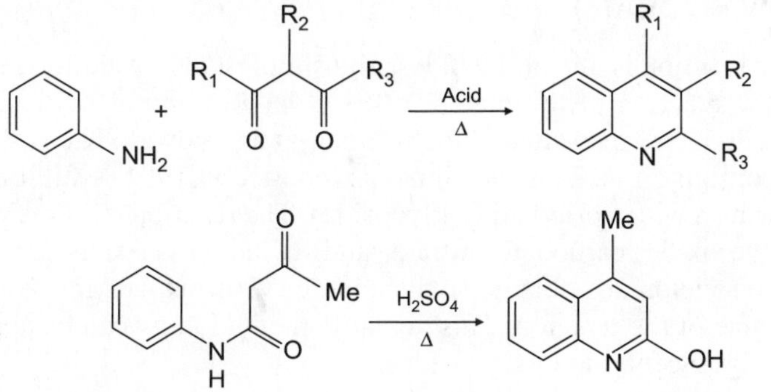

217. KOCH-HAAF CARBOXYLATIONS

Koch-Haaf carboxylation refers to strong acid-catalyzed tertiary carboxylic acid formation from alcohols or olefins and CO. In the reaction, tertiary carboxylic acids or esters are formed by the treatment of alcohols or alkenes with anhydrous formic acid, which behaves as both an acidic catalyst and a source of carbon monoxide. This reaction is found useful in production of carboxylic acids.

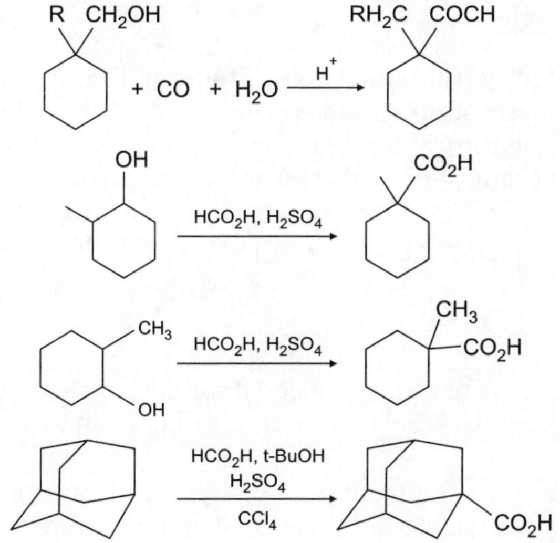

218. KOCHI REACTION

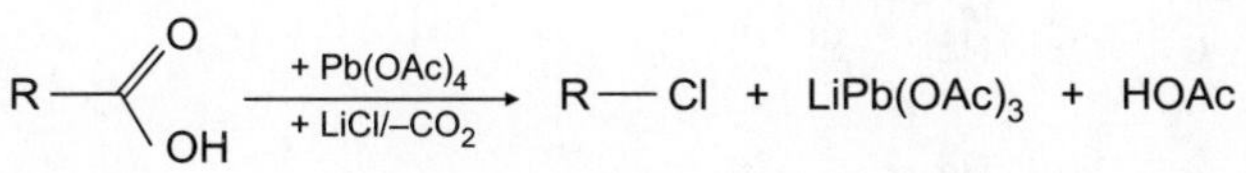

$$R\text{—}C(=O)OH \xrightarrow[\text{+ LiCl/–CO}_2]{\text{+ Pb(OAc)}_4} R\text{—Cl} + \text{LiPb(OAc)}_3 + \text{HOAc}$$

The Kochi reaction involves conversion of carboxylic acids into corresponding halides via one-carbon oxidative degradation of carboxylic acid by lead tetraacetate (LTA) accompanied by a simultaneous replacement with a halogen under free-radical conditions in the presence of a stoichiometric amount of metal halide (lithium chloride or other lithium salts). This reaction is suitable for synthesis of secondary and tertiary chlorides. The reaction is a valuable alternative to the Hunsdiecker reaction.

219. KOENIGS-KNORR SYNTHESIS (FORMATION OF THE β-GLYCOSIDE FROM α-HALOCARBOHYDRATE)

The Koenigs-Knorr reaction is one of the oldest and simplest glycosylation reactions which yields a β-glycoside by the reaction of acetylated glycosyl halides and alcohols or phenols under the influence of silver salt. It is named after Wilhelm Koenigs and Edward Knorr. The reaction proceeds with inversion of configuration. This reaction was reported in 1901 is still one of the most useful reactions for preparing a wide variety of O-glycosides. The reaction requires silver salts as catalyst and among them the oxide, carbonate, nitrate, and triflate silver salts are the most commonly employed. Improved yields are obtained with iodide, vigorous stirring, and protection against light during the course of the reaction. This versatile methodology can be applied for preparation of alky, aryl, and oligosaccharide O-glycosides.

(i) Ag_2O or Ag_2CO_3/PhH, drierite, I_2. (ii) MeONa/MeOH

220. KOLBE ELECTROLYSIS

The reaction was discovered by **Hermann Kolbe** (Professor of chemistry in Marburg and Leipzig) in 1849 during the electrolysis of potassium acetate.

In Kolbe electrolysis symmetrical hydrocarbons or dimmers are formed through the coupling of radicals generated from carboxylic acid at an anode via electrolysis.

This reaction is an effective and simple method for the coupling of carboxylates. The *Kolbe* synthesis and the Kolbe electrolysis are two different reactions and should not to be confused with each other.

The reaction is useful for the preparation of hydrocarbons e.g. alkanes are formed, on electrolysis of concentrated aqueous solution of sodium or potassium salt of saturated mono carboxylic acids.

Sodium acetate   Ethane

221. KOLBE-SCHMITT REACTION

In 1860, German chemist Hermann Kolbe prepared salicylic acid by heating a mixture of phenol and sodium in the presence of carbon dioxide at atmospheric pressure. Later on in 1884, the reaction was modified under enhanced pressure by Schmitt. This modification significantly increased the yield of salicylic acid.

The carboxylation reaction of alkali metal phenoxides with carbon dioxide to hydroxybenzoic acids is known as the Kolbe-Schmitt reaction. In the Kolbe *synthesis*, sodium phenoxide is heated with CO_2 under pressure and the reaction mixture is subsequently acidified to yield salicylic acid.

Sodium phenoxide   Sodium salicylate   Salicylic acid (79%)

The Kolbe synthesis

The Kolbe-Schmitt reaction is highly dependent on the metal cation used.

The main product of the reaction between sodium phenoxide and carbon dioxide is salicylic acid (94–97%), though p-hydroxybenzoic and 4-hydroxy-isophthalic acids are also identified at low yields (2–4%). If potassium phenoxide is used instead of sodium phenoxide, p-hydroxy benzoic acid is formed in excess, imposing that the distribution of products in the Kolbe-Schmitt reaction is highly dependent on the metal cation used.

Alkyl derivatives of phenol behave very much like phenol itself. Phenols that bear strongly electron-withdrawing substituents usually give low yields of carboxylated products; their derived phenoxide anions are less basic.

p-cresol   2-hydroxy-5-methyl benzoic acid
                             (78%)

The Kolbe-Schmitt carboxylation reaction is an excellent commercially feasible method to introduce the carboxylic acid functionality onto an aromatic substrate. The Kolbe-Schmitt reaction has been used in industry since 1874. Most of the commercial applications of this carboxylation reaction are associated with aromatic compounds having an additional hydroxyl group present on the ring. Salicylic acid is used in manufacture of *aspirin and methyl salicylate* while 4-hydroxybenzoic acid methyl esters are used as preservative in food, beverages and cosmetics.

Salicylic acid undergoes acetylation by heating with acetic anhydride to form aspirin which is used as an analgesic and antipyretic.

Salicylic acid → (CH₃CO)₂O → Acetyl salicylic acid or aspirin + CH₃COOH

The mechanism of the Kolbe-Schmitt is not clearly understood, but it is postulated that a transition complex is formed between the reactants, which polarizes the carbon dioxide molecule and makes the positively charged carbon atom more susceptible to addition onto the ring. The ortho isomer product predominates because of the stabilizing influence of chelation on the transition state. Potassium phenoxide, which is less likely to form such a complex, mainly leads to the para isomer. The actual reaction mechanism appears to be more complicated than this simple picture. At least a part of the potassium p-hydroxybenzoate that forms comes from a rearrangement of the initially formed potassium salicylate. It is significant to note that the sodium salicylate does not exhibit any rearrangement.

222. KOSTANECKI ACYLATION

The Kostanecki reaction also known as Kostanecki-Robinson reaction involves the formation of coumarins or chromones from readily available 2-hydroxyarylketones and an aliphatic anhydride in the presence of the corresponding sodium salt via O-acylation and aldol condensation. Depending on the actual reaction conditions and substituents, either coumarin or chromone or both can be produced.

223. KRAFFT DEGRADATION

Krafft degradation involves conversion of carboxylic acids into the next lower homolog. This conversion is carried out by dry distillation of the alkaline earth salt of carboxylic acids with the corresponding acetate, followed by chromic acid oxidation of the methyl ketone. Reaction is especially applicable for high molecular weight carboxylic acids.

M = Alkali earth metal

224. KRAPCHO DECARBOXYLATION

Krapcho decarboxylation involves an alkali halide promoted nucleophilic decarboxylation of active esters, i.e. esters having an electron-withdrawing group in the beta position (e.g. β-keto esters, β-diesters, α-cyano esters) in polar or dipolar aprotic solvents (e.g. DMF, DMSO, HMPA).

The alkali halides, such as $MgCl_2$, $Na_3PO_4 \cdot 12H_2O$ and lithium acetate (LiAc) are found effective for this decarboxylation.

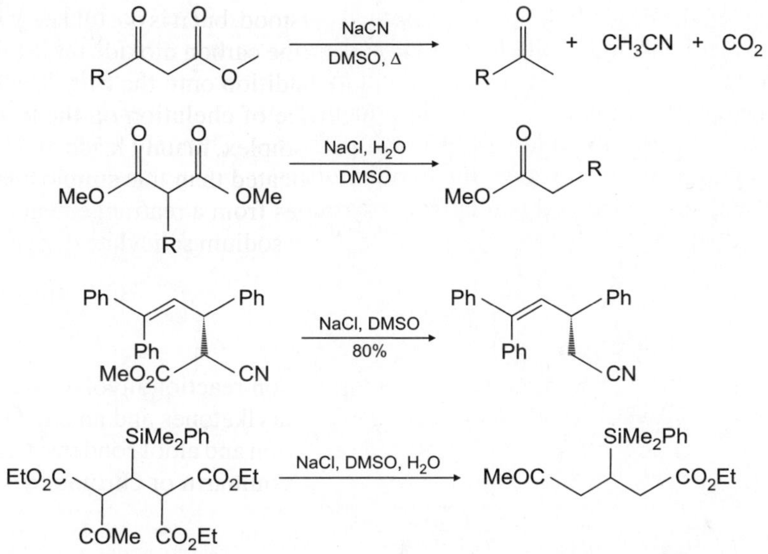

225. KRÖHNKE OXIDATION

The Kröhnke reaction uses combination of pyridine and p-nitrosodimethylaniline as oxidizing reagent for aldehyde synthesis. In the reaction, the activated halides are transformed into aldehydes via their pyridinium salts, which yield nitrones upon treatment with p-nitrosodimethylaniline. Aldehydes or ketones are then generated by hydrolysis of nitrones. This reaction is limited to active halides and requires very high temperature.

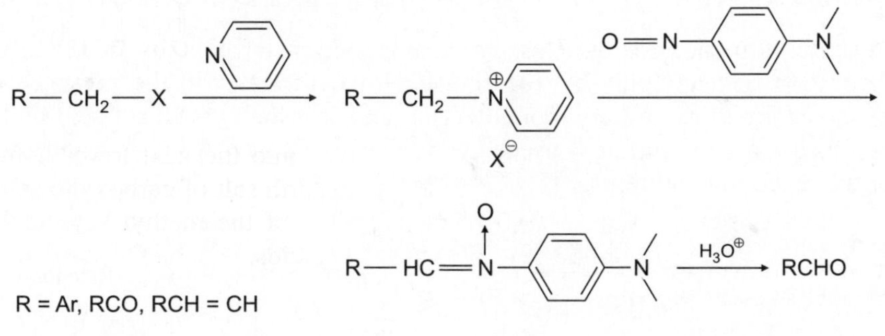

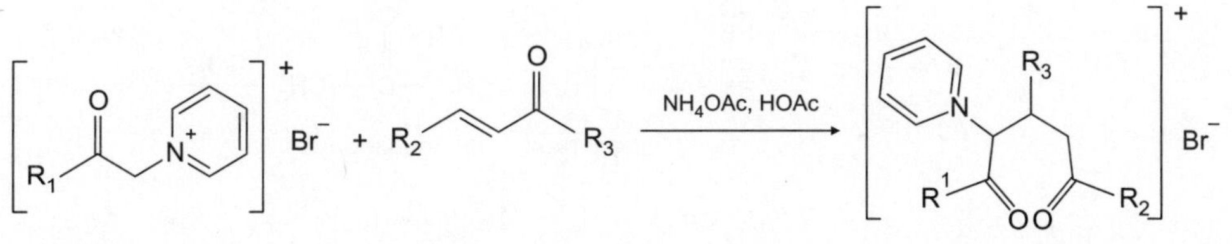

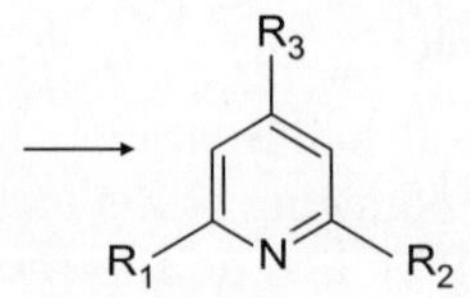

226. KRÖHNKE PYRIDINE SYNTHESIS

Kröhnke pyridine synthesis yields 2,4,6-trisubstituted pyridine derivatives by 1,4-Michael addition of α-pyridinium methyl ketone salts to α,β-unsaturated ketones, generating the 1,5-dicarbonyl compounds followed by cyclization with ammonium acetate in acetic acid. In other words, this reaction involves condensation of α-pyridinium methyl ketone salts and eneones that proceeds through a 2,3-ene-1,5-dione to yield trisubstituted pyridines. Three different substituents can be introduced into pyridine ring. However, the pyridinium salts and the unsaturated ketones have to be synthesized first, so this method is relatively expensive.

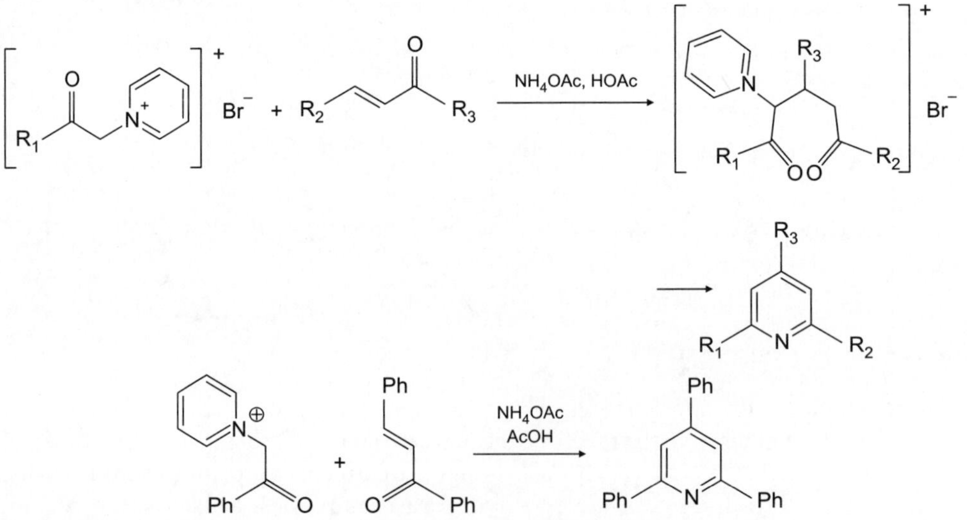

227. KUCHEROV REACTION

Kucherov reaction is named for the Russian chemist Mikhail Kucherov in 1881. The reaction describes hydration of acetylenic hydrocarbons with with water in the presence of acid and mercury(II) salts form carbonyl compounds. The acid is usually sulfuric acid and among the mercury(II) salts usually the sulfate is used, e.g. mercuric sulphate. Alkynes readily combine with water under the reaction conditions.

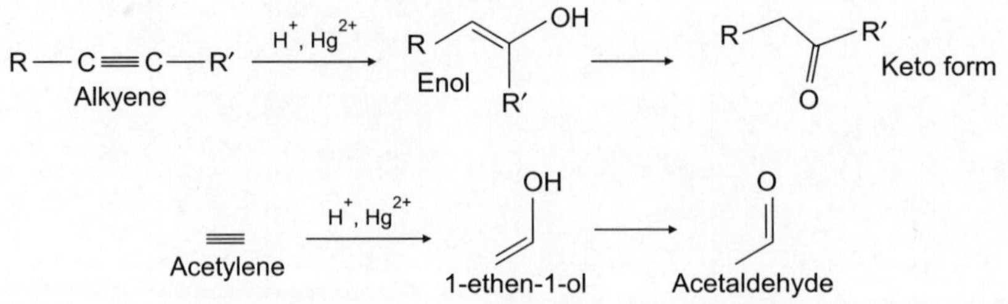

The reaction provides products according to the type of alkynes used:

- Acetylene gives acetaldehyde
- Alkynes other than acetylene give ketones.
- Terminal alkynes form methyl ketones
- Symmetric internal alkynes give single ketone
- Unsymmetrical internal alkynes give mixtures of ketones

An enol intermediate is formed during the reaction which transforms to its keto tautomer through keto-enol tautomerization.

228. KUHN-WINTERSTEIN REACTION

The Kuhn-Winterstein reaction describes conversion of the 1,2-glycols by reaction with diphosphotetraiodide (P_2I_4) into corresponding alkenes. In most cases, the major product is trans-alkenes. This reaction is useful in the preparation of polyenes.

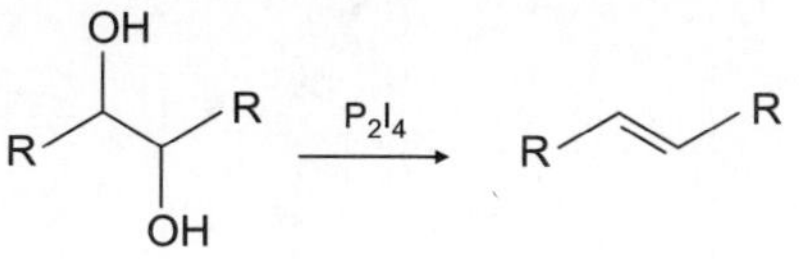

229. LADENBURG REARRANGEMENT

Ladenburg rearrangement describes thermal transformations of N-alkylpyridinium salts to C-alkylpyridine derivatives.

R = Alkyl group; X⁻ = Halide

230. LEBEDEV PROCESS

Lebedev process is the process of formation of butadiene from ethanol is a single-step process developed by Sergei Lebedev. The process involves production of butadiene from ethanol at temperature of 400–450°C in the presence of a catalyst. Ethanol in the process is converted to butadiene, hydrogen and water by catalytic pyrolysis. A variety of metal oxide catalysts used for the process such as mixtures of sililcates and aluminum and zinc oxides, silica gel impregnated with tantalum oxide (SiO_2–Ta_2O_5), silica-magnesium-tantalum, and silica-magnesium-chromium. This process has been used for the production of butadiene for synthetic rubber.

231. LEHMSTEDT-TANASESCU REACTION

The Lehmstedt-Tanasescu reaction (or Lehmsted-Tanasescu-acridone-synthesis) involves formation of acridones from the reaction of o-nitrobenzaldehyde (2-nitrobenzaldehyde) and halobenzene in the presence of concentrated sulfuric acid containing nitrous acid as catalyst.

The reaction is named after the German chemist Kurt Lehmstedt and the Romanian chemist Ioan Tănăsescu. Among the several methods for the preparation of acridones, the Lehmstedt-Tanasescu reaction is found to be the most acceptable.

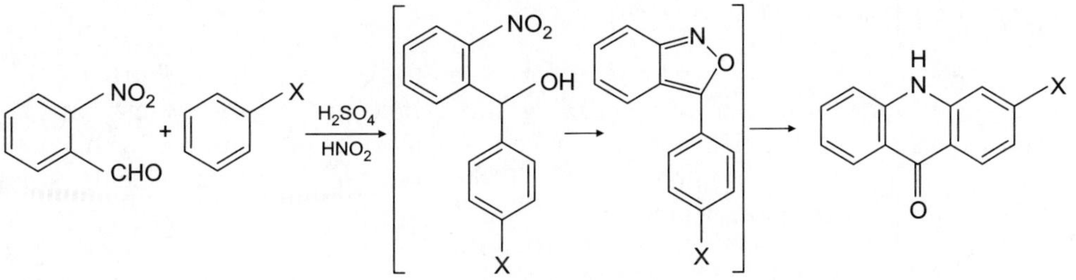

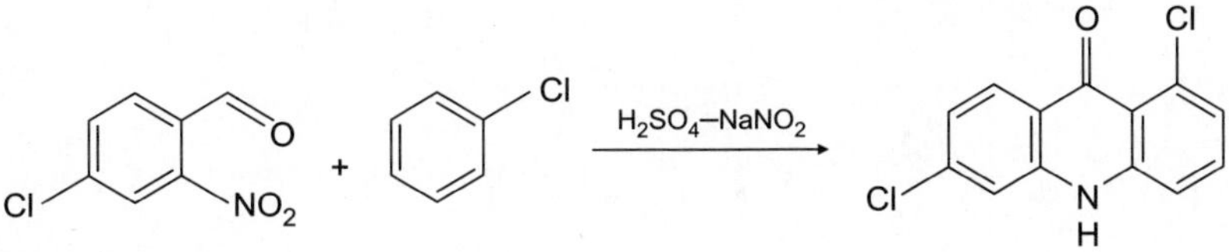

232. LETTS NITRILE SYNTHESIS

The Letts nitrile synthesis involves heating of aromatic carboxylic acids with metal thiocyanates to form nitriles.

$$RCOOH + KSCN \xrightarrow{\Delta} RCN + CO_2 + KHS$$

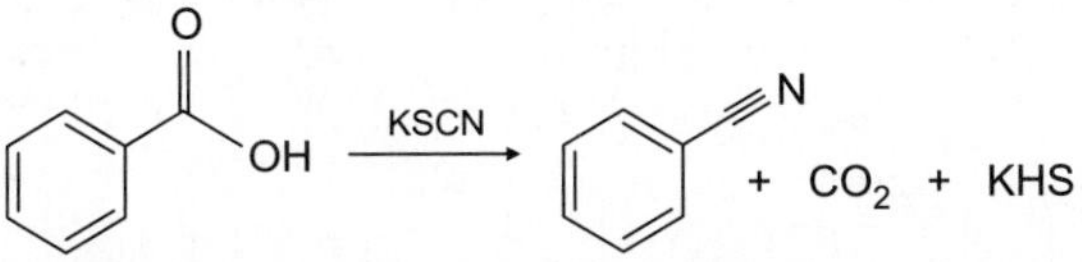

233. LEUCKART (LEUKART) REACTION; LEUCKART-WALLACH REACTION; ESCHWEILER-CLARKE REACTION

The Leuckart reaction is a process for reductive alkylation of ammonium (or amine) salts of formic acid or formamides by aldehydes or ketones. The reaction is named after Rudolf Leuckart. Formic acid acts as the reducing agent in the reaction which reduces the iminium into an amine. With this reaction ketones can convert to their corresponding amines without the employment of halogen or nitro groups.

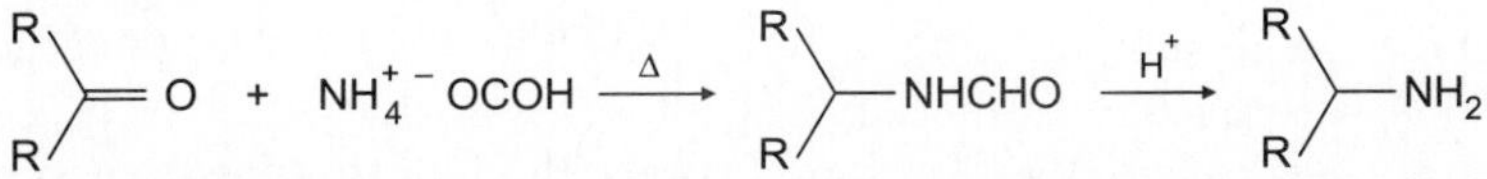

When excess formic acid is used in the reaction it is referred to as the Leuckart-Wallach reaction. The reaction is named after Rudolf Leuckart and Otto Wallach. The Leuckart-Wallach reaction has been used for synthesis of tetrahyddro-1,4 benzodiazipin-5-ones, a molecule that is part of the benzodiazepine family.

The reductive methylation of primary or secondary amines employing formaldehyde and formic acid is known as the Eschweiler-Clarke reaction.

Amine synthesis from reductive amination of a ketone and an amine in the presence of excess formic acid, which serves as the reducing reagent by delivering a hydride. When the ketone is replaced by formaldehyde, it becomes Eschweiler-Clarke reductive alkylation of amines.

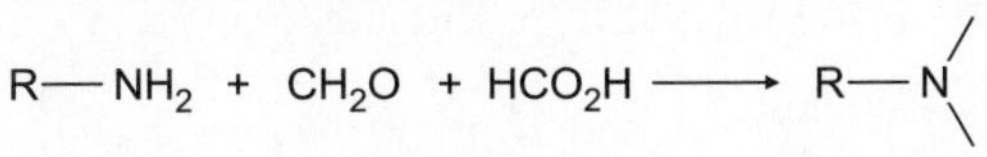

$$R-NH_2 + CH_2O + HCO_2H \longrightarrow R-N\big\langle$$

234. LEUCKART THIOPHENOL REACTION

Leuckart thiophenol reaction was first reported by Rudolf Leuckart in 1890. The reaction allows the preparation of thiophenols and corresponding thioethers from anilines or their corresponding diazonium salts (ArN_2X). The first step is the reaction of an aryl diazonium salt with a potassium alkyl xanthate to give an aryl xanthate, which afford aryl thiols (aryl mercaptan) on alkaline hydrolysis and aryl thioethers on warming.

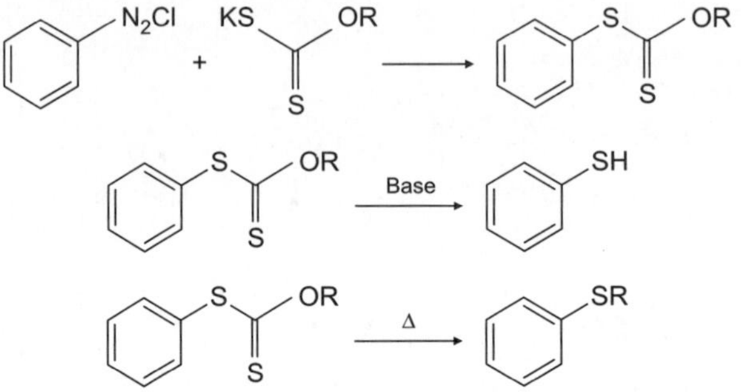

235. LIEBEN IODOFORM REACTION (HALOFORM REACTION)

The haloform reaction also said to be oxidative cleavage of methyl ketones is a reaction of a methyl ketone (a 2-alkanone) in basic solution in the presence of excess halogen. The organic products of this reaction are the haloform (α,α,α-trihalomethane (CHX_3, where X is a halogen)) and a carboxylate anion. The overall reaction is shown below. The reaction can be used to produce chloroform ($CHCl_3$), bromoform ($CHBr_3$), or iodoform (CHI_3).

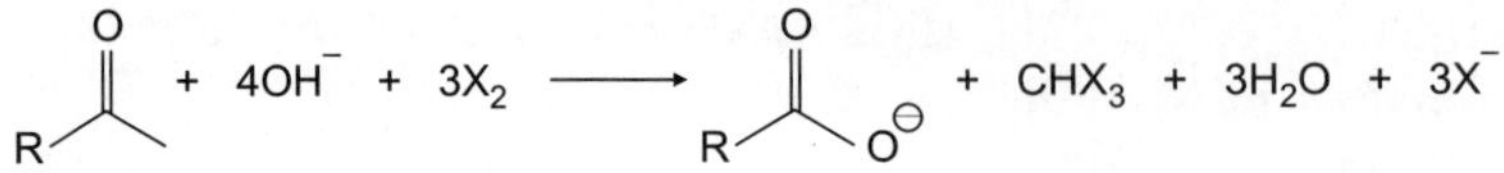

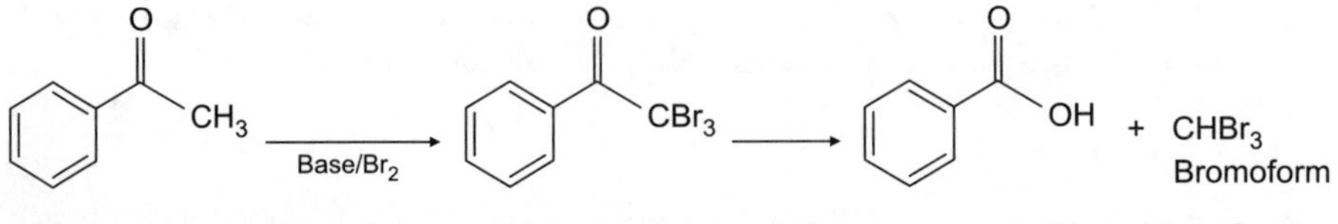

The haloform reaction is unique to methyl ketones. Other ketones will undergo base-promoted α-halogenation, but will react no further.

If the halogen used is iodine, the reaction becomes the iodoform test reaction. Iodoform is a bright yellow solid, and its formation can be used as a qualitative test for the presence of a methyl ketone. The iodoform is easily identified by its yellow colour, its characteristic odour and the melting point.

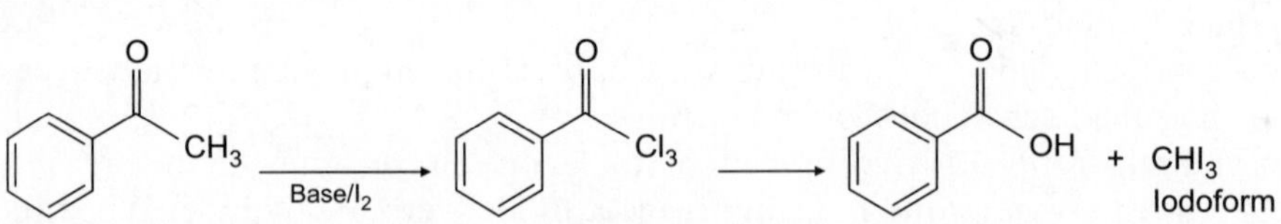

Fluorine cannot be used or fluoroform (CHF_3) cannot be prepared from a methyl ketone by the haloform reaction due to the instability of hypofluorite.

Substrates that successfully undergo the haloform reaction are:

Compounds having the group shown will give a positive test for haloform reaction

- methyl ketones
- secondary alcohols oxidizable to methyl ketones, such as isopropanol
- primary alcohols such as ethanol
- aldehydes such as ethanal
- 1,3-diketones such as acetylacetone also give the haloform reaction
- β-keto acids such as acetoacetic acid will also give the test upon heating

The halogen used in the reaction may be chlorine, bromine, or iodine.

This reaction was traditionally used to determine the presence of a methyl ketone, or a secondary alcohol oxidizable to a methyl ketone through the iodoform test. The reaction has some synthetic utility in the oxidative demethylation of methyl ketones if the other substituent on the carbonyl groups bears no enolizable α-protons.

236. LOBRY DE BRUYN-VAN EKENSTEIN TRANSFORMATION

The Lobry de Bruyn-Alberda van Ekenstein transformation also known as enolization reaction, produces enediol anion species. Through these reactions of sugar transformation aldoses are converted into ketoses and vice versa.

This transformation involves both epimerization of aldoses and ketoses and aldose-ketose interconversion, when they are exposed to base. Therefore used for the reciprocal interconversion of carbohydrates into their isomers in an alkaline solution. This glycochemical conversion is usually catalyzed by base but can also take place under acid or neutral conditions.

Aldose   Enolate   Enediol   Different enolate   Ketose

| Aldose carbohydrate | Recovered starting material | Epimer at alpha carbon | Ketose form | + other compounds |

The Lobry de Bruyn-van Ekenstein transformation was discovered in 1885 by Cornelis Adriaan Lobry van Troostenburg de Bruyn and Willem Alberda van Ekenstein. The reaction mechanism takes place via the transformation reactions through a tautomeric enediol as reaction intermediate.

237. LOSSEN REARRANGEMENT

The Lossen rearrangement classically is described as the thermal decomposition of hydroxamic acids, their salts, or esters, in a manner resulting in migration of the group attached to the carbon atom of the hydroxamic acid group to the nitrogen atom.

The rearrangement of acyl nitrenes to isocyanates is the crux of this rearrangement. It involves thermal or alkaline conversion of hydroxamic acid into an isocyanate. The mechanism of the reaction is considered to proceed by formation of an intermediate isocyanate which, in dilute aqueous solution, may hydrolyze with production of an amine; on the other hand, if amines or alcohols are present, the isocyanate may react with them to form ureas or urethanes respectively. Thus the nature of the product or products isolated depends on the conditions under which the rearrangement is performed and on the types of available reactive groups. As the starting material hydroxamic acid is difficult to obtain the reaction has limited synthetic application.

This reaction occurs via a similar mechanism as of Curtius or Hofmann rearrangement i.e. by means of the formation of a univalent nitrogen intermediate (known as nitrene). Reaction is generally carried out starting with esters (O-acyl, sulfonyl, or phosphoryl derivative) of hydroxamic acid because of higher yield of products compared to hydroxamic acid itself.

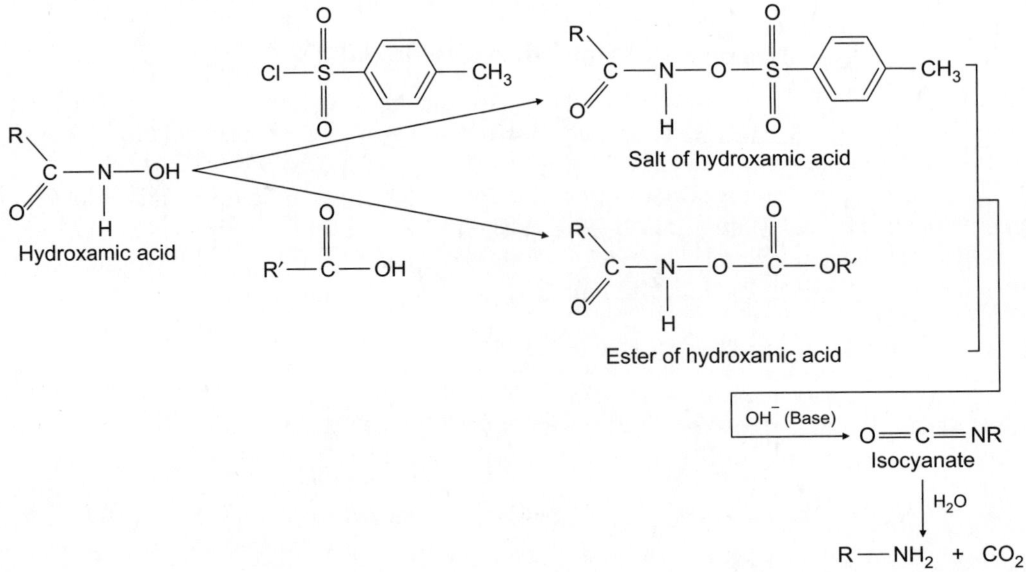

238. McFADYEN-STEVENS REACTION

McFadyen-Stevens reaction describes base-catalyzed thermal decomposition of acylbenzene-sulfonylhydrazines to aldehydes. Both aromatic or heterocyclic aldehydes can be synthesized from this reaction.

239. McLAFFERTY REARRANGEMENT

The fragmentation of radical cations of carbonyl compounds by transfer of a hydrogen atom from the γ-position to the carbonyl-O and cleavage of the α,β-bond is known as the McLafferty rearrangement.

Molecular ion

EI = Electron ionization (formerly known as electron impact)

The McLafferty rearrangement is a characteristic fragmentation of the molecular ion of a carbonyl compound containing at least one gamma hydrogen. This rearrangement is found to be reliable and useful in mass spectrometric fragmentation mechanisms for structural analysis of organic compounds.

240. McMURRY COUPLING REACTION

McMurry coupling reaction describes reductive coupling of carbonyls with low valent transition metals, Ti(0) or Ti(II), to give olefins.

McMurry has started the application of the low-valent titanium as reducing agent for several functional group and studied the scope of coupling of the carbonyl groups using $LiAlH_4$, $LiBH_4$, etc. in combination with $TiCl_3$ or $TiCl_4$ is generally known as the McMurry coupling or McMurry reaction. The combination of $TiCl_3$ and Zn/Cu couple in DME or $TiCl_3$-Li in DME is known as the McMurry reagent. This reaction generally gives trans-olefins but sometimes contaminated with pinacol. The study finds the formation of the predominant cis-stilbene from acetophenone under the McMurry coupling condition is due to the π-π preassociation. This coupling reaction has been further modified to enhance the chemoselectivity and diastereoselectivity. This reaction is useful in the preparation of olefins. Deoxygenative coupling of carbonyl compounds to alkenes induced by low-valent titanium.

The McMurry reaction is an organic reaction in which two ketone or aldehyde groups are coupled to an alkene using titanium chloride compound such as titanium(III) chloride and a reducing agent. The reaction is named after its co-discoverer, John E. McMurry. The McMurry reaction originally involved the use of a mixture $TiCl_3$ and $LiAlH_4$, which produces the active reagent(s). This reaction is related to the Pinacol coupling reaction which also proceeds by reductive coupling of carbonyl compounds.

This reductive coupling involves two steps. The coupling is induced by single electron transfer to the carbonyl groups from alkali metal, followed by deoxygenation of the 1,2-diol with low-valent titanium to yield the alkene.

The McMurry reaction works well to produce symmetric products or rings:

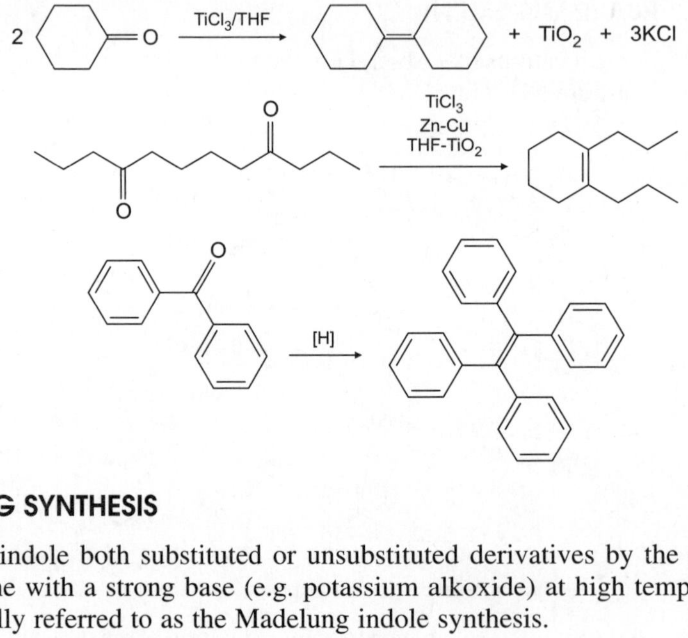

241. MADELUNG SYNTHESIS

The synthesis of indole both substituted or unsubstituted derivatives by the reaction of ortho-alkyl N-acylaniline with a strong base (e.g. potassium alkoxide) at high temperature (e.g. 300–400°C), is generally referred to as the Madelung indole synthesis.

Classically, this reaction involves treatment of N-benzoyl-o-toluidines with alkoxides at high temperatures (360–380°C). The harsh conditions used kept the reaction far from wide application. Although the classical Madelung synthesis is rarely employed nowadays, recent modification, which utilizes BuLi or LDA as bases under milder conditions than the original Madelung harsh conditions, has been extended in several ways. This reaction is useful for the preparation of unsubstituted indole and indoles with substitutents at 2-position.

242. MAILLARD REACTION (BROWNING REACTION)

Non-enzymatic browning, also known as the Maillard reaction, occurs between carbonyl compounds, especially reducing sugars, and compounds with free amino groups, such as amines, amino acids, and proteins. This reaction usually requires heat. The reaction in aqueous solution is usually vital to foods as well as to general biochemistry and even to medicine.

✓ Maillard browning

Reducing sugar + amine ⟶ Brown pigments + flavors

Effects of Maillard Reaction
Desirable:
Color—bread crust, syrup, meat
Flavor—coffee, cocoa, meats
Antioxidants
Undesirable:
Color—changes in color during storage
Flavor—changes during processing and storage
Nutritional loss—essential amino acids
Vitamins (vit C), palatability and digestibility
Toxicity/mutagenicity

243. MALAPRADE REACTION (PERIODIC ACID OXIDATION)

Compounds containing two hydroxyl groups (vicinal diols or glycol), or a hydroxyl and an amino group (aminoalcohols), attached to adjacent carbon atoms, undergo cleavage of the carbon-carbon bond when treated with periodic acid in aqueous solution to yield aldehydes.

The cleavage of 1,2-diols by periodic acid is associated with the name of the French chemist Malaprade (first reported by Léon Malaprade in 1934). The reaction mechanism is found to involve a five-membered ring periodate ester intermediate.

This reaction is found to be extended to the cleavage of α-hydroxy carbonyl compounds, 1,2 dicarbonyl compounds, α-amino alcohols, α-amino acids, and polyhydroxy alcohols and successfully applied for structural analysis.

244. MALONIC ESTER SYNTHESIS

The malonic ester sythesis has proven to be versatile method for building the desired carbon skeletons of carbonyl-containing molecules. This method rely on the alkylation of an enolate stabilized by two neighboring carbonyl groups, followed by hydrolysis of one or two ester groups and decarboxylation of the resulting carboxylic acid.

After alkylation, β-carboxy esters can be hydrolyzed (saponified). The resulting β-diacids will decarboxylate upon gentle heating. This alkylation-saponification-decarboxylation sequence is termed the "malonic ester synthesis". This sequence is used for the synthesis of substituted carboxylic acids.

Malonic esters are more acidic than simple esters, so that alkylations can be carried out via enolate formation promoted by relatively mild bases such as sodium alkoxide, and subsequent alkylation with halides.

A malonic ester
R = Me or Et

A substituted
acetic acid

The major drawback of malonic ester synthesis is that the alkylation stage can also produce dialkylated structures. This makes separation of products difficult and yields lower.

A carboxylic acid group with a carbonyl group in the β-position can undergo decarboxylation on heating, through a cyclic transition state. The net result is that a –CO_2H group β to the carbonyl is replaced with a hydrogen.

245. MANNICH REACTION

Mannich reaction (aminomethylation of CH-acidic compounds) is a multi-component condensation reaction, which involves the compounds having an active hydrogen with non-enolizable aldehydes and ammonia or primary or secondary amines to give aminomethylated products. The final product is a β-amino-carbonyl compound also known as a Mannich base. This reaction can also operate under basic conditions.

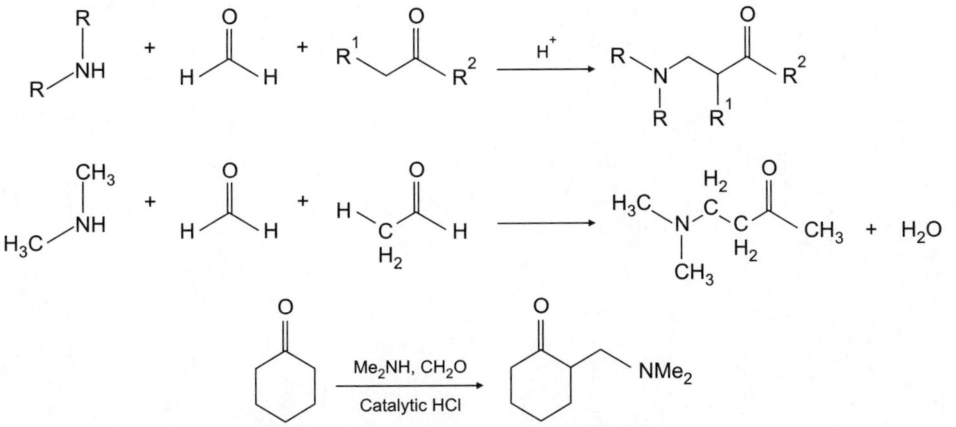

The Mannich reaction has been proposed in many biosynthetic pathways, especially for alkaloids. The reaction is named after the chemist Carl Mannich.

The Mannich reaction is an example of nucleophilic addition of an amine to a carbonyl group followed by dehydration to the Schiff base. The Schiff base is an electrophile which reacts in the second step in a electrophilic addition with a compound containing an acidic proton (which is, or had become an enol). The Mannich reaction is also considered a condensation reaction.

246. MARSCHALK REACTION

The Marschalk reaction is the sodium dithionite promoted reduction of 1-hydroxy- or aminoanthraquinones to their leuco-forms, followed by condensation with aldehydes to yield the 2-alkylated anthraquinones. 2-hydroxyanthraquinones yield 1-alkylated products.

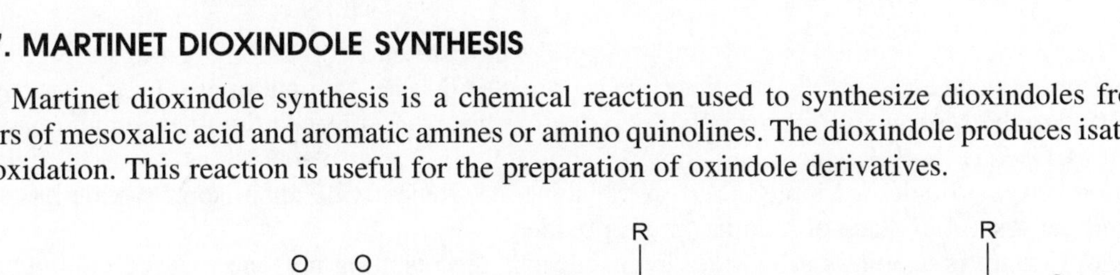

X = OH, Y = O
X = NH₂, Y = NH

247. MARTINET DIOXINDOLE SYNTHESIS

The Martinet dioxindole synthesis is a chemical reaction used to synthesize dioxindoles from esters of mesoxalic acid and aromatic amines or amino quinolines. The dioxindole produces isatins on oxidation. This reaction is useful for the preparation of oxindole derivatives.

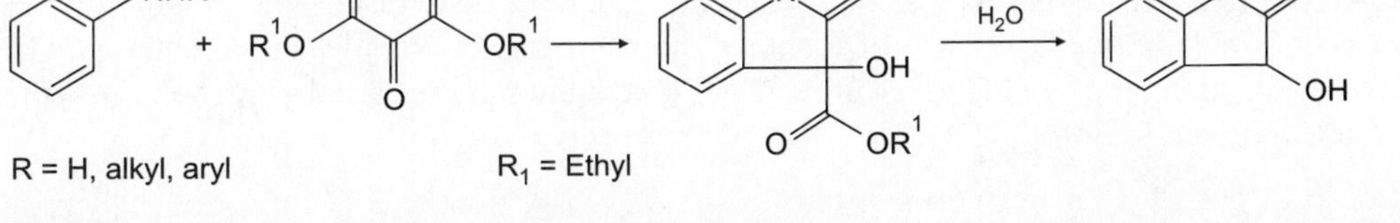

R = H, alkyl, aryl   R₁ = Ethyl

248. MEERWEIN ARYLATION

The Meerwein arylation involves formation of arylated olefins on treatment of olefins with aryl diazonium salt (ArN_2X) in the presence of cupric salts. The reaction product is an alkylated arene compound. The reaction is named after Hans Meerwein.

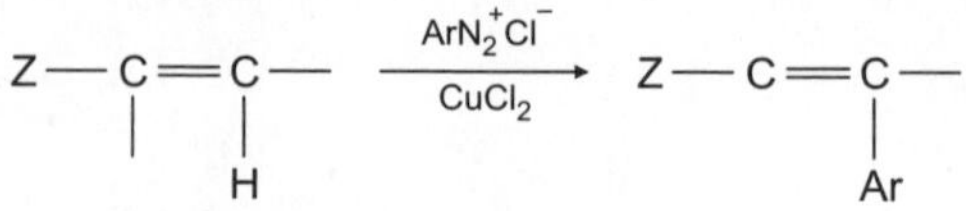

$$Z-C=C- \xrightarrow[CuCl_2]{ArN_2^+Cl^-} Z-C=C-$$

$$Z=C=C, C=O, Ar, CN, H$$

Example:

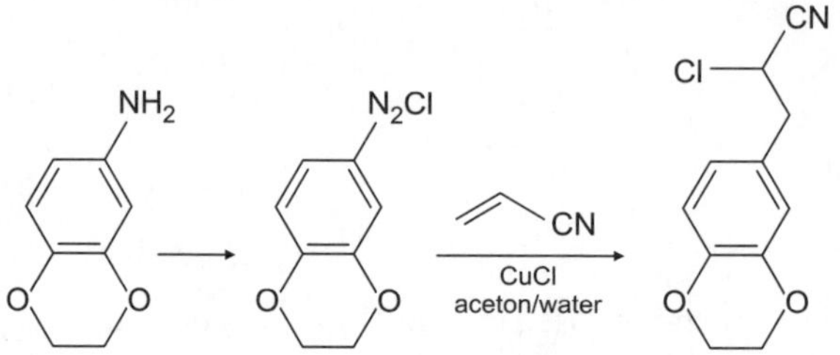

249. MEERWEIN-PONNDORF-VERLEY REDUCTION (ALUMINUM ALKOXIDE REDUCTION/ REDUCTION OF ALDEHYDES AND KETONES WITH ALUMINUM ISOPROPOXIDE)

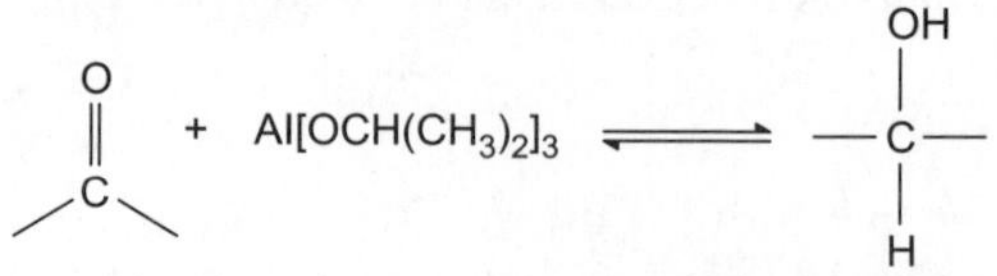

$$\overset{O}{\underset{C}{\|}} + Al[OCH(CH_3)_2]_3 \rightleftharpoons \overset{OH}{\underset{H}{-C-}}$$

Reduction of aldehydes or ketones to the corresponding alcohols with aluminum alkoxides. The aluminium-catalyzed hydride shift from the α-carbon of an alcohol component to the carbonyl carbon of a second component, which proceeds via a six-membered transition state, is referred to as the Meerwein-Ponndorf-Verley Reduction (MPV) or the Oppenauer Oxidation, depending on which component is the desired product. If the alcohol is the desired product, the reaction is viewed as the Meerwein-Ponndorf-Verley Reduction.

$$\overset{R}{\underset{R'}{>}}C=O \xrightarrow[HOi-Pr]{Al(Oi-Pr)_3} \overset{R}{\underset{R'}{>}}CHOH + \overset{}{\underset{}{>}}=O$$

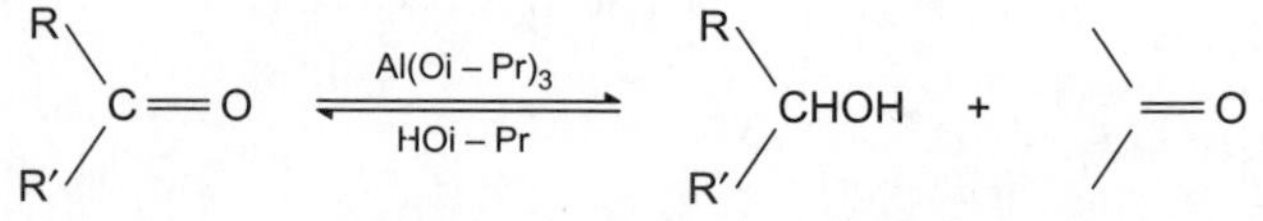

The reaction is specific for carbonyl group. The case when two carbonyl groups are present, one group is reduced and other ketonic group is protected by acetal formation from getting reduced.

Though the Meerwein-Ponndorf-Verley reduction has been replaced by other reducing agents such as $LiAlH_4$, $NaBH_4$ etc. but is known for its mildness and selectivity as it is not affecting carbon-carbon double or triple bonds of the substrate. Recently, lanthan isopropoxide has been found purposeful in place of aluminium propoxide.

Isopropanol is useful as a hydride donor because the resulting acetone may be continuously removed from the reaction mixture by distillation. The popularity of the MPV reduction lies in its high chemoselectivity, and its use of a cheap environmentally friendly metal catalyst. Grignard reagents will sometimes yield the result of an MPV reduction if the carbonyl carbon is too hindered for nucleophilic addition.

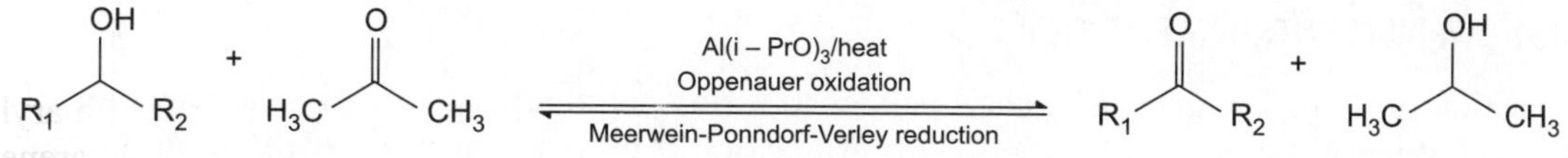

The MPV reduction was discovered by Meerwein and Schmidt and separately by Verley in 1925.

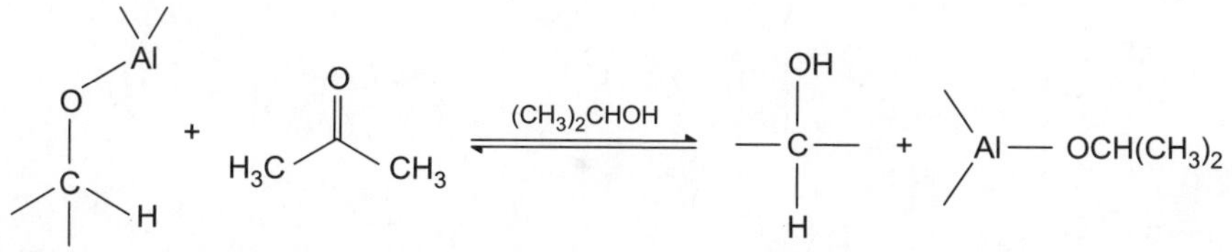

250. MEISENHEIMER REARRANGEMENTS

In Meisenheimer rearrangement the tertiary aliphatic amine oxides (tertiary N-oxides) are undergoe thermal rearrangement into substituted N-alkoxylamines or N-hydroxylamines, *via* [1,2]-R group migration, or [2,3]- sigmatropic rearrangement. For example, the N-oxides ($R_1R_2R_3N^+O^-$) rearrange to hydroxylamines ($R_2R_3N^-O^-R_1$). The migrating group in the reaction is usually benzylic or allylic. Aromatic N-oxides do not respond to this rearrangement.

In a 1,2-rearrangement:

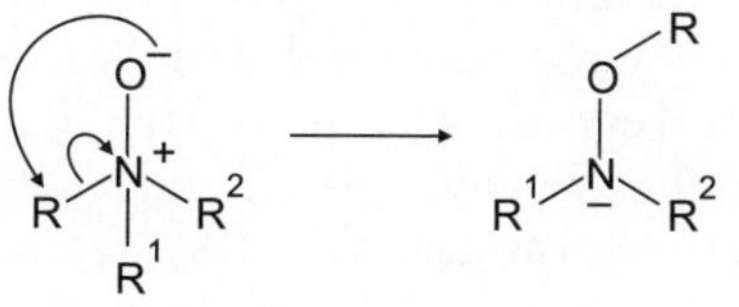

or a 2,3-rearrangment:

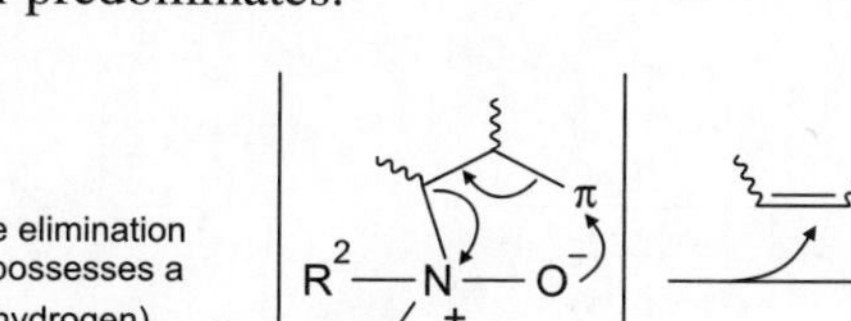

This rearrangement has limited synthetic utility because if any of the groups possess a β-hydrogen, Cope elimination predominates.

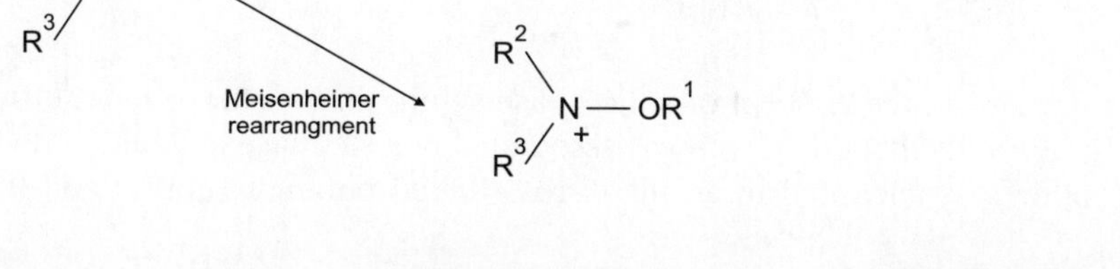

251. MENSCHUTKIN REACTION

Reaction of tertiary amines with alkyl halides to form quaternary salts or the quaternization of a trialkylamine is known as the Menschutkin reaction. The reaction is formally an alkyl group transfer, a process of central importance in biochemistry and a widely used strategy in organic synthesis. The reaction has been named after its discoverer, Nikolai Menshutkin, who described the procedure in 1890.

In Menschutkin reaction, two neutral molecules undergo an S_N2 reaction to produce an ion pair. Although the reaction can also proceed via either an S_N1 or an S_N2 mechanism. The process is accelerated by polar aprotic solvents, increased pressure, elevated temperature, and increased leaving-group ability. This reaction is widely used as a model reaction to study the substituent and solvent effects on the reaction mechanism and for the synthesis of quaternary ammonium salt.

252. MERRIFIELD SOLID PHASE PEPTIDE SYNTHESIS (SPPS)

Synthesis of long peptides and small proteins by first attachment of the C-terminal amino acid to an insoluble polymeric support resin, then elongation of the peptide chain, and finally the cleavage of the peptide from the resin until the desired peptide or protein is assembled, is generally referred to as the Merrifield solid phase peptide synthesis (or SPPS) and the polymeric resin is known as the Merrified resin.

Elongation step (P = support resin)

Merrifield introduced the concept of solid phase synthesis to achieve more efficient synthesis of peptides. In SPPS, the peptide chain was assembled in a stepwise manner while the C-terminal end of the peptide was anchored to an inert cross-linked polymer support and the peptide was grown from C-terminal to N-terminal residue. Merrifield demonstrated the feasibility of the idea by synthesizing a model—tetra peptide L-leucyl-L-alanyl-glycyl-L-valine.

Merrifield's method employs an insoluble and filterable polymeric support such as cross-linked polystyrene that function as the carboxyl-protecting group for the C-terminal amino acid of the peptide. The target peptide sequence was formed in a stepwise manner by attaching temporary N-protected C-terminal amino acid to the chloromethylated PS-DVB resin. After the removal of N-protection, the next N-protected amino acid is coupled and the process is repeated until the entire desired peptide is assembled on the polymer support. Dicyclohexyl carbodiimide (DCC) is used as the coupling agent, and all the reactions are carried out under non-aqueous conditions in organic solvents. The target peptide was deprotected and cleaved from the polymer matrix by acidolysis with HF or anhydrous TFA in the presence of suitable scavenger.

Solid phase peptide synthesis has the following advantages over the classical solution phase method.

(i) The peptide is synthesized while its C-terminus is covalently attached to an insoluble polymeric support. This permits the easy separation of the growing peptide from any by-products or excess unused amino acid components.

(ii) The reactions are driven to completion by using an excess of reactants and reagents.

(iii) No mechanical loss occurs because the growing peptide is retained on the polymer in a single reaction vessel throughout the synthesis.

(iv) The final peptide is detached from the polymer support by a single cleavage step at the end of synthesis. The side chain protecting groups can also be cleaved in the same reaction in order to simplify the workup and the isolation of the final peptide. The cleavage step does not degrade the assembled peptide.

(v) The physical operations involved in the synthesis are simple, rapid and amenable to automation.

(vi) The spent resin can be recycled.

In spite of many advantages given above, Merrifield's solid phase method has a number of limitations. They are:

(i) Non-compatibility of resin and growing peptide chain.

(ii) Lack of stability of peptide-resin linkage under the conditions of synthesis.

(iii) Non-equivalence of functional groups attached to the polymer support.

(iv) Formation of error peptides due to truncated and failure sequences.

(v) Peptide conformation changes in macroscopic environments inside the polymer matrix and also due to peptide resin linkage.

A number of modifications have been introduced to overcome the difficulties associated with Merrifield's SPPS which includes:

(i) Development of new supports with high swelling properties permitting improved solvation of both matrix and growing peptide chain.

(ii) Introduction of multi-detachable anchoring groups improving the flexibility of synthetic strategy.

(iii) Development of newer separation method (e.g. preparative and semi-preparative HPLC) and characterization techniques in peptide synthesis.

253. MEYER REACTION

In Meyer reaction, alkali stannite reacts with an alkyl halides to form alkylstannonic acids. When applied to alkali arsenites reacts with an alkyl halides yields alkylarsonic acid.

$$Na_2SnO_2 \xrightarrow{\text{RX}} RSnO_2Na$$

$$Na_2AsO_3 \xrightarrow{\text{RX}} RAsO_3Na_2$$

254. MEYER-SCHUSTER REARRANGEMENT; RUPE REARRANGEMENT

The Meyer-Schuster and the Rupe rearrangements are traditionally the best known reactions of alkynols. Both reactions afford α,β-unsaturated ketones or aldehydes from the starting alkynols.

The Meyer-Schuster rearrangement mechanism, involves acid catalysed rearrangement of both secondary and tertiary alkynols (propargyl alcohols) affording α,β-unsaturated enones as the products. In this reaction, the type of product, whether it is an enal or enone, depends on the type of group at the terminal end of the acetylene group. In the case of terminal acetylenic hydrogen, the reaction gives an α,β-unsaturated aldehyde (an enal) as the end product (when R' = H). When R' is an alkyl group, the end product of the Meyer-Schuster rearrangement is an α,β-unsaturated ketone (an enone). In short, aldehydes result when the acetylenic group is terminal, ketones when it is internal.

The reaction appears to involve an 1,3-hydroxyl shift.

The reaction may be catalyzed with Lewis or protic acids and is not sensitive to moisture in that it may be conducted in either aqueous or anhydrous conditions. However, the Meyer-Schuster rearrangement is but one possible fate of the propargyl cation and selecting for the Meyer-Schuster pathway has been a long lasting challenge. The most significant competing pathway is the Rupe rearrangement.

In the **Rupe rearrangement**, alkynols, upon acid catalysis, are converted to only α,β-unsaturated ketones. Even if the terminal acetylenic group is a hydrogen, the reaction does not yield an aldehyde but proceeds to rearrange the alkynol to α,β-unsaturated ketone (when $R' = H$). The study finds that in many cases, this reaction is competed by the Meyer-Schuster rearrangement, which also leads to the formation of α,β-unsaturated ketones or aldehydes.

$$R_1CH_2 - \overset{\overset{\displaystyle OH}{|}}{\underset{\underset{\displaystyle R_2}{|}}{C}} - C \equiv C - R' \xrightarrow[\substack{R_1; R_2 = H_3, alkyl \\ R' = Alkyl\ or\ H}]{HCOOH} \underset{R_2}{\overset{R_1}{{}}}C = C\underset{R_2}{\overset{COCH_2R'}{{}}}$$

The mechanism for the Meyer-Schuster reaction, though related to the Rupe's, deviates from the latter in that an allene or an allenol is thought to be the key intermediate and not the enyne as described for the Rupe rearrangement.

α,β-unsaturated carbonyl compounds, important intermediates in the manufacture of fragrances, carotenoids, and vitamins, are accessible by *Meyer-Schuster-* and *Rupe-Kambli*-type rearrangement of α-acetylenic alcohols. The acid catalyzed *Meyer-Schuster*-rearrangement starts from α-alkynols to yield α,β-unsaturated aldehydes. In the *Rupe-Kambli*-type rearrangement α,β-unsaturated ketones are obtained as products. While the traditional Meyer-Schuster rearrangement uses harsh conditions with a strong acid as the catalyst, this introduces competition with the Rupe reaction if the alcohol is tertiary. Milder conditions have been used successfully with transition metal-based and Lewis acid catalysts (for example, Ru- and Ag-based catalysts).

Iodo Meyer-Schuster rearrangement

In 1991 Angara and Mc Nelis reported novel formations of α-iodoenones and α-bromoenones from a reaction involving secondary alkynols and halonium producing systems.

$$CH_3 - \overset{\overset{\displaystyle OH}{|}}{\underset{\underset{\displaystyle H}{|}}{C}} - C \equiv C - R' \xrightarrow[C_6H_5I(OH)\ (OTS)]{\substack{NXS \\ catalytic}} \text{product}$$

$$R' = C_2H_5 \text{ or } C_6H_5 \qquad\qquad X = I,\ Br$$

255. MEYER SYNTHESIS (VICTOR MEYER SYNTHESIS)

$$R - X + AgNO_2 \xrightarrow[X = I\ or\ Br]{(C_2H_5)_2O} R - NO_2 + AgX$$

The formation of a nitroalkanes (aliphatic nitrites) from the reaction between an alkyl halide and metal nitrites (silver nitrite) is known as the Victor Meyer reaction. This reaction is a convenient method for the preparation of the homologues of primary nitroalkanes higher than 1-nitropropane. Primary alkyl iodides and bromides are excellent substrates for this reaction on the other hand reaction with primary alkyl chlorides is too slow to be synthetically useful. Secondary alkyl halides as well as substrates with branching on the carbon chain give much lower yields of nitro compounds.

256. MEYERS ALDEHYDE SYNTHESIS

The reaction, named after Albert I. Meyers, describes aldehyde synthesis starting from a dihydro-1,3-oxazine with an alkyl group in the 2 position. In the reaction a dihydro-1,3-oxazine with an alkyl group (e.g. 4,4,6-trimethyl-5,6-dihydro-1,3-oxazines) first undergo deprotonation by a strong base such as butyl lithium and subsequently alkylated by an alkyl halide (haloalkane). In the next step the nitrogen to carbon double bond (imine) is reduced with sodium borohydride and the resulting oxazine (a hemiaminal) hydrolyzed with water and oxalic acid to the aldehyde. t-butyl lithium is found suitable reagent for the deprotonation and alkyl bromides and iodides are good as alkylating reagents.

This reaction can be used to prepare α, β-unsaturated aldehyde, cycloalkyl aldehyde, γ-hydroxy or γ-oxo aldehyde and ketones, etc.

Z = H, C_6H_5, $COOC_2H_5$

257. MICHAEL REACTION (ADDITION, CONDENSATION)

The Michael addition involves the addition of a nucleophile, also called a 'Michael donor,' to an activated electrophilic olefin, the 'Michael acceptor', resulting in a 'Michael adduct'

Although, the Michael addition is generally considered the addition of enolate nucleophiles to activated olefins, a wide range of functional groups possess sufficient nucleophilicity to perform as Michael donors. Reactions involving non-enolate nucleophiles such as amines, thiols, and phosphines are typically referred to as 'Michael-type additions'.

The Michael acceptor possesses an electron withdrawing and resonance stabilizing activating group, which stabilizes the anionic intermediate. Michael addition acceptors are far more numerous and varied than donors, due to the plethora of electron withdrawing activating groups that enable the Michael addition to olefins and alkynes. Acrylate esters, acrylonitrile, acrylamides, maleimides, alkyl methacrylates, cyanoacrylates and vinyl sulfones serve as Michael acceptors and are commercially available. Less common, but equally important, vinyl ketones, nitro ethylenes, α,β-unsaturated aldehydes, vinyl phosphonates, acrylonitrile, vinyl pyridines, azo compounds and even b-keto acetylenes and acetylene esters also serve as Michael acceptors.

The Michael addition, named for Arthur Michael, is a facile reaction between nucleophiles and activated olefins and alkynes in which the nucleophile adds across a carbon-carbon multiple bond. The Michael addition benefits from mild reaction conditions, high functional group tolerance, a large host of polymerizable monomers and functional precursors as well as high conversions and favorable reaction rates.

The Michael reaction is a "1,4 addition" or "conjugate addition" of a carbanion or another nucleophile to an α,β-unsaturated carbonyl compound. However, there is the obvious competitive reaction, where we see the 1,2-addition reaction.

Michael additions are conducted in a suitable solvent in the presence of a strong base either at room temperature or at elevated temperatures. Due to the presence of the strong base, side reactions such as multiple condensations, polymerizations, rearrangements and retro-Michael additions are common. These undesirable side reactions decreases the yields of the target adduct and render their purification difficult. Better results can be obtained by employing weaker bases such as piperidine, quaternary ammonium hydroxide, tertiary amines etc.

The Michael reaction has been widely used in organic synthesis for its C-C bond forming ability. It is employed in the traditional sense where an enolate reacts with an α,β-unsaturated carbonyl. This reaction is also used in tandem with other reactions. The best known of these is the Robinson annulations where the Michael addition occurs as the first step.

Cortisone

Michael reaction employing non-enolic carbon nucleophile

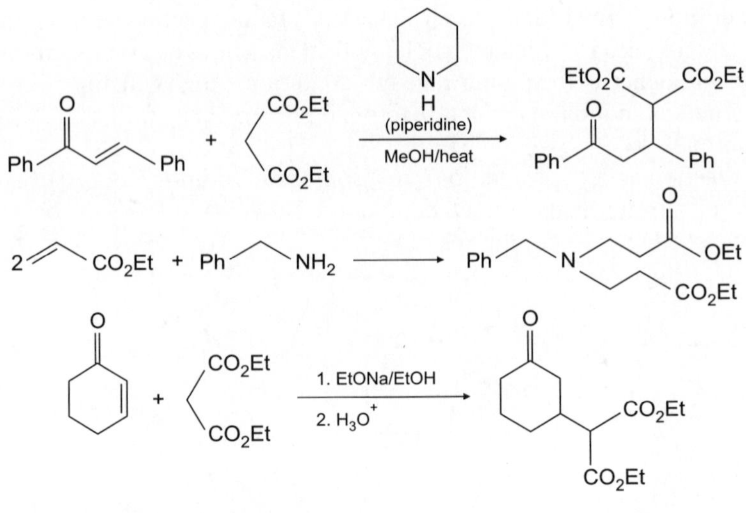

258. MICHAELIS-ARBUZOV REACTION

Formation of carbon-phosphorus bonds is one of the greatest challenges in organophosphorus chemistry. Michaelis-Arbuzov reaction (also called the Arbuzov reaction) is the most common C-P bond forming reaction is the, which involves a reaction between an alkyl halide and a trialkyl or dialkyl phosphate via the intermediate phosphonium salt resulting trialkoxyphosphonium salt.

However, this reaction usually gives acceptable yields when primary alkyl halides are used as substrates similarly alkyl bromides are more reactive than the corresponding chlorides. In other words, the Michaelis-Arbuzov reaction of trialkyl phosphites and α-halogenoketones leads to β-keto phosphonates.

Michaelis-Arbuzov reaction is one of the methods commonly used for the preparation of phosphonates but this method is restricted to highly reactive α-halogenoketones, taking into account competition with the Perkow reaction which gives enol phosphates. This reaction sees extensive application in the preparation of phosphonate esters for use in the Horner-Emmons reaction.

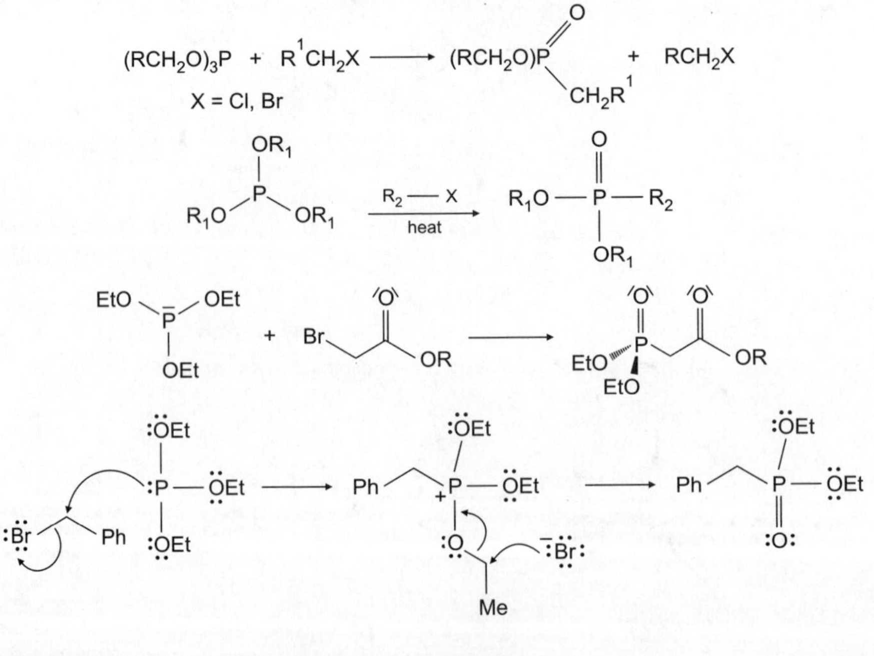

259. MIESCHER DEGRADATION

Miescher degradation is generally referred as a multistep cleavage of the side chain of steroid, such as bile acid methyl ester to the stage of methyl ketone involving conversion of the methyl ester of the bile acid to the tertiary alcohol, followed by dehydration, bromination, dehydrohalogenation and oxidation of the diene. This reaction is carried out by sequential reactions with the phenyl Grignard reagent in presence of chromium trioxide. The reaction is useful primarily in steroid chemistry.

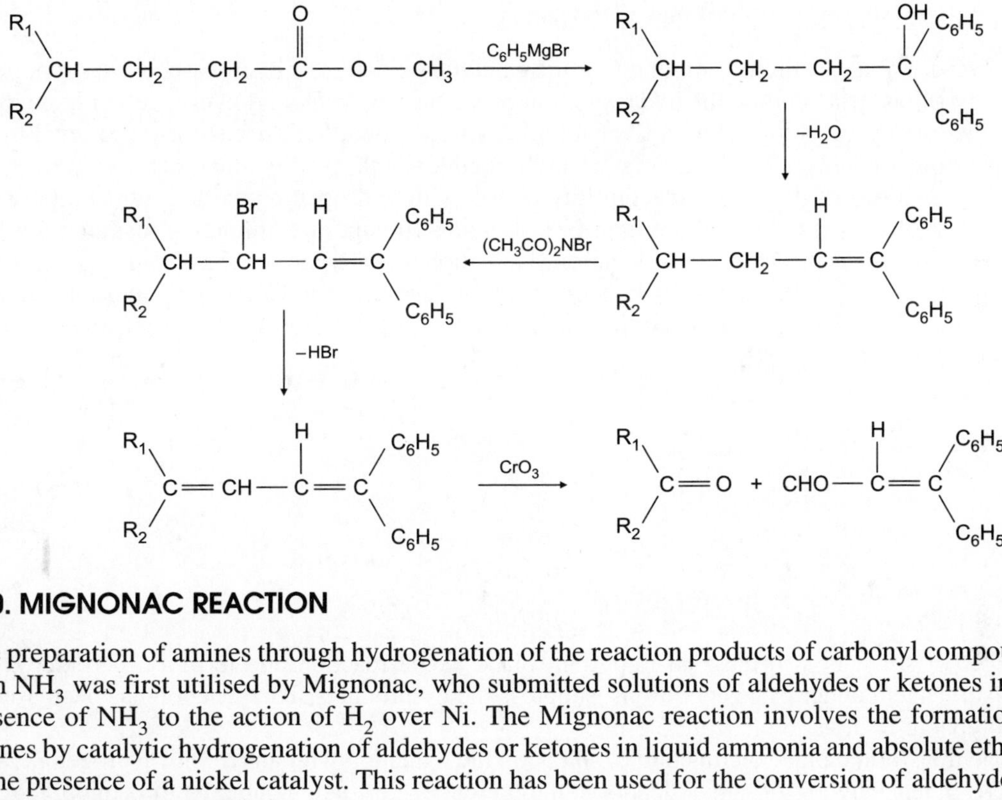

260. MIGNONAC REACTION

The preparation of amines through hydrogenation of the reaction products of carbonyl compounds with NH_3 was first utilised by Mignonac, who submitted solutions of aldehydes or ketones in the presence of NH_3 to the action of H_2 over Ni. The Mignonac reaction involves the formation of amines by catalytic hydrogenation of aldehydes or ketones in liquid ammonia and absolute ethanol in the presence of a nickel catalyst. This reaction has been used for the conversion of aldehydes or ketones into primary amines.

The initially formed primary amines can, in their turn, behave as aminating agents for carbonyl compounds to afford secondary amines.

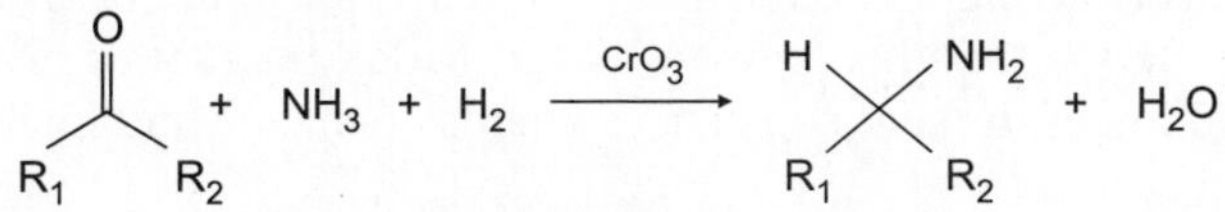

Synthesis of 1-phenylethylamine starting from acetophenone

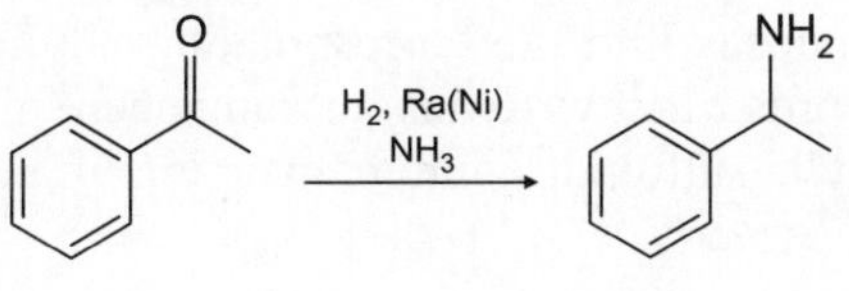

$$RCHO + NH_3 \rightleftharpoons RCH(OH)NH_2 \xrightarrow{-H_2O}$$

$$\downarrow\uparrow -H_2O \qquad\qquad \xrightarrow{H_2} RCH_2NH_2$$

$$RCH = NH \xrightarrow{H_2}$$

$$R_2NH + O = C\underset{CHR^2R^3}{\overset{R^1}{\diagup}} \rightleftharpoons R_2NC(OH)R^1CHR^2R^3 \xrightarrow{-H_2O} R_2NCR^1 = CR^2R^3 \xrightarrow{H_2} R_2NCR^1HCHR^2R^3$$

261. MILAS HYDROXYLATION OF OLEFINS

The first use of a vanadium compound for the catalytic oxidation of hydrocarbons was reported in 1937 by Milas, who studied the hydroxylation of alkenes by V_2O_5 and hydrogen peroxide. Since this date, many vanadium (V) and vanadium (IV) compounds have been used to catalyze the oxidation of alcohols, alkenes, aromatics and thioethers with hydrogen peroxide.

Formation of cis-glycols by reaction of alkenes with hydrogen peroxide in tert-butyl or tert-amyl alcohol and either ultraviolet light or a catalytic amount of osmium e.g. osmium tetroxide (OsO_4) is known as the Milas hydroxylation. Transition metal oxides of vanadium, e.g. pentoxide, or of chromium, e.g. chromium trioxide are also found effective under similar conditions. In other words, the Milas hydroxylation converts an alkene to a vicinal diol. The reaction was developed by N.A. Milas in the 1930s.

$$\diagup C = C \diagdown + H_2O \xrightarrow[\text{or } h\nu]{\text{Metal oxide}} \overset{HO \quad OH}{-C - C -}$$

$$\diagup\diagdown = \diagup\diagdown \xrightarrow[^tBuOH, H_2O_2]{OsO_4(cat.)} \overset{HO \quad OH}{\diagup\diagdown}$$

Role of solvents also have an important place in the reaction and tert-buyl peroxide as the oxidation agent under basic conditions, suppress the by-products. This reaction has been used for the preparation of cis-diols from olefins.

The transformation of olefins into cis-diols by the oxidation with anhydrous hydrogen peroxide (H_2O_2) in tert-butyl or tert-amyl alcohol in the presence of a catalytic amount of osmium tetroxide (OsO_4) is generally referred to as the Milas hydroxylation, and hydrogen peroxide is known as the Milas reagent. Transition metal oxides such as vanadium pentoxide and chromium trioxide are also found effective under similar conditions. The cis-diol is formed by reaction of alkenes with hydrogen peroxide and either ultraviolet light or a catalytic osmium, vanadium, or chromium oxide.

V_2O_5 and aqueous H_2O_2 together known as the Milas reagent is an effective catalyst for hydroxylation of organic unsaturated substances such as benzene and many types of alkenes.

262. MISLOW-EVANS REARRANGEMENT

This reaction is a base-promoted [2,3]-sigmatropic rearrangement of allylic sulfoxides to allylic sulfenates which then decomposes into the corresponding allylic alcohol in the presence of a thiophilic reagent. In other words Mislow-Evans rearrangement given by Kurt Mislow in 1968 is a thermal racemization of allylic sulfoxides, interconversion of achiral allylic sulfenates to chiral allylic sulfoxides.

Sulfoxide-Sulfenate Rgt.: (Mislow-Evans)

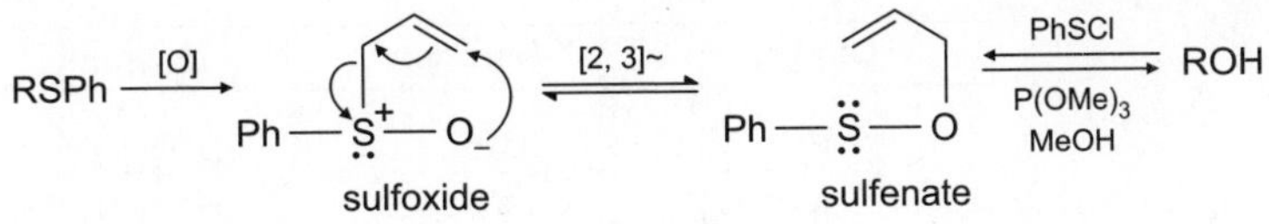

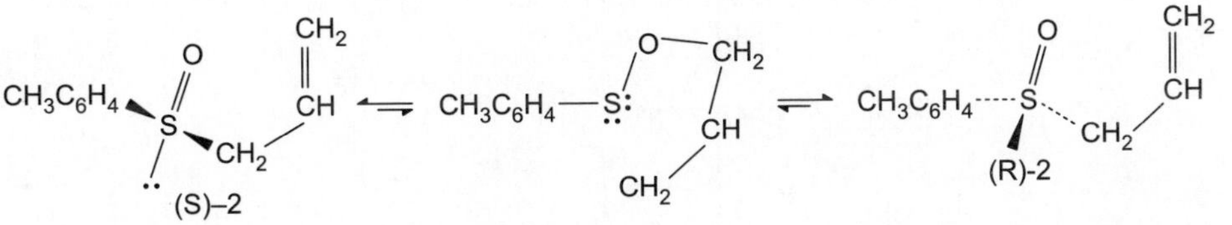

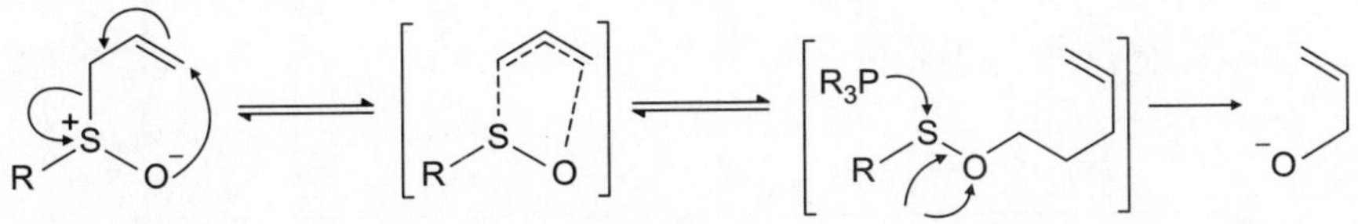

The Mislow-Evans rearrangement is believed to proceed exclusively through a concerted mechanism ([2,3]-*sigmatropic rearrangement with intramolecular* a, g *shift of the allyl group*) and there is no dissociation into kinetically free allyl and p-toluenesulfinyl radicals.

The reaction is reversible and equilibrium lies largely to the left, sulfonate not detectable by NMR. The reverse process is accomplished by treating the alcohol with arylsulfenyl chloride, followed by thermal rearrangement of the sulfenate to generate the allylic sulfoxide. It has been found that increased heating in the reaction can result in 1,3-shift of sulfoxide. Driving force in the reaction is the formation of the S=O sulfoxide bond (90 kcal/mol).This reaction is highly stereoselective and sensitive to salvation and found general application in the preparation of trans-allylic alcohols.

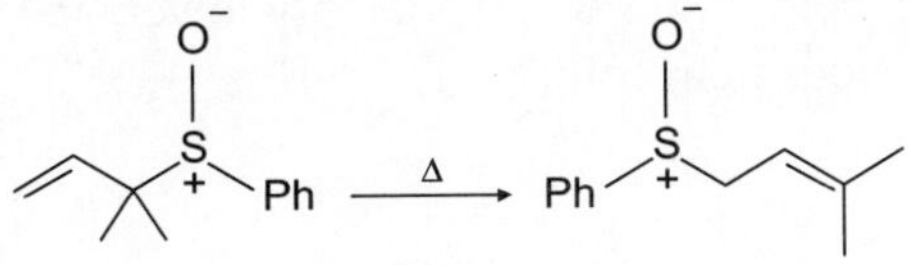

263. MITSUNOBU REACTION

In the Mitsunobu reaction, a unique dehydration occurs between alcohols and various Brønsted-Lowry acids (HA) utilizing a combination of diethyl azodicarboxylate (DEAD) - triphenylphosphine (TPP).

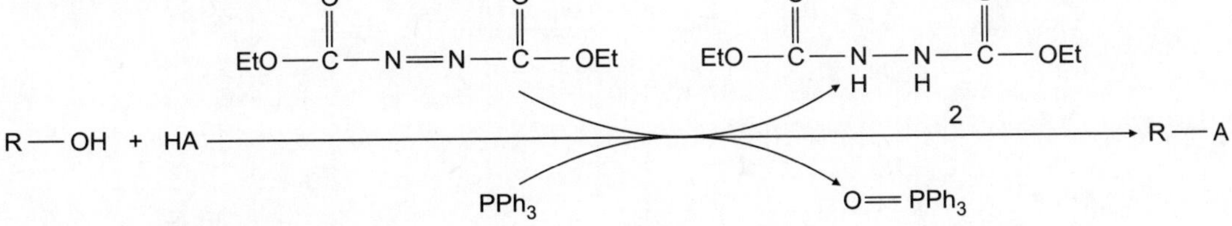

ex.
Nucleophiles (HA)

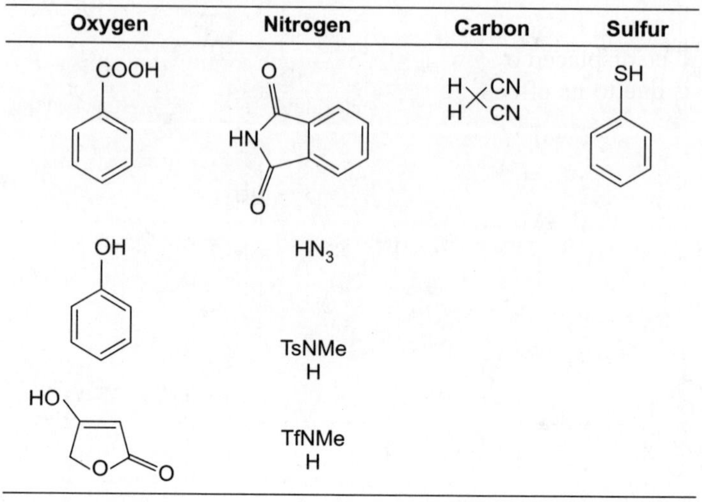

Oxygen	Nitrogen	Carbon	Sulfur
COOH		H CN / H CN	SH
OH	HN$_3$		
HO (furanone)	TsNMe / H		
	TfNMe / H		

Without any prerequisite activation of the alcohol, this redox condensation reaction proceeds under mild conditions with complete Walden inversion of stereochemistry (for example: Scheme 2), while DEAD is reduced to dihydro-DEAD and TPP is oxidized to triphenylphosphine oxide.

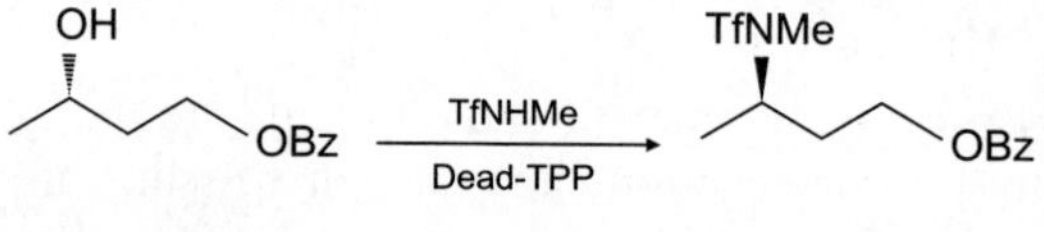

$$\text{(OH ... OBz)} \xrightarrow[\text{Dead-TPP}]{\text{TfNHMe}} \text{(TfNMe ... OBz)}$$

Organic chemists have enjoyed these advantages of the Mitsunobu reaction in organic synthesis. However, the reaction has a serious limitation (the so-called "the restriction of pKa"); the acidic hydrogen in HA has to have a pKa of less than 11 for the reaction to proceed satisfactorily. If HA has a pKa higher than 11, the yield of RA is considerably lower, and with HA having a pKa higher than 13, the desired reaction does not occur.

(i) pk$_a$ < 11

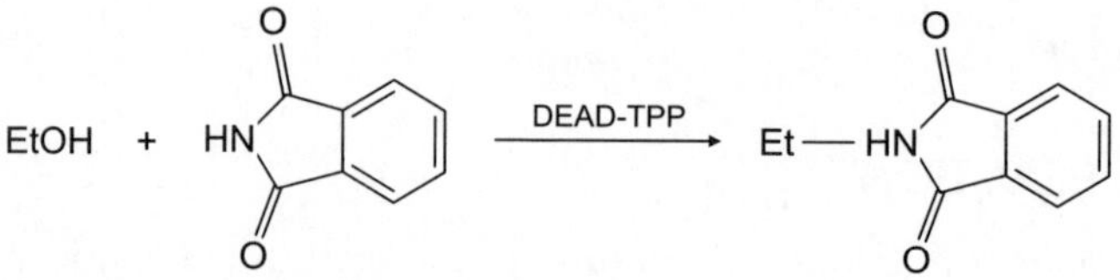

$$\text{EtOH} + \text{HN(phthalimide)} \xrightarrow{\text{DEAD-TPP}} \text{Et—HN(phthalimide)}$$

(pK$_a$ = 8.3)

(ii) 11 < pk$_a$ < 13

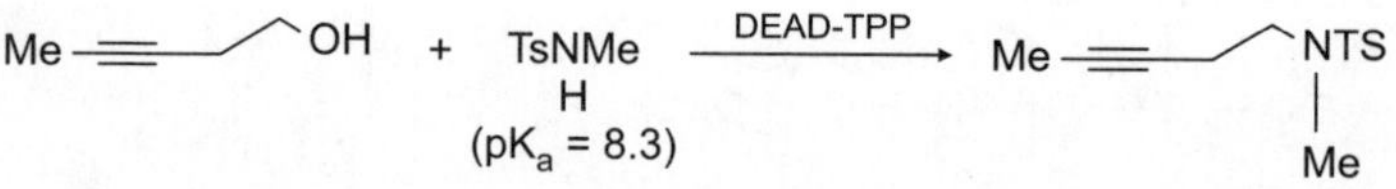

$$\text{Me}\equiv\!\!-\text{OH} + \text{TsNMe} \atop \text{H} \xrightarrow{\text{DEAD-TPP}} \text{Me}\equiv\!\!-\text{NTS} \atop \text{Me}$$

(pK$_a$ = 8.3)

(iii) pk$_a$ > 13

$$\text{(OH)} + \text{H}-\text{C(CO}_2\text{Et)}_2-\text{H} \xrightarrow{\text{DEAD-TPP}} \text{(product)}$$

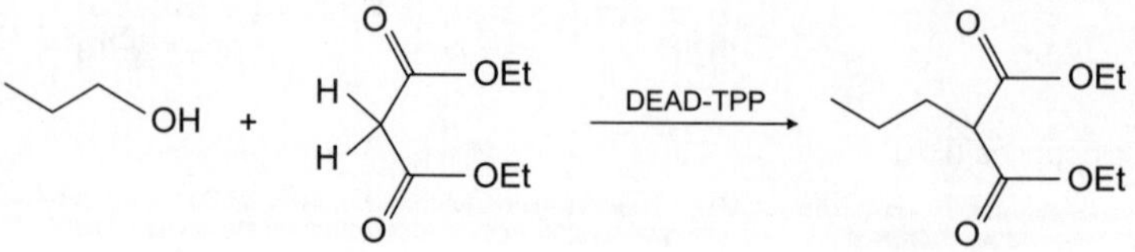

(pK$_a$ = 13.3)

The Mitsunobu reaction (Discovered in 1967) converts a hydroxyl group into a potent leaving group that is able to be displaced by a wide variety of nucleophiles. It has gained wide acceptance in organic synthesis due to its effectiveness and versatility.

$$ROH \xrightarrow[\text{DEAD, PPh}_3]{\text{NuH}} RNu$$

By far the largest use of the Mitsunobu reaction is the inversion of secondary alcohols. This is accomplished by the conversion of a secondary alcohol to an ester followed by reduction of the ester. Acetic and benzoic acids have typically been used for this procedure but provide poor results with hindered secondary alcohols.

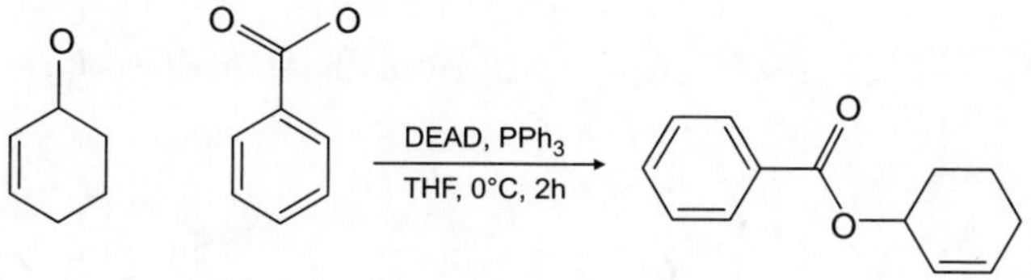

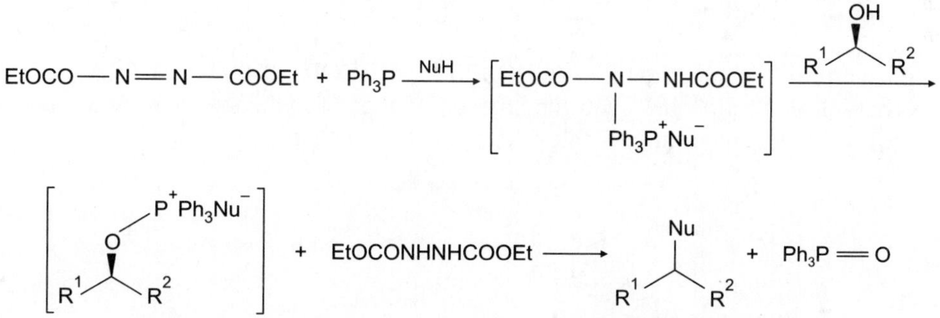

NuH = phosphoric mono- and diesters, carboxylic acids, phenols, imides, oximes, hydroxymates, heterocycles, thiols, thioamides, β-keto esters

264. MOORE MYERS CYCLIZATION; MOORE CYCLIZATION; MYERS CYCLIZATION

The synthesis of phenol (or quinone) derivatives via thermal generation of a aryl/phenoxy biradical by cyclization of enyne-ketenes is known as the Moore cyclization. the cylcization occurs C_2 and C_7 via an aryl/phenoxy biradical intermediate.

The similar process when proceed by cyclization of enyne-allenes is referred to as the Myers cyliczation.

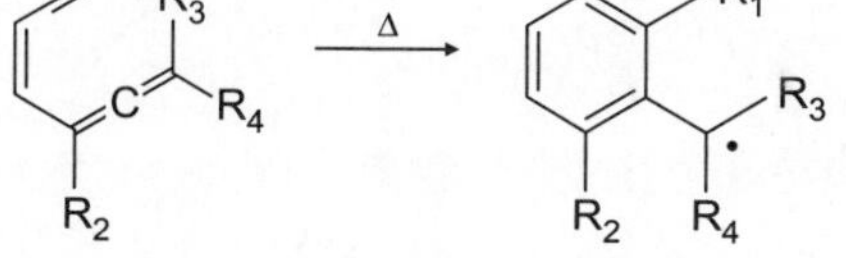

Myers cyclization

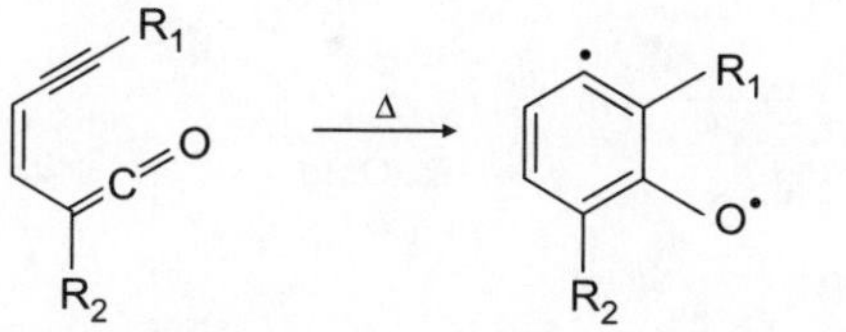

Moore cyclization

265. MUKAIYAMA ALDOL REACTION

The Lewis-acid catalyzed addition of silyl enol ethers to aldehydes is known as the Mukaiyama Aldol reaction. The reaction is a type of aldol reaction and used extensively in organic synthesis because it allows for a crossed aldol reaction between an aldehyde and a ketone or a different aldehyde. Some of the Lewis-acids used for the reaction are titanium tetrachloride, tin tetrachloride or boron trifluoride etherate. This reaction is highly sensitive to the solvent and to reactant concentrations.

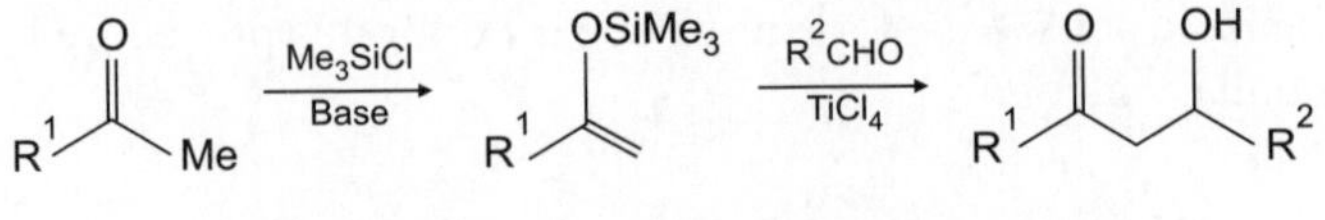

Lewis acid catalysis: 1, 2-addition, Mukaiyama-aldol

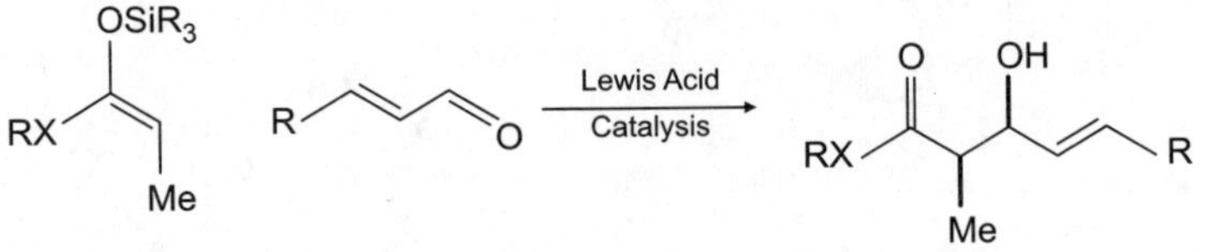

Reactivity order as an electrophile: RCHO (78°C) > RCOR′ (0°C) >> $RCO_2R′$

The archetypical reaction involved the silyl enol ether of cyclohexanone with benzaldehyde with one equivalent of titanium tetrachloride in dichloromethane.

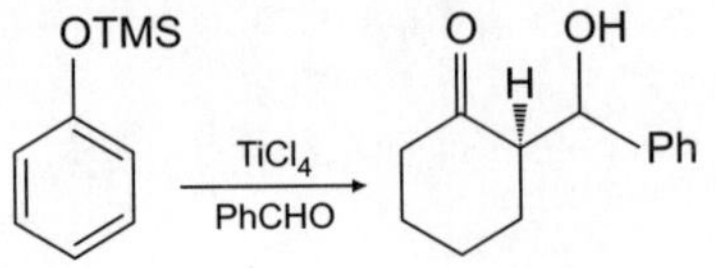

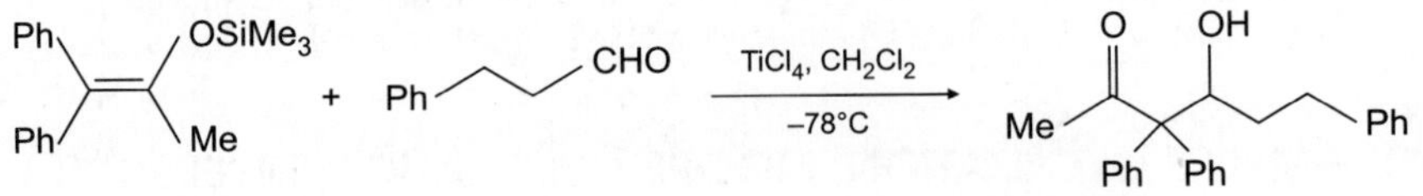

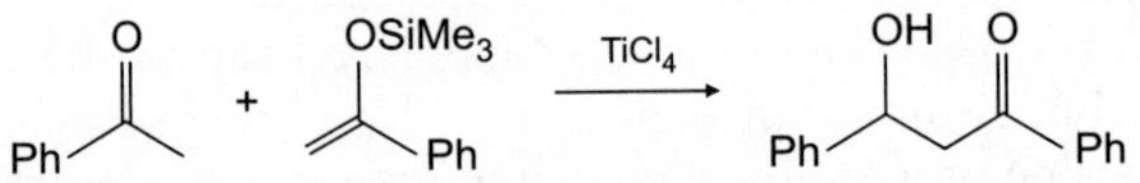

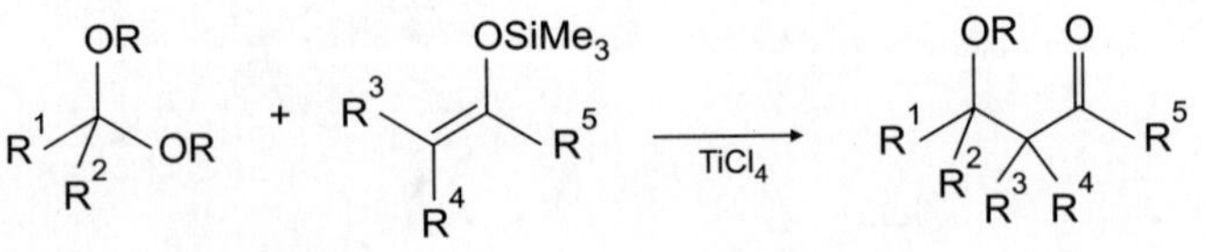

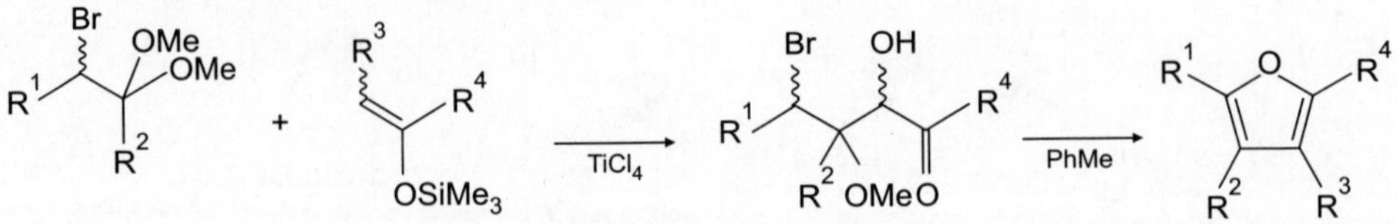

Mukaiyama aldol reaction has a number of advantages like mild condition, directed donor and acceptor involved in reaction and catalytic asymmetric synthesis, etc. However, the reaction has disadvantages of being that it is indirect and many factors decide the stereochemistry of the product.

266. MUKAIYAMA-MICHAEL REACTION

The conjugate addition of O-silylated ketene acetals to α,β-unsaturated ketones (the Mukaiyama-Michael reaction) is a well documented and important method for carbon-carbon bond formation. The utility of the reaction has been demonstrated by numerous applications in organic synthesis. The reaction has been performed at high temperatures or high pressures. However, a strong Lewis acid is normally used to catalyze the reaction under moderate conditions. Typical Lewis acids reported to be useful are $TiCl_4$, $TiCl_4/Ti(i\text{-}PrO)_4$, $SnCl_4$ SmI_2 and lanthanum triflates. Cobalt-bis-dicarbollide $[LiCo(B_9C_2H_{11})_2]$, $Mg(ClO_4)_2$ and $LiClO_4$ in diethyl ether have also been reported as useful catalysts for this reaction. Uncatalyzed Mukaiyama-Michael reactions (solvent assisted reaction) have also been reported in highly polar solvents such as acetonitrile, nitromethane and in DMSO.

In 1974, Mukaiyama and co-workers reported the first examples of Lewis acid-catalyzed Michael reactions between enolsilanes and α,β-unsaturated carbonyl acceptors. This reaction variant is an attractive alternative to the conventional metalloenolate process due to the mild reaction conditions and frequently superior regiocontrol (1,4- versus 1,2-addition).

Organocatalysis: 1,4-addition, Mukaiyama-Michael

Since its discovery, the Mukaiyama-Michael reaction has become a powerful chemical tool for carbon-carbon fragment couplings with the accompanying formation of vicinal carbon-sp^3 stereochemistry.

267. NAGATA HYDROCYANATION

In 1962, Nagata and co-workers developed two new hydrocyanation methods using organoaluminum compounds. One involving a combination of hydrogen cyanide and an alkylaluminum and the other one involves an alkylaluminum cyanide, both in an aprotic solvent. In other words, alkylaluminum-mediated 1,4-addition (conjugated addition) of hydrogen cyanide to α,β-unsaturated carbonyl compounds is generally referred to as the **Nagata hydrocyanation.**

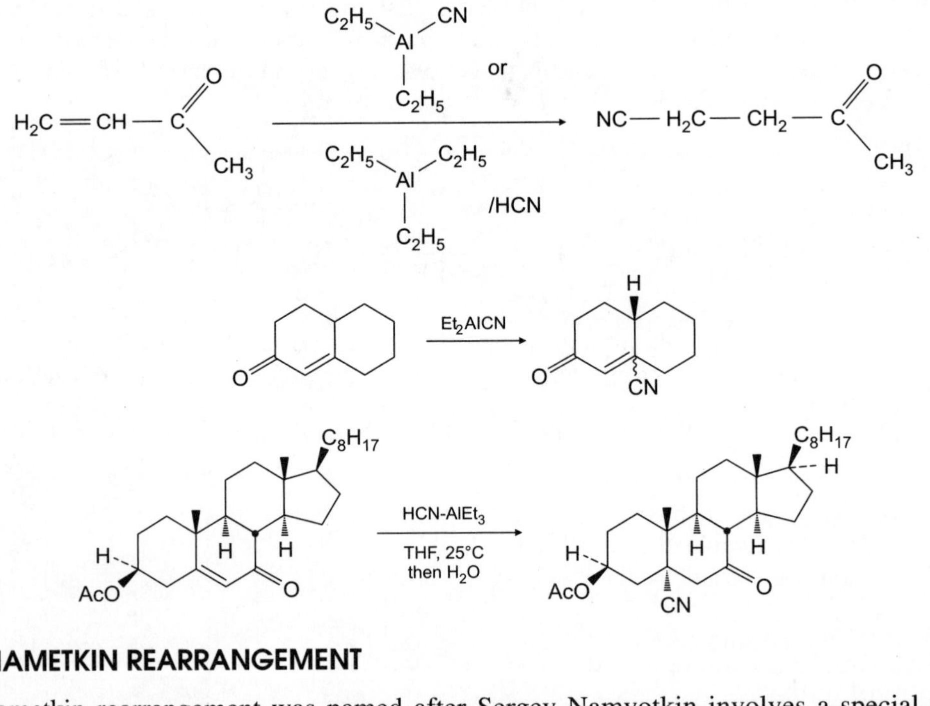

268. NAMETKIN REARRANGEMENT

The Nametkin rearrangement was named after Sergey Namyotkin involves a special case of carbonium ion rearrangement in camphene hydrochloride derivatives involving the migration of a methyl group e.g. rearrangement of methyl groups in certain terpenes. In some cases the reaction type is also called a retropinacol rearrangement. A "Nametkin rearrangement"can also be defined as a W-M double rearrangement" [Wagner-Meerwein rearrangement (W-M rearrangement)].

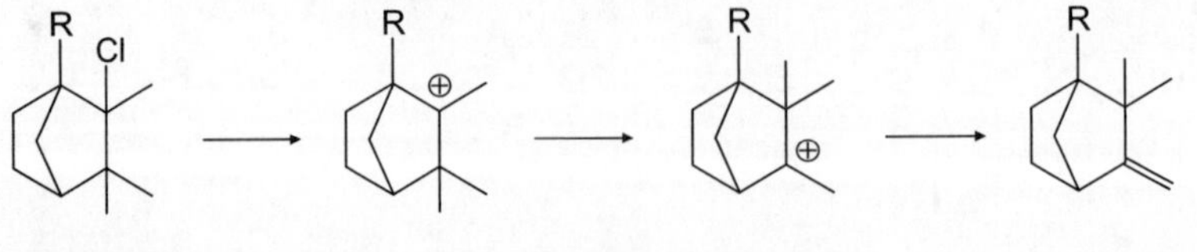

A successive Nametkin rearrangement takes place in the example given below where flower petals are formed by expansion of the cyclobutyl rings.

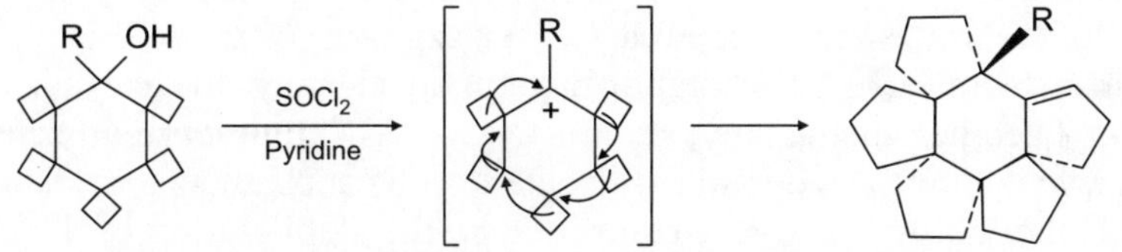

269. NAZAROV CYCLIZATION REACTION

The classical Nazarov reaction is acid (protic or Lewis) induced cationic 4π conrotatory electrocyclic ring closure reaction of divinyl ketone to form cyclopentenones. Thereby, this reaction allows the synthesis of cyclopentenones from divinyl ketones. The reaction was named after the eminent Russian chemist I. N. Nazarov in 1942.

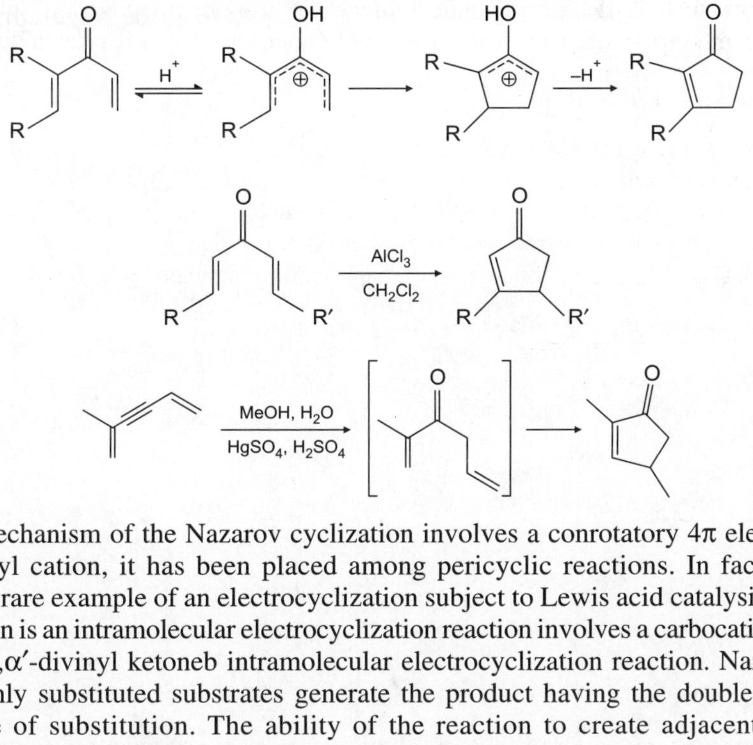

Since the mechanism of the Nazarov cyclization involves a conrotatory 4π electrocyclization of a pentadienyl cation, it has been placed among pericyclic reactions. In fact, the Nazarov cyclization is a rare example of an electrocyclization subject to Lewis acid catalysis. The classical Nazarov reaction is an intramolecular electrocyclization reaction involves a carbocation intermediate proceeds via α,α'-divinyl ketoneb intramolecular electrocyclization reaction. Nazarov reactions with more highly substituted substrates generate the product having the double bond with the highest degree of substitution. The ability of the reaction to create adjacent stereocenters stereospecifically, should make it a valuable synthetic tool, but various factors can complicate the efficiency of the reaction.

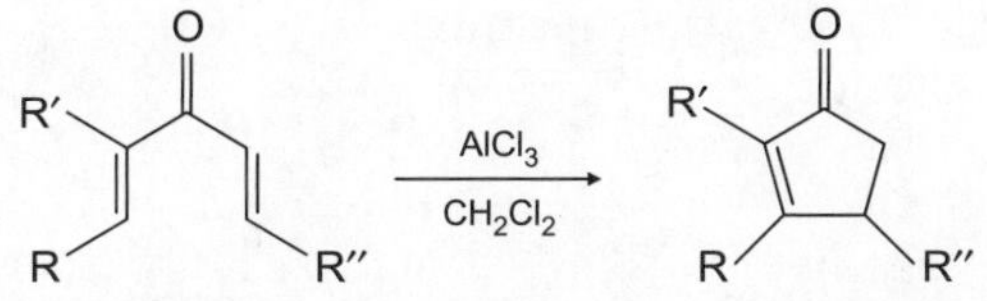

The Nazarov cyclization is one of the most versatile methods for the synthesis of five-membered carbocycles. It has been used in the construction of numerous complex target molecules, including polyquinane natural products and prostanoids. Modern variants are based on the interception of cationic intermediates or the use of highly reactive allene substrates, and even include examples of the reverse reaction.

A wide variety of substrates are susceptible to the Nazarov cyclization. Biscyclic, monocyclic, and acyclic precursors with varying substitution patterns are all amenable to cyclization. In addition to traditional divinyl ketones, allene vinyl ketones, aryl vinyl ketones, and allyl vinyl ketones can undergo Nazarov cyclizations. Heteroatom and heterocyclic substituents are also well tolerated.

Reaction rate is thought to be largely controlled by the stabilization of the two carbocation intermediates. It is thought that the electrocyclization is the slow step, so substituents that stabilize the pentadienyl intermediate are predicted to slow the reaction rate. Substituents that stabilize the oxallyl cation intermediate are believed to speed the reaction rate due to the lowering of the transition state energy as the reaction progresses from the pentadienyl cation intermediate to the oxallyl cation.

Traditionally, the reaction conditions of the Nazarov cyclization have been somewhat harsh. Usually one or more equivalent of a strong Lewis acid ($AlCl_3$, $BF_3 \cdot OEt_2$, $TiCl_4$) or Brønsted acid (HCl, H_2SO_4, H_3PO_4) is needed to promote the reaction. The cyclization can proceed in a wide variety of solvents, such as dichloromethane, toluene, THF, and methanol. Typically, reactions are conducted at room temperature or below, but it is not uncommon to see elevated temperatures.

270. NEBER REARRANGEMENT

The Neber rearrangement involves conversion of an oxime (ketoxime O-arylsulfonate) into an alpha-aminoketone in a rearrangement reaction. This rearrangement involves treatment of sulfonic esters of ketoximes with potassium ethoxide, followed by hydrolysis. This rearrangement is generally carried out in ethanol by treatment of the ketoxime tosylate with sodium ethoxide, followed by the acidic hydrolysis. This reaction has general application in the preparation of α-amino ketones.

271. NEF REACTION

The Nef reaction is the acid hydrolysis of a salt of a primary or secondary nitroalkane to yield an aldehyde or a ketone, respectively or corresponding carbonyl compound, and nitrous oxide.

$$\left[\begin{array}{c} \underset{R}{\overset{R}{\diagdown}} \overset{}{\underset{}{C}} - \overset{+}{N} \overset{O}{\diagdown}_{O^-} \quad M^+ \end{array} \longleftrightarrow \begin{array}{c} \underset{R}{\overset{R}{\diagdown}} C = \overset{+}{N} \overset{O^-}{\diagdown}_{O^-} \quad M^+ \end{array} \longleftrightarrow \begin{array}{c} \underset{R}{\overset{R}{\diagdown}} C = \overset{+}{N} \overset{O^-}{\diagdown}_{O} \quad M^+ \end{array} \right]$$

$$\xrightarrow[H_2O]{H^+} \quad \underset{R}{\overset{R}{\diagdown}} C = O \; + \; N_2O \; + \; M^+$$

R and R′ are hydrogen, alkyl, or aryl; M^+ is a cation.

This reaction was reported by the Swiss chemist J. U. Nef in 1894 who treated the sodium salt of nitroethane with sulfuric acid resulting in nitrous oxide and acetaldehyde. Hydrochloric and sulfuric acid give the same result. The Nef reaction is one of the better examples of "umpolung" reactivity in which the original nitro compound anion functions as an acyl anion equivalent. The Nef reaction should not be confused with the Nef synthesis.

272. NEF SYNTHESIS

In Nef synthesis, acetylenic carbinols are obtained by the addition of sodium acetylides are with aldehydes or ketones. This reaction is often erroneously referred to as the Nef reaction.

$$O = C \underset{R_2}{\overset{R_1}{\diagdown}} \quad \xrightarrow[\text{2. } H_2O/H^+]{\text{1. Na} - C \equiv CH} \quad HO - \underset{R_2}{\overset{R_1}{\underset{|}{\overset{|}{C}}}} - C \equiv CH$$

273. NEGISHI CROSS COUPLING

The reaction is named after Ei-ichi Negishi who received the 2010 nobel prize in chemistry for the discovery and development of this reaction. The Negishi coupling is the nickel- or palladium-catalyzed cross coupling reaction of organometallic species containing zinc or zirconium with organoelectrophiles. The reaction involves an organozinc compound, an organic halide and a nickel or palladium catalyst creating a new carbon-carbon covalent bond.

The Negishi coupling allows the preparation of unsymmetrical biaryls in good yields. The versatile nickel- or palladium-catalyzed coupling of organozinc compounds with various halides (aryl, vinyl, benzyl, or allyl) has broad scope, and is not restricted to the formation of biaryls.

$$R - X \; + \; R' - Zn - X' \xrightarrow{ML_n} R - R'$$

The halide X can be chloride, bromine or iodine but also a triflate or acetyloxy group with as the organic residue R alkenyl, aryl, allyl, alkynyl or propargyl.

The halide X′ in the organozinc compound can be chloride, bromine or iodine and the organic residue R′ is alkenyl, aryl, allyl or alkyl.

The metal M in the catalyst is nickel or palladium.

The ligand L in the catalyst can be triphenylphosphine, dppe, BINAP or chiraphos.

Palladium catalysts, in general, have higher chemical yields and higher functional group tolerance.

Synthesis of alkenes by palladium-catalyzed Negishi coupling:

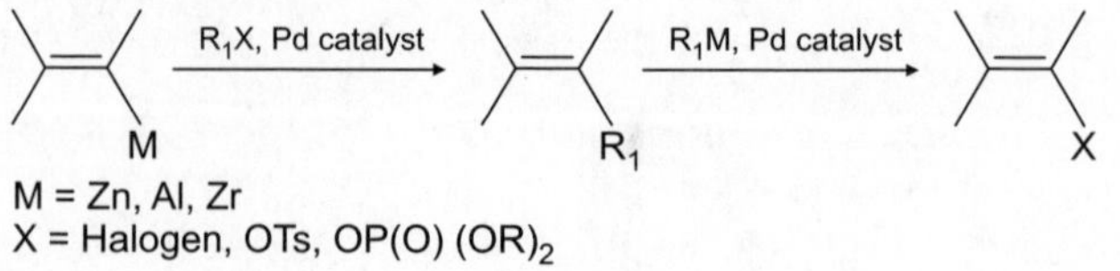

M = Zn, Al, Zr
X = Halogen, OTs, OP(O) (OR)$_2$

There are quite a few examples of heterocylic substrates are also known, mainly nitrogen containing that participate in the Negishi reaction.

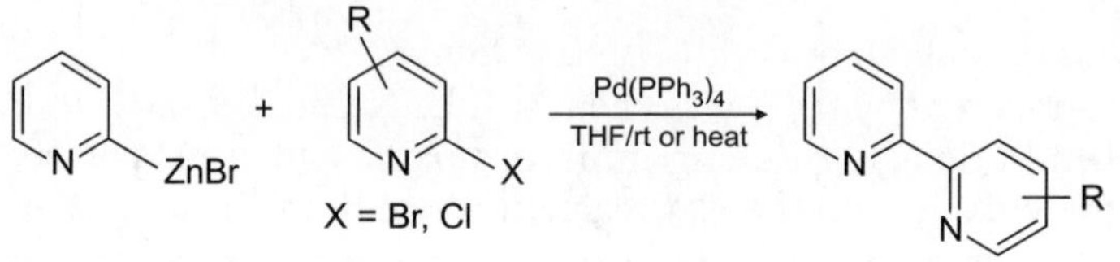

X = Br, Cl

274. NENCKI REACTION

Nencki reaction describes preparation of an acyl phenol derivative by the ring acylation of phenols with acids in the presence of zinc chloride. Acyl chloride, anhydride, or carboxylic acid can be employed for acylation in the reaction.

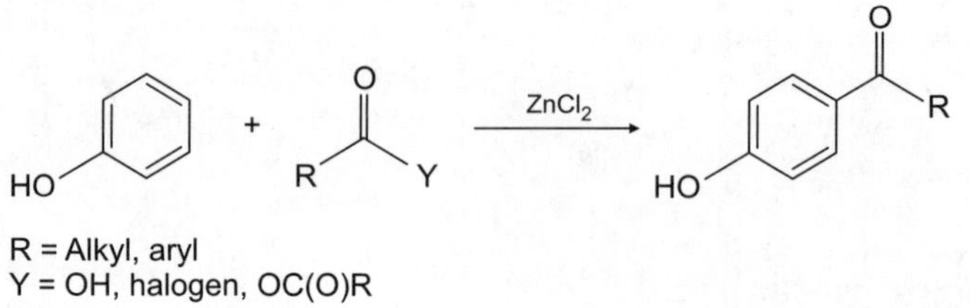

R = Alkyl, aryl
Y = OH, halogen, OC(O)R

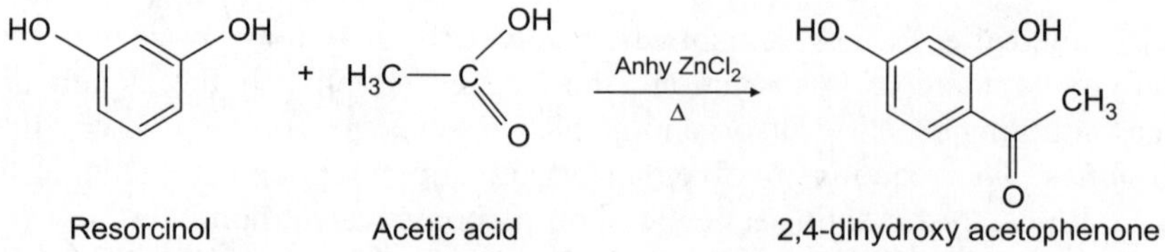

| Resorcinol | Acetic acid | 2,4-dihydroxy acetophenone |

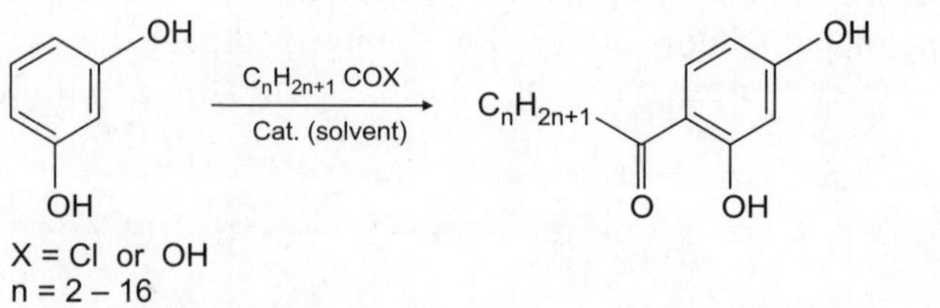

X = Cl or OH
n = 2 – 16

275. NENITZESCU INDOLE SYNTHESIS

The synthesis of a 5-hydroxyindole derivative involving the condensation between a 1,4-benzoquinone and α-amino-α,β-unsaturated compound and subsequent cyclization is generally known as the Nenitzescu indole synthesis.

Synthesis of 5-hydroxyindole derivatives by condensation of p-benzoquinone with β-amino-crotonic esters.

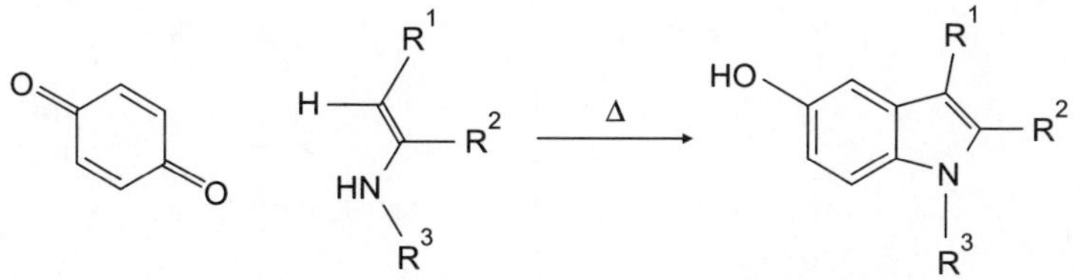

The combination of the benzoquinone with the resin-bound enamine gave, after release from the resin, the indole.

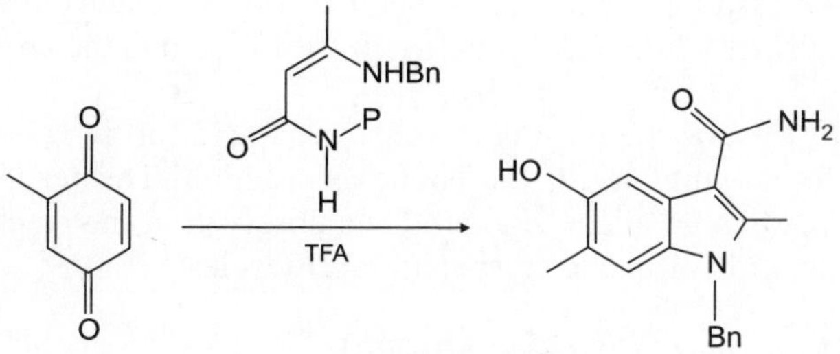

This reaction is particularly useful for the preparation of 5-hydroxyindole derivatives.

276. NENITZESCU REDUCTIVE ACYLATION

The Nenitzescu reductive acylation (or Darzens-Nenitzescu synthesis of ketones or Darzens-Nenitzescu reaction or Nenitzescu acylation) involves the hydrogenative acylation of cycloolefins with acid chlorides, e.g. acylation of cyclohexene with acetyl chloride to methylcyclohexenyl-ketone.

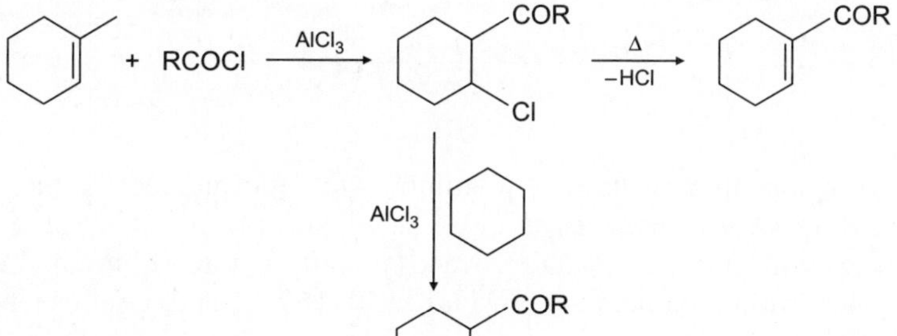

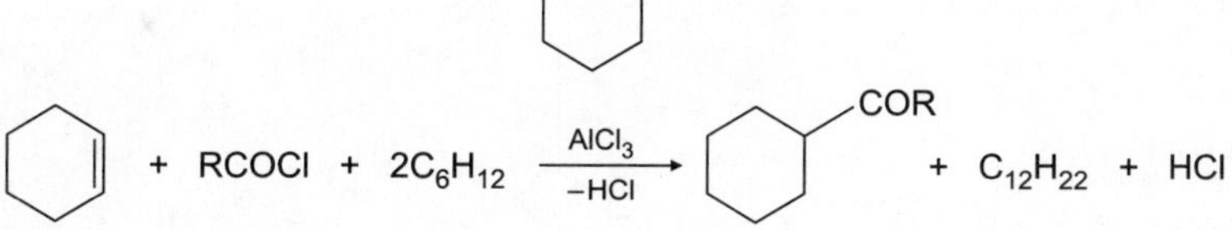

277. NICHOLAS REACTION

The Nicholas reaction, or chemistry of hexacarbonyl (m-propargylium) cobalt cations, has been known since 1972. The reaction of cobalt complexed propargylic alcohols with HBF_4 provides a cobalt-stabilized carbocation that can be treated with a variety of carbon nucleophiles to provide alkylated products or, in other words, the Nicholas reaction is the reaction of dicobalthexacarbonyl-stabilized propargyl cations with nucleophiles, followed by oxidative demetalation to yield propargylated products.

The Nicholas reaction is a highly useful inter- or intramolecular propargylic substitution reaction. Intermolecular reactions with a variety of nucleophiles are well known and the most common intramolecular variations involve exocyclic cyclizations with carbon nucleophiles. Intramolecular Nicholas reactions are classified as exocyclic when the cobalt-complexed alkyne ends up outside of the newly generated ring, while endocyclic cyclizations include the cobalt-alkyne complex in the newly formed ring.

The Lewis acid $BF_3 \cdot OEt_2$ is the most commonly used acid for this reaction. C, O, N, and S-nucleophiles are all found suitable for the Nicholas reaction. The reaction can be conducted intramolecularly and there is no allene side products observed in the reaction. This reaction has been used for the preparation of cyclic ethers and macrocycles.

278. NIEMENTOWSKI QUINAZOLINE SYNTHESIS

The so-called Niementowski synthesis involves preparation of fused pyrimidines from the anthranilic acid derivatives. For first time, this type of synthesis was described by Niementowski in 1895. He obtained quinazolin-4(3H)-one by fusion of anthranilic acid with formamide. In general, the Niementowski quinazoline synthesis is the cyclization reaction of anthranilic acids with amides to form 4-oxo-3,4-dihydroquinazolines.

The Niementowski reaction is the most common synthetic method for the synthesis of 3H-quinazolin-4-one ring which involves the fusion of anthranilic acid (or a derivative, e.g. 2-aminobenzonitrile) with formamide and proceeds usually via an o-amidine intermediate. This procedure usually needs high temperatures (usually 130–150°C) and requires lengthy and tedious (average time 6 hrs) conditions.

279. NIEMENTOWSKI QUINOLINE SYNTHESIS

The Niementowski quinoline synthesis describes the chemical reaction of anthranilic acids and carbonyl compounds (ketones or aldehydes) to form γ-hydroxyquinoline derivatives.

280. NIERENSTEIN REACTION

The synthesis of haloketones by treatment of an aliphatic or aromatic acyl chloride with diazomethane is known as the Nierenstein reaction.

$$ArCOCl + CH_2NH_2 \longrightarrow ArCOCH_2Cl + N_2$$

It is an insertion reaction in that the methylene from the diazomethane is inserted into the carbon-chlorine bond of the acid chloride.

281. NORRISH TYPE CLEAVAGE

Carbonyl compounds when irradiated can undergo cleavage of the carbonyl C-α bond. In the gas phase, the radicals are generated and detected (longer lifetime). In solution, the radicals undergo further reactions to give products.

Hydrogen Abstraction. The radicals can abstract a hydrogen atom from a donor. The resulting radicals can then undergo further reactions (e.g. dimerization, photopinacolization).

Intramolecular hydrogen abstraction followed by cleavage = *Norrish type II cleavage*.

The radicals can abstract a hydrogen atom from a donor.

The resulting radicals can then undergo further reactions.

Norrish type I reaction (α-cleavage reaction)

The Norrish type I reaction is the photochemical cleavage or homolysis of aldehydes and ketones into two free radical intermediates. In this type of cleavage, the carbonyl group is excited by the absorption of a photon to a photochemical singlet state and the homolysis of the (C=O)–R bond may result. α-cleavage of an excited carbonyl compound leading to two radical fragments evolve according to their intrinsic reactivity, an acyl-alkyl radical pair (from an acyclic carbonyl compound) or an acyl-alkyl diradical (from a cyclic carbonyl compound) as a primary photoproduct. Several secondary reaction modes are now open to these fragments depending on the exact molecular structure.

Decarbonylation of the primary diradical (by extrusion of carbon monoxide) and subsequent recombination of the biradical afford a new carbon carbon bond.

The fragments can simply recombine to the original carbonyl compound

When the carbon fragment has an α-proton available, it gets abstracted forming a ketene and a saturated hydrocarbon in path C.

When the alkyl fragment contains a β-proton, it gets abstracted with formation of an aldehyde and an alkene. Such homolytic cleavage of aldehydes and ketones originate from their excited nπ* state. This process is described as Norrish type I reaction. This reaction is synthetically useful for the ring cleavage of cyclic ketones.

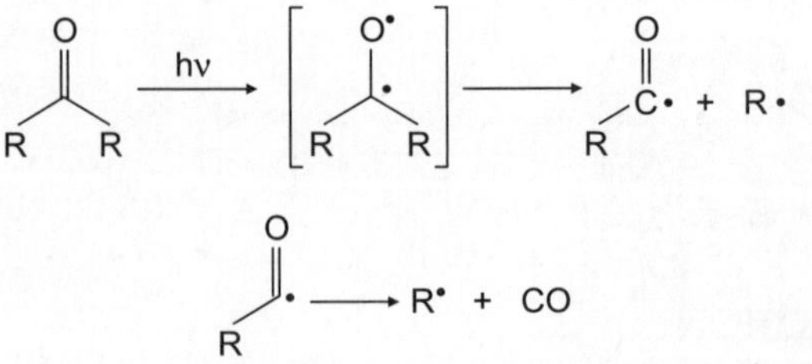

The three products of the below reaction scheme show that the initial pair of radicals formed can combine to give a photoaddition product, or the radical coming from the ketone can dimerize, or abstract another hydrogen atom.

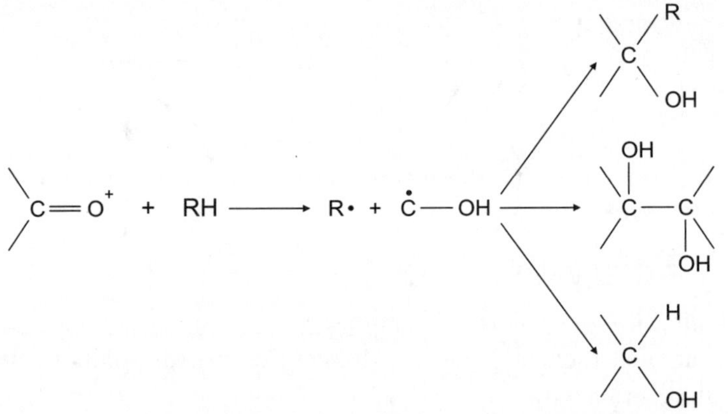

Several secondary reaction modes are open to these fragments depending on the exact molecular structure.

The fragments can simply recombine to the original carbonyl compound (path A).

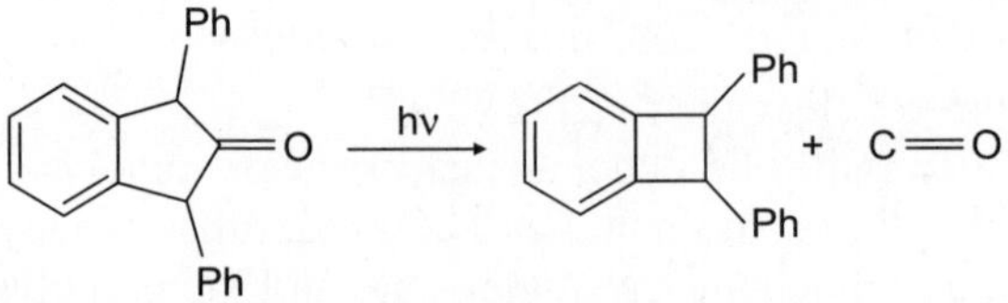

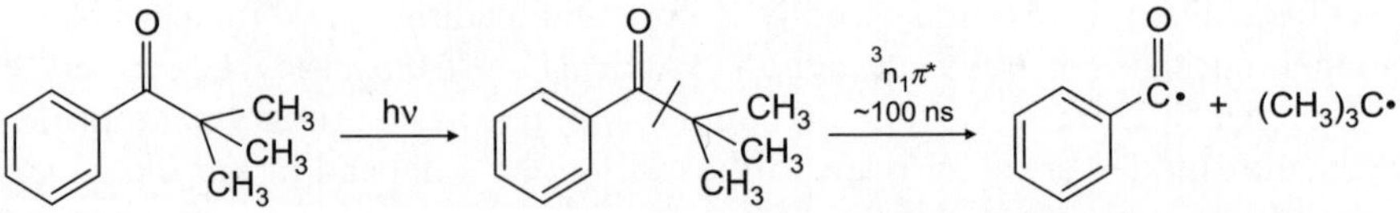

Norrish type II photoelimination reaction (Norrish Type II cleavage):

Reaction originating from the $n\pi^*$ excited state of aldehydes and ketones that involves intramolecular γ-hydrogen abstraction followed by cleavage of the resulting diradical to an olefin and an enol which tautomerizes to the carbonyl compound.

The excitation of an organic chromophore can lead to the formation of a highly reactive diradical species. Among all its possible reaction pathways, hydrogen abstraction in the γ- position is quite common, and was identified very early on by Norrish. This process is now commonly described for carbonyl compounds as a Norrish type II reaction.

Beside carbon monoxide extrusion acyl radicals formed in a α-cleavage can be stabilized by subsequent hydrogen migration.

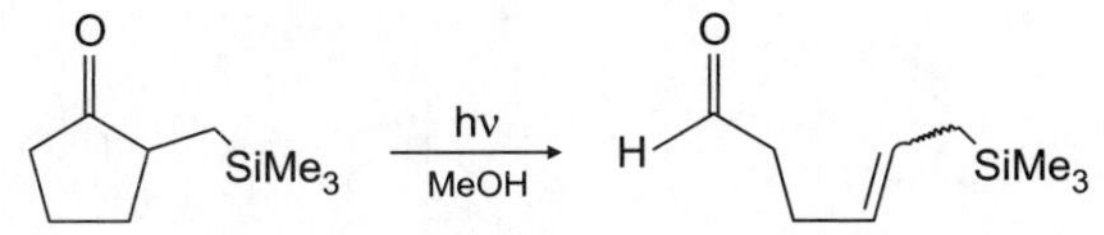

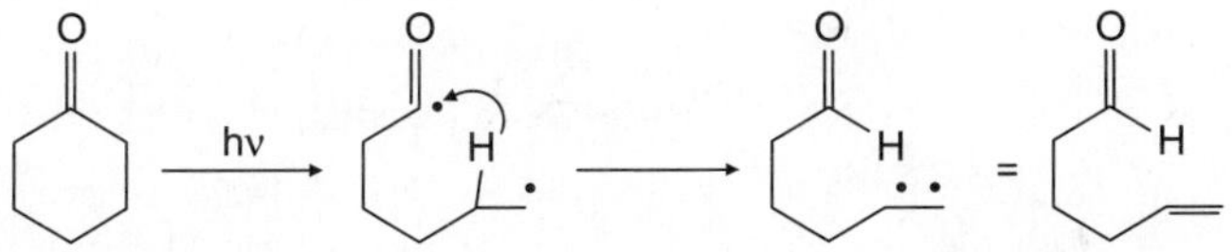

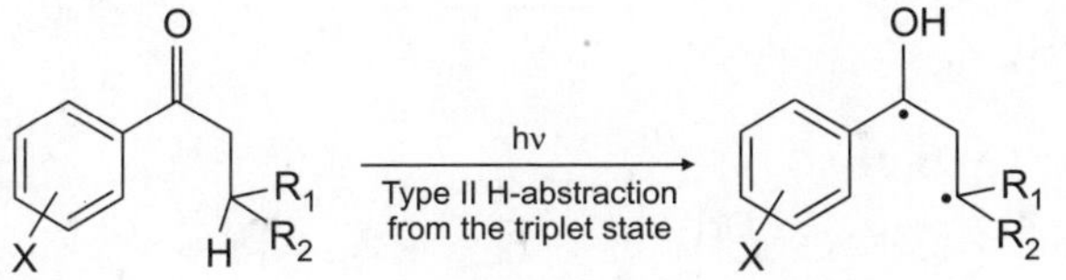

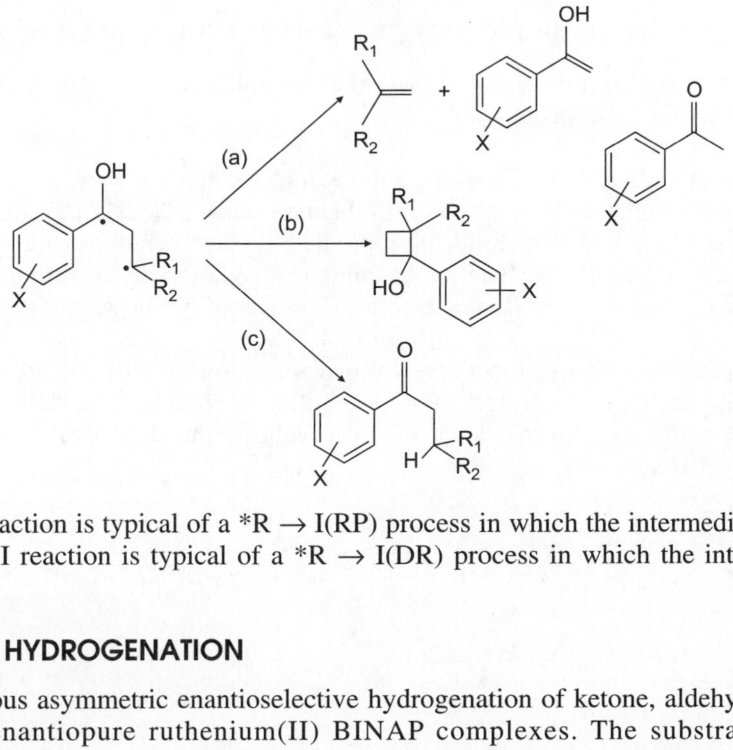

The type I reaction is typical of a *R → I(RP) process in which the intermediate I is a radical pair. The type II reaction is typical of a *R → I(DR) process in which the intermediate I is a diradical.

282. NOYORI HYDROGENATION

The homogeneous asymmetric enantioselective hydrogenation of ketone, aldehydes, and imines mediated by enantiopure ruthenium(II) BINAP complexes. The substrates must have coordinating functionalities in neighbouring positions which serve as directing groups during the transformation:

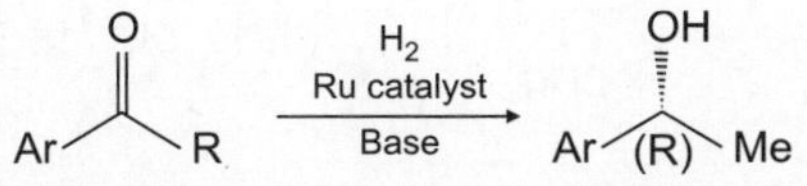

BINAP-Ru catalyst is used for the asymmetric hydrogenation of functionalized ketones and BINAP/diamine-Ru catalyst is used for the asymmetric hydrogenation of simple ketones.

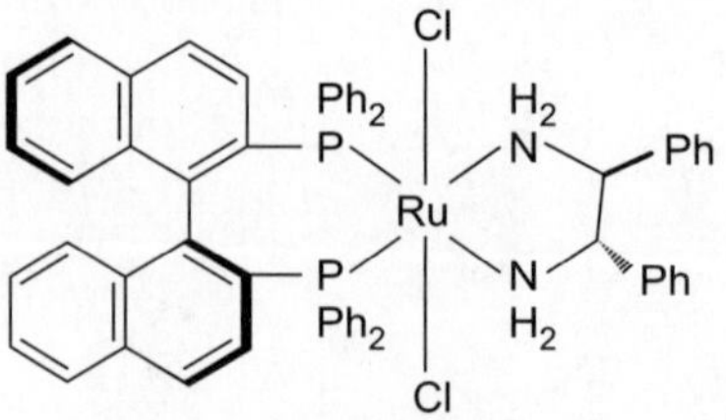

(S)-BINAP/(S, S)-DPEN-Ru(II) catalyst

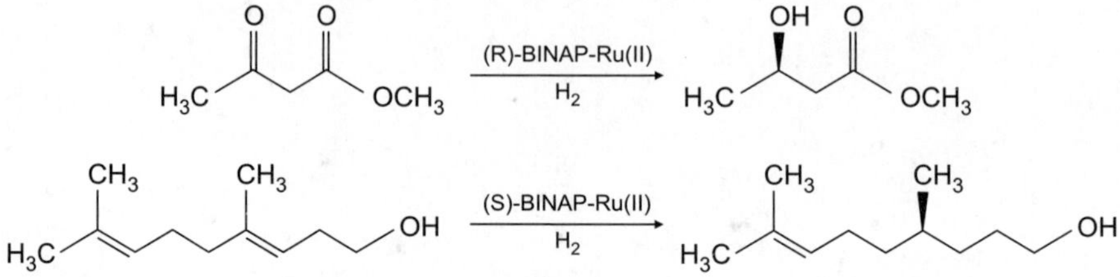

283. NOZAKI-HIYAMA COUPLING REACTION (NOZAKI-HIYAMA-KISHI REACTION)

Organochromium-mediated allyl and vinyl additions to aldehydes are commonly known as the Nozaki-Hiyama-Kishi (NHK) reaction.

Or

The chromium (Cr(II), Cr(III), Cr(VI)) catalyzed redox additions between an organic halide and a carbonyl compound usually aldehyde, is often known as the Nozaki-Hiyama-Kishi reaction.

These transformations are very mild, give predictable stereochemical outcomes and show exceptional chemoselectivity. This reaction has high chemoselectivity toward aldehydes but use of excess toxic chromium salts is disadvantageous. This reaction is useful for the synthesis of the natural products.

The NHK reaction is tolerant of functional group diversity on both the organochromium species as well as on the aldehyde, resulting in its wide utility in complex natural product synthesis. An intramolecular version of this reaction to afford a cyclic product is generally called the Nozaki-Hiyama-Kishi cyclization.

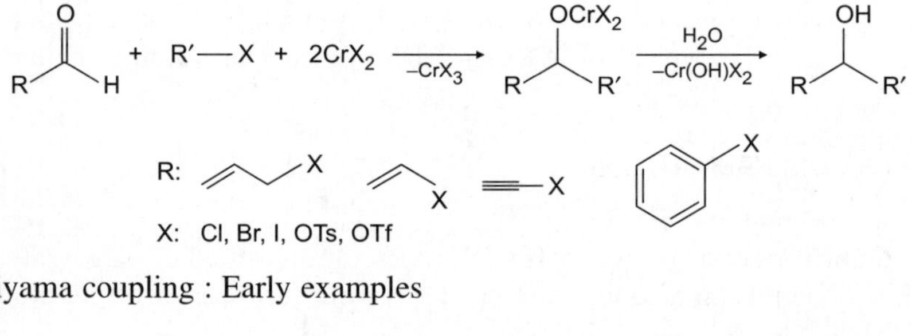

X: Cl, Br, I, OTs, OTf

Nozaki-Hiyama coupling : Early examples

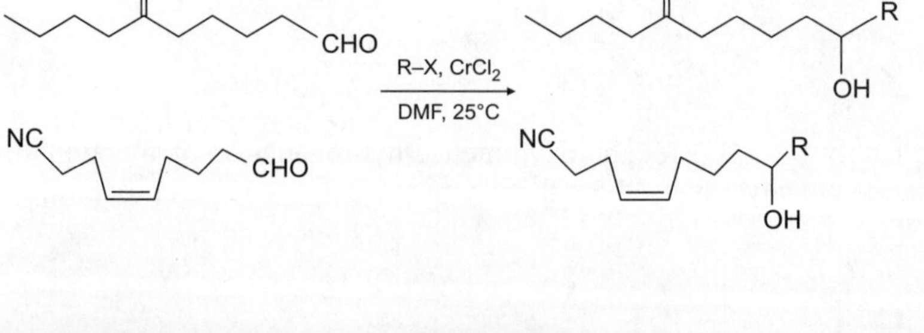

284. OLEFIN METATHESIS

This reaction was discovered serendipitously by Banks and Bailey, when they were seeking an effective heterogeneous catalyst to replace the HF acid catalyst for converting olefins into high-octane gasoline via olefin–isoparaffin alkylation.

Olefin metathesis ('metathesis' from the Greek meaning 'change of position, transposition') reorganizes the carbon atoms of two C=C bonds (olefins or alkenes), generating two new ones; it promotes unique skeletal rearrangements.

Olefin metathesis allows the exchange of substituents between different olefins, i.e. a transalkylidenation. Some useful procedures include ring closure between terminal vinyl groups, cross metathesis—the intermolecular reaction of terminal vinyl groups and ring opening of strained alkenes.

Ring-closing metathesis

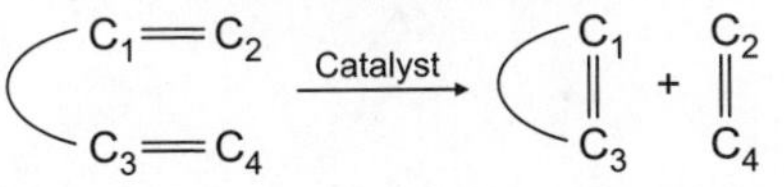

$$C_1=C_2 \quad \xrightarrow{\text{Catalyst}} \quad C_1 \parallel C_3 \quad + \quad C_2 \parallel C_4$$

Ring-opening metathesis

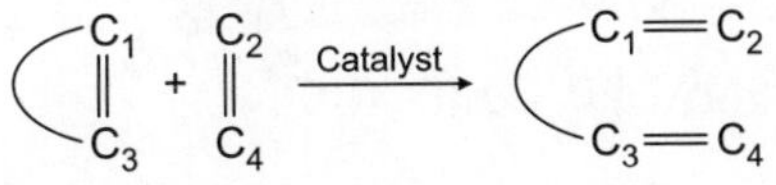

$$C_1 \parallel C_3 \quad + \quad C_2 \parallel C_4 \quad \xrightarrow{\text{Catalyst}} \quad C_1=C_2$$
$$C_3=C_4$$

Cross metathesis

$$C_1=C_2 \quad \xrightarrow{\text{Catalyst}} \quad C_1 \parallel C_3 \quad + \quad C_2 \parallel C_4$$
$$C_3=C_4$$

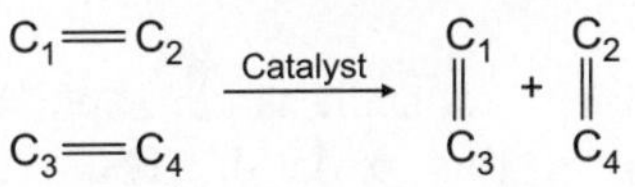

Acyclic Diene Metathesis Polymerization (ADMET)

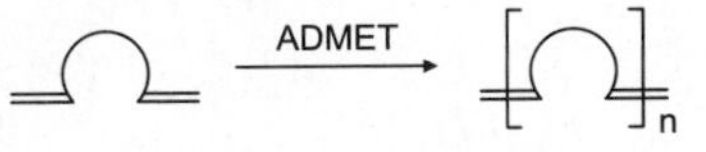

$$\xrightarrow{\text{ADMET}}$$

Catalysts for this reaction:

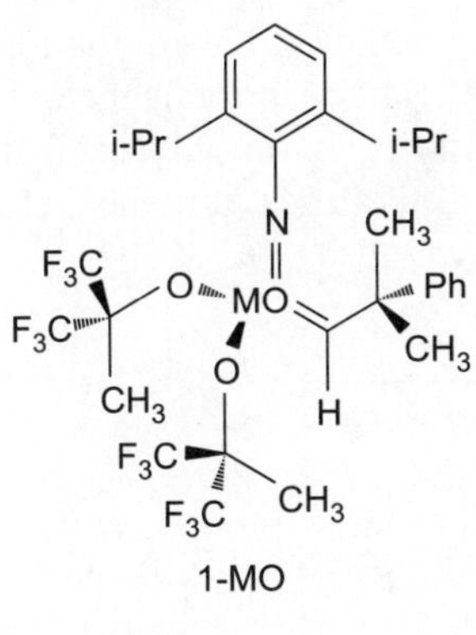

1-MO

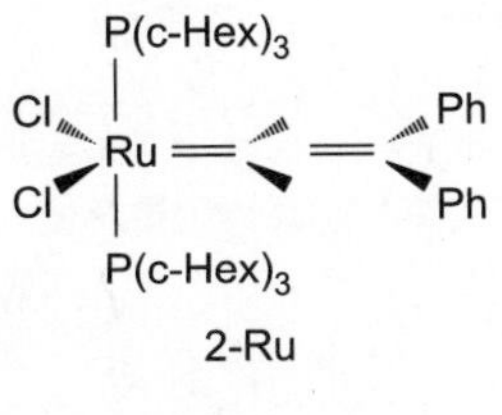

2-Ru

191

3-Ru
(Grubbs'1st generation catalyst)

4-Ru
(Grubbs'2nd generation catalyst)

- The well-defined catalysts shown above have been used widely for the olefin metathesis reaction. Titanium- and tungsten-based catalysts have also been developed but are less used.
- Schrock's alkoxy imidomolybdenum complex 1-Mo is highly reactive toward a broad range of substrates; however, this Mo-based catalyst has moderate to poor functional group tolerance, high sensitivity to air, moisture or even to trace impurities present in solvents, and exhibits thermal instability.
- Grubbs' Ru-based catalysts exhibit high reactivity in a variety of ROMP, RCM, and CM processes and show remarkable tolerance toward many different organic functional groups.
- The electron-rich tricyclohexyl phosphine ligands of the d^6 Ru(II) metal center in alkylidenes 2-Ru and 3-Ru leads to increased metathesis activity. The NHC ligand in 4-Ru is a strong - donor and a poor -acceptor and stabilizes a 14 e– Ru intermediate in the catalytic cycle, making this catalyst more effective than 2-Ru or 3-Ru.
- Ru-based catalysts show little sensitivity to air, moisture or minor impurities in solvents. These catalysts can be conveniently stored in the air for several weeks without decomposition. All of the catalysts above are commerically available, but 1-Mo is significantly more expensive.

285. OPPENAUER OXIDATION

Oppenauer oxidation describes aluminum or potassium alkoxide-catalyzed oxidation of a secondary alcohol to the corresponding ketone which involves hydride shift from the α-carbon of an alcohol component to the carbonyl carbon of a second component and proceeds over a six-membered transition state.

Non-enolizable ketones with a relatively low reduction potential, such as benzophenone, can serve as the carbonyl component used as the hydride acceptor in this oxidation. The oxidation is highly selective for secondary alcohols and does not oxidize other sensitive functional groups such as amines and sulfides.

Oppenauer oxidation, named after Rupert Viktor Oppenauer. This oxidation is a gentle method for selectively oxidizing secondary alcohols to ketones. The reaction is the opposite of Meerwein-Ponndorf-Verley reduction. The Oppenanuer oxidation is commonly used in various industrial processes such as the synthesis of steroids, hormones, alkaloids, terpenes, etc.

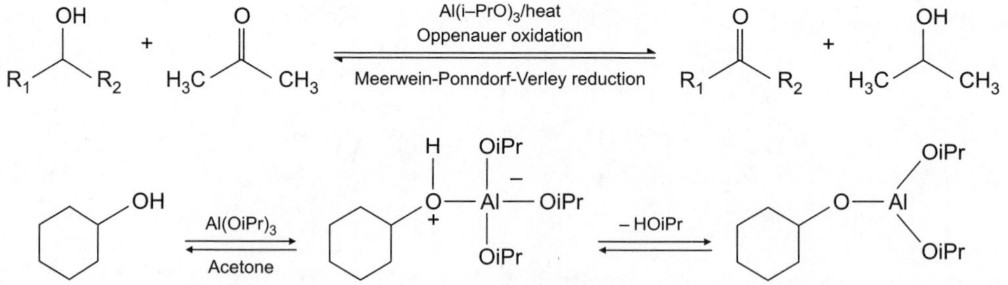

Six-membered transition state in Oppenauer oxidations,

286. OVERMAN REARRANGEMENT

Rearrangement of allylic imidates was discovered in 1937 by Mumm and Moller which required harsh reaction conditions, hence not useful synthetically. In 1974, Overman discovered that trichloroacetimidates rearranged thermally or catalyzed by a transition metal under more mild reaction conditions.

The aza-Claisen rearrangement of allylic trichloroacetimidates to allylic trichloroacetamides (Overman rearrangement) is a powerful and attractive strategy for the synthesis of allylic amines from readily available allylic alcohols. This transformation can be conducted thermally at high temperatures or by transition metal catalysis under very mild conditions.

Allylic alcohol can be 1°, 2° or 3°.

- Z = CCl_3, aryl; X = H, alkyl, aryl
- Base: NaH, NaOR, etc
- Addition of weak base (e.g. K_2CO_3) can improve reproducibility in the reaction, especially when scaling up.

287. OXO PROCESS (HYDROFORMYLATION REACTION)

Hydroformylation, also known as oxo synthesis or oxo process, is an important homogeneously catalyzed industrial process for the production of aldehydes from alkenes. This chemical reaction entails the addition of a formyl group (CHO) and a hydrogen atom to a carbon-carbon double bond. This process has undergone continuous growth since its invention in 1938; production capacity reached 6.6×106 tons in 1995. It is important because the resulting aldehydes are easily converted into many secondary products. For example, the resulting aldehydes are hydrogenated to alcohols that are converted to plasticizers or detergents. Hydroformylation is also used in specialty chemicals, relevant to the organic synthesis of fragrances and natural products. The development of hydroformylation, which originated within the German coal-based industry, is considered one of the premier achievements of 20th-century industrial chemistry.

The process typically is accomplished by treatment of an alkene with high pressures (between 10 to 100 atmospheres) of carbon monoxide and hydrogen at temperatures between 40 and 200°C. Transition metal catalysts are required.

Hydroformylation was discovered by Otto Roelen in 1938 during an investigation of the origin of oxygenated products occurring in cobalt catalyzed Fischer-Tropsch reactions. Roelen's observation that ethylene, H_2 and CO were converted into propanal, and at higher pressures, diethyl ketone, marked the beginning of hydroformylation.

Cobalt catalysts completely dominated industrial hydroformylation until the early 1970's when rhodium catalysts were commercialized. In 2004, ~75% of all hydroformylation processes are based on rhodium triarylphosphine catalysts, which excel with C8 or lower alkenes and where high regioselectivity to linear aldehydes is critical.

Most aldehydes produced are hydrogenated to alcohols or oxidized to carboxylic acids. Esterfication of the alcohols with phthalic anhydride produces dialkyl phthalate plasticizers that are primarily used for polyvinyl chloride plastics – the largest single end-use. Detergents and surfactants make up the next largest category, followed by solvents, lubricants and chemical intermediates.

288. PAAL-KNORR PYRROLE SYNTHESIS

The Paal-Knorr synthesis is a reaction where 1,4-diketones are converted to either furans, thiophenes or pyrroles. The Paal-Knorr pyrrole synthesis refers to the formation of pyrroles by condensation of 1,4-dicarbonyl compounds with an excess of ammonia or primary amines.

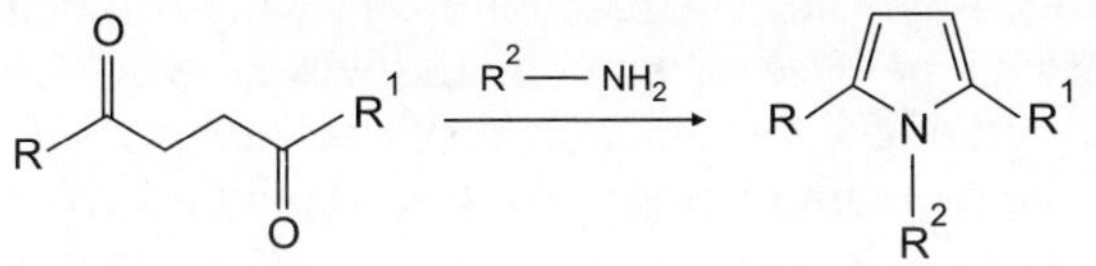

Therefore, the 2,4-disubstituted or 1,2,4-trisubstituted pyrroles can be synthesized by the Paal-Knorr Pyrrole synthesis as given below:

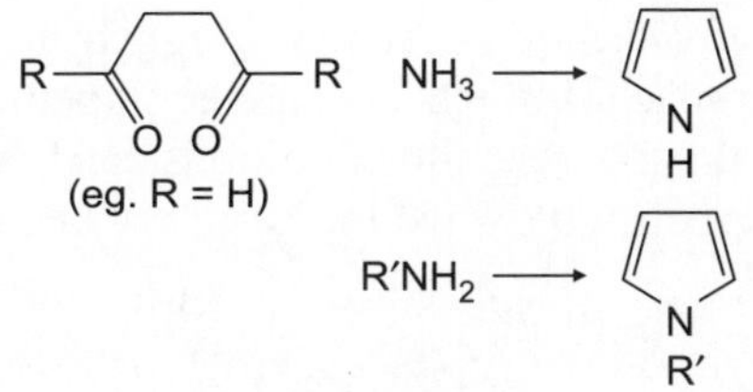

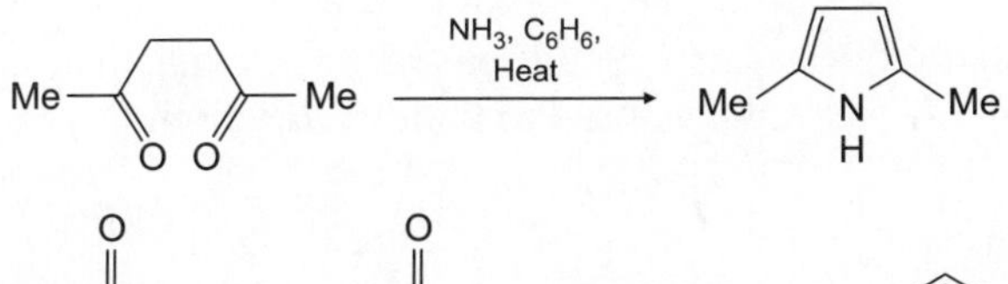

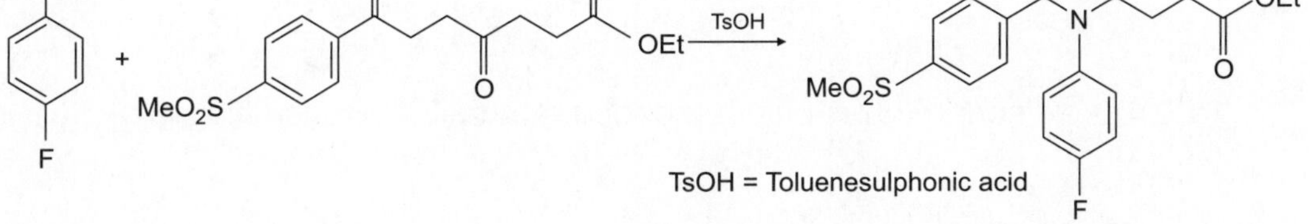

(eg. R = H)

TsOH = Toluenesulphonic acid

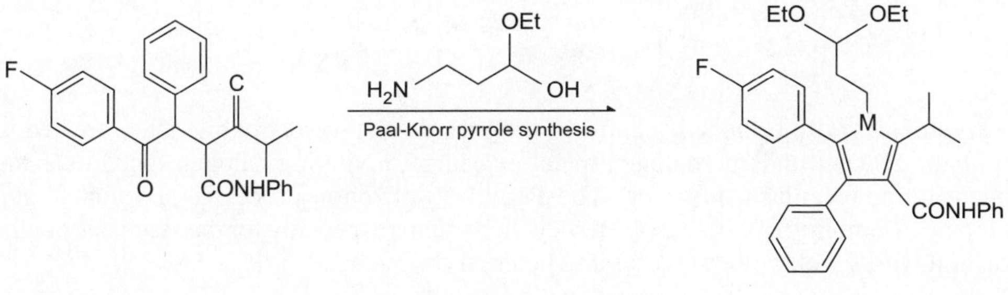

Paal-Knorr pyrrole synthesis

Synthesis of atorvastatin also involves the Paal-Knorr pyrrole synthesis.

This reaction was reported independently by Carl Paal and Ludwig Knorr in 1884 as a method for the synthesis of furans and soon be adapted for pyrroles and thiophenes. Reaction proceeds via formation of a 2,5-dihydroxytetrahydropyrrole derivative which undergoes dehydration to give the corresponding substituted pyrrole. The reaction is typically run under protic or Lewis acidic conditions, with a primary amine. Use of ammonium hydroxide or ammonium acetate gives the N-unsubstituted pyrrole. although, the reaction can also be carried out in neutral or weakly acidic conditions. Use of ammonium hydroxide or ammonium acetate gives the N-unsubstituted pyrrole.

Traditionally, the synthetic applications of this reaction were limited by the availability of the appropriate 1,4-diketone as synthetic precursors. However, 1,4-Diketones can now be accessible easily, e.g. by the Nef reaction. Apart from this, variations of the Paal-Knorr now allow for different precursors to be used. Additionally, newer modifications (such as Microwave-assisted Paal-Knorr reactions) of the Paal-Knorr approaches to pyrroles are equipped with milder reaction conditions which avoid the other traditional harsh reaction conditions, such as prolonged heating in acid for cyclization. With the application of microwave the can now be carried out at room temperature within even few minutes.

289. PARHAM CYCLIZATION

Aromatic lithiation, usually carried out by lithium-halogen exchange, and subsequent reaction with an internal electrophile is known as the **Parham cyclization.**

X = Br, I

E = COOH, CONR$_2$, epoxide, CH$_2$Br, CH$_2$Cl, OCONR$_2$, NCHArCONRCOCH$_2$R, POPh$_2$

The aryl halides with ortho side chains bearing a side chain electrophile when treated with an organolithium reagent undergo halogen-metal exchange. involving halogen-metal exchange and subsequent intramolecular cyclisation. The Parham cyclization process is a valuable synthetic tactics for the assembling of carbo and heterocyclic systems especially for the synthesis of alkaloids. The reaction works well with both aryl and heteroaryl systems.

290. PASSERINI REACTION

This reaction was discovered by Mario Passerini in 1921 that now bears his name. It involves a three component reaction to prepare an α-acyloxy amide by the treatment of an isocyanide with a carboxylic acid and an aldehyde or ketone.

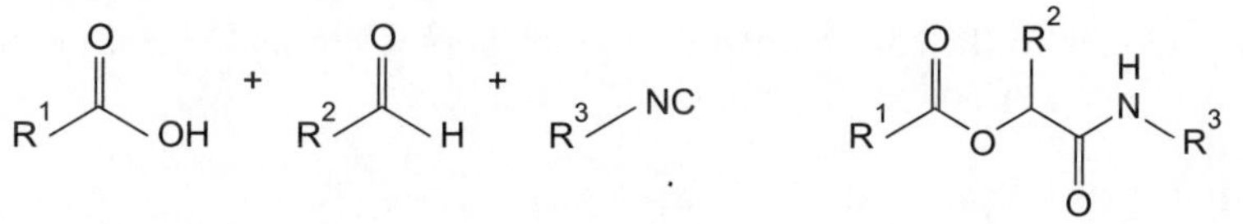

In other words, this reaction refers to coupling of a carbonyl compound and an isocyanide with a carboxylic acid to form an α-acyloxycarboxamide.

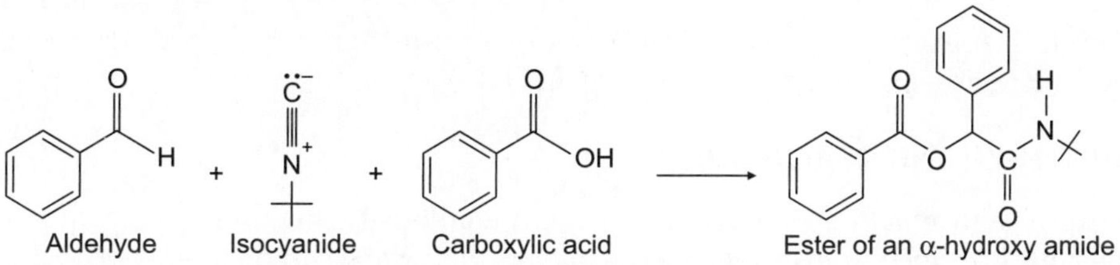

| Aldehyde | Isocyanide | Carboxylic acid | | Ester of an α-hydroxy amide |

The Passerini reaction is preferably carried out in aprotic solvents where it found to proceed rapidly at room temperature indicating a non-ionic mechanism for the reaction. One drawback of the Passerini reaction is its sensitivity to the steric hindrance of the carbonyl compounds. Therefore, aldehydes perform much better than ketones. The Passerini, have been used extensively in the field of combinatorial chemistry, which seeks to synthesize large libraries of similar molecules in a parallel fashion.

291. PATERNO-BÜCHI REACTION

The Paterno-Büchi reaction involves formation of oxetanes (four-membered cyclic ethers) by photochemical [2 + 2] cycloaddition of carbonyl compounds to olefins. The reaction was first described by Emanuele Paternò and George Hermann Büchi.

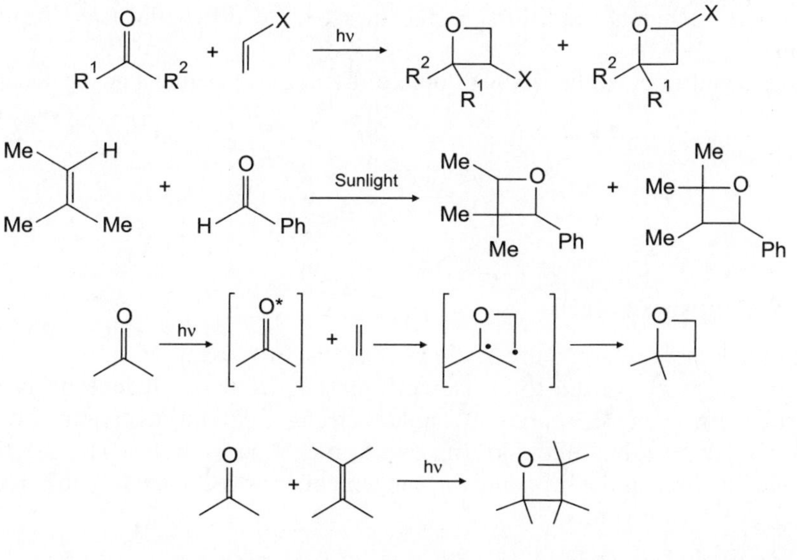

This reaction belongs to the more general class of photochemical [2 + 2] cycloadditions and believed to proceed via a diradical intermediate.

This reaction found useful for the synthesis of four-membered oxygen heterocyclic rings (oxetanes) in a regio- and stereoselective manner.

292. PAUSON-KHAND REACTION

The Pauson-Khand reaction is a widely utilized method for making cyclopentenones. The cyclopentenone is formed by cyclization of an alkyne, olefin, and carbon monoxide in the presence of $Co_2(CO)_8$ in a formal [2 + 2 + 1] cycloaddition.

Pauson and Khand first reported the reaction in detail in 1973. General reaction conditions involved heating the premade alkyne-$Co_2(CO)_6$ with the alkene to receive moderate yields of cyclopentenones. The Pauson-Khand reaction is tolerant of a wide variety of functionality such as esters, ethers, thioethers, tertiary amines, amides, sulfonamides, nitriles, and alcohols, which makes it an attractive reaction for organic synthesis.

The traditional Pauson-Khand reaction suffered from many drawbacks like stoichiometric amount of $Co_2(CO)_8$, harsh conditions were employed to effect the transformation, lack of regioselectivity, etc.

In 1981, Schore introduced the first example of the intramolecular Pauson-Khand reaction.

293. PAYNE REARRANGEMENT

The payne rearrangement refers to the base-promoted isomerization of 2,3-epoxyalcohols, where a 2,3-epoxy alcohol is converted to an isomeric one under the influence of an aqueous base. In payne rearrangement, the 2,3-epoxy alcohols rearrange with an inversion of configuration at the C-2 carbon of the original epoxide ring as of an S_N2 substitution. The reaction, originally referred to in the literature as the β-oxanol rearrangement, is now exclusively referred to as epoxide migration or payne rearrangement.

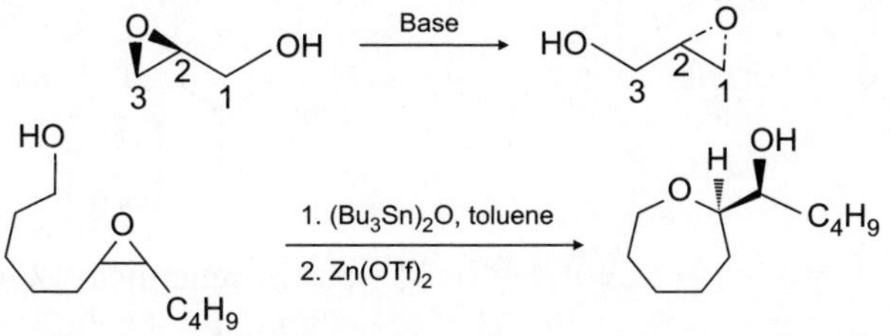

Aza- and Thia-Payne Rearrangements

Related rearrangements in which either the epoxide or hydroxy oxygen has been replaced with nitrogen or sulfur is referred to as aza-payne and thia-payne rearrangements respectively.

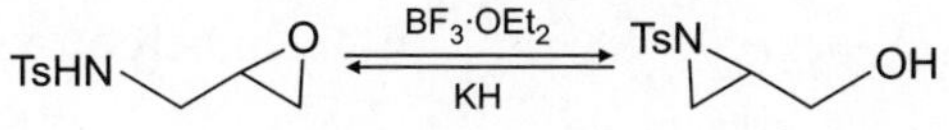

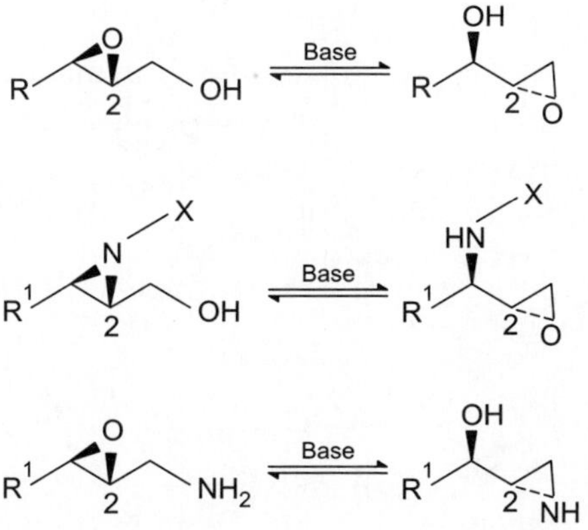

R^1 = Alkyl or aryl   X = H, mesyl, tosyl, etc.

294. PECHMANN CONDENSATION

The Pechmann condensation involves the condensation of phenols with β-ketonic esters in the presence of a variety of acidic condensing agents and gives good yields of coumarins. The reaction is conducted with a strong Brønstedt acid such as methanesulfonic acid or a Lewis acid such as $AlCl_3$.

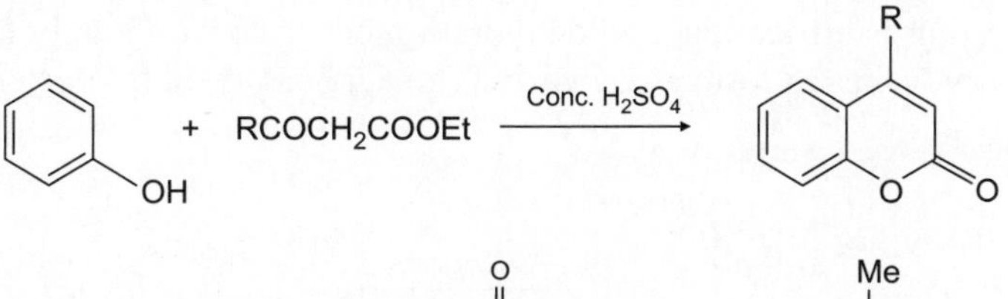

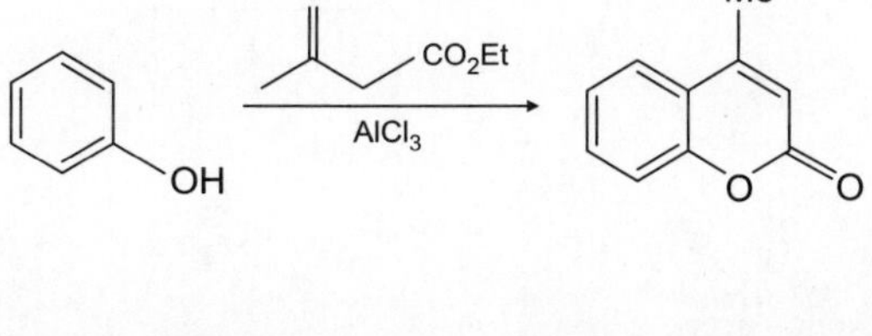

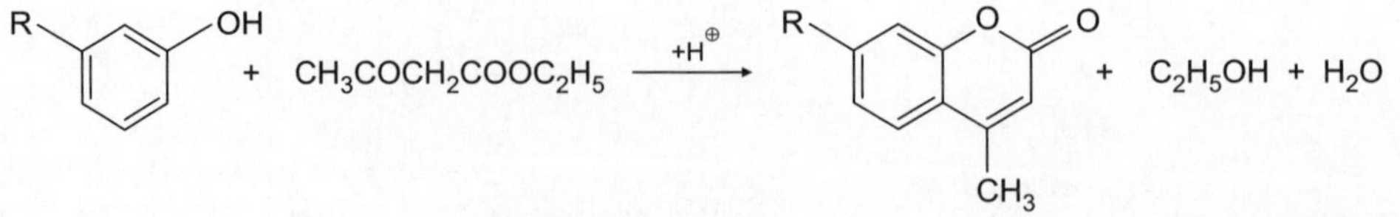

The Pechmann condensation is facilitated by the presence of hydroxyl (OH), dimethylamino (NMe$_2$) and alkyl groups meta to the hydroxyl of the phenol. For example, activated phenols such as resorcinol, the reaction can be performed under much milder conditions while with simple phenols, the conditions are harsh.

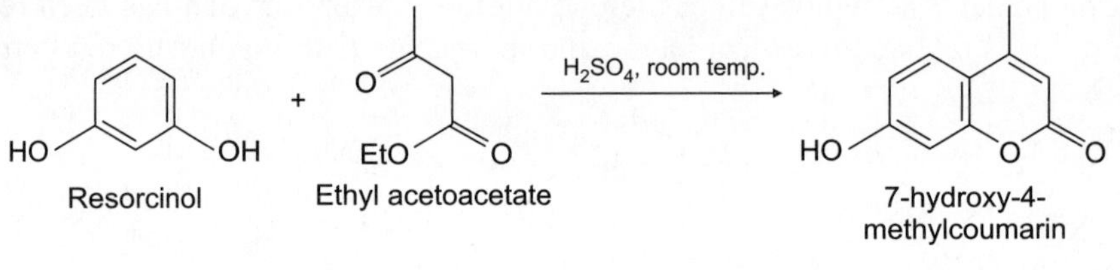

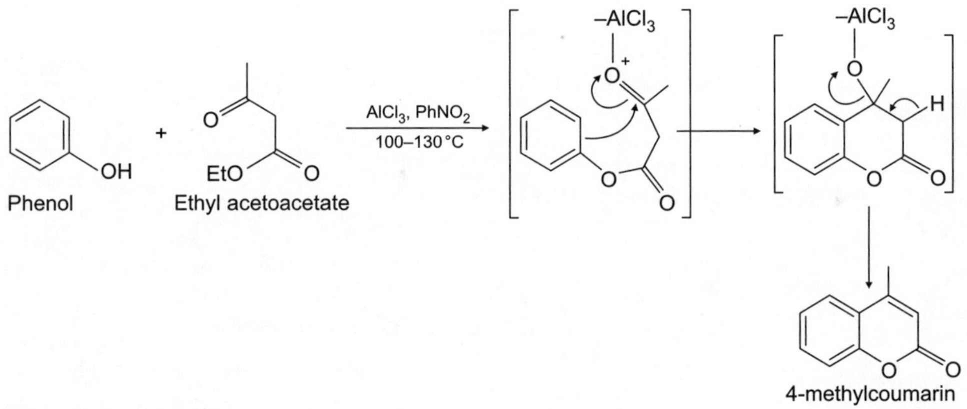

A variety of condensing reagents such as concentrated sulphuric acid, methanesulfonic acid, hydrogen fluoride, Lewis acids (AlCl$_3$, ZnCl$_2$, ZnCl$_2$/Al$_2$O$_3$, ZrCl$_4$, etc.), dehydrating agents (P$_2$O$_5$) or montmorillonite clay, etc. are used in the reaction.

The main disadvantages of the processes using these catalysts are longer reaction time, large amount of the catalyst, tedious purification process after completion of the reaction and some of the catalysts are highly expensive, some of the existing methods are that the catalysts are destroyed in the workup procedure and cannot be recovered or re-used.

When the phenols react with the ester in the presence of an acid catalyst, three reactions—hydroxyalkylation, transesterification, and dehydration occur concomitantly condensing together the two reactants at two sites to form a coumarin. The acid catalyses transesterification as well as keto-enol tautomerisation.

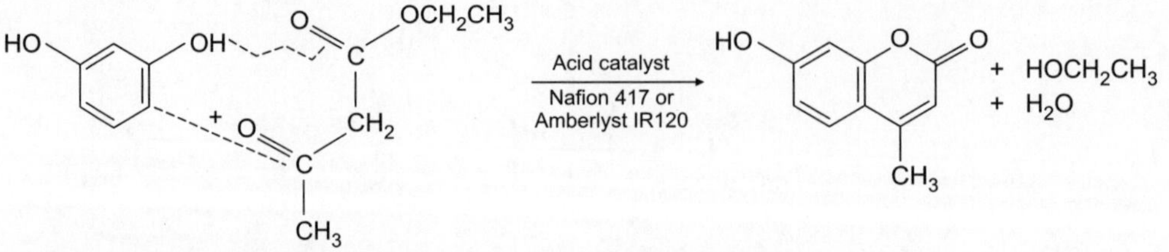

This reaction has found applications in the synthesis of various coumarins substituted on the pyrone or benzene ring or both.

The Simonis chromone cyclization (a variation to the **Pechmann** reaction) involves condensation of phenols, beta-keto esters and phosphorus pentoxide to yield a chromone.

295. PECHMANN PYRAZOLE SYNTHESIS

It describes preparation of pyrazole derivatives by 1,3-dipolar cycloaddition between a diazomethane (or other diazonium salt) and a molecule with carbon-carbon double bonds to pyrazoline and subsequent oxidation.

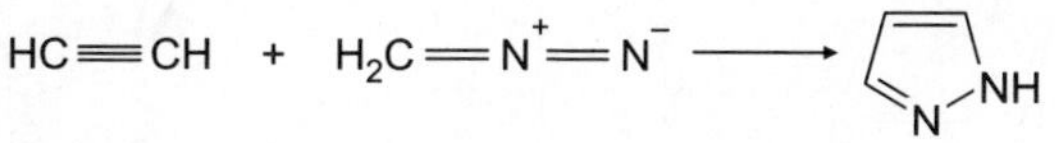

$$HC \equiv CH \ + \ H_2C = \overset{+}{N} = \overset{-}{N} \longrightarrow$$

296. PELLIZZARI REACTION

In Pellizzari reaction, substituted 1,2,4-triazoles are formed by the condensation of amides and acyl hydrazines.

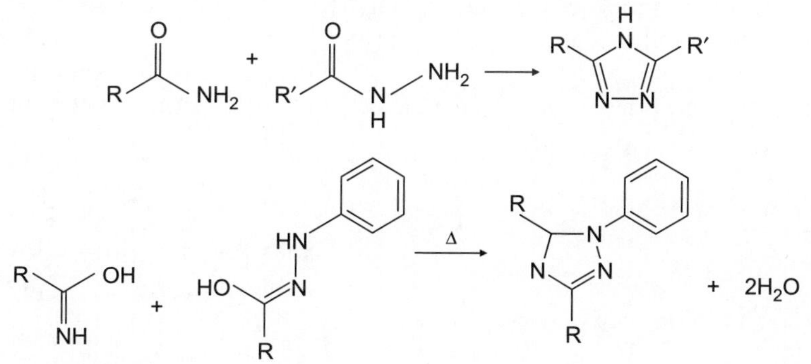

297. PELOUZE SYNTHESIS

Pelouze synthesis involves preparation of nitriles from alkali cyanides by alkylation with alkyl sulfates or alkyl phosphates.

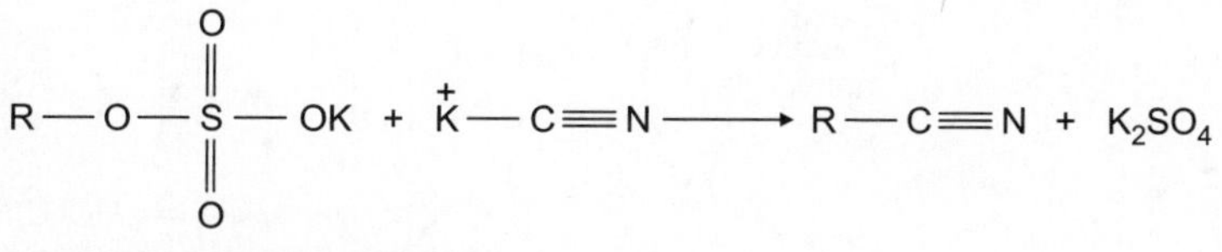

$$R - O - \overset{\overset{O}{\|}}{\underset{\underset{O}{\|}}{S}} - OK \ + \ \overset{+}{K} - C \equiv N \longrightarrow R - C \equiv N \ + \ K_2SO_4$$

298. PERKIN ALICYCLIC SYNTHESIS

This reaction was discovered by William Henry Perkin, Jr. He described synthesis of alicyclic compounds from α,ω-dihaloalkanes and compounds containing active methylene groups in the presence of sodium ethoxide.

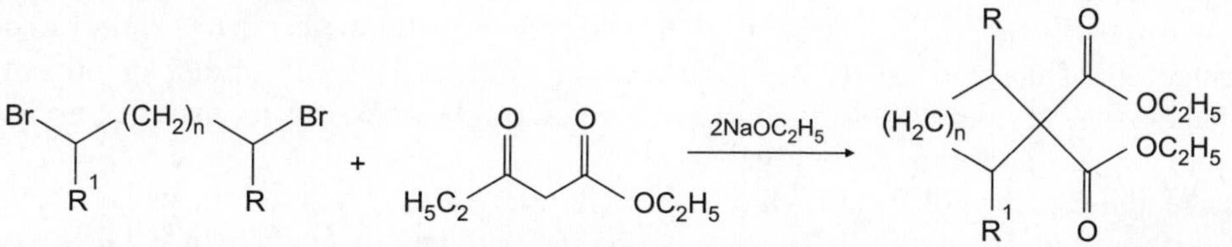

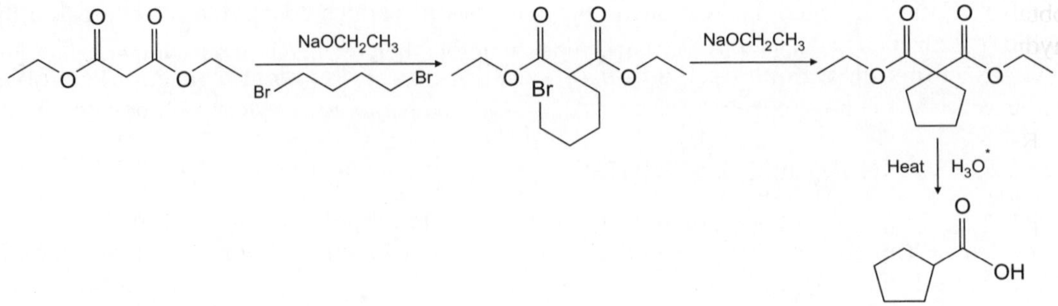

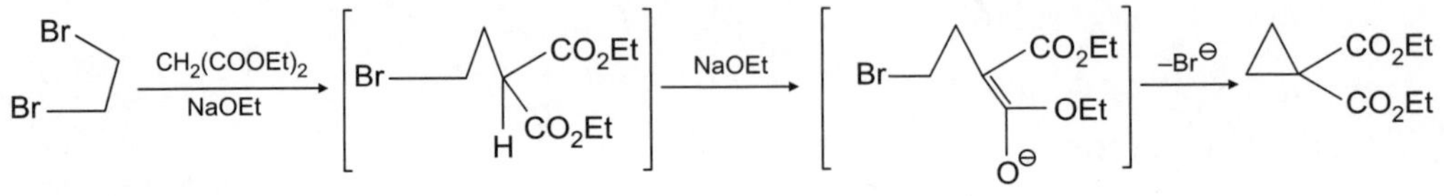

299. PERKIN REACTION

The base-catalyzed condensation of aromatic aldehydes with aliphatic acid anhydrides is called the Perkin reaction. It is a classical method for the synthesis of α,β-unsaturated aromatic acids. This reaction is a type of aldol condensation of aromatic aldehydes and the anhydride of an aliphatic acid in the presence of sodium salt of the same acid, to give on heating an α,β-unsaturated acid.

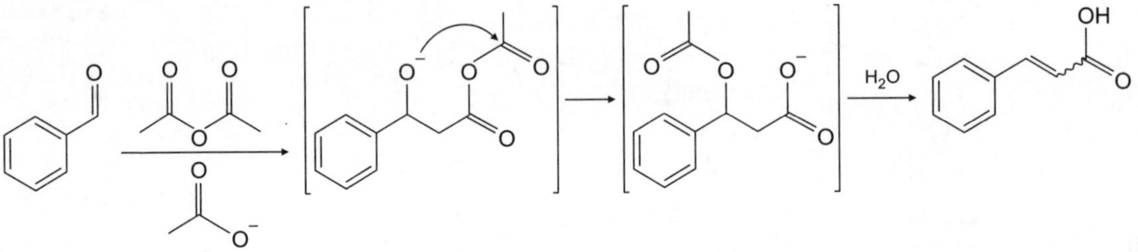

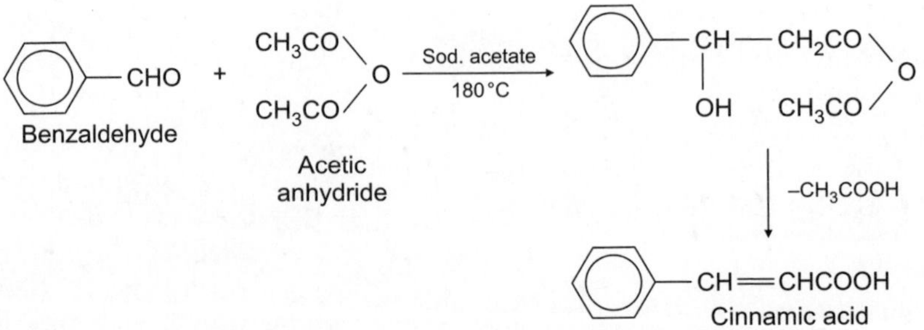

Benzaldehyde

Acetic anhydride

Cinnamic acid

Several modifications and extensions of the Perkin reaction, such as the paraconic acid synthesis of Fittig and the azlactone synthesis of Erlenmeyer, have served to broaden the scope and usefulness of the original process. The synthesis of coumarin from salicylaldehyde and acetic anhydride is of commercial importance.

α,β-unsaturated aromatic acids from the Perkin reaction can be subjected to a variety of chemical transformations. Thus, Perkin reaction gives access indirectly to a number of other types of compounds such as arylethylenes and acetylenes, arylacetaldehydes, arylethylamines, arylpropionic and propiolic acids, and their derivatives.

In the reaction, deprotonation of the carboxylic anhydride gives anion, which then adds to aldehyde. If the starting anhydride bears only one α-hydrogen, then a β-hydroxy carboxylic acid is obtained as the reaction product. However, if the anhydride used bears two α-hydrogens, then a β-hydroxy carboxylic acid will not be isolated as product.

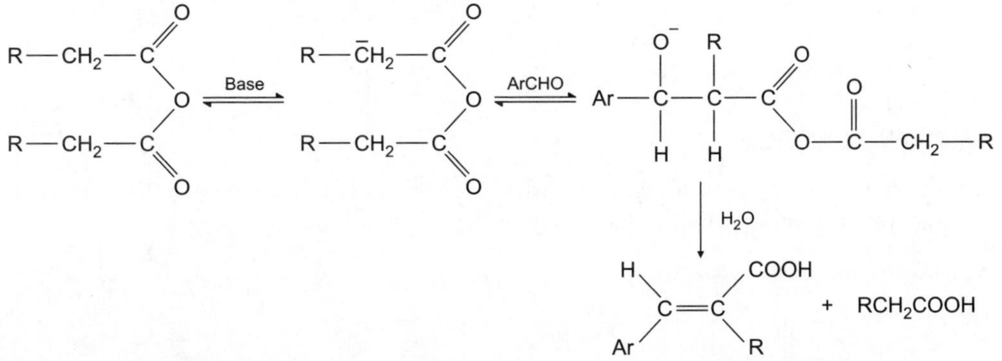

The Erlenmeyer-Plochl-azlactone synthesis is a variant of the Perkin reaction where an azlactone is obtained via condensation of an aromatic aldehyde with an N-acyl glycine in the presence of sodium acetate and acetic anhydride.

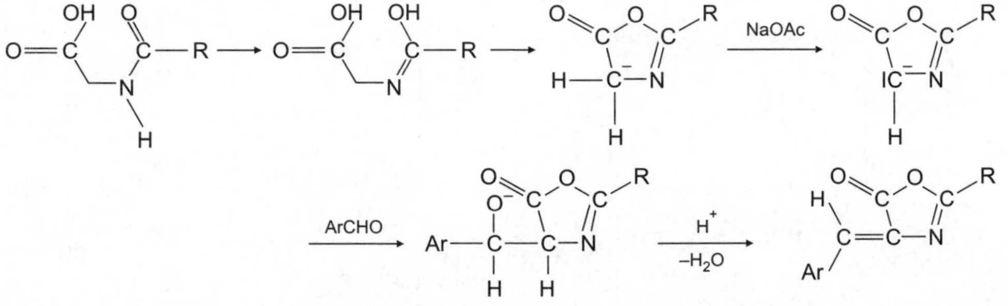

300. PERKIN REARRANGEMENT (COUMARIN-BENZOFURAN RING CONTRACTION)

Perkin rearrangement refers to the formation of benzofuran-2-carboxylic acids and benzofurans by heating 3-halocoumarins with alkali.

In another method called the Perkin rearrangement, a coumarin is reacted with a hydroxide:

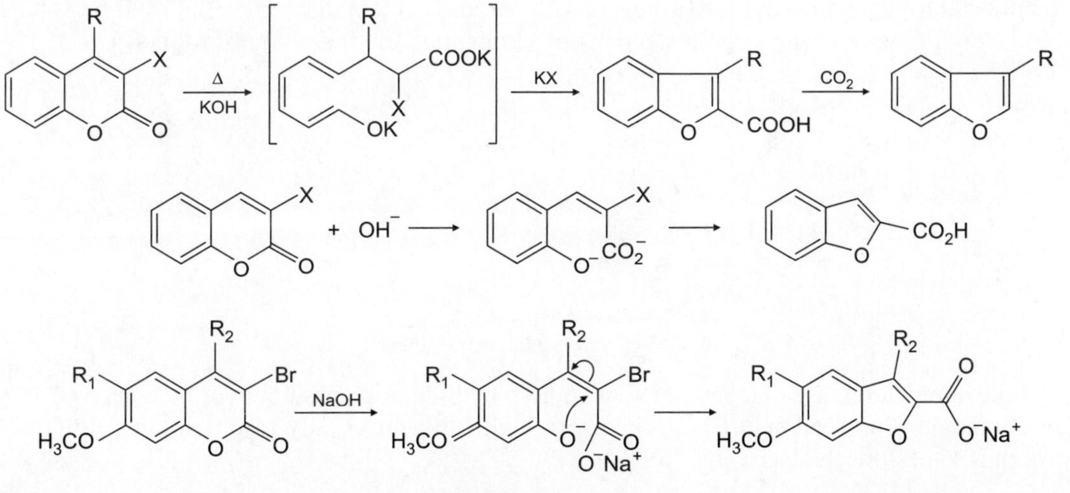

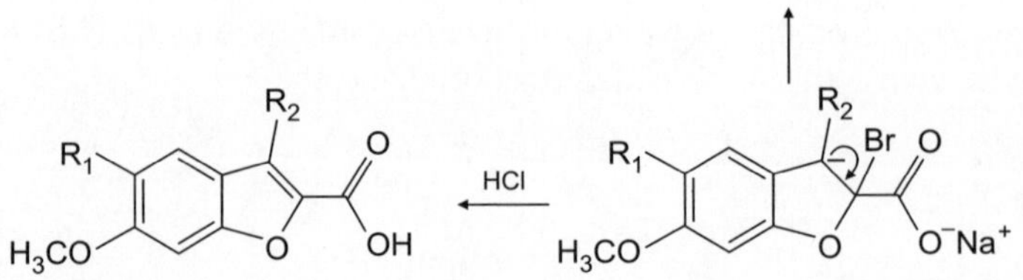

301. PERKOW REACTION

The Perkow reaction involves preparation of vinyl (or enol) phosphates by the reaction of a **trialkyl phosphite ester** or with α-halo carbonyl compounds.

R^1, R^2 = Alkyl   A = Alkyl, aryl, functional group

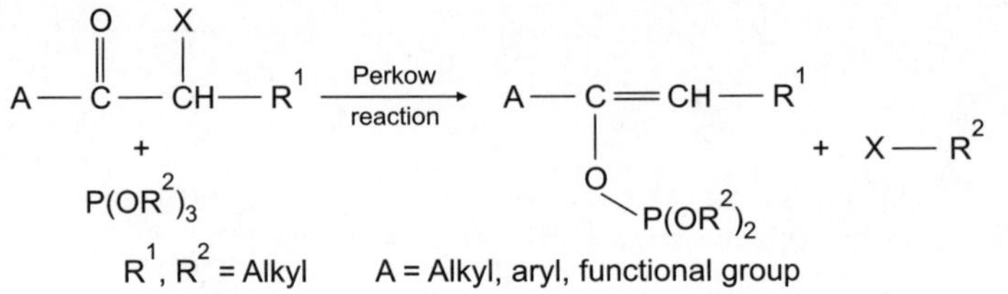

Order of reactivity of α-halo carbonyl compounds in the Perkow reaction is found as below:

α-halo aldehydes > α-halo ketones > α-halo esters (α-halo amides do not react at all)

It has also been found that the reaction rate increases with the number of α-halo atoms in the carbonyl compound.

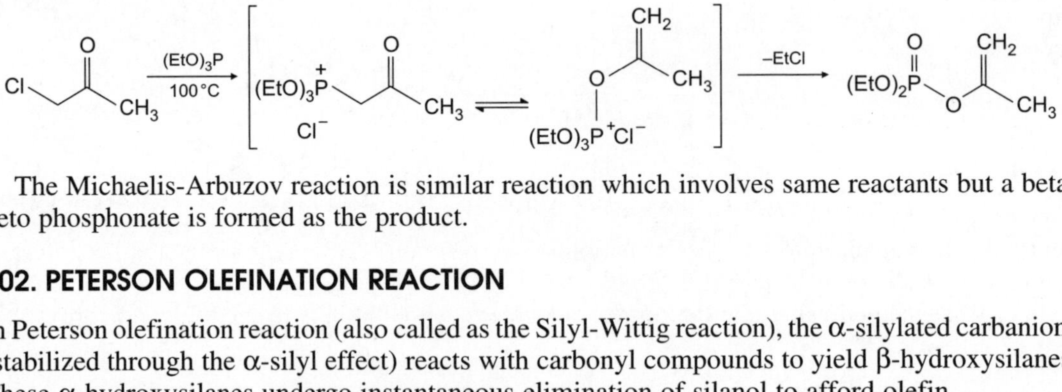

The Michaelis-Arbuzov reaction is similar reaction which involves same reactants but a beta-keto phosphonate is formed as the product.

302. PETERSON OLEFINATION REACTION

In Peterson olefination reaction (also called as the Silyl-Wittig reaction), the α-silylated carbanions (stabilized through the α-silyl effect) reacts with carbonyl compounds to yield β-hydroxysilanes. These α-hydroxysilanes undergo instantaneous elimination of silanol to afford olefin.

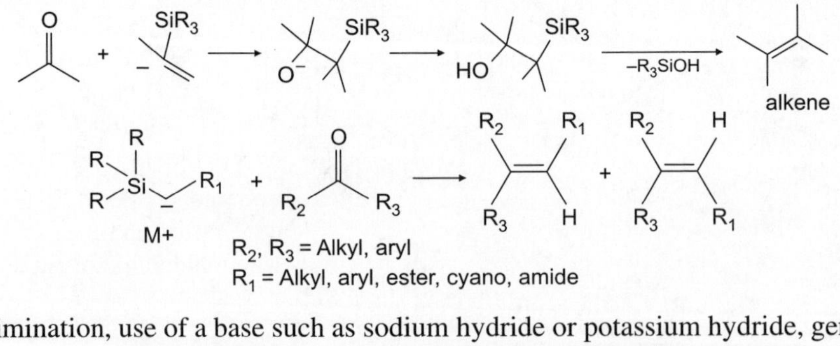

R_2, R_3 = Alkyl, aryl
R_1 = Alkyl, aryl, ester, cyano, amide

While elimination, use of a base such as sodium hydride or potassium hydride, generally gives syn elimination (proceeds via E1 mechanism), while an acid usually results in anti elimination (proceeds via E2 mechanism).

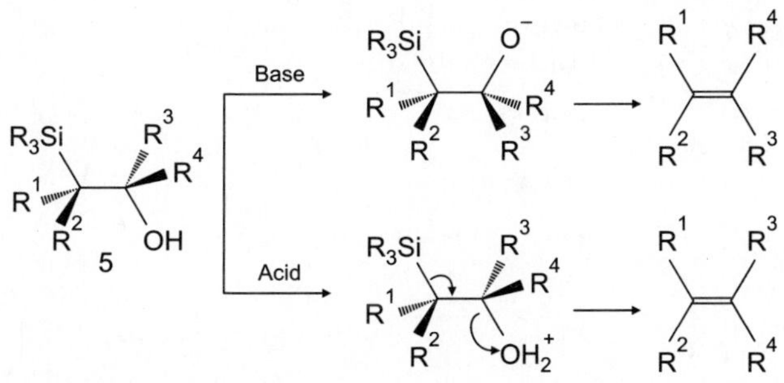

Synthesis of β-gorgonene, a non-isoprenoid sesquiterpene:

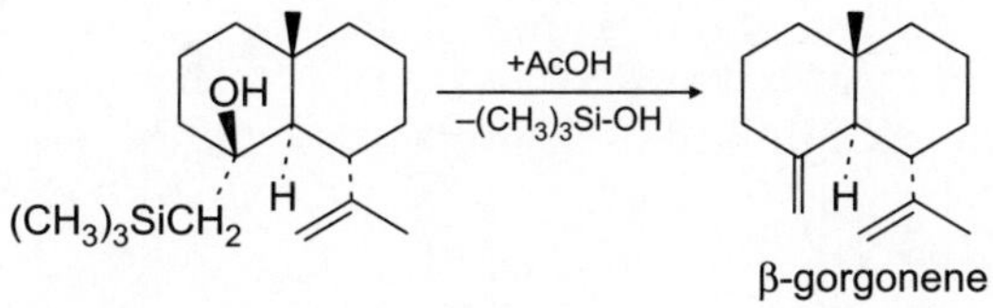

β-gorgonene

The Peterson olefination is a valuable alternative to the Wittig reaction and is advantageous to allow for a simple control of the alkene geometry. However, its applicability in synthesis depends on the availability of the required silanes.

303. PETRENKO-KRITSCHENKO PIPERIDONE SYNTHESIS

The Petrenko-Kritschenko reaction is a multicomponent reaction which involves formation of piperidones via cyclization of two equivalents of aldehyde and one equivalent each of acetonedicarboxylic ester and ammonia (or a primary amine) in a double Mannich reaction.

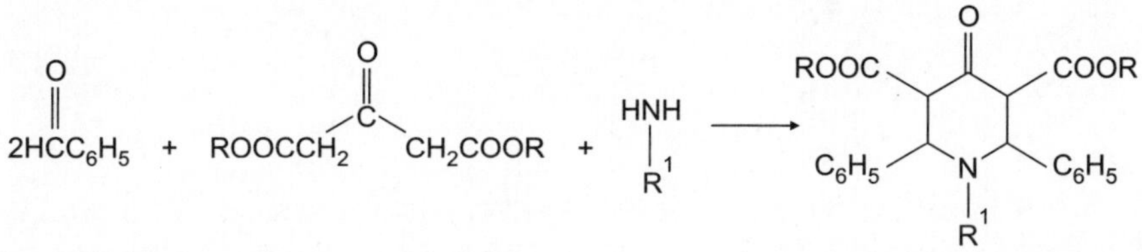

In its original publication, this reaction was shown to be carried with diethyl-α-ketoglurate in combination with ammonia and benzaldehyde.

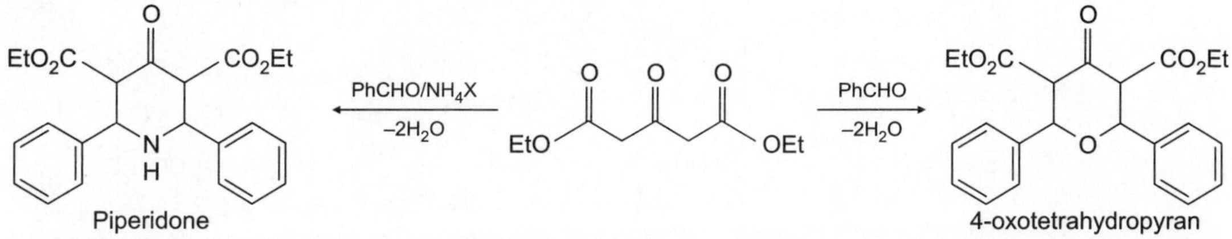

Piperidone

4-oxotetrahydropyran

This reaction is closely related to the Robinson-Schöpf tropinone synthesis, but in contrast to the Robinson synthesis, the Petrenko-Kritschenko reaction employ simpler aldehydes like benzaldehyde (Robinson synthesis employ dialdehydes like succinaldehyde or glutaraldehyde).

304. PFAU-PLATTNER AZULENE SYNTHESIS

Pfau-Plattner azulene synthesis involves a multistep preparation for the formation of azulenes which are synthesized by ring expansion of indanes on addition of diazoacetic ester, hydrolysis, dehydrogenation and decarboxylation of the resulting acid.

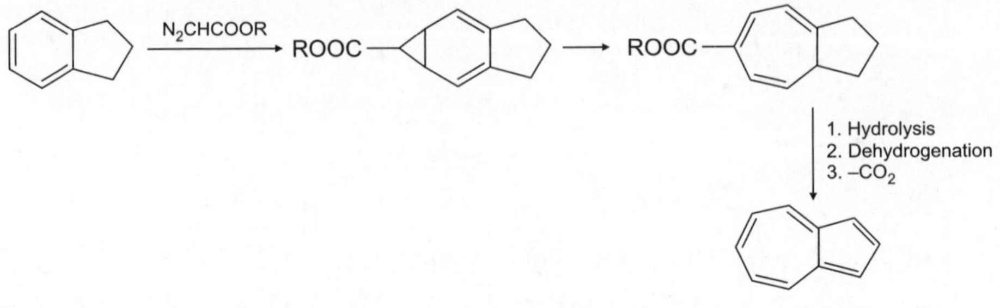

In 1939, St. Pfau and Plattner reported synthesis of azulenes starting from indane and ethyl diazoacetate. This reaction has a limited use in organic synthesis.

305. PFITZINGER REACTION

The formation of 4-quinolinecarboxylic (cinchoninic) acid derivatives as a result of the reaction of isatin or its derivatives with ketones containing the —CH$_2$CO—group (α-methylene carbonyl compounds or enolizable carbonyl compound) in the presence of sodium hydroxide or potassium hydroxide was first discovered at the end of the nineteenth century by Pfitzinger and is known as the Pfitzinger reaction.

Subsequent decarboxylation of quinoline-4-carboxylic acids yields quinolines.

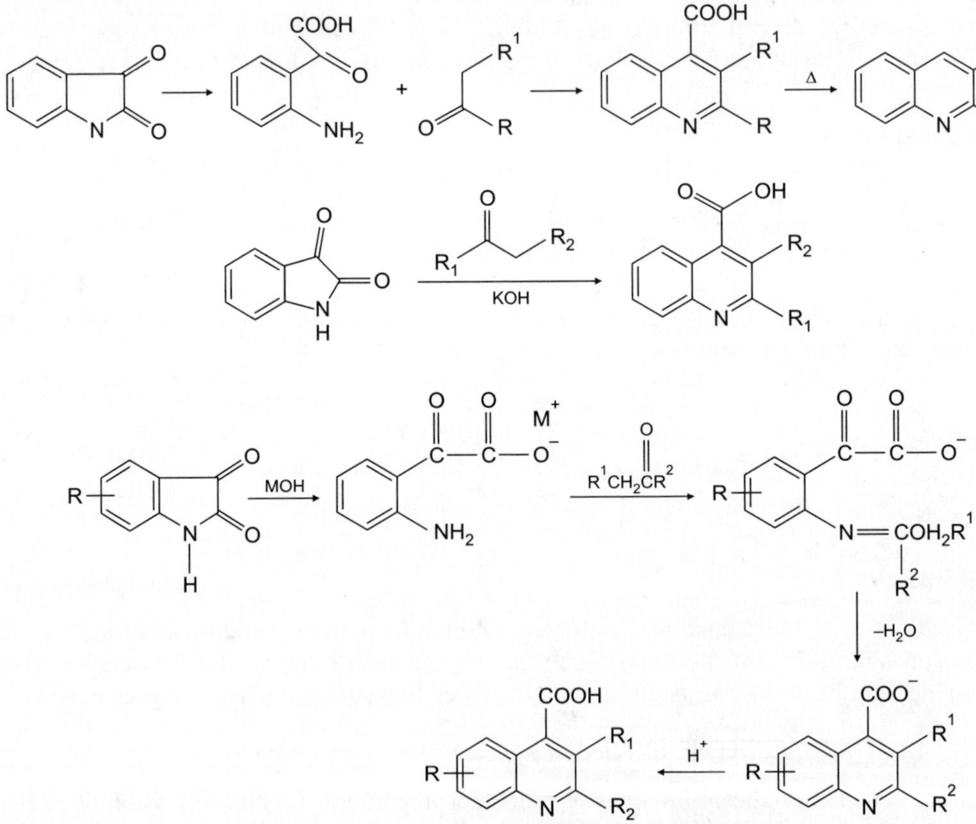

306. PFITZNER-MOFFATT OXIDATION (MOFFATT OXIDATION)

The Pfitzner-Moffatt oxidation also referred to as Moffatt oxidation, describes the mild oxidation of primary and secondary alcohols by dimethyl sulfoxide (DMSO) activated with a carbodiimide, such as dicyclohexylcarbodiimide (DCC). The resulting alkoxysulfonium ylide rearranges intramolecularly to generate corresponding aldehydes and ketones. In this way, a primary alcohol can be converted to the aldehyde with no carboxylic acid being produced. The combination of DMSO and DCC is known as the Pfitzner-Moffatt reagent.

The Moffatt oxidation is the first reported DMSO-based oxidation procedure where dicyclohexyl-carbodiimide (DCC) functions as the electrophilic activating agent in conjunction with a Brønsted acid promoter. Oxidations are carried out with an excess of DCC at or near 23 °C.

The strong acid conditions are sometimes a problem, and complete removal of the dicyclohexyl-urea and methylthiomethyl (MTM) ether formation can limit usefulness. Alternatively, application of carbodiimides that yield water-soluble by-products, (e.g. 1-ethyl-3-(3-dimethylaminopropyl) carbodiimide (EDC)) can simplify workup procedures.

307. PICTET-GAMS ISOQUINOLINE SYNTHESIS

The reaction involves formation of isoquinolines by cyclization of acylated aminomethyl phenyl carbinols or their ethers with a dehydrating agent (such as phosphorus pentaoxide, or phosphorus oxychloride), under reflux conditions and in an inert solvent such as decalin.

This synthesis is a variation on the Bischler-Napieralski reaction. Both the reactions work similarly, the only difference being that an elimination of the hydroxy group forms isoquinoline.

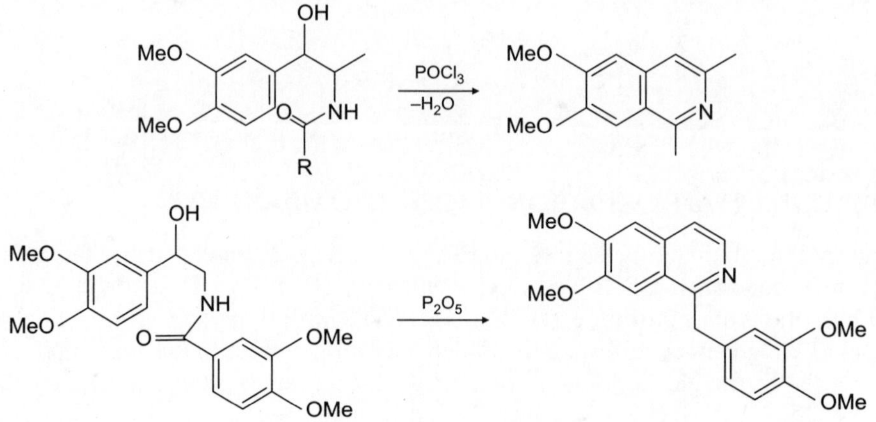

308. PICTET-HUBERT REACTION; MORGAN-WALLS REACTION

The synthesis of phenanthridine derivatives by dehydrative ring closure of N-acyl ortho-aminobiphenyls on heating with zinc chloride at high temperature (250–300°C) is referred to as the Pictet-Hubert phenanthridine synthesis, while with phosphorus oxychloride in boiling nitrobenzene is known as the Morgan-Walls reaction or Morgan-Walls cyclization.

Pictet-Hubert reaction was discovered in 1899. The reaction conditions of the **Pictet-Hubert reaction** were improved by Morgan and Walls in 1931, replacing the metal by phosphorus oxychloride and using nitrobenzene as a reaction solvent.

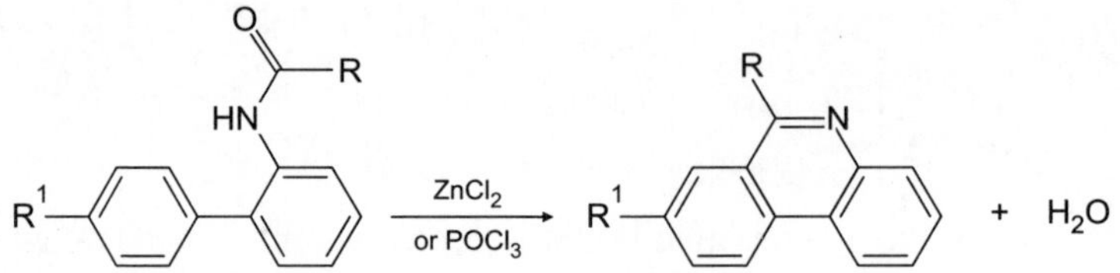

309. PICTET-SPENGLER ISOQUINOLINE SYNTHESIS

The reaction was discovered in 1911 by Amé Pictet and Theodor Spengler. In Pictet-Spengler reaction, a condensation of a β-arylethylamine, such as tryptamine and an aldehyde or ketone forms an imine, which undergoes a cyclization to form a tetrahydroisoquinoline instead of the dihydroisoquinoline.

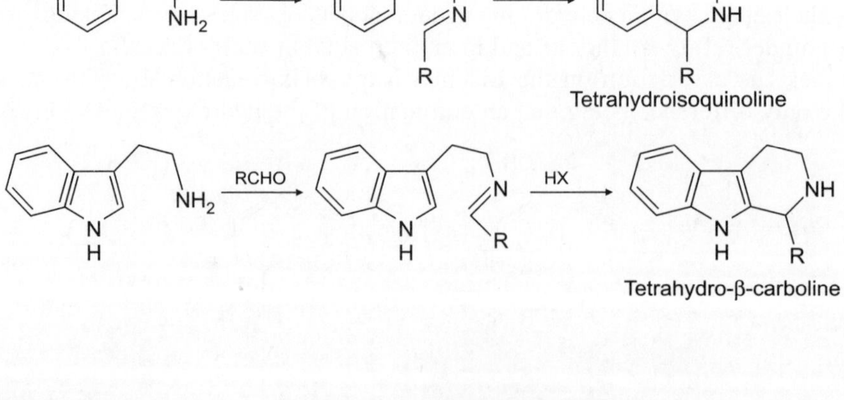

Tetrahydroisoquinoline

Tetrahydro-β-carboline

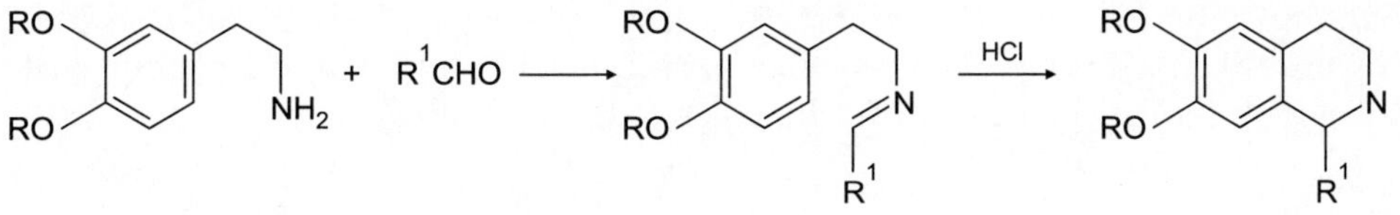

Synthesis of Tadalafil

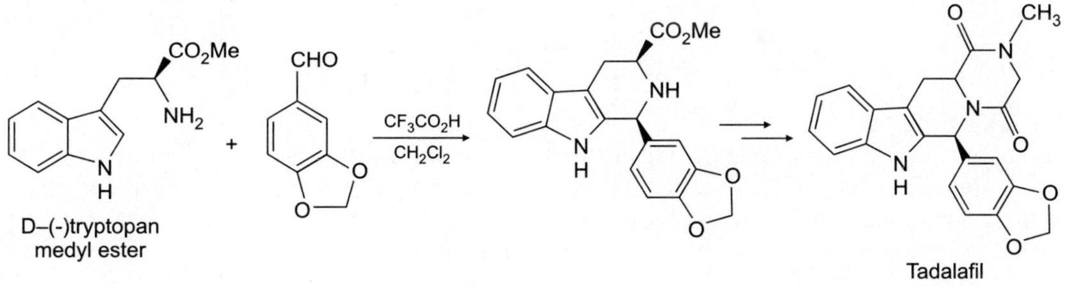

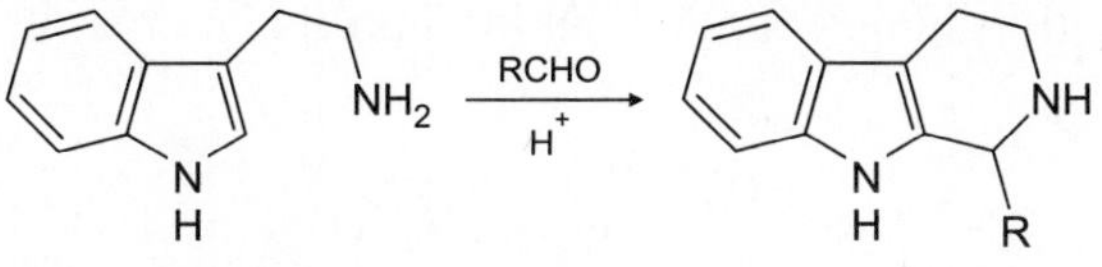

D–(-)tryptopan
medyl ester

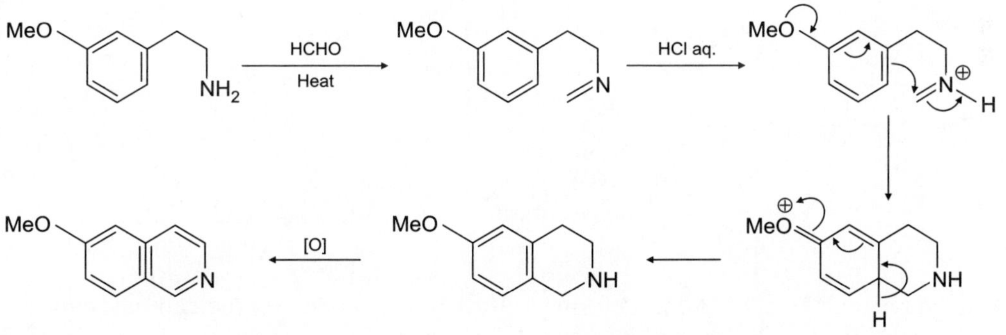

Tadalafil

310. PILOTY-ROBINSON SYNTHESIS

Formation of pyrroles by heating azines of enolizable ketones with acid catalysts (Lewis acid or Brønsted acid usually zinc chloride or hydrogen chloride).

The starting materials in the Piloty-Robinson pyrrole synthesis are 2 equivalents of an aldehyde and hydrazine. The product is a pyrrole with specific substituents in the 3rd and 4th positions. The aldehyde reacts with the diamine to an intermediate di-imine (R–C=N–N=C–R), which with added hydrochloric acid, gives ring-closure and loss of ammonia to the pyrrole. This thermal transformation of enolizable ketazines into pyrrole derivatives required a catalytic amount of Lewis acid or Brønsted acid.

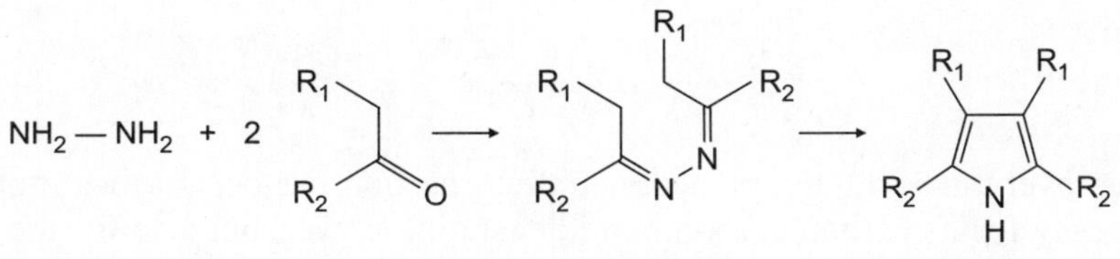

The Piloty-Robinson pyrrole synthesis is a less well-known hydrazine-based heterocycle synthesis that, like the Fischer indole synthesis, utilizes a [3,3]-sigmatropic rearrangement. This synthesis allows for the conversion of ketone-derived azines into symmetrical pyrroles.

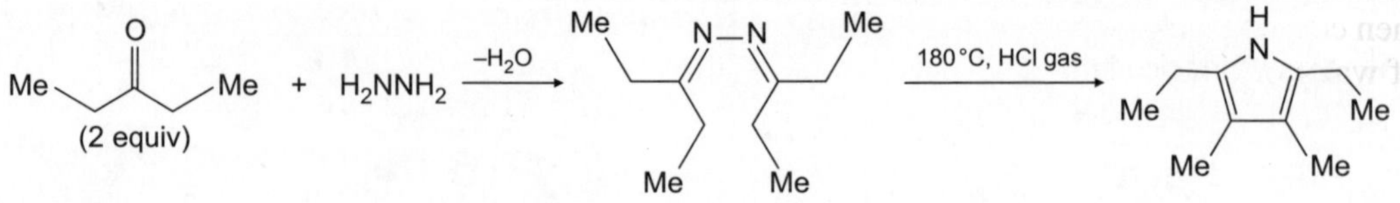

311. PINACOL COUPLING REACTION

The reductive radical-radical coupling of carbonyl compounds leading to 1,2-diols, known as pinacol coupling, is one of the oldest and best-studied carbon-carbon bond forming reactions. The reaction is named after pinacol (also known as 2,3-dimethyl-2,3-butanediol or tetramethylethylene glycol), which is the product of this reaction with acetone.

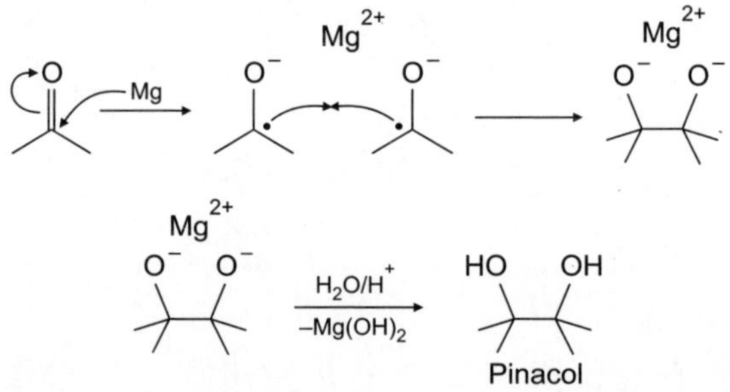

A carbon-carbon covalent bond is formed in the reaction between the carbonyl groups of an aldehyde or a ketone in presence of an electron donor in a free radical process to produce a vicinal diol. The reaction is usually a homocoupling but intramolecular cross-coupling reactions are also possible.

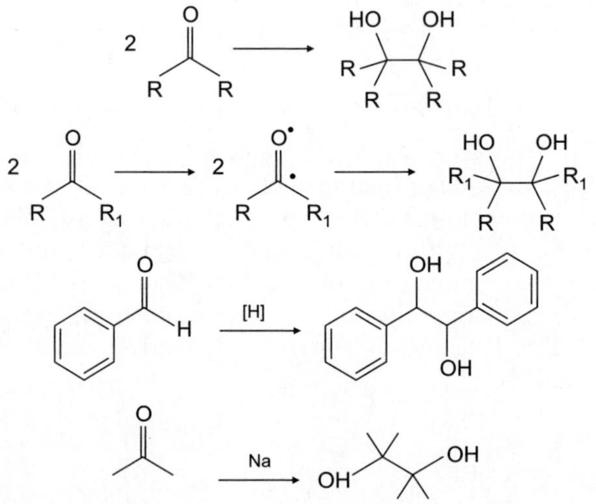

Since its discovery in 1859, the pinacol coupling reaction has been a focal point in synthetic chemistry not only for its carbon-carbon bond generating power, but also for the wide utility of the 1,2-diols obtained from these reactions. These 1,2-diols are valuable synthons in the synthesis of biologically active natural products or molecular fragments.

The first step in the reaction mechanism is that the carbonyl group is reduced to a ketyl radical anion species in one-electron reduction by a reducing agent such as magnesium. Two ketyl groups then coupled yielding a vicinal diol with both hydroxyl groups deprotonated. Subsequent addition of water or other proton donor gives the diol.

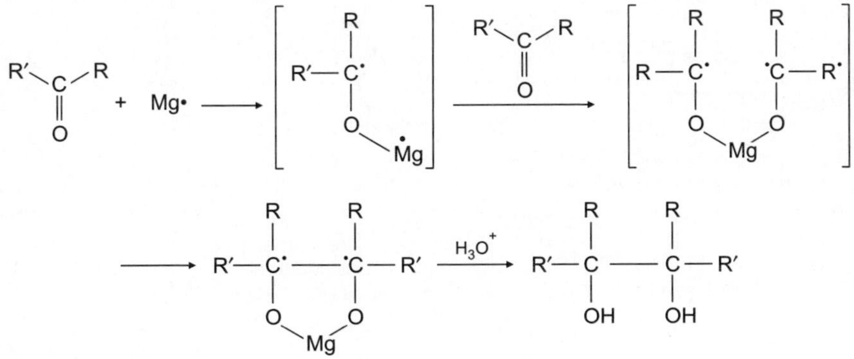

312. PINACOL REARRANGEMENT

In 1860, Rudolph Fittig reported one of the earliest rearrangements in organic chemistry. It consisted of the transformation of pinacol to pinacolone under acidic treatment. In other words, it is the acid-catalyzed elimination of water from pinacol which gives t-butyl methyl ketone (conversion that gave its name to this reaction).

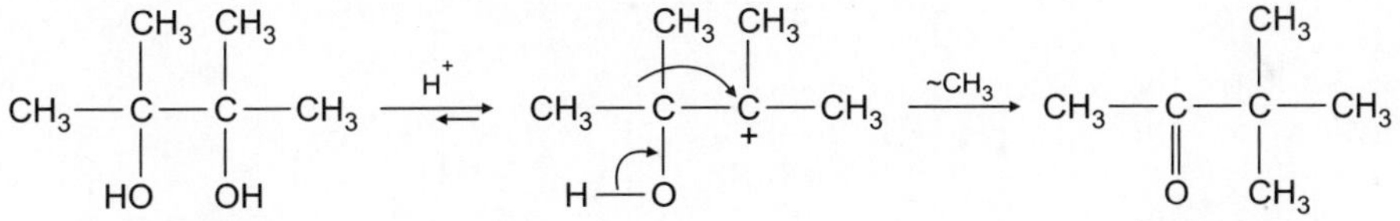

In recent time, the pinacol rearrangement encompasses all α-glycols other than methyl-containing ones. It proceeds via a 1,2 alkyl shift, and the overall reaction is:

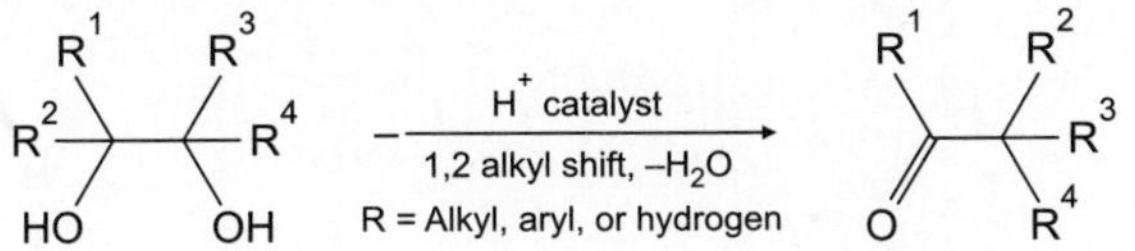

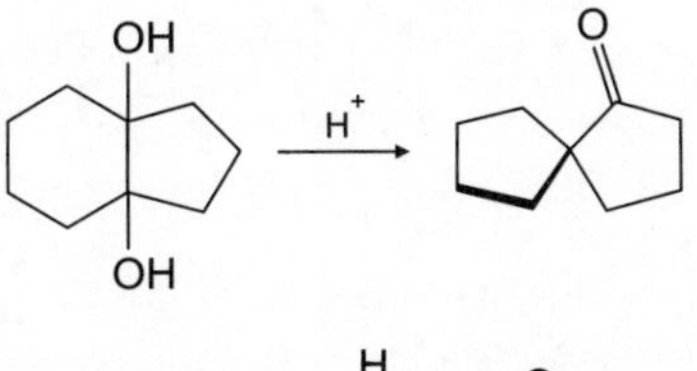

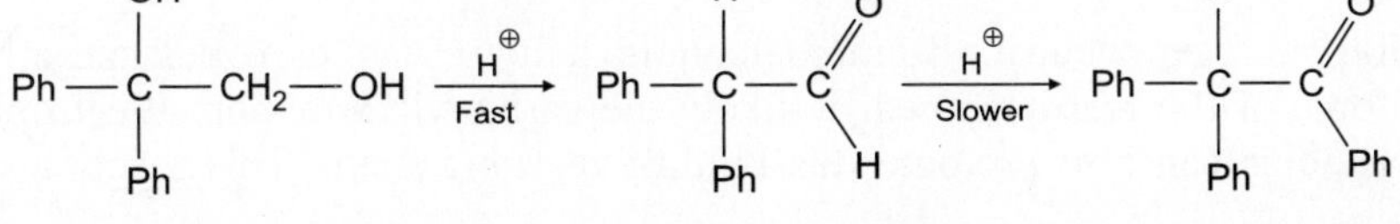

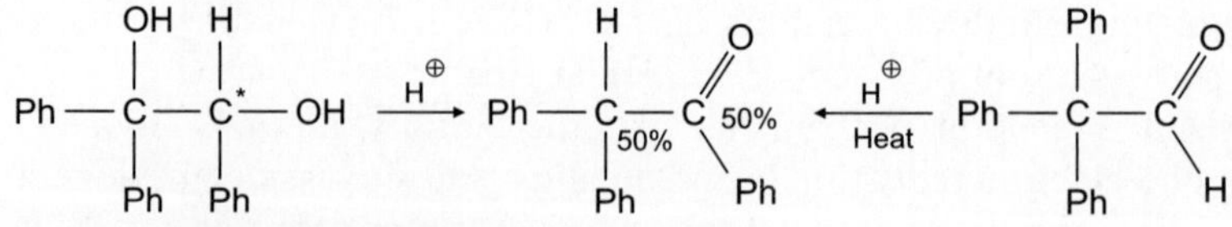

Rate of racemization = Rate of rearrangement
isotropic label C is scrambled C

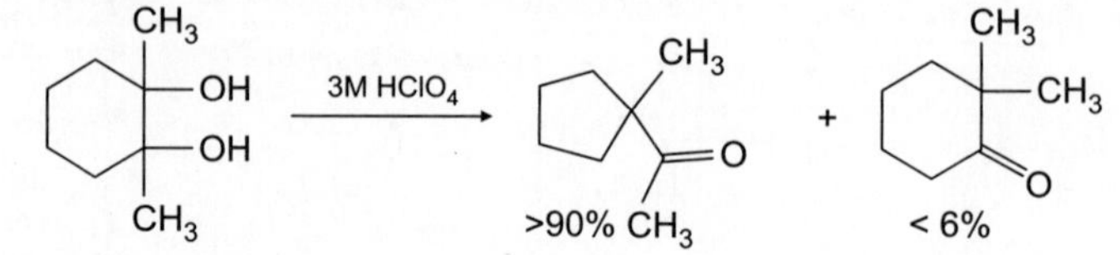

cis and trans-isomers
are interconverted

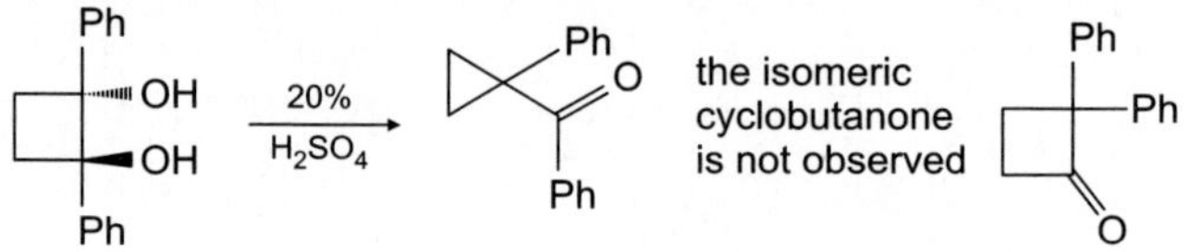

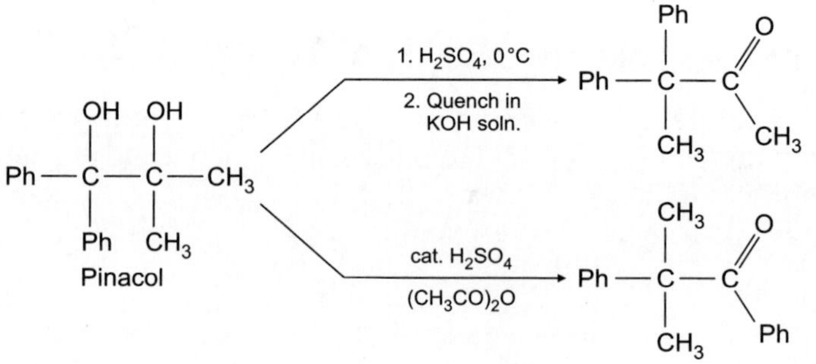

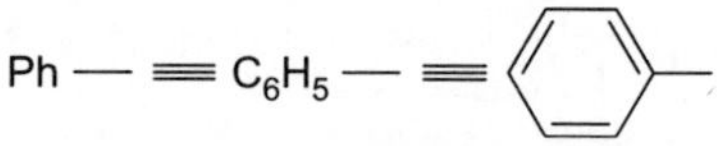

$$Ph — \equiv C_6H_5 — \equiv \text{(tolyl)}$$

*The reaction is a useful method of preparing **spirocyclic compounds**:*

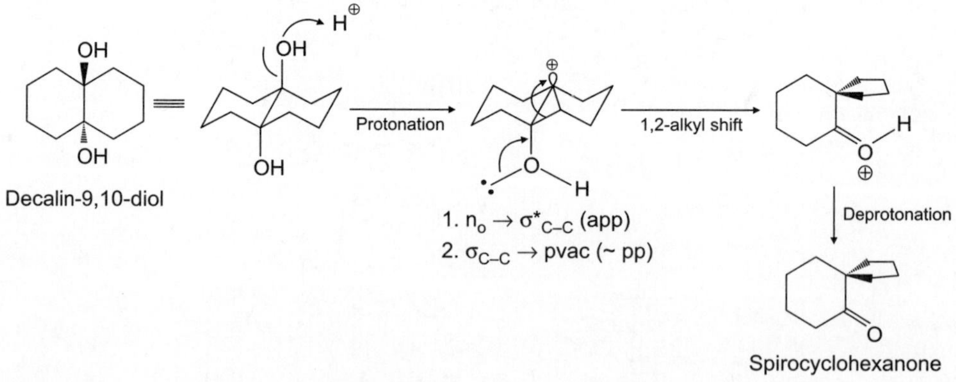

Decalin-9,10-diol

1. $n_o \rightarrow \sigma^*_{C-C}$ (app)
2. $\sigma_{C-C} \rightarrow$ pvac (~ pp)

Spirocyclohexanone

The reaction can also occur on β-aminoalcohols, halohydrins, epoxides and allylic alcohols. The driving force of the rearrangement is likely the carbonyl formation. Electrophilic reagents such as Lewis acids can also promote this kind of rearrangement. This reaction occurs with a variety of fully substituted 1,2-diols, and can be understood to involve the formation of a carbenium ion intermediate that subsequently undergoes a rearrangement.

This reaction involves formation of a carbenium ion intermediate that subsequently undergoes a rearrangement. The first generated intermediate, an α-hydroxycarbenium ion, rearranges through a 1,2-alkyl shift to produce the carbonyl compound. If two of the substituents form a ring, the Pinacol Rearrangement can constitute a ring-expansion or ring-contraction reaction.

Reaction mechanism involves protonation of one of the –OH groups and a carbocation is formed. Subsequently, an alkyl group from the adjacent carbon migrates to the carbocation center. The driving force for this rearrangement step is believed to be the relative stability of the resultant oxonium ion. The elimination of water to give an alkene may be observed as a side-reaction. Substituents R_1, R_2, R_3, R_4 can be alkyl or aryl; single substituents can even be hydrogen.

The pinacol rearrangement occurs when a 1,2 diol is exposed to acidic conditions. The reaction begins when one of the OH groups is protonated. Following departure of water, a carbocation is formed. The interesting part of this reaction is that an adjacent alkyl group or H atom migrates to the carbocation center to create a more stable cation. Loss of a proton gives the carbonyl product.

When the starting diol is not symmetric, the initial carbocation forms on the side that makes the more stable carbocation.

1,2 diol starting material   Intermediate A

Intermediate B   Intermediate C

Carbonyl products

The migratory aptitude of the substituents occurs with the concepts that which group stabilises carbo cation more effectively is migrated, i.e.

$$Ar >>>> \text{hydride} > Ph- > R_3C > R_2CH > RCH_2 > CH_3 > H$$

Reaction with an unsymmetrical diol as starting material may give rise to formation of a mixture of products. If the migrating alkyl group has a chiral center as its key atom, the configuration at this center is *retained* even after migration takes place. The reaction is strictly intramolecular; the migrating group R is never completely released from the substrate.

313. PINNER REACTION (AMIDINE AND ORTHO ESTER SYNTHESIS)

The addition of dry HCl to a mixture of a nitrile and an alcohol in the absence of water leads to the hydrochloride salt of an imino ester (imino esters are also called imidates and imino ethers). This reaction is called the Pinner synthesis. The imino ethers can be converted to either an ester or an amidine. The Pinner reaction or the Pinner synthesis is named after Adolf Pinner.

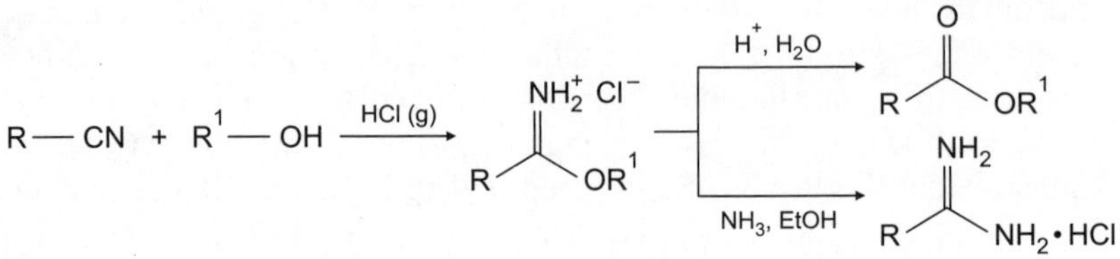

314. PINNER TRIAZINE SYNTHESIS

The Pinner triazine synthesis was named after Adolf Pinner in 1890. It involves preparation of 2-hydroxy-4,6-diaryl-sym-triazines (or sym-triazines) by reaction of aryl amidines and phosgene via the intermediate of bisimidyl urea. The reaction may be extended to halogenated aliphatic amidines. The yields of the reaction depends on the temperature however, this reaction might not be suitable for aliphatic amidines.

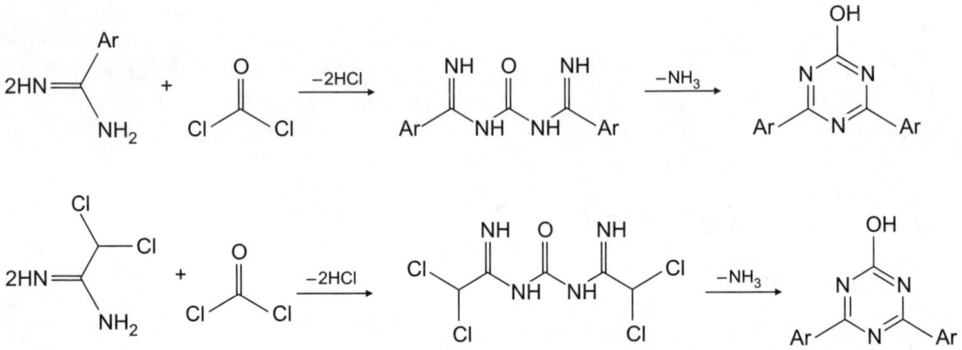

315. PIRIA REACTION

In Piria reaction (R. Piria, 1851) the aromatic nitro compounds, e.g. nitrobenzene reacts with a metal bisulfite forming an aminosulfonic acid as a result of combined nitro group reduction and sulfonation. This reaction has been applied for the preparation of aromatic amines.

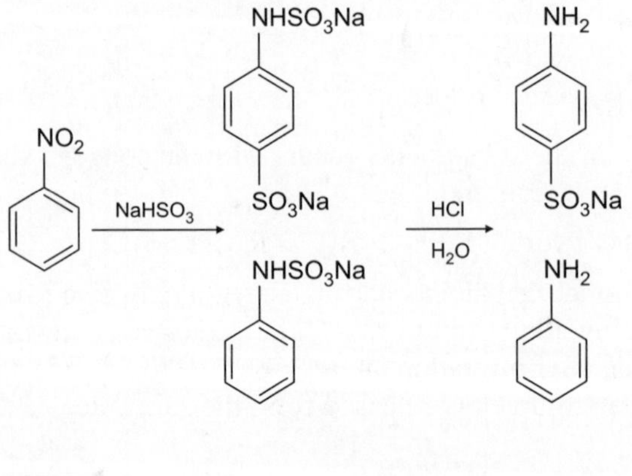

316. POLONOVSKI REACTION; POTIER-POLONOVSKI REACTION

In the Polonovski reaction, a tertiary N-oxide is cleaved by acetic acid anhydride to the corresponding acetamide and aldehyde. In this reaction one of the alkyl groups attached to the nitrogen is cleaved, generating the N,N-disubstituted acetamide and aldehyde.

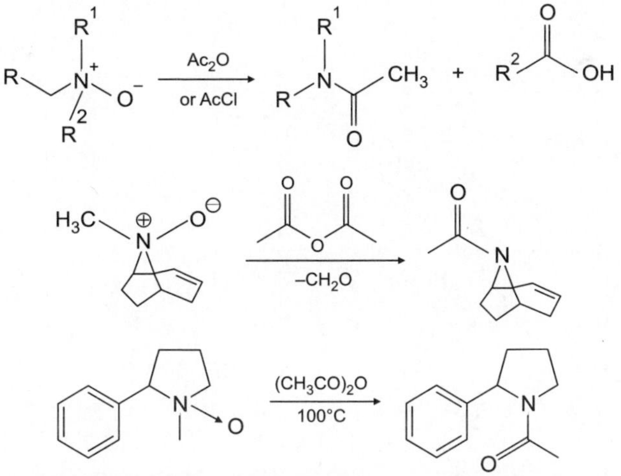

The **Potier-Polonovski reaction** is a modification of the Polonovski reaction where trifluoro-acetic anhydride is used in place of acetic anhydride. The reaction proceeds via an iminium ion intermediate which becomes the stable reaction product when trifluoroacetic anhydride is employed. Because the reaction conditions for the Polonovski-Potier reaction are mild, it has largely replaced the Polonovski reaction.

Tertiary N-oxide

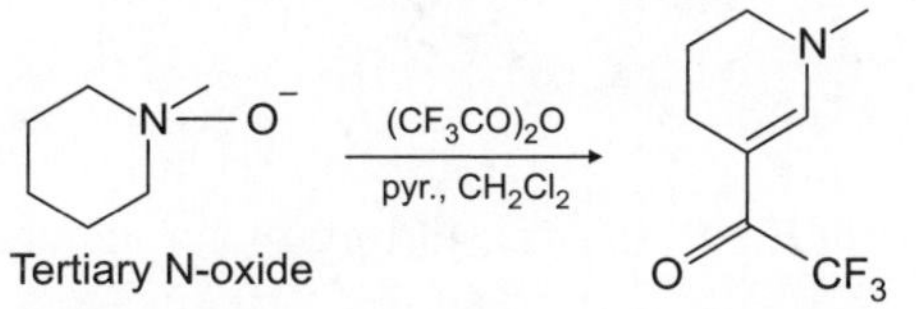

317. POMERANZ-FRITSCH REACTION (SCHLITTLER-MÜLLER MODIFICATION)

Pomeranz-Fritsch reaction involves formation of isoquinolines by the acid-catalyzed (e.g. sulphuric acid) electrophilic cyclization of benzalaminoacetals prepared from an aromatic aldehyde with an amino acetal. This reaction is generally favored by electron-donating substituents on the benzalaminoacetal.

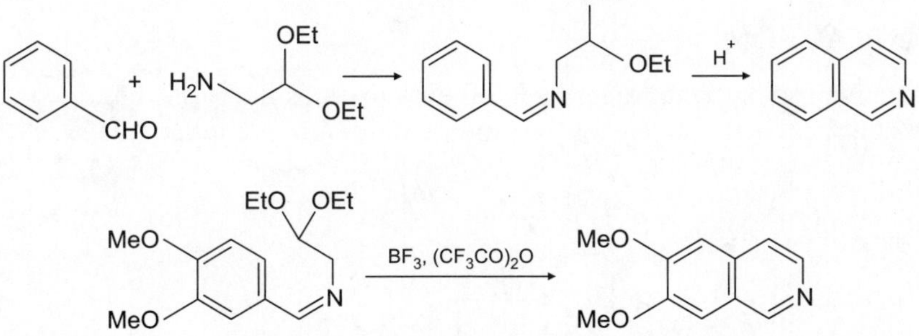

A benzaldehyde is the starting material, and it is reacted with an amino-acetaldehyde dialkyl acetal to form an imine, which is then cyclized directly under relatively severe acidic conditions (e.g. conc H_2SO_4 at 100°C) to give the isoquinoline. Although the Pomeranz-Fritsh ring-closure conditions permit the cyclization of unsubstitued imines, the reaction is accelerated greatly if electron-donating groups are present in the benzene ring.

In the **Schlittler-Müller modification**, the starting materials are benzyl amines and glyoxal semiacetal.

318. PONZIO REACTION

The Ponzio reaction refers to the transformation of ketoximes into *gem*-dinitro compounds via a pseudonitrole intermediate by means of oxidation with dinitrogen tetraoxide.

319. PRÉVOST REACTION

In Prévost reaction, an alkene is converted to a vicinal diol with anti stereochemistry by silver(I) carboxylate and iodine in an anhydrous solvent. Hydrolysis of the 1,2-dicarboxylate intermediates gives the desired diol. In this method, the alkene is treated with iodine and silver benzoate in a 1:2 molar ratio. The reaction was discovered by the French chemist Charles Prévost. This reaction has been applied for the conversion of alkenes into adjacent trans-diols.

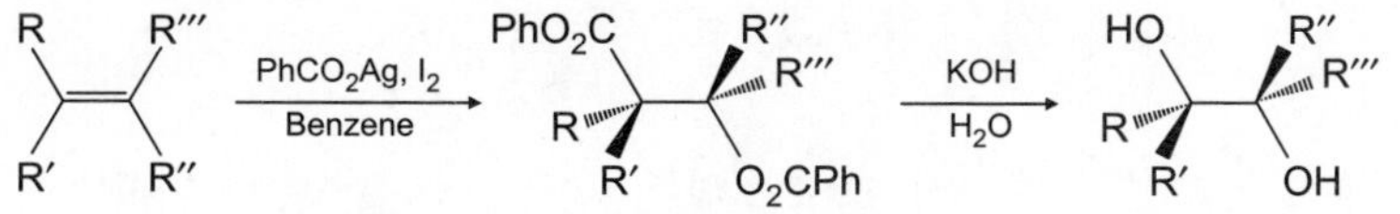

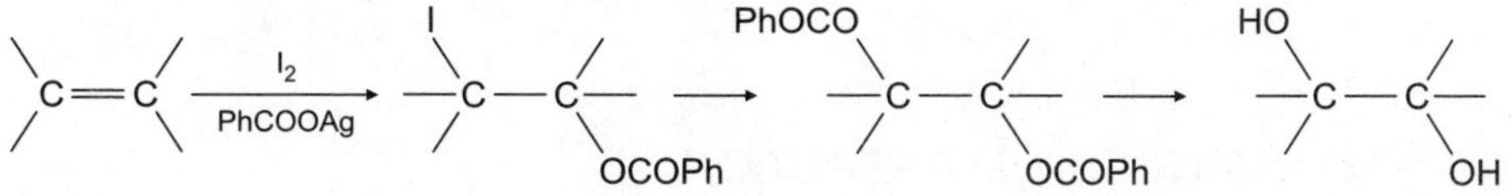

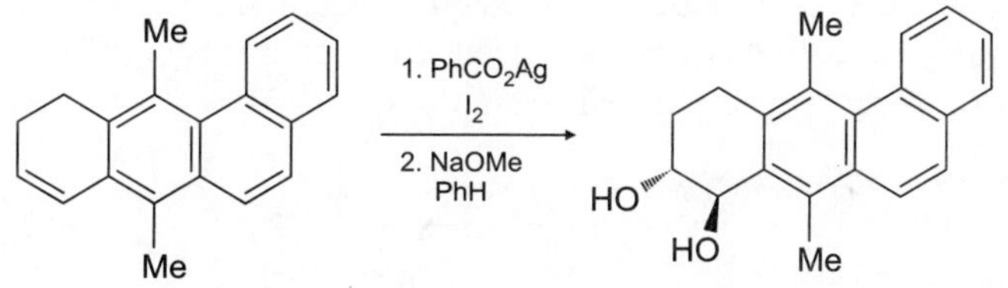

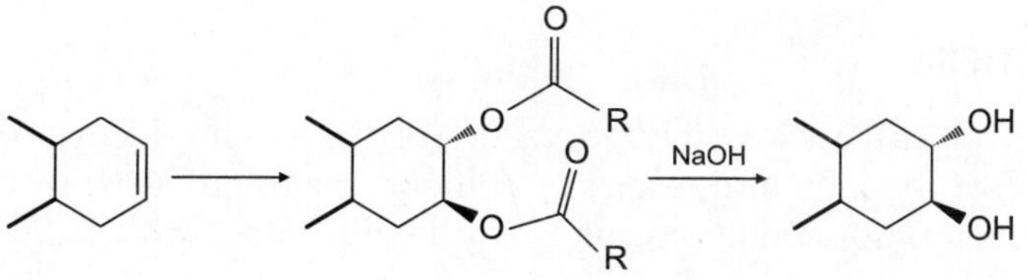

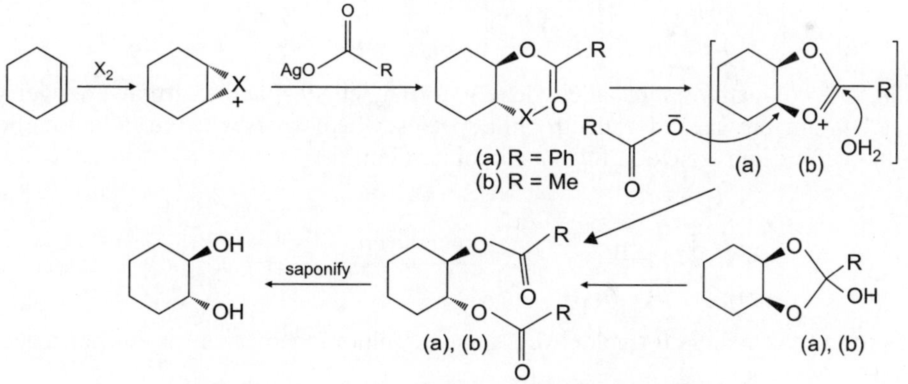

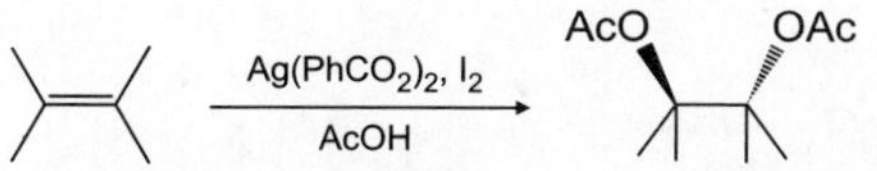

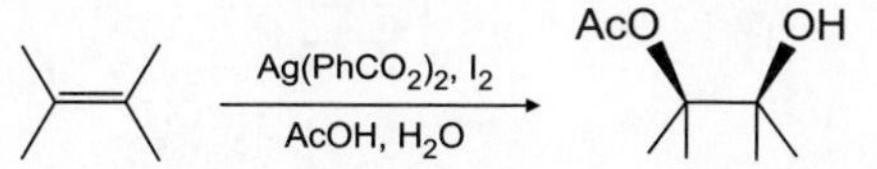

The Woodward modification of this reaction results in overall syn hydroxylation (syn-diols) where added water decomposes the benzoxonium intermediate directly to a syn-substituted diol. The alkene is treated with iodine and silver acetate in a 1:1 molar ratio in acetic acid containing water.

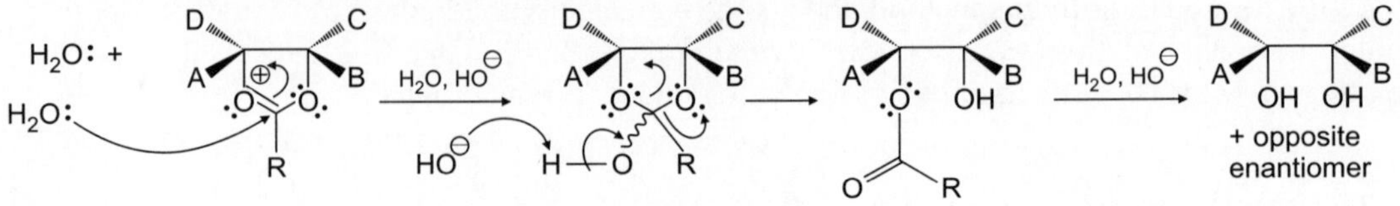

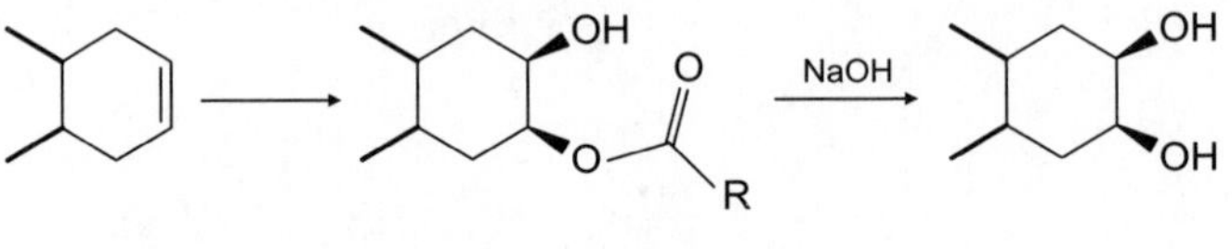

320. PRILEZHAEV (PRILESCHAJEW) REACTION

Formation of epoxides by the reaction of alkenes (with isolated olefinic double bond) with peracids. The commercial available mCPBA is a widely used reagent for this conversion, while magnesium mono-perphthalate and peracetic acid are also employed.

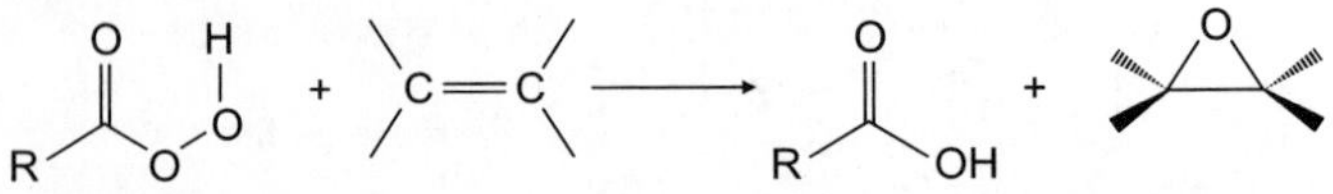

321. PRINS REACTION

The Prins reaction involves the acid-catalyzed reaction of olefins with aldehyde (often formaldehyde) usually leading to the formation of dioxanes as common major products. However, depending on substrate structure and reaction conditions, a 1,3-diol, allylic alcohol or a 1,3-dioxane may be formed.

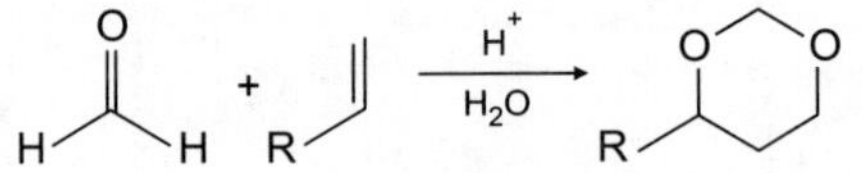

Initially, the protonation of the aldehyde, e.g. formaldehyde—at the carbonyl oxygen produces hydroxycarbenium ion as reactive electrophile species, which reacts with the carbon-carbon double bond of the olefinic substrate and form a carbenium ion.

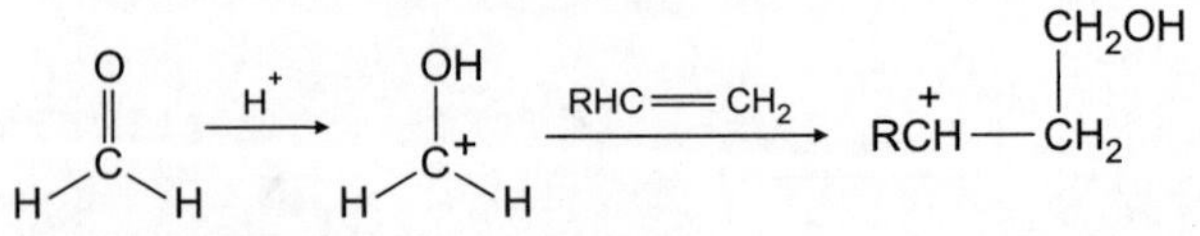

In the presence of excess formaldehyde, the carbenium ion species can further react to give a 1,3-dioxane.

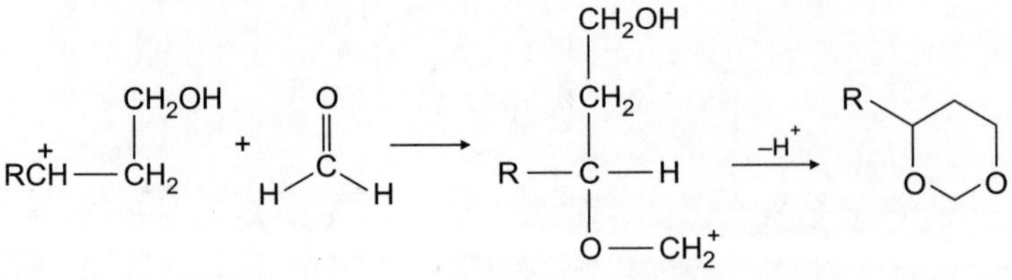

If only one equivalent of formaldehyde is used, however, 1,3-diol is formed as the major product:

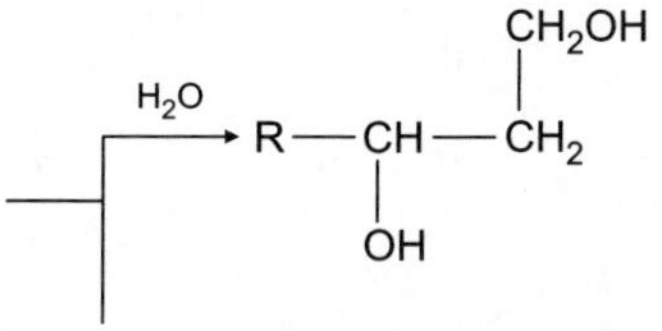

The Prins reaction suffers from the drawback of formation of mixtures of products. For example, the reaction of aqueous formaldehyde with cyclohexene under acid catalysis:

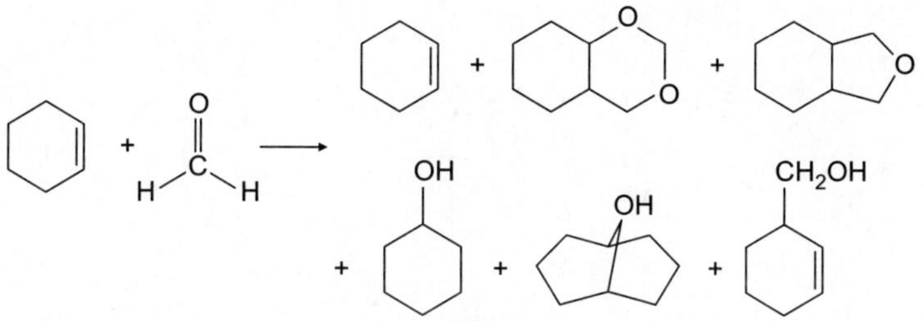

As a catalyst sulfuric acid is most often used; phosphoric acid, boron trifluoride or an acidic ion exchange resin have also found application. 1,1-disubstituted alkenes are especially suitable substrates, since these are converted to relatively stable tertiary carbenium ion species upon protonation.

322. PSCHORR REACTION

The Pschorr reaction involves intramolecular carbon-carbon bond formation between two aromatic rings of an aryl or heteroaryl diazonium salt through a carbonium or radical intermediate. This radical is generated *in situ* from an aryl diazonium salt by copper catalysis. Although excess copper salts are used, the yield is normally moderate.

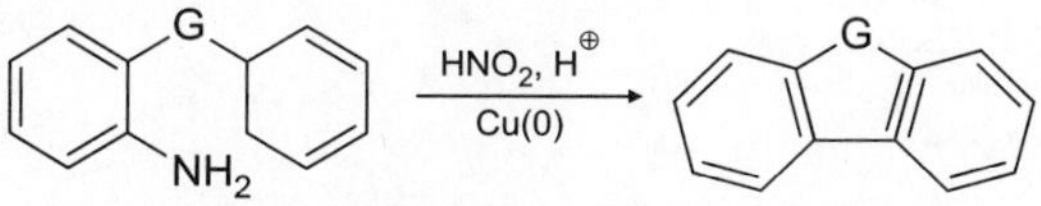

$G = CH = CH, CH_2 - CH_2, CO, NH, CH_2$, others

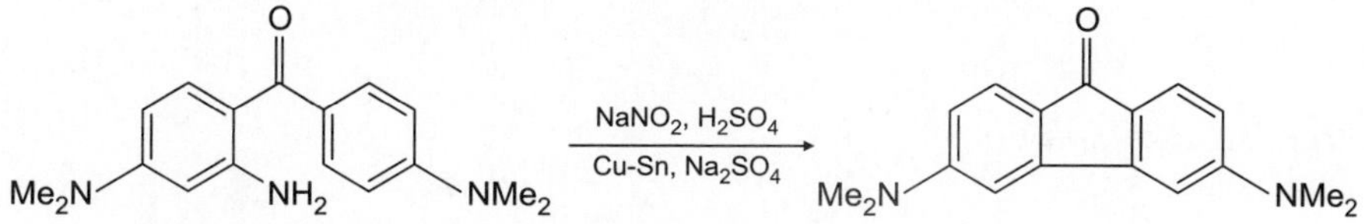

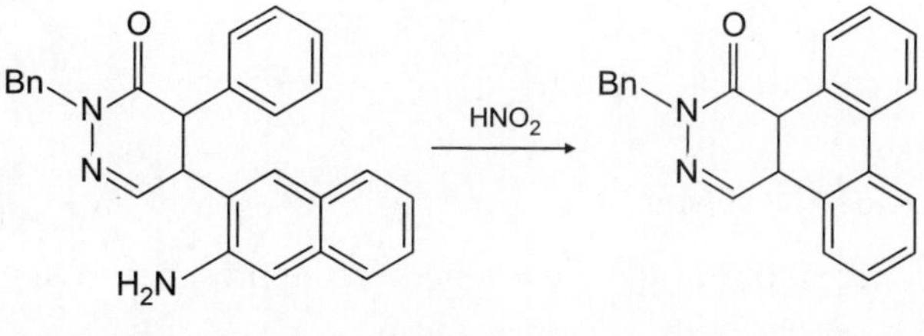

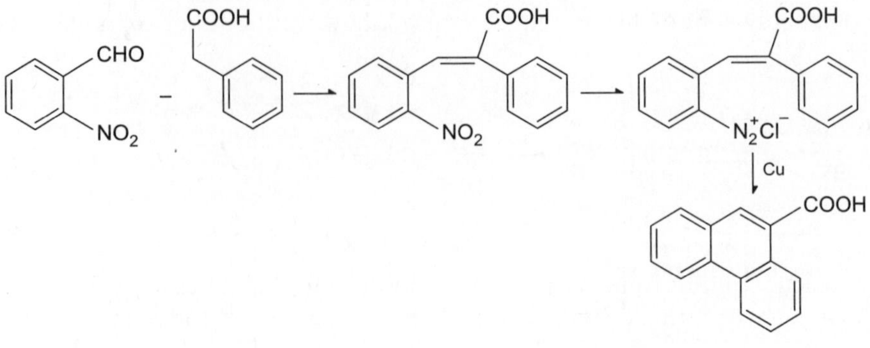

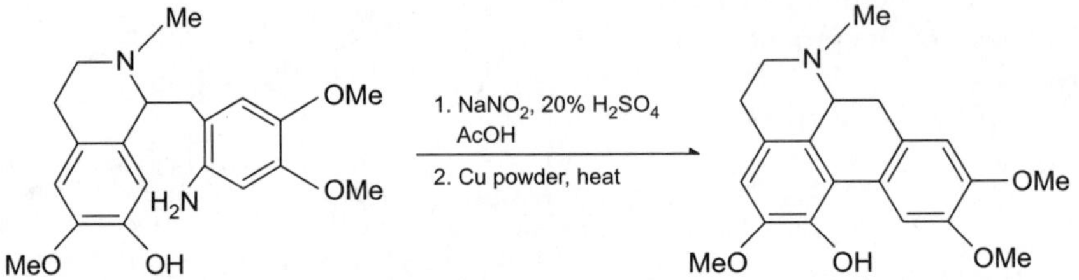

Mechanism: The reactive intermediate depends on the reaction conditions. Under acidic conditions, the diazonium salt is believed to decompose into aryl cation and nitrogen. The aryl cation is highly reactive and attacked by the aryl ring that leads to cyclization. On the other hand, under neutral and basic conditions, the diazonium salt is reduced by single electron transfer to give aryl radical which proceeds reaction intramolecularly with benzene ring to give the cyclized product.

In Acid Solution:

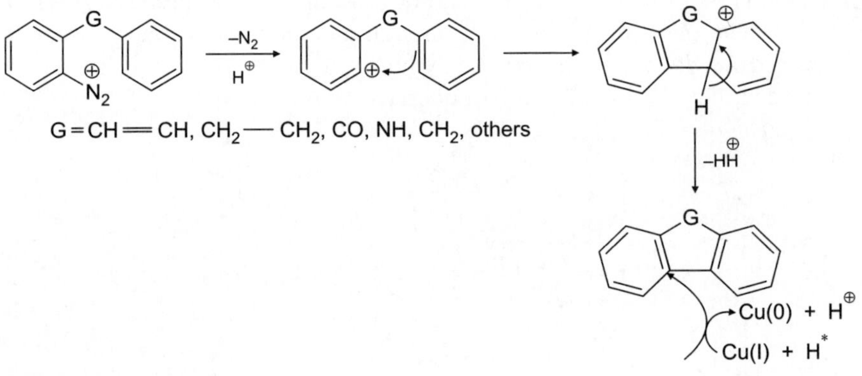

$$G = CH = CH, \ CH_2 - CH_2, \ CO, \ NH, \ CH_2, \ others$$

In Neutral or Basic Solution:

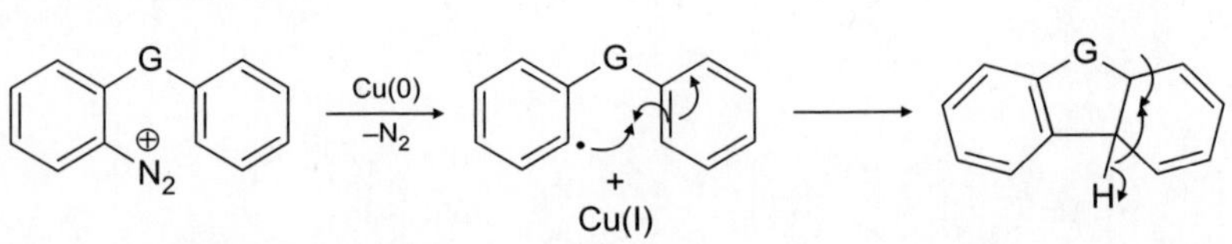

323. PUMMERER REARRANGEMENT

Rearrangement of sulfoxides to α-acyloxythioethers in the presence of acyclic anhydrides. When nucleophiles other than those derived from the anhydride are present, different functionalization is achieved:

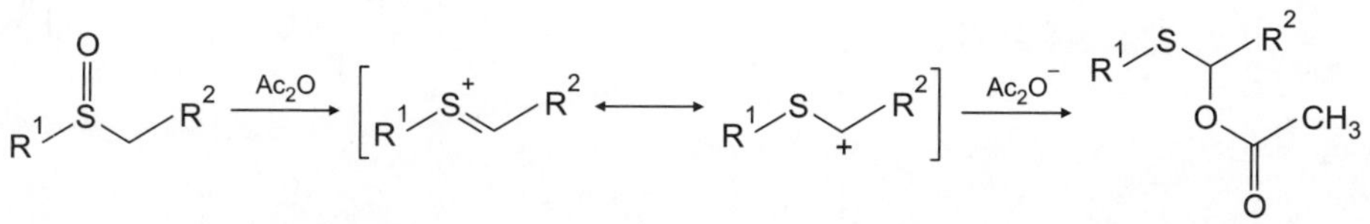

Pummerer rearrangements are well-known to yield internal redox products from C,H-acidic sulfoxides including deoxygenation of sulfur and oxygenation of carbon-hydrogen bond.

Pummerer rearrangements can be used to introduce an "aldehyde" on the carbon next to the sulfur, the product of the pummerer rearrangement is in the same oxidation state as an aldehyde and with hydrolysis of the product the aldehyde can be obtained.

324. PURDIE METHYLATION (IRVINE-PURDIE METHYLATION)

In Purdie methylation, alkyl glycoside on repeated treatment with methyl iodide and silver oxide leads to exhaustive methylation followed by hydrolysis with dilute acid to yield the anomeric hydroxyl group. This reaction is applicable for the structural analysis of polysaccharide and the methylation of sugars.

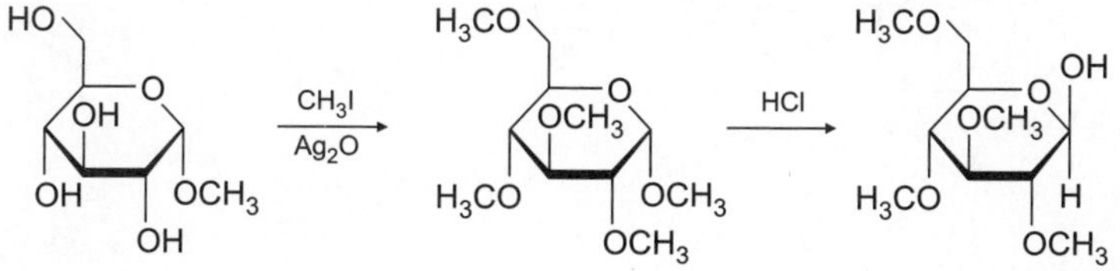

325. QUELET REACTION

The Quelet reaction or the Blanc-Quelet reaction involves a substitution reaction to yield α-chloroalkyl aromatic derivatives when dry hydrochloric acid passes through the solution of a phenolic ether and an aliphatic aldehyde in the presence or absence of a Lewis acid reagent (or a dehydration catalyst, e.g. $ZnCl_2$). In the reaction, substitution occurs on the para position to the alkoxy group or on the ortho position in the case of para-substituted phenolic ethers. This reaction is useful for the preparation of different styrene derivatives.

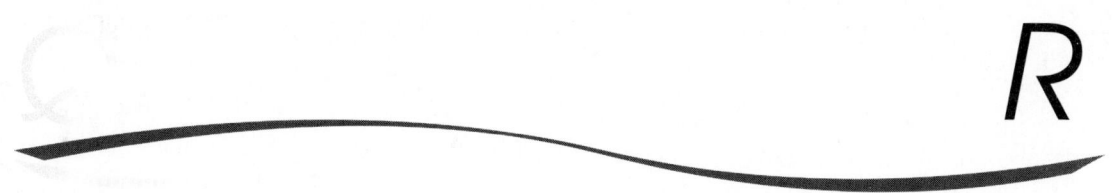

326. RAMBERG-BÄCKLUND REACTION

The base-mediated conversion of α-halosulfones into alkenes with extrusion of sulfur dioxide is known as the Ramberg-Bäcklund reaction. The reaction was named after Ludwig Ramberg and Birger Bäcklund in 1940. In the reaction, *Z* alkenes are often observed with weak bases, whereas strong bases give predominantly *E* alkenes.

Mechanistically, it is indicated that abstraction of a proton from the non-halogenated α-centre leads to formation of an episulfone, which extrudes sulfur dioxide to provide the product alkene.

This reaction was found to be a convenient procedure for preparing alkenes, because

- The starting material sulfones can be accessed easily. For example, they can be prepared from a sulfide by reaction with thionyl chloride (or with N-chlorosuccinimide) to give an α-chlorosulfide, followed by oxidation to the sulfone, e.g. using m-chloroperbenzoic acid.
- Location of the double bond can be clearly defined as alkene rearrangements do not occur under basic conditions.
- The broad range of substitution patterns that can be prepared including fully substituted alkenes, cyclobutenes and polyenes.
- Polyfunctional molecules can also be converted as long as base-sensitive groups are absent.

Base for the reaction can be alkoxides, e.g. potassium t-butoxide in an ethereal solvent, as well as aqueous alkali hydroxide. The RBR has become a versatile method for the preparation of alkene π bonds within a variety of structural motifs, including strained cyclic systems. For example, synthesis of ene-diyne,

The utility of this reaction was greatly augmented by Meyers' development of a one-pot chlorination/RBR sequence, which overall converts a sulfone into an alkene using carbon tetrachloride, potassium hydroxide, water and *t*-butanol. However, the formation of dichlorocarbene as a by-product of the reaction can lead to undesired reactivity in some instances.

Meyers' extension

327. RASCHIG PHENOL PROCESS

Raschig phenol process or Raschig-Hooker process is an industrial process for the production of phenol by the hydrolysis of chlorobenzene. Chlorobenzene for the reaction is produced by the chlorination of benzene with hydrochloric acid in the presence of air at 200 to 260°C over a copper catalyst on an alumina base. Chlorobenzene is then hydrolyzed with steam at about 425 °C temperature.

The Raschig phenol process should not be confused with two similarly named processes— Raschig process which is used for producing hydroxylamine and the Olin Raschig process used for producing hydrazine.

328. REED REACTION OR REED PROCESS

Reed process describes a photochemical reaction of hydrocarbons (paraffins and cycloparaffins) with sulfuryl chloride or sulfur dioxide and chlorine under irradiation with ultraviolet light to yield sulfonyl chlorides.

The reaction proceeds via the free radical mechanism starting with the generation of chlorine atom by the dissociation of chlorine molecule in presence of light which then attacks the hydrocarbon chain to form hydrogen chloride what results in the formation of alkyl free radical. Then SO_2 as an electron donor bonds to the reaction center, forming a sulfonyl radical. Both sulfochlorination and chlorination products are obtained in the reaction. This reaction can also be done by the use of gamma-ray illumination.

Chain initiation:

$$Cl_2 \xrightarrow{h\nu} 2Cl\cdot$$

Chain propagation steps:

$$R\!-\!H \ + \ \cdot Cl \longrightarrow R\cdot \ + \ HCl$$

$$R\cdot \ + \ :SO_2 \longrightarrow R\!-\!\dot{S}O_2$$

$$R\!-\!\dot{S}O_2 \ + \ Cl_2 \longrightarrow R\!-\!SO_2\!-\!Cl \ + \ Cl\cdot$$

329. REFORMATSKY (REFORMATSKII) REACTION

The Reformatsky reaction involves condensation of ester-derived zinc enolates with aldehydes or ketones to give corresponding β-hydroxy esters. The reaction was named after its discoverer Sergey Nikolaevich Reformatsky.

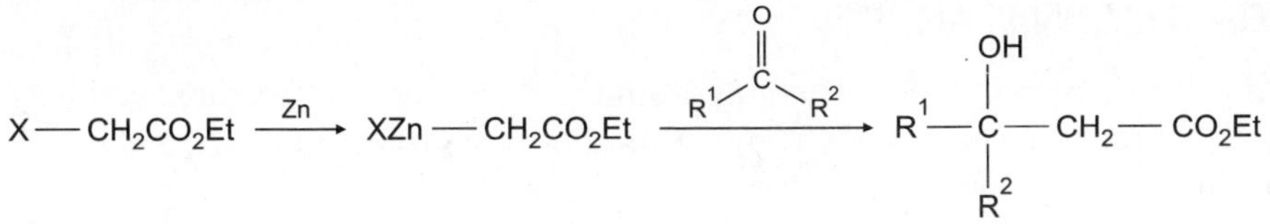

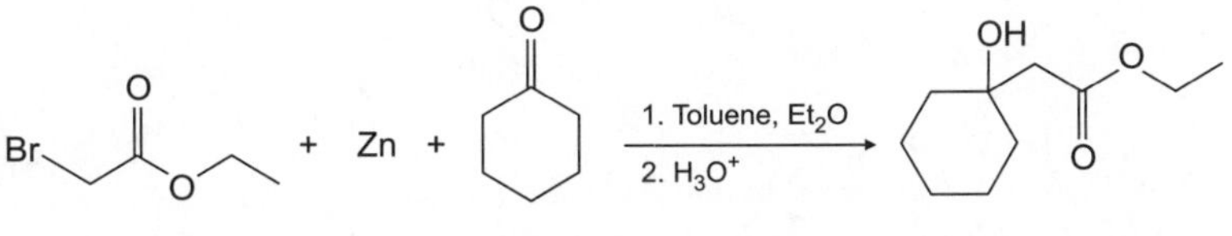

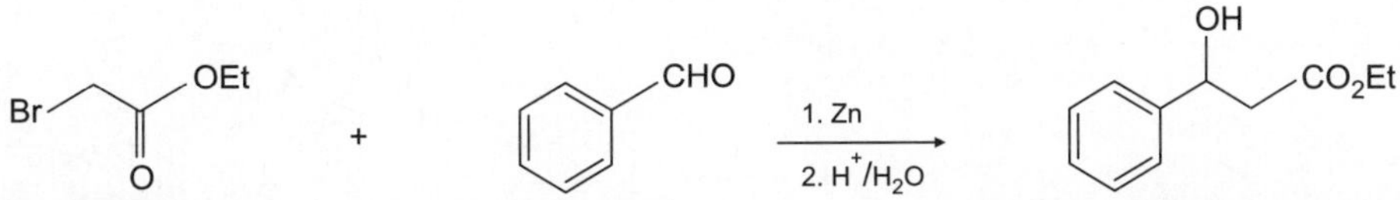

Ethyl-2-bromoacctato   Benzaldehyde   Ethyl-3-hydroxy-3-phenylpropanoate

Zinc enolates are generated by addition of an α-haloester in THF, DME, Et$_2$O, benzene, or toluene to an **activated zinc**, such as a Zn-Cu couple or zinc obtained by reduction of zinc halides with potassium (**Rieke zinc**). Better yields are obtained if using Zn-Cu couple or *in situ* preparation of zinc by reduction of zinc halides by potassium (also known as Rieke zinc).

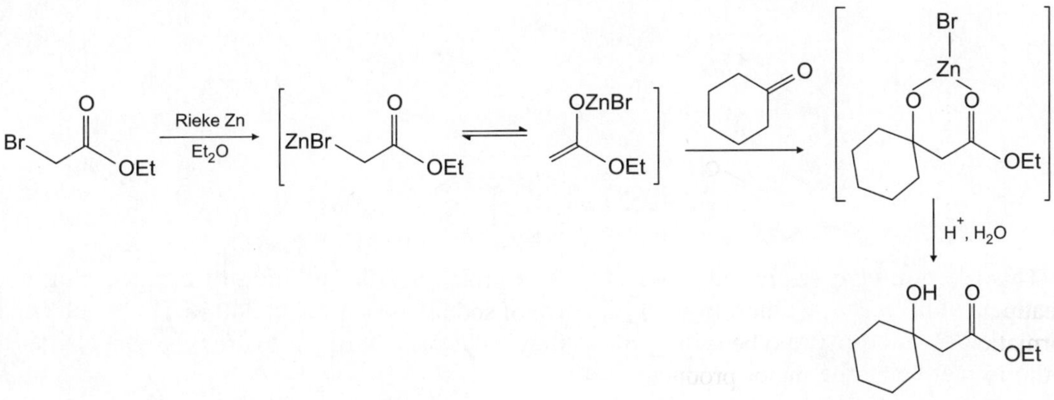

Organozinc compounds for the reaction are prepared from α-halogenesters, which react as the nucleophilic partner with ketones or aldehydes in an addition to give β-hydroxy esters. The carbonyl substrate in the reaction can be an aldehyde or ketone. Moreover, these aldehydes or ketones can be aliphatic, aromatic, or heterocyclic or contain various functional groups. Solvents such as ethers, including diethyl ether, THF, and 1,4-dioxane can be used, although the reaction can also be done in water using dibenzoyl peroxide and $MgClO_4$.

Several modified Reformatsky reactions using other metals have been reported. Also, the scope of the Reformatsky reaction has been extended with special techniques for the activation of the metal (e.g. for removal of the oxide layer, and the preparation of finely dispersed metal). Much more effective protocol is the use of special alloys, e.g. Zn-Cu couple, or the reduction of zinc halides using potassium (the so-called Rieke procedure) or potassium graphite. Although, the application of ultrasound has also been reported for zinc metal, promoters such as iodine and potassium iodide were needed sometimes to obtain the desired products in good yields.

330. REIMER-TIEMANN REACTION

The reaction was discovered by Karl Ludwig Reimer and Ferdinand Tiemann. Reimer-Tiemann reaction is an electrophilic substitution reaction used for the ortho-formylation of aromatic rings with chloroform and hydroxide ion.

This reaction involves introduction of –CHO group at ortho position of benzene ring upon treatment of phenol with chloroform in presence of sodium hydroxide at 340 K. This results in the formation of ortho-hydroxybenzaldehyde (salicylaldehyde) and para-hydroxybenzaldehyde, the ortho isomer being the major product.

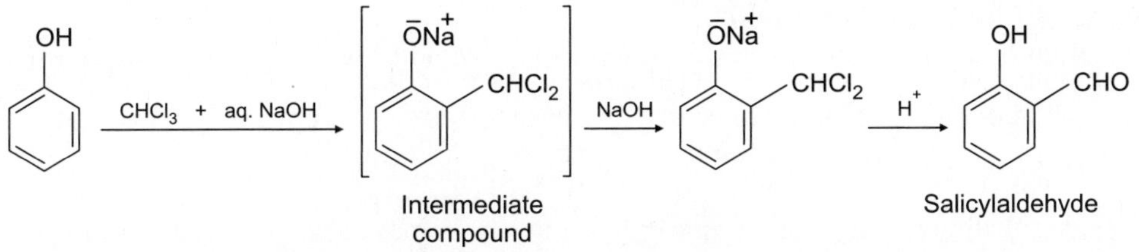

Intermediate
compound

Salicylaldehyde

Ortho formylated phenols, e.g. salicylic aldehyde are preferentially formed in Reimer-Tiemann reaction contrast to the other formylation reactions, e.g. the *Gattermann reaction* where para-formyl derivative is obtained as a major product.

The actual formylation process is preceded by the formation of dichlorocarbene as the reactive species. Chloroform is deprotonated in strongly alkaline solution to form the chloroform carbanion (trichloromethide anion), which quickly alpha-eliminate or decomposes into dichlorocarbene and a chloride.

In alkaline solution, the phenol is also deprotonated to the phenolate and reacts at the ortho-position with dichlorocarbene to form an initial addition reaction product, which isomerizes to the aromatic o-dichloromethyl phenolate. Under the reaction conditions, aromatic o-dichloromethyl phenolate is hydrolyzed to the o-formyl phenolate.

(a) Carbene generation:

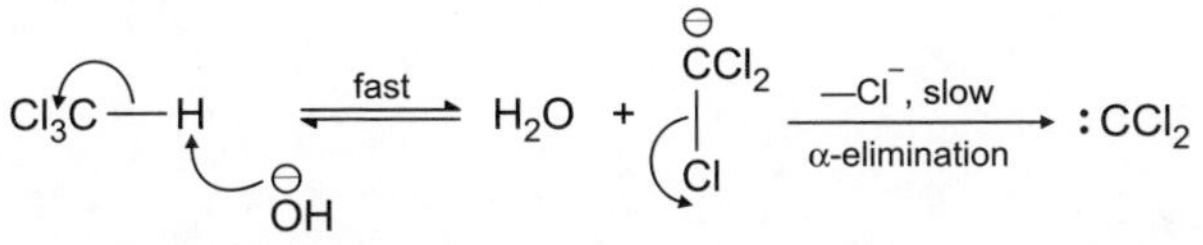

(b) Addition of dichlorocarbene and hydrolysis:

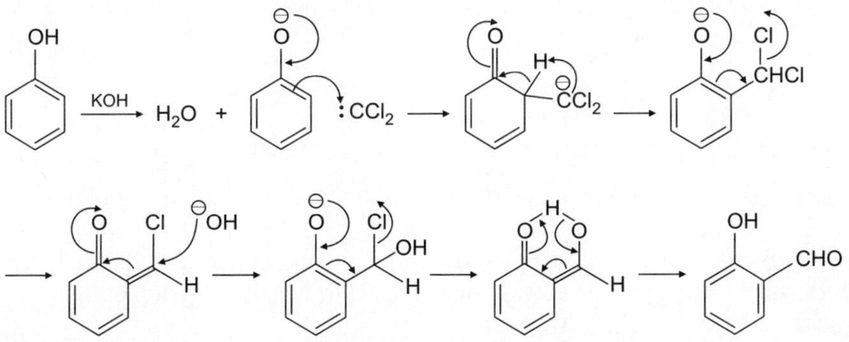

This reaction is limited to the formylation of phenols and certain reactive heterocycles like pyrroles and indoles.

The Reimer-Tiemann reaction, utilized in the formylation of activated aromatic rings, suffers from two major drawbacks:

Firstly, the biphasic nature of the system acts to limit the reaction interface.

Inefficient transfer of reagents between these layers may stifle the rate of reaction and decrease yields.

Secondly, to obtain a satisfactory reaction rate, the process should be performed at reflux.

331. REISSERT INDOLE SYNTHESIS

Reissert Indole synthesis is a multistep synthesis of indole derivatives, where an ortho-nitrotoluene reacts first with oxalic ester in the presence of a base to form o-nitrophenylpyruvic ester, then the nitro group of the resulting α-keto ester is reduced to amino followed by cyclization to indole-2-carboxylic acid.

The ester substituent in the resulting indole may be removed, if required by hydrolysis and thermal decarboxylation.

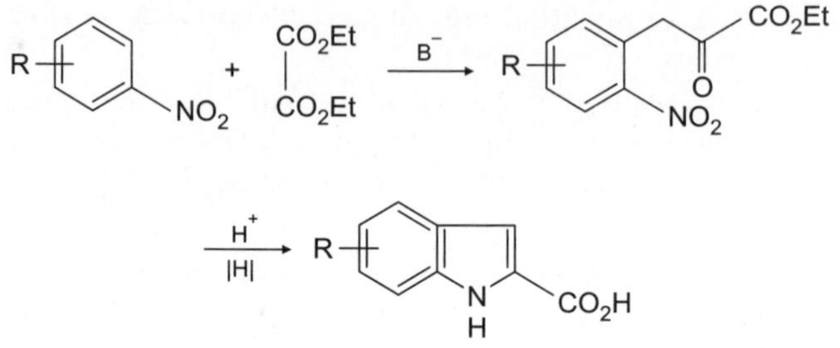

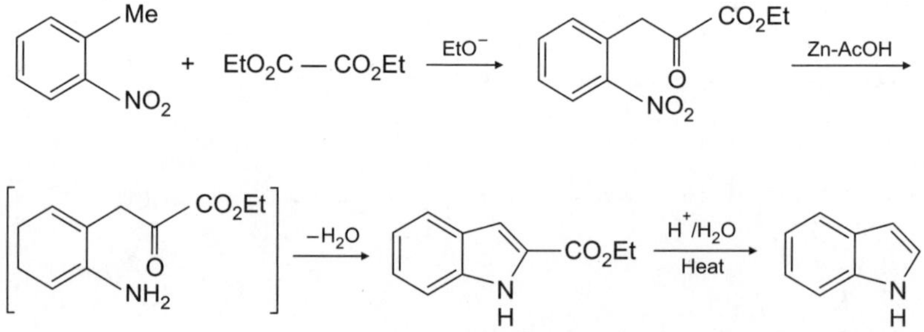

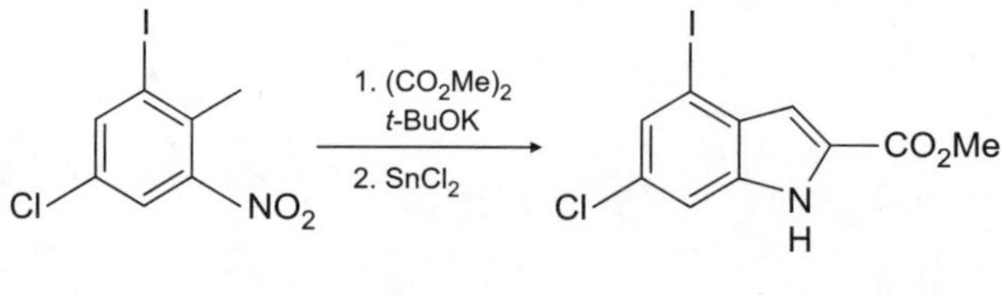

The Reissert synthesis is suitable for the preparation of 2-substituted indoles. For example, when 1-methyl-2-nitrobenzene is treated with dimethyl oxalate in the presence of sodium methoxide as a base, it gave methyl 3-(2-nitrophenyl)-2-oxopropanoate, which upon catalytic hydrogenation (H_2, Pd-C) reduced the nitro group to an amino group followed by a spontaneous cyclodehydration to give indole-2-carboxylic acid methyl ester.

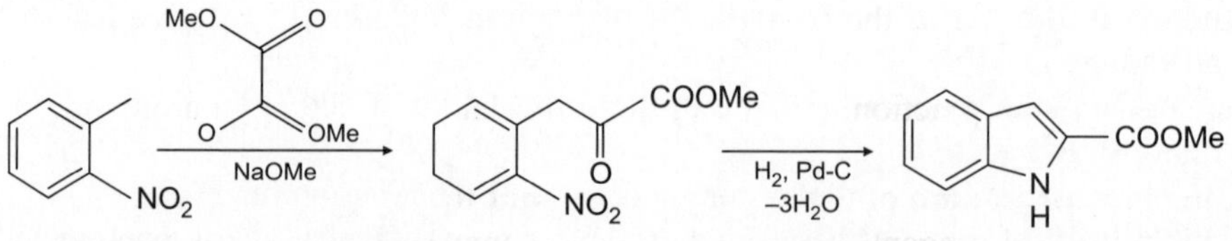

332. REISSERT REACTION (GROSHEINTZ-FISCHER-REISSERT ALDEHYDE SYNTHESIS)

The Reissert reaction is a series of chemical reactions that transforms quinoline to quinaldic acid. It involves formation of 1-acyl-2-cyano-1,2-dihydroquinoline derivatives, also known as Reissert compounds, by reaction of acid chlorides with quinoline and potassium cyanide; hydrolysis of these compounds yield aldehydes and quinaldic acid.

333. REPPE CHEMISTRY

During the 1930s, I.G. Farben embarked on a completely novel area of diversification based on high-pressure acetylene chemistry. This had previously been considered far too dangerous, since acetylene under pressure is inherently unstable and explodes with extreme violence. The basic research for safely reacting the highly flammable gas with other chemicals under high pressures was done by J. Walter Reppe.

The term **Reppe chemistry** involves the use of acetylene at high pressures in the presence of suitable catalysts to carry out vinylation, ethynylation, cyclopolymerization and carbonylation.

334. RETRO-DIELS–ALDER REACTION

The retro-Diels–Alder reaction is essentially the microscopic reverse of the Diels–Alder reaction. It is a concerted, pericyclic, single-step process accomplished spontaneously with heat, or with acid or base mediation. It involves the formation of a diene and dienophile from a cyclohexene.

Like the Diels–Alder reaction, the rDA preserves configuration in the diene and dienophile. Thermal dissociation of Diels-Alder adducts occur most readily when one or both fragments are particularly stable. Because the Diels–Alder reaction exchanges two π bonds for two σ bonds, it is intrinsically thermodynamically favored in the forward direction, and due to the high reactivity or volatility of the emitted dienophile only a few rDA reactions occur spontaneously at room temperature. Most, however, require either thermal or chemical activation.

335. RETROPINACOL REARRANGEMENT

The retropinacol rearrangement involves conversion of alcohols with a tertiary β-carbon into olefin along with the migration of an alkyl group from the tertiary β-carbon in the presence of an acid. This reaction allows conversion of an alcohol to the rearranged olefin on treatment with acid.

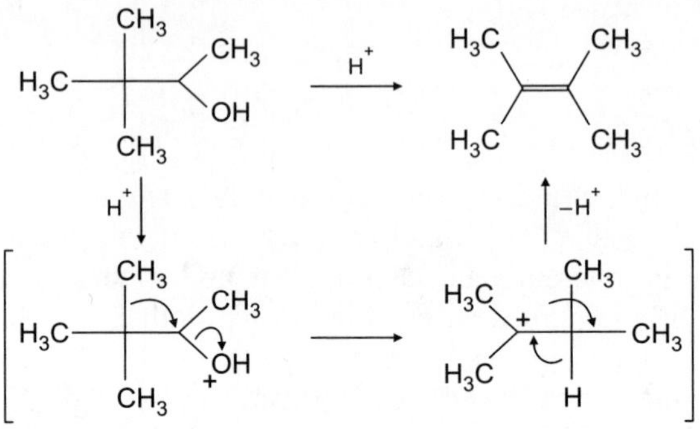

336. REVERDIN REACTION OR REVERDIN REARRANGEMENT

Reverdin reaction describes migration of iodine during nitration of iodophenolic ethers. The reaction involves nitrosode-iodination. Migration of iodine occurs from the 4-position to the 2-position during the nitration and 2-iodo-4-nitro-anisole to give 2-iodo-4-nitroanisol.

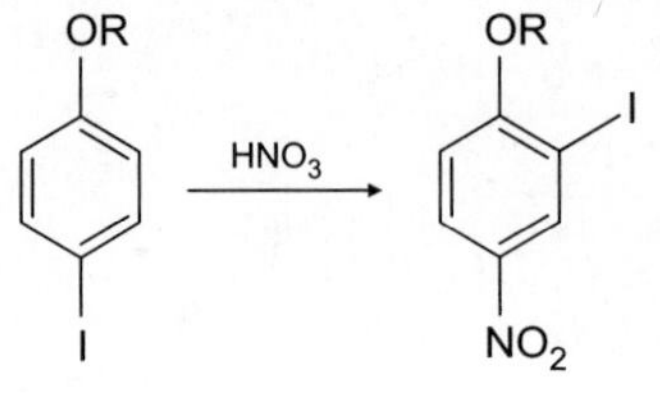

337. RIEHM QUINOLINE SYNTHESIS

Synthesis of quinolines by arylamine hydrochlorides with ketones with or without use of aluminum chloride or phosphorus pentachloride is generally referred to as the Riehm quinoline synthesis.

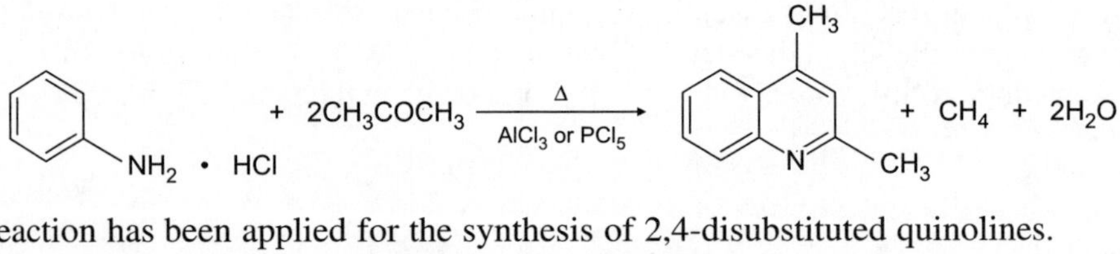

This reaction has been applied for the synthesis of 2,4-disubstituted quinolines.

338. RIEMSCHNEIDER THIOCARBAMATE SYNTHESIS

The acidic (e.g. concentrated sulfuric acid) transformation of alkyl or aryl thiocyanates into the corresponding thiocarbamates is known as the Riemschneider reaction. In the reaction, the thiocyanate is first treated with sulfuric acid and then hydrolyzed with ice water. The reaction was discovered by the German chemist Randolph Riemschneider.

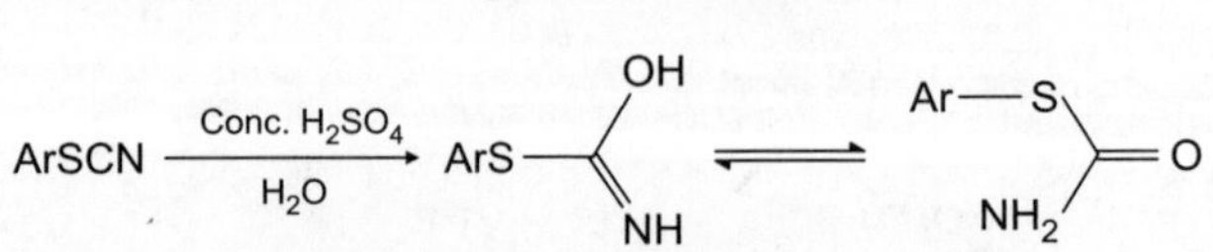

339. RILEY OXIDATIONS (SELENIUM DIOXIDE OXIDATION)

The oxidation of active methyl or methylene group into a carbonyl or alkoxyl group using selenium dioxide as the mild oxidant is generally known as the Riley oxidation. The active methylene or methyl group usually refers to when it is present adjacent to the carbonyl group, an adjacent double bond, or aromatic, etc.

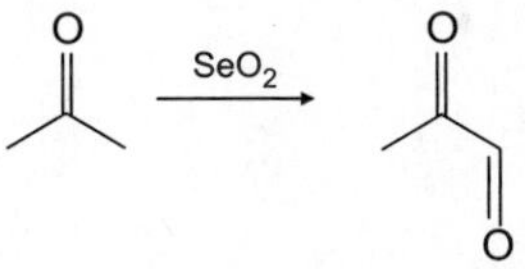

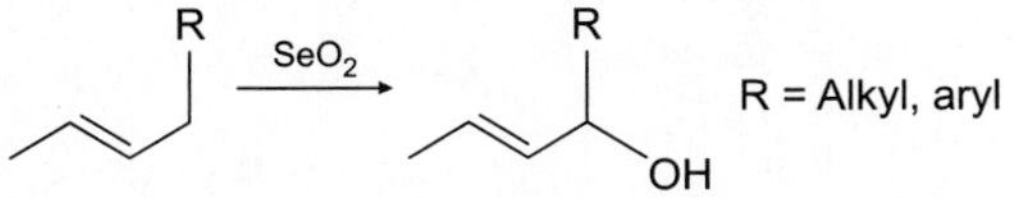

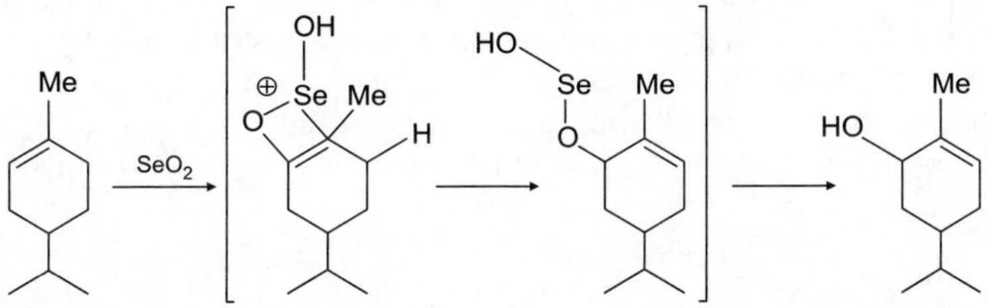

R = Alkyl, aryl

This reaction is especially useful for the preparation of allyl alcohols, natural products and biologically active compounds.

340. RITTER REACTION

The ritter reaction is a general method for the amidation of alcohols or alkenes with nitriles. The reaction typically requires a strongly ionizing solvent and a stoichiometric amount of a strong acid (usually sulfuric acid), thus limiting its applicability to compounds with groups that survive such harsh conditions.

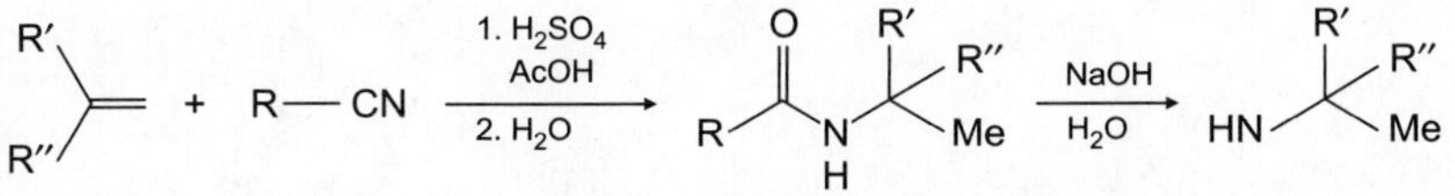

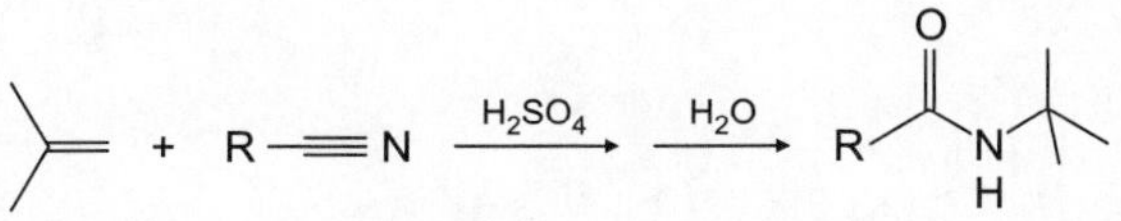

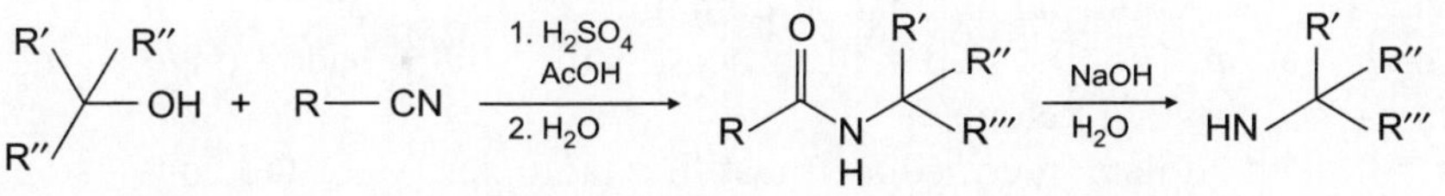

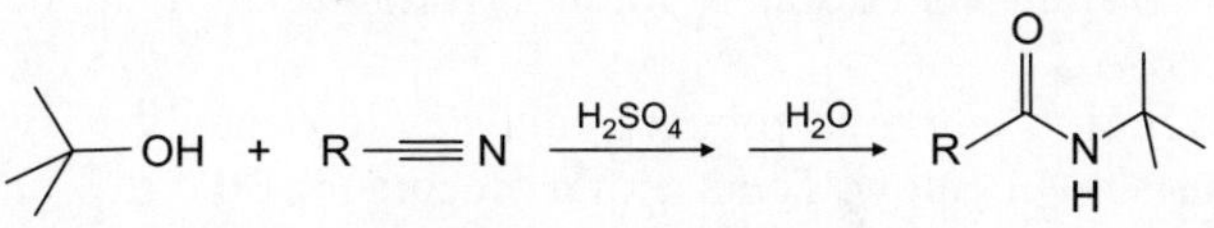

Acidification of the appropriate alcohol or alkene generates a carbenium ion, which reacts with nitrile. While the successful course of the reaction certainly depends on the reactivity of nitrile, a major factor is the stability and reactivity of the carbocationic intermediates.

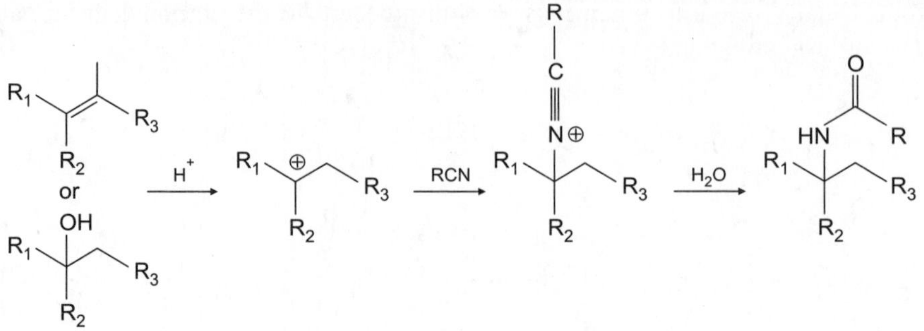

Any substrate capable of generating a stable carbenium ion is a suitable starting material; primary alcohols do not react under these conditions with exception of benzylic alcohols.

Ritter reaction is a well-known reaction in the synthetic organic processes, which provides versatile synthetic method for preparation of a variety of amides. It is also used in industrial processes as it can be effectively scaled up from laboratory experiments to large-scale applications while maintaining high yield.

341. ROBINSON ANNULATION

Robinson annulation is an important extension of the intramolecular aldol condensation and refers to the Michael addition of a cyclic ketone, e.g. cyclohexanone with methyl vinyl ketone resulting in a ring closure to yield a bicyclic α,β-unsaturated ketone.

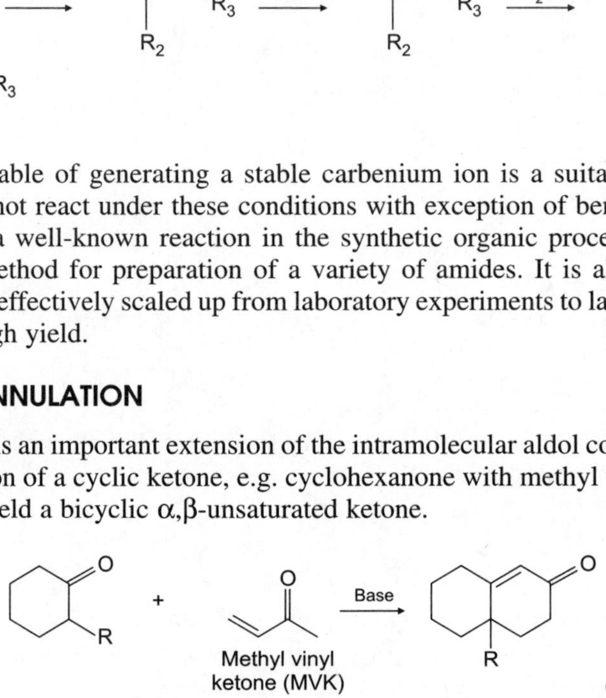

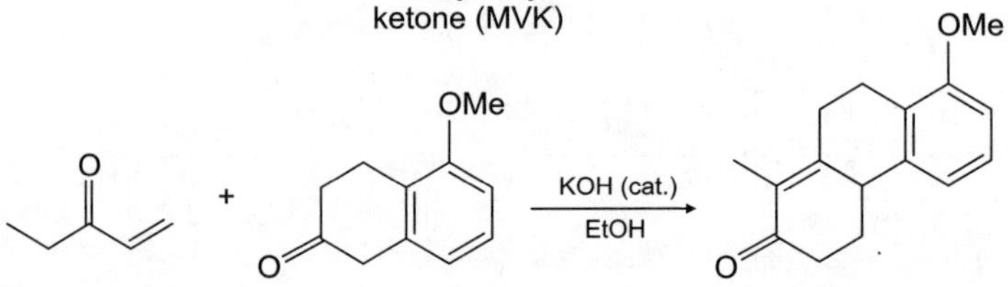

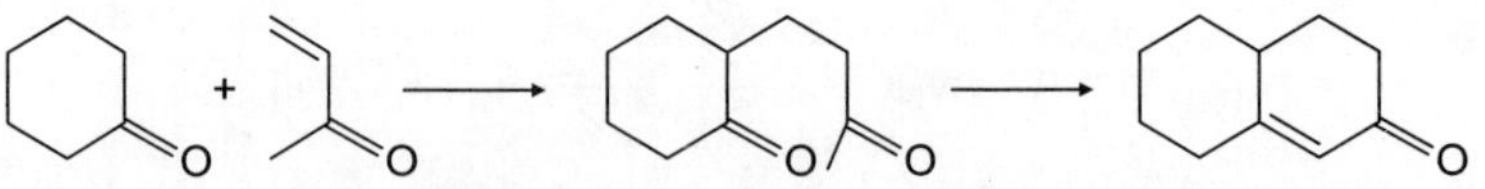

Mechanistically the Robinson annulation combines two reactions: the Michael Addition and the Aldol Condensation. The first step in the process is the Michael addition to an α,β-unsaturated ketone, such as methyl vinyl ketone.

The newly formed enolate intermediate must first tautomerize for the conversion to continue:

The subsequent cyclization via Aldol Addition is followed by a condensation to form a six-membered ring enone:

The Robinson Annulation can also proceed under acidic catalysis, with the entire process occurring in one pot, as shown below. The use of a precursor of the α,β-unsaturated ketone, such as a β-chloroketone, can reduce the steady-state concentration of enone and decrease the side reaction of polymerization.

Initially, upon treatment with a base, the cyclic ketone is deprotonated to give an enolate, which undergoes a conjugate addition to the methyl vinyl ketone, i.e. a Michael addition, to give a 1,5-diketone which undergoes an intramolecular aldol reaction leading to closure of a six membered ring in the next step. Subsequent dehydration yields the bicyclic enone.

Because methyl vinyl ketone has a tendency to polymerize in the presence of a strong base, the yield of annulation product is therefore often low. Precursors are often used instead, that is from which the Michael acceptor is generated *in situ*, upon treatment with a base. The quaternary ammonium salt for example, $MeCOCH_2CH_2NEt_2Me^+ I^-$ can be obtained by quaternization reaction of the tertiary amine, which in turn is prepared from acetone, formaldehyde and diethylamine in a *Mannich reaction*.

This reaction has found wide application in the synthesis of terpenes, and for the formation of six-membered rings in polycyclic compounds, such as steroids.

342. ROBINSON-SCHOPF REACTION

The Robinson tropinone synthesis refers to the one-pot multicomponent synthesis of tropinone from succindialdehyde, acetonedicarboxylic acid and methylamine in aqueous solution involves a double Mannich reaction.

343. ROSENMUND REDUCTION

In Rosenmund reduction the acid chlorides are reduced to aldehydes by hydrogen in presence of palladium suspended in barium sulphate as catalyst. $BaSO_4$ is used as a catalytic poison, to stop the reduction at the stage of aldehyde because the untreated catalyst may cause over reduction.

Therefore the aldehyde formed will be further reduced to primary alcohol. Side products are also formed in the reaction due to presence of water and can be avoided if the reaction is conducted in anhydrous solvents. Rosenmund reduction is named after Karl Wilhelm Rosenmund. It is one of the most useful methods for preparation of a large variety of aldehydes from acids. However, the original Rosenmund reduction was considered unsafe and inefficient because it requires bubbling gaseous hydrogen though the reaction mixture at high temperatures.

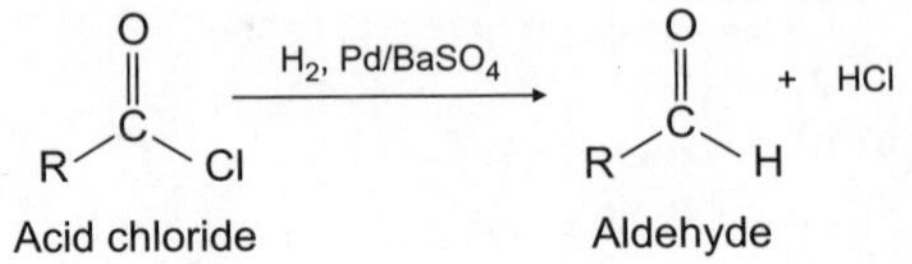

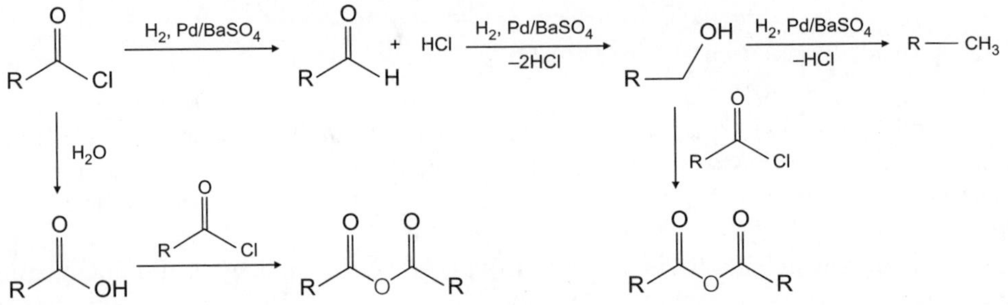

The reaction mechanism presumably involves an organopalladium species as an intermediate, which then reacts with the hydrogen. This reduction is usually applied for the conversion of a carboxylic acid into the corresponding aldehyde via the acyl chloride.

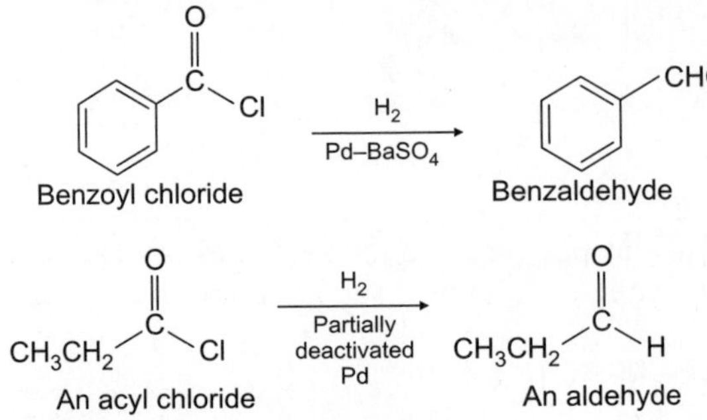

344. ROSENMUND-VON BRAUN SYNTHESIS

Rosenmund-von Braun synthesis gives rise to aryl nitriles from the coupling of copper cyanide (CuCN) with aryl halides at reflux temperature. By this reaction aryl nitriles can be prepared by the cyanation of aryl halides with an excess of copper(I) cyanide in a polar high-boiling solvent such as DMF, nitrobenzene, or pyridine at reflux temperature.

$$ArX + CuCN \xrightarrow{\Delta} ArCN + CuX$$

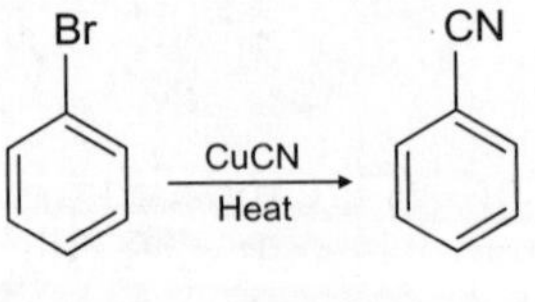

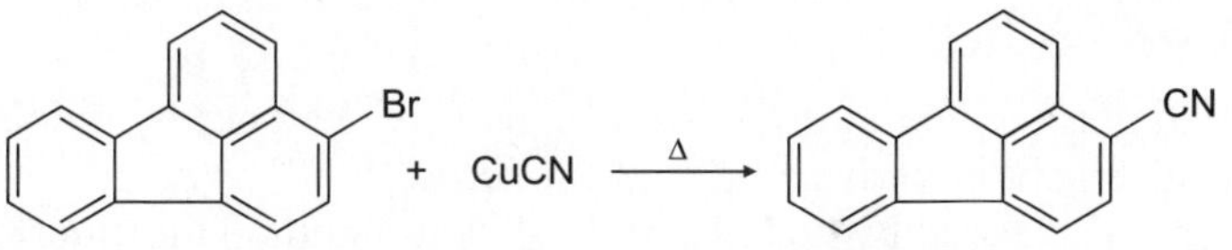

This reaction is usually carried out in polar and high boiling solvent solvents such as nitrobenzene, DMF and pyridine using excess of CuCN at reflux temperature.

Because of excess of copper cyanide and the use of a polar, high-boiling point solvent, the product isolation in this reaction is a tedious process, involving washings with water followed by extensive silica gel column chromatography. Apart from these, elevated temperatures (up to 200°C) also a drawback for the reaction since it loweres the functional group tolerance.

Modern methodologies using catalytic amounts of transition-metal complexes, together with less toxic cyanide sources have made the chemistry of this reaction more efficient, applicable and safer. Since the reactions often still require the use of elevated temperatures the application of microwave heating has proven valuable.

345. ROTHEMUND REACTION OR ROTHEMUND PORPHYRIN SYNTHESIS

Rothemund reaction refers to the preparation of meso-tetrasubstituted porphyrins by the thermal condensation of pyrrole with an aldehyde:

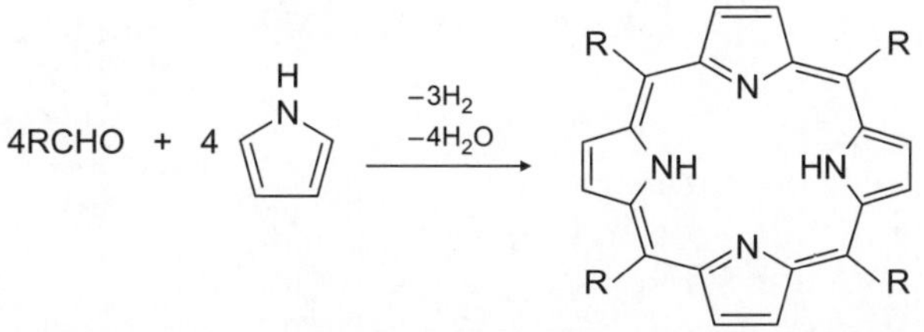

The chemistry of meso-substituted porphyrins has its foundation in the work of Rothemund in 1935. Rothemund performed the synthesis of meso-tetramethylporphyrin by utilizing the condensation reaction between acetalaldehyde and pyrrole in methanol at various temperatures. Various aldehydes like propionaldehyde, benzaldehyde, n-butyraldehyde, α-furaldehyde were utilized using this methodology

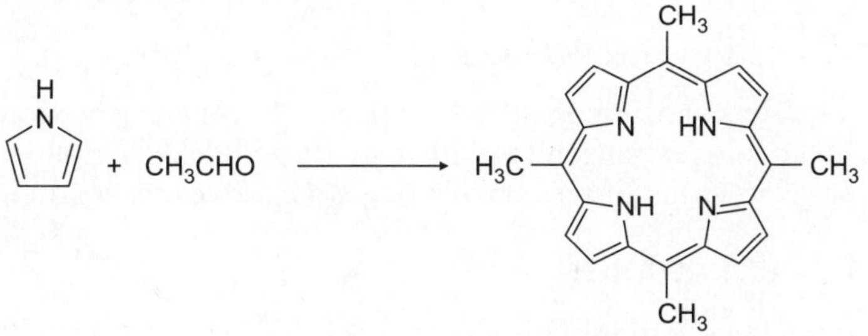

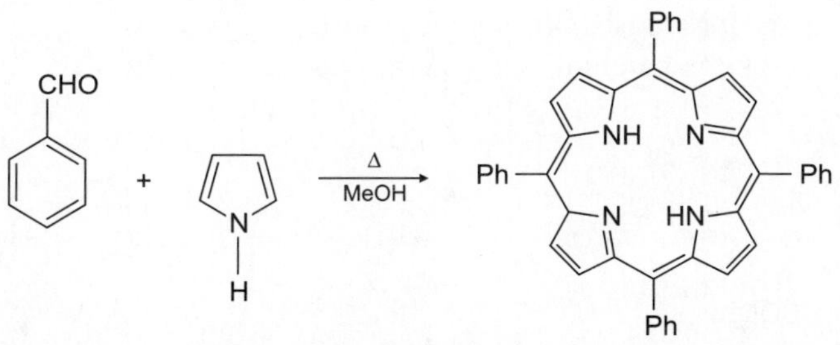

346. RUBOTTOM OXIDATION

The Rubottom oxidation refers to the synthesis of α-hydroxy ketones by oxidation of enolsilanes (silyl enol ethers) with m-chloroperbenzoic acid (m-CPBA).

The silyl group can be cleaved by means of tetra-N-butylammonium fluoride (TBAF) or with aqueous workup. This reaction is an excellent method for α-hydroxyl incorporation using silyl enol ethers, involving an epoxidation.

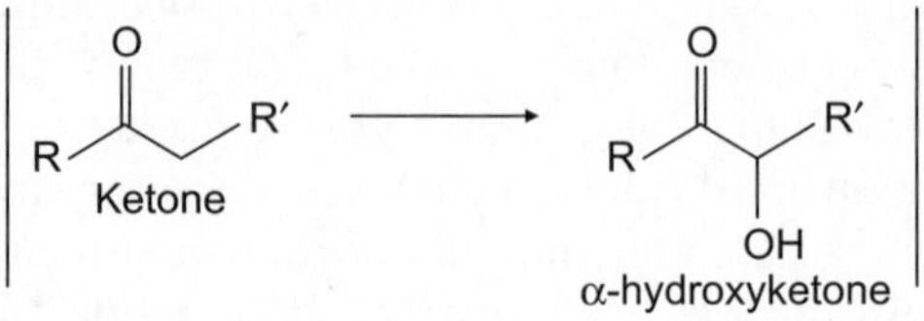

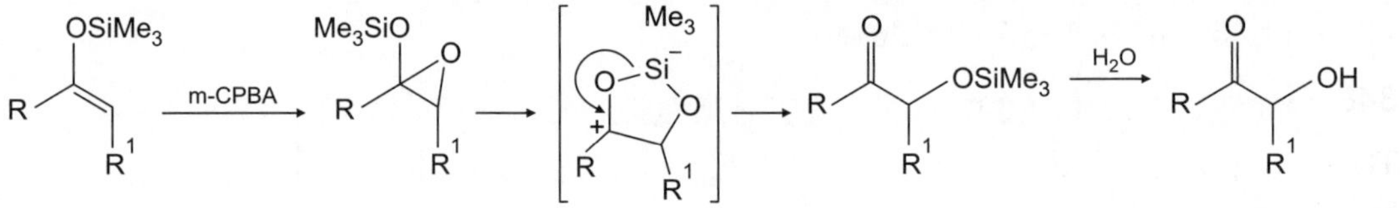

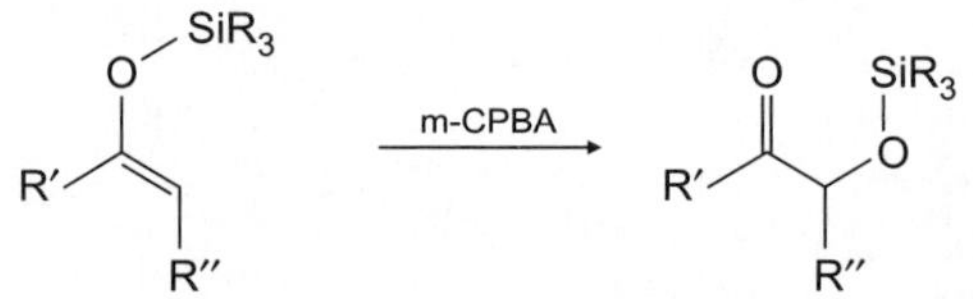

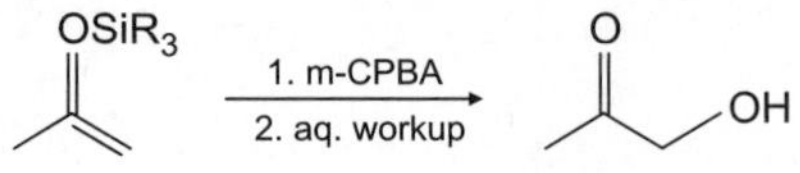

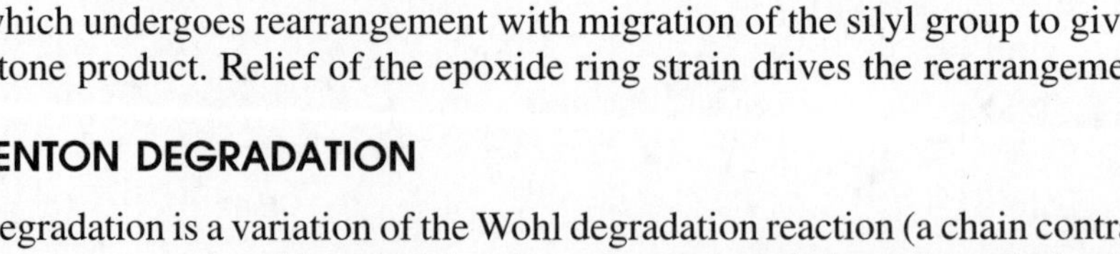

TMSCl trimethylsily chloride/LDA lithium diisopropylamide

The enol ether double bond is epoxidized with m-chloroperoxybenzoic acid to give an epoxysilane which undergoes rearrangement with migration of the silyl group to give the silylated α-hydroxy ketone product. Relief of the epoxide ring strain drives the rearrangement.

347. RUFF-FENTON DEGRADATION

Ruff-Fenton degradation is a variation of the Wohl degradation reaction (a chain contraction method in carbohydrate chemistry for aldoses). It involves shortening of the carbon chain of sugars by the oxidation of aldonic acids (as calcium salts) with hydrogen peroxide and ferric salts (Iron(III) sulfate).

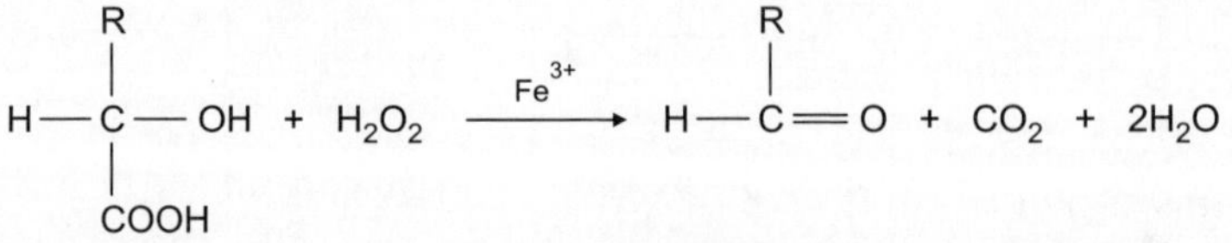

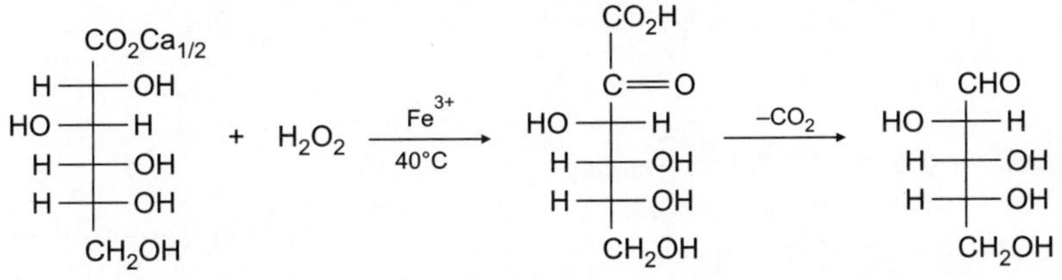

$$Fe^{2+} + H_2O_2 \longrightarrow Fe^{3+} + HO^- + HO^-$$

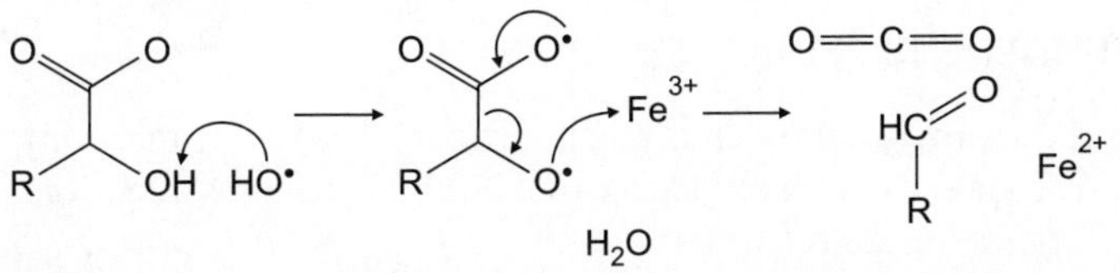

348. RUZICKA CYCLIZATION OR RUZICKA LARGE RING SYNTHESIS

The reaction is named after Lavoslav Ruzicka who invented it in 1926. The Ruzicka large ring synthesis or Ruzicka cyclization involves formation of large ring alicyclic ketones from dicarboxylic acids by thermal decomposition of salts with metals of the second and fourth groups of the periodic table (Ca, Th, Ce) such as thorium oxide. This reaction has been applied in the synthesis of Exaltone (a synthetic Muscone).

Ruzicka cyclization works well for the preparation of six and seven membered rings but results in lower yields of C_8 and C_{10}–C_{30} cyclic ketones. A free radical mechanism has been suggested for the reaction.

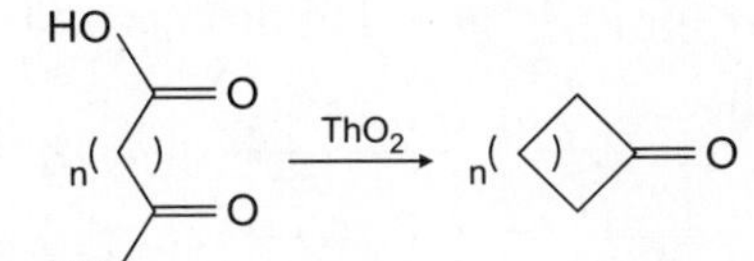

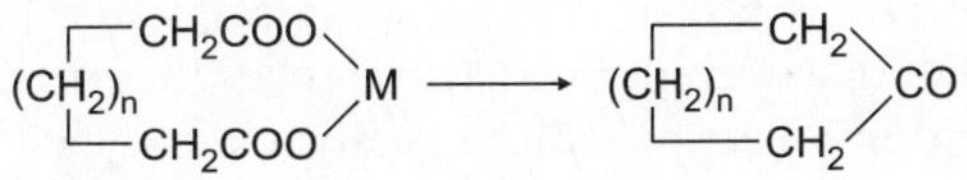

349. SABATIER-SENDERENS REDUCTION

Sabatier-Senderens reduction involves catalytic hydrogenation of organic compounds in the vapour phase by passage over finely divided, heated nickel catalyst. This reaction is named after Paul Sabatier and Jean-Baptiste Senderens.

$$H_2C = CH_2 + H_2 \xrightarrow[300°C]{Ni} H_3C - CH_3$$

$$H_3C - HC = CH_2 + H_2 \xrightarrow[300°C]{Ni} H_3C - CH_2 - CH_3$$

Purity of catalyst and temperature of the reaction are important parameters for this reaction. The course of the **Sabatier-Senderens Reduction** is different and does not involve nascent hydrogen as the reducing agent. This reaction is especially useful in the transformation of carbon monoxide or carbon dioxide into organic compounds.

$$CO + H_2 \xrightarrow{Ni}{\Delta} CH_4$$

It is employed commercially for hydrogenating unsaturated vegetable oils to make margarine.

$$\text{Cotton seed oil} \xrightarrow[\Delta]{H_2/Ni \text{ on silica black}} \text{Partial saturated oil}$$

350. SAEGUSA OXIDATION (SAEGUSA–ITO OXIDATION)

Takeo Saegusa and Yoshihiko discovered this reaction in 1978. The reaction is also known as the Saegusa enone synthesis and involves palladium catalyzed conversion of silyl enol ethers (enol silanes) to enones.

The reaction uses palladium acetate to generate α,β-unsaturated carbonyls from an intermediate silyl enol ether. This reaction has importance for the preparation of α,β-unsaturated ketones.

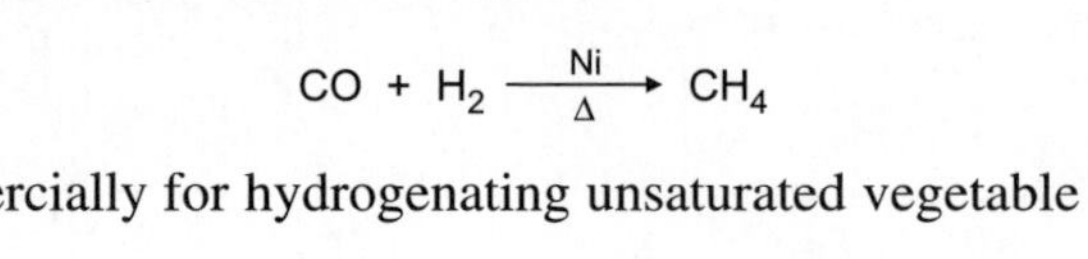

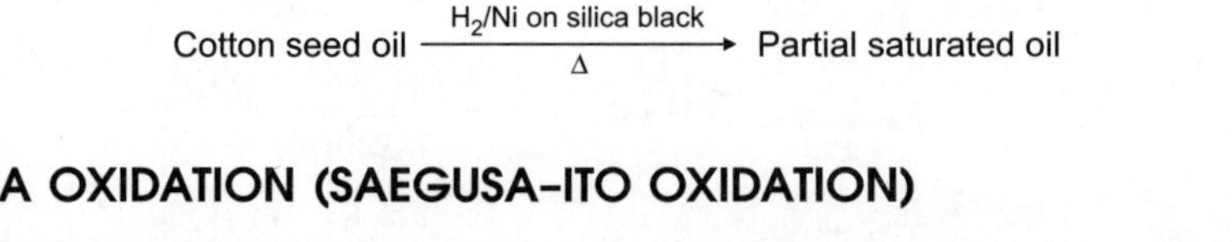

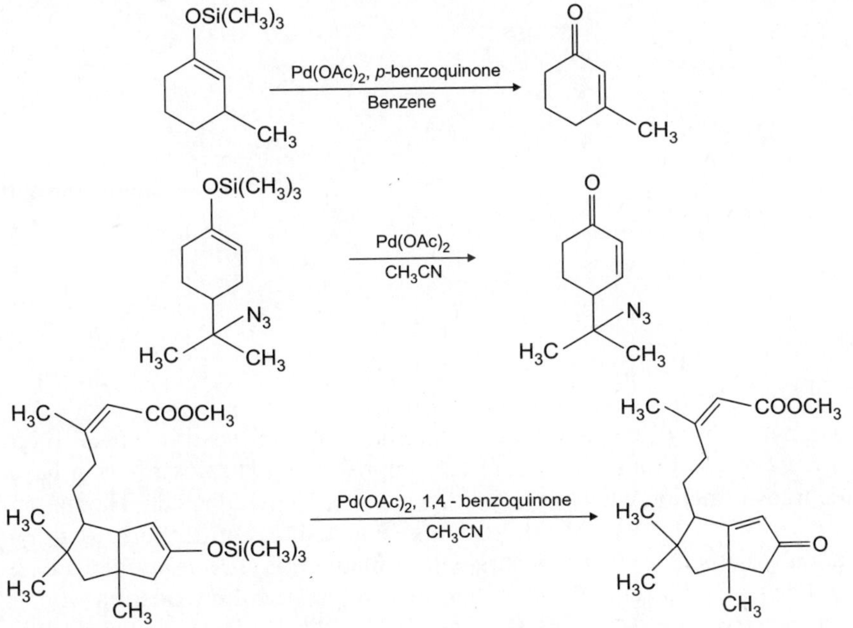

The mechanism of the Saegusa–Ito oxidation is similar to that of the Wacker oxidation. It involves coordination of palladium to the enol olefin followed by loss of the silyl group and formation of an oxoallyl-palladium complex. β-hydride elimination yields the palladium hydride enone complex which upon reductive elimination yields the product along with acetic acid and Pd^0.

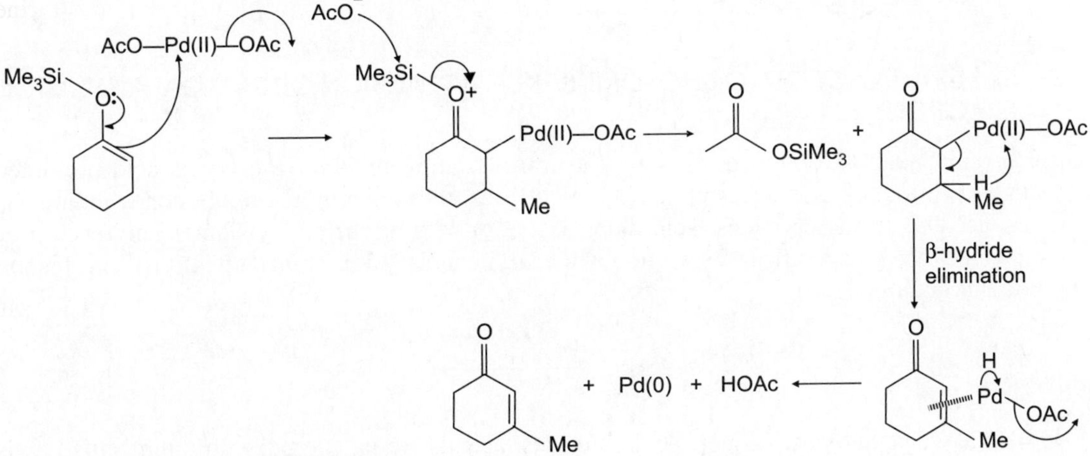

Regenerating the Pd(II) oxidant:

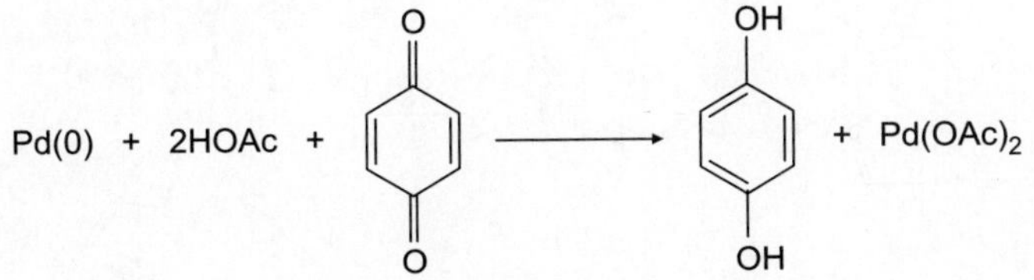

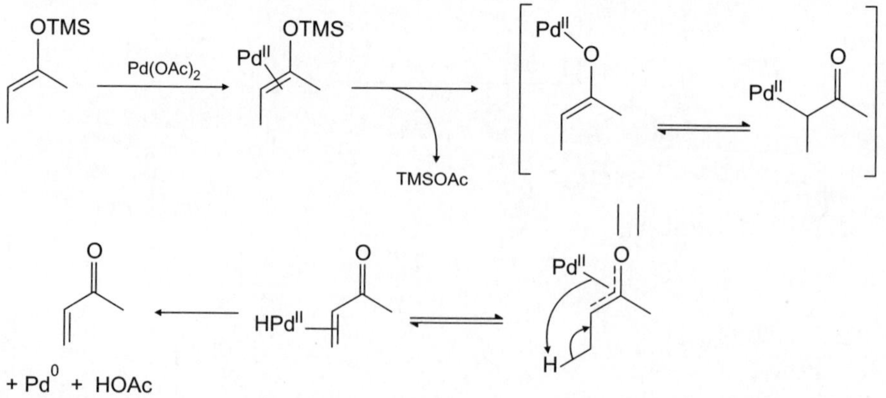

This reaction found too expensive for industrial usage because it typically employs near-stoichiometric amounts of palladium therefore, improvements to this reaction have focused on rendering the transformation catalytic with respect to the palladium salt. The majority of efforts were focused towards the use of co-oxidants (e.g. using atmospheric oxygen as well as stoichiometric allylcarbonate) that regenerate the palladium(II) species effectively. For example, Larock (1998) developed a catalytic substitute for the Saegusa–Ito oxidation where the Pd(II) is regenerated using oxygen.

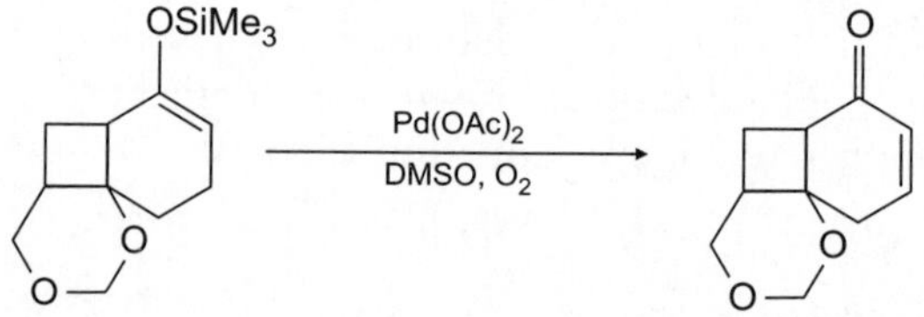

351. SAKURAI REACTION (OR HOSOMI-SAKURAI REACTION OR SAKURAI ALLYLATION REACTION)

The Hosomi Sakurai reaction or Hosomi-Sakurai allylation involves the Lewis acid promoted allylation of various carbon electrophiles (e.g. aldehydes, ketones, iminium salts, enones, aldimines, alkenes, acetals, ketals, epoxides, acid chlorides, etc.) with allyltrimethylsilane. The reaction is accompanied by regiospecific transposition of the allylic moiety. It is named after the Akira Hosomi and Hideki Sakurai.

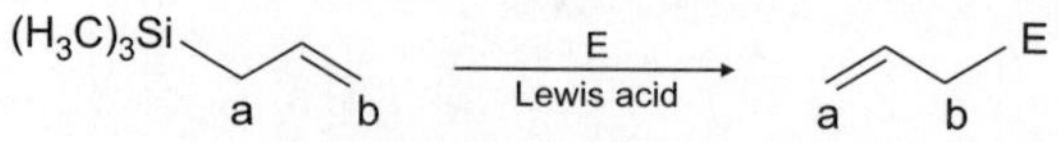

E = Aldehydes, ketones, enones, acid chlorides, acetals, ketis, epoxides, iminium salts Lewis acid = $TiCl_4$, $AlCl_3$, $BF_3 \cdot O(CH_2CH_3)_2$, $SnCl_4$, $(CH_3CH_2)_2AlCl$, cat. $(CH_3)_3SiOSO_2CF_3$

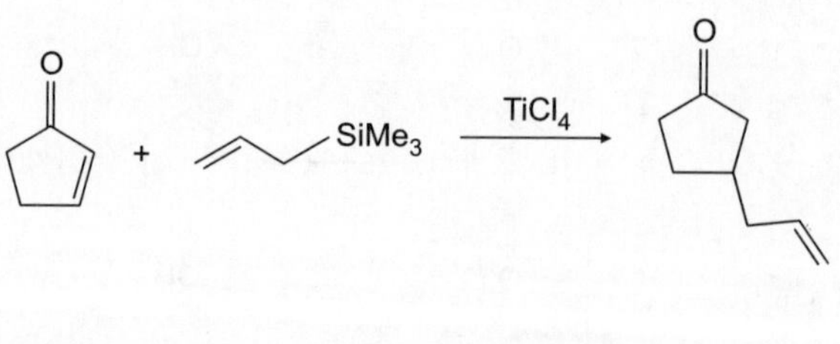

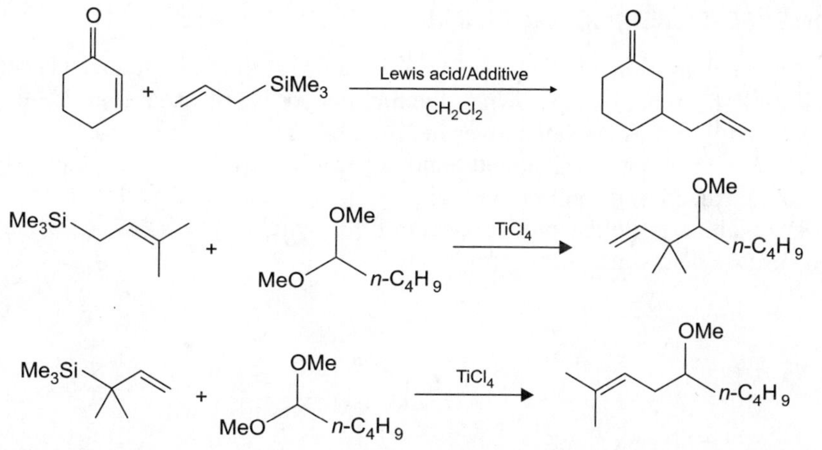

Activation by Lewis acids is critical for an efficient allylation. The range of Lewis acids employed in Sakurai reactions is extensive, among which titanium tetrachloride ($TiCl_4$), $AlCl_3$, and $BF_3:OEt_2$, tin tetrachloride, and $AlCl(Et)_2$ are in general the most effective for the allylations. In all cases, however, the procedures require a stoichiometric amount or even an excess of the Lewis acid to obtain reasonable reaction rates and acceptable yields of products. The reaction is a type of electrophilic allyl shift with formation of an intermediate beta-silyl carbocation. Driving force is the stabilization of said carbocation by the beta-silicon effect.

Sakurai reactions have been extensively applied in organic synthesis, in natural product synthesis, and in the preparation of some heterocyclic compounds. They are now considered to be one of the most efficient means of C-C bond formation, and examples of both inter- and intramolecular reactions have been reported. Sakurai reactions are regiospecific with regard to the newly formed C-C bond.

352. SANDMEYER DIPHENYLUREA ISATIN SYNTHESIS

The Sandmeyer diphenylurea isatin synthesis starts from diphenylthiourea, potassium cyanide, and lead carbonate to form a cyanoformamidine, followed by reduction with ammonium sulphide and finally ring-closure with concentrated sulfuric acid to isatin-2-anil.

Ring closure can also be done with aluminum chloride in benzene or carbon disulfide.

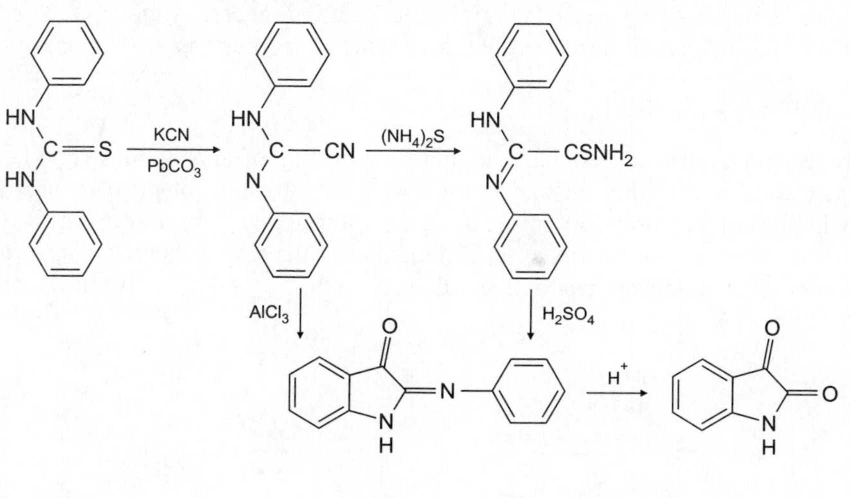

353. SANDMEYER ISONITROSOACETANILIDE ISATIN SYNTHESIS

Cyclicization of condensation product of chloral hydrate, aniline and hydroxylamine in sulfuric acid affords isatin. This reaction is known as the Sandmeyer isonitrosoacetanilide Isatin Synthesis and was discovered by Traugott Sandmeyer in 1919.

This method was developed by Traugott Sandmeyer in 1919 and is most frequently used for the synthesis of isatin. Reaction of aniline with chloral hydrate and hydroxylamine hydrochloride in aqueous sodium sulphate yield isonitrosoacetanilide, which after isolation, when treated with concentrated sulfuric acid, furnishes isatin

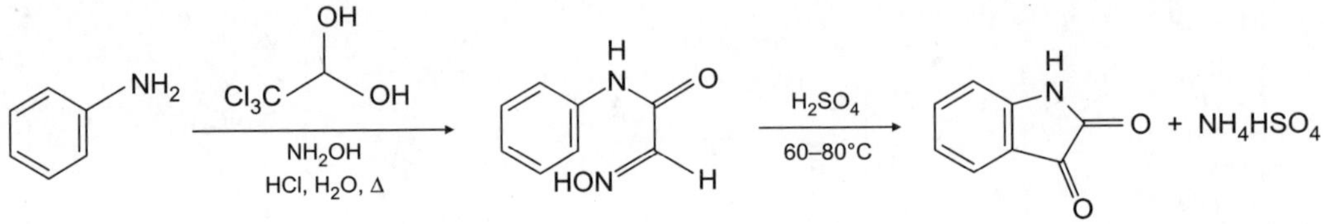

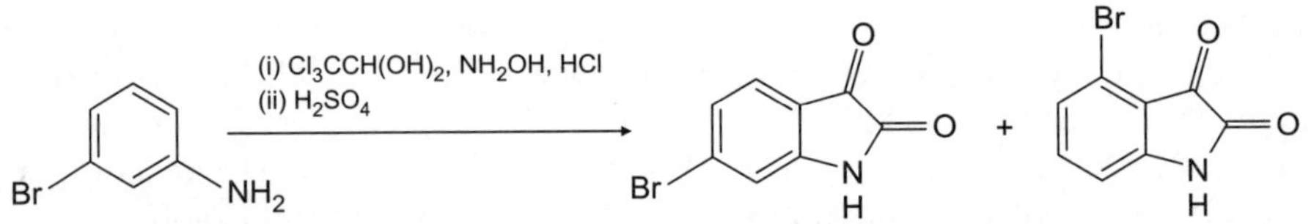

The method applies well to anilines with electron-withdrawing substituents, such as 2-fluoroaniline, and to some heterocyclic amines, such as 2-aminophenoxathine as given below:

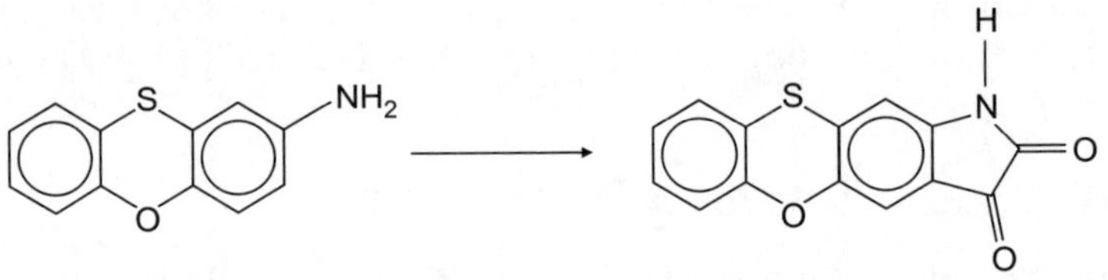

As the reagents for the reaction are cheap and readily available, and the yields are usually high therefore this method has economic advantages also. Despite its efficiency, the Sandmeyer method is limited by: (a) harsh conditions; (b) fails if the isonitrosoacetanilides bear electron-donating groups; (c) the formation of a mixture of regioisomers; and (d) generally moderate yields. However, this latter shortcoming can be improved under microwave conditions.

354. SANDMEYER REACTION

The Sandmeyer reaction is a chemical method to synthesis of aryl halides (F, Cl. Br and I) as well as aryl cyanides from primary aryl amines by diazotization of primary aryl amines with $NaNO_2/HCl$. Subsequent transformation of diazo intermediates into aryl halides or cyanides with cuprous halides or cyanide. Reaction is named after the Swiss chemist Traugott Sandmeyer. In other words the Sandmeyer reaction involves the cuprous salts catalyzed decomposition of diazonium salts.

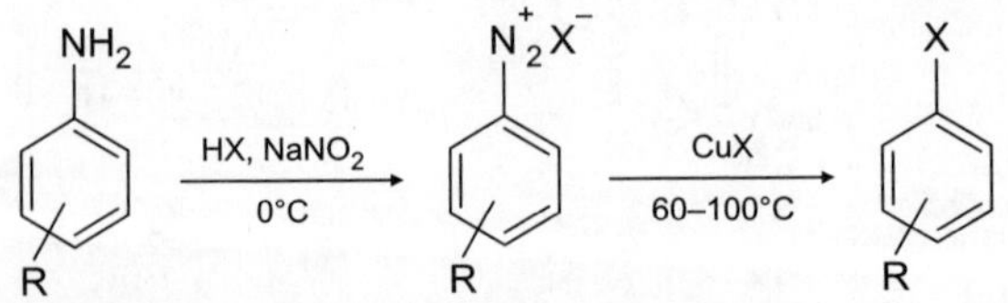

X = CN, Br, Cl, SO$_3$H

The Sandmeyer reaction suffers from the drawback of the formation of some side products which in turn affects yield of aryl halide. The Sandmeyer-type reactions with thiols, water and potassium iodide do not require catalysis.

The Sandmeyer reaction is a versatile means of replacing the amine group of a primary aromatic amine with a number of different substituents. It is done via preparation of its diazonium salt and subsequent displacement with a nucleophile (Cl$^-$, I$^-$, CN$^-$, RS$^-$, OH$^-$). This reaction is also important because it can result in some substitution patterns that are not achievable by direct substitution.

The diazonium salt is formed by the reaction of nitrous acid with the amine in acid solution. Nitrous acid is not stable and must be prepared in situ:

$$NaNO_2 + HCl \rightleftharpoons HONO + Na^+ Cl^-$$

Sodium   Nitrous
nitrite   acid

In strong acid, nitrous acid dissociates to form nitroso ions, +NO, which attack the nitrogen of the amine.

$$H_3O^+ + HONO \rightleftharpoons H_2O + H_2\overset{+}{O}NO \rightleftharpoons 2H_2O + \overset{+}{N}=O$$

The intermediate so formed loses a proton, rearranges, and finally loses water to form the resonance-stabilized diazonium ion.

355. SARETT OXIDATION; COLLINS OXIDATION

The Collins/Sarett oxidation is referred as the mild oxidation of alocohols with chromium trioxide-pyridine complex.

The Sarett oxidation is named after the American chemist Lewis Hastings Sarett. It involves oxidation of primary and secondary alcohols to corresponding aldehydes and ketones by means of CrO_3-pyridine complex as oxidant. In this reaction the primary alcohols are not further oxidized to carboxylic acids.

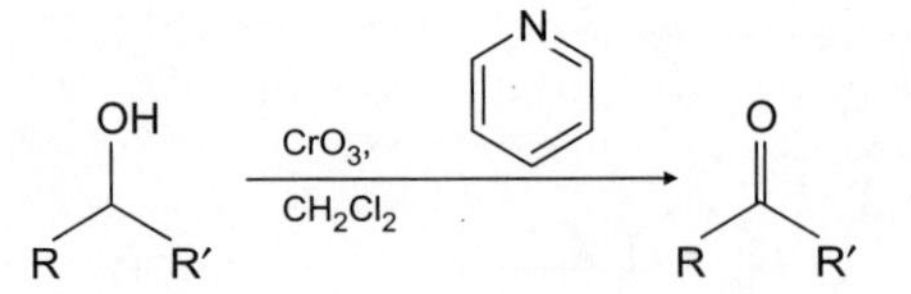

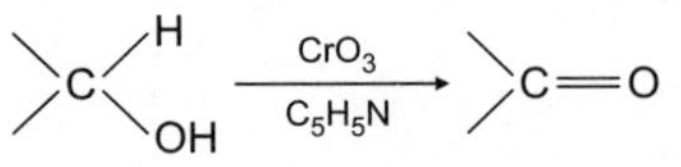

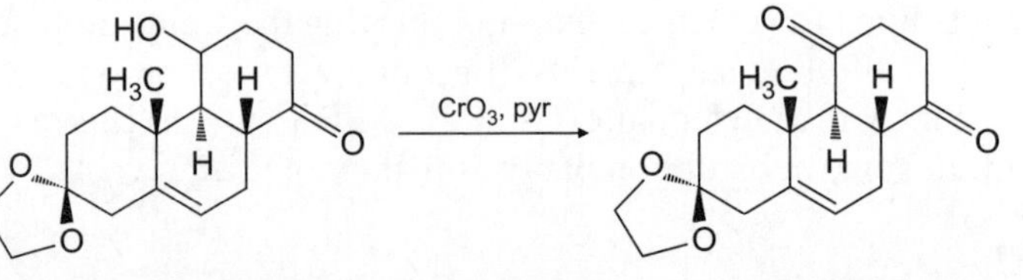

The Collins oxidation is characterized by a modified procedure (dichloromethane used as solvent while the Sarett oxidation usually requires pyridine as a solvent) that reliably oxidizes primary alcohols to aldehydes. This modification was developed to deal with the problem of poor yields in the oxidation of primary alcohols to aldehydes, and to improve the isolation of the carbonyl products.

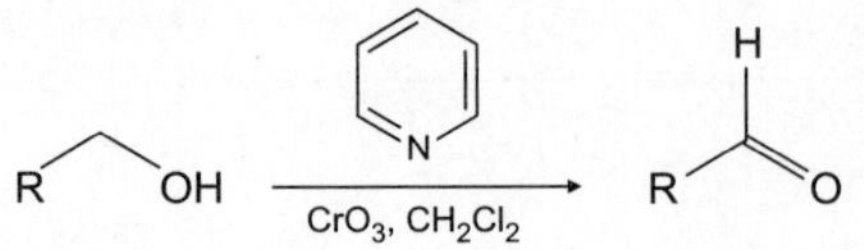

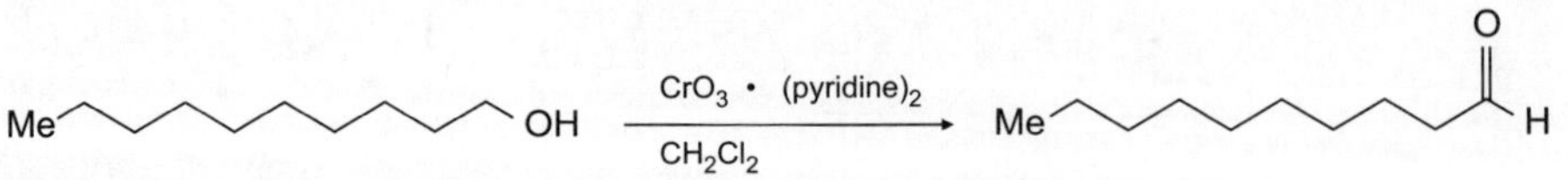

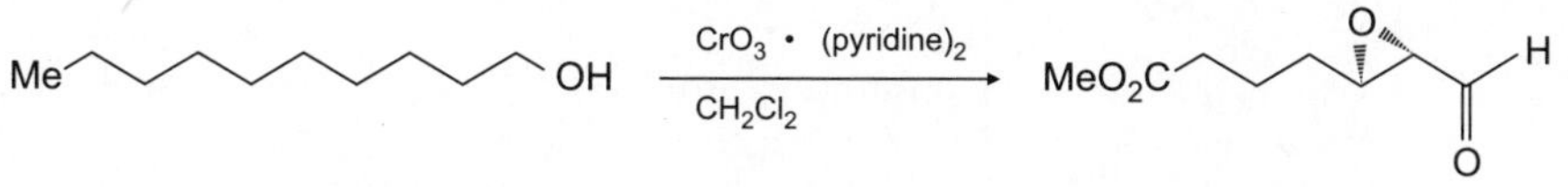

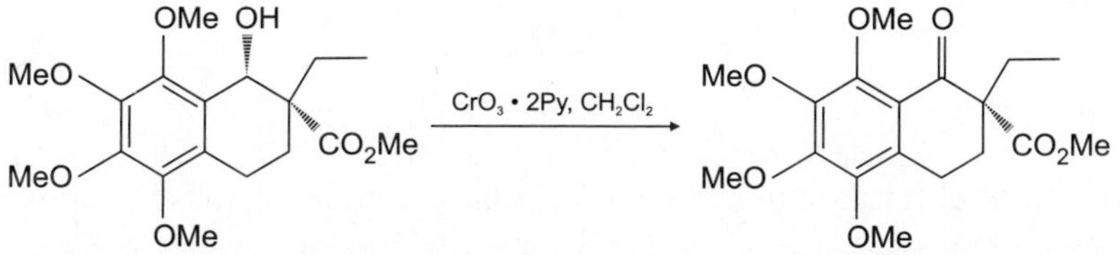

Sarett reagent (CrO_3 • 2Py complex) is formed when chromium trioxide is added to an excess of pyridine. This reagent is highly hygroscopic and requires special care in its preparation.

The Collins reagent is CrO_3 • 2Py complex diluted in dichloromethane or it can be said CrO_3 • 2Py complex in dichloromethane. The use of pyridine as solvent in Sarett reagent (CrO_3 • 2 Py complex) does not permit the oxidation of base-sensitive substrates. The Collins reagent can tolerate many other functional groups within the molecule therefore, broader substrate scope. Moreover, the Collins reagent is especially useful for oxidations of acid sensitive compounds.

The chief drawbacks in using the Collins reagent are the excess use of the reagent, the nuisance involved in preparing pure dipyridine-chromium(VI) oxide, its hygroscopic nature [easily hydrolyzed to the yellow dipyridinium dichromate ($[Cr_2O_7]{-}2(pyrH^+)_2$)], its propensity to inflame during preparation.

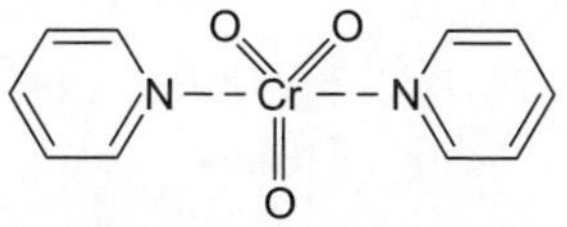

Collins/Sarett

356. SCHIEMANN REACTION (BALZ-SCHIEMANN REACTION)

The Schiemann reaction (also called the Balz-Schiemann reaction) involves formation of aryl fluorides from arenediazonium fluoroborates. In the reaction the arylamines transformed to aryl fluorides by the formation and isolation of the diazonium tetrafluoroborate, followed by thermal decomposition. The reaction is named after the German chemists Günther Schiemann and Günther Balz.

$$R{-}NH_2 \xrightarrow[HBF_4]{NaNO_2} R{-}\overset{+}{N}{\equiv}N\ BF_4^- \xrightarrow{\Delta} R{-}F$$

$$Ar{-}NH_2 \xrightarrow[HBF_4]{HNO_2} Ar{-}\overset{+}{N}{\equiv}N\ BF_4^- \xrightarrow[-N_2]{\Delta} Ar{-}F + BF_3$$

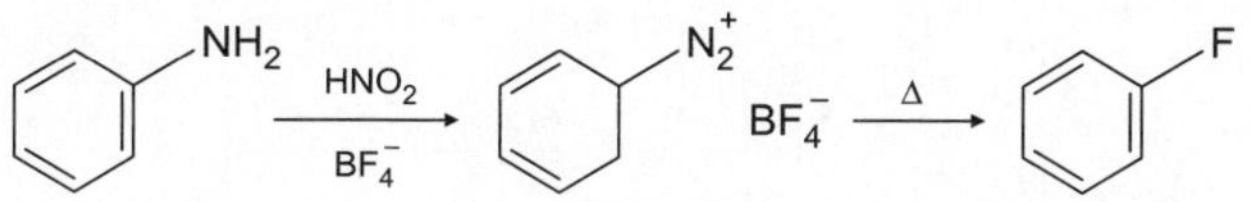

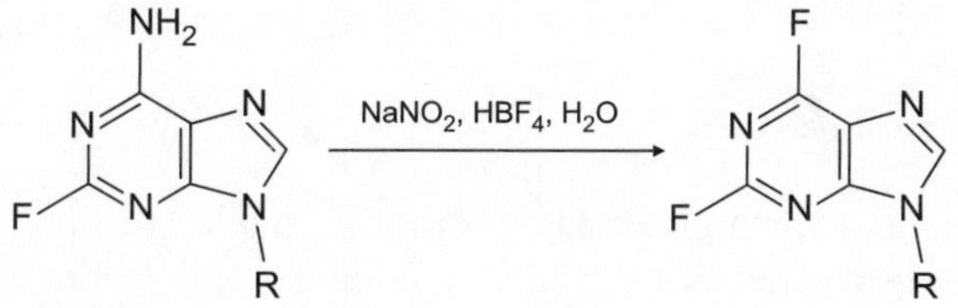

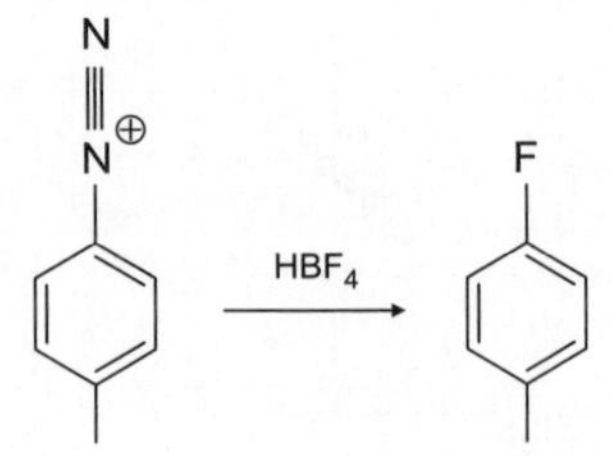

This reaction is useful for the selective introduction of a fluorine substituent onto an aromatic ring and works with condensed aromatic amines also. The reaction is similar to the Sandmeyer reaction.

357. SCHMIDT REACTION OR SCHMIDT REARRANGEMENT

The reaction of carbonyl group containing compounds with hydrazoic acid in the presence of acid catalysts proceeds via 1, 2 - migration is known as the Schmidt reaction. Reaction involves alkyl migration over the carbon-nitrogen bond in an azide with expulsion of nitrogen. The reaction product depends on the type of reactant: carboxylic acids form amines through an isocyanate intermediate, ketones form amides and aldehydes form cyanides. Schmidt reaction is related to the Curtius rearrangement except that in this reaction the azide is protonated and hence with different intermediates.

Reaction with aldehydes gives cyanides

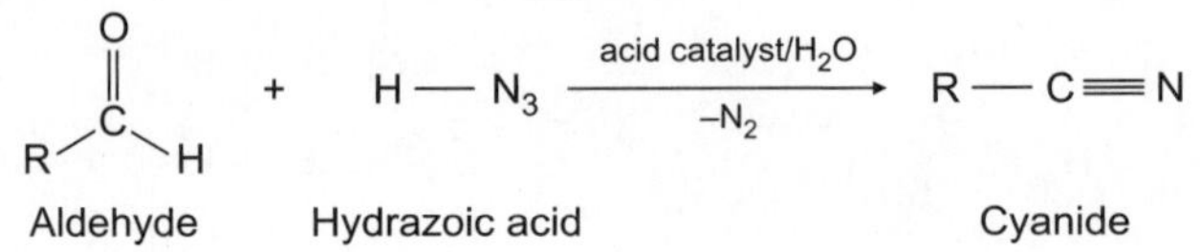

Reaction with ketones gives amides

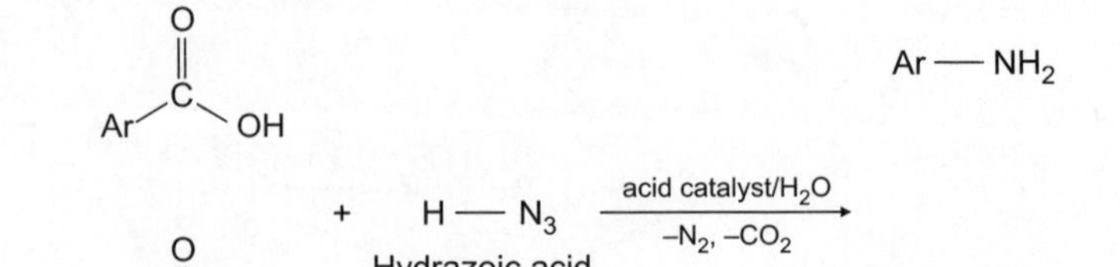

Reaction with carboxylic acids gives amines

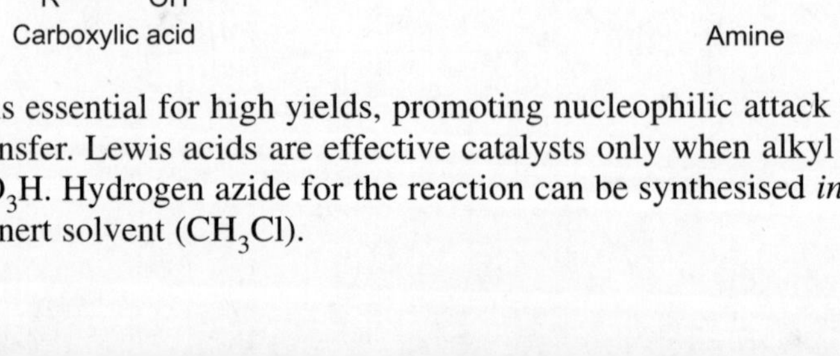

Acidic solution is essential for high yields, promoting nucleophilic attack at the carbonyl and allowing proton transfer. Lewis acids are effective catalysts only when alkyl azides are used,e.g. $TiCl_3$, TFA, CH_3SO_3H. Hydrogen azide for the reaction can be synthesised *in situ* (NaN_3 + acid) or as a solution in inert solvent (CH_3Cl).

$$nC_5H_{11} \!-\! COOH \ + \ NH_3 \ \xrightarrow[\text{2. H}_2\text{O}]{\text{1. H}_2\text{SO}_4\text{/benzene}} \ nC_5H_{11} \!-\! NH_2$$

Hexanoic acid   Pentylamine

358. SCHOLL REACTION

The Scholl reaction or Scholl condensation (occasionally referred as the Scholl oxidation) was discovered by Roland Scholl. It involves coupling of two arene molecules by treatment with a Lewis acid and a protic acid. The reaction is thought to goes through an arenium ion and activating groups like methoxy found to facilitate the reaction.

$$2ArH \ \xrightarrow[\text{H}^+]{\text{AlCl}_3} \ Ar \!-\! Ar \ + \ H_2$$

The Scholl reaction suffers from drawbacks like requirement of high reaction temperature, strong-acid catalysts therefore the reaction fails for substrates that are destroyed by these conditions. These drawbacks are also associated with low yield therefore the method is not a popular one and is seldom useful.

9-fenylfluorene

359. SCHÖLLKOPF BIS-LACTIM AMINO ACID SYNTHESIS

The Schöllkopf method or Schöllkopf Bis-Lactim Amino Acid Synthesis was established in 1981 by Ulrich Schöllkopf, is used for the asymmetric synthesis of chiral amino acids. It involves diastereoselective alkylation of the lithiated bis-lactim ether (derived from L-Val and Gly or Ala) by an electrophile. Acidic hydrolysis leads to formation of the alkylated amino acid.

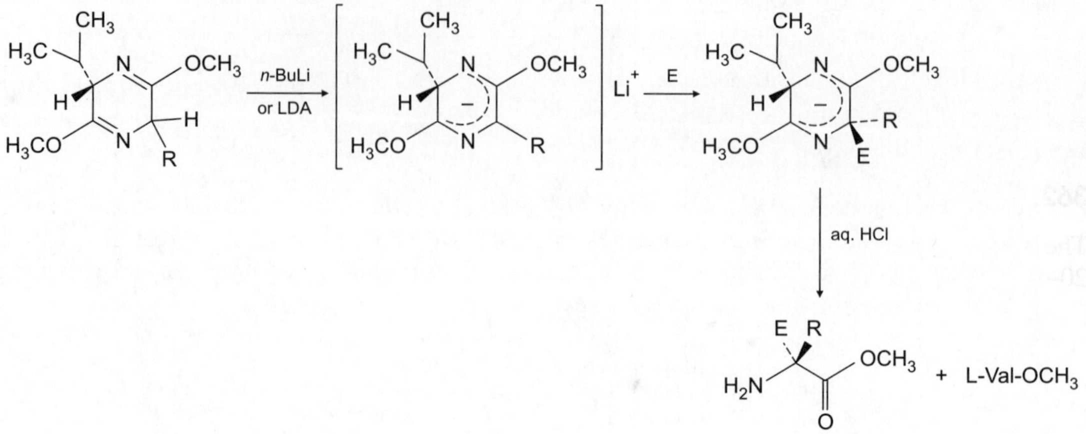

E = Alkyl halides, aldehydes, ketones, thioketones, acid chlorides, epoxides, acrylates.

360. SCHOTTEN-BAUMANN REACTION

The Schotten-Baumann reaction involves synthesise of amides from amines and acid chlorides in aqueous alkaline solution (usually aqueous sodium hydroxide).

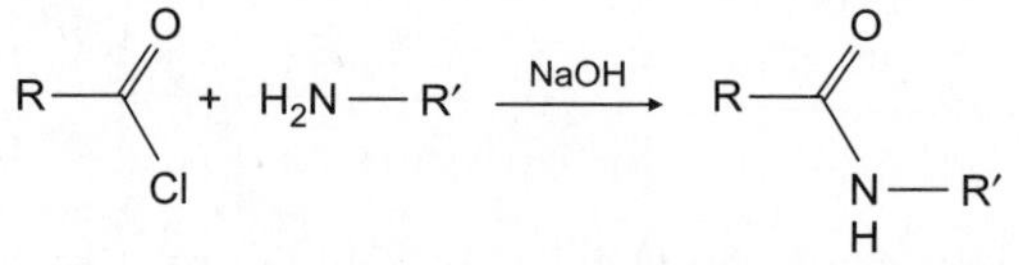

$C_6H_5NH_2$ + $ClCOC_6H_5$ $\xrightarrow{\text{NaOH}}$ $C_6H_5NHCOC_6H_5$

Benzoyl chloride   Benzanilide

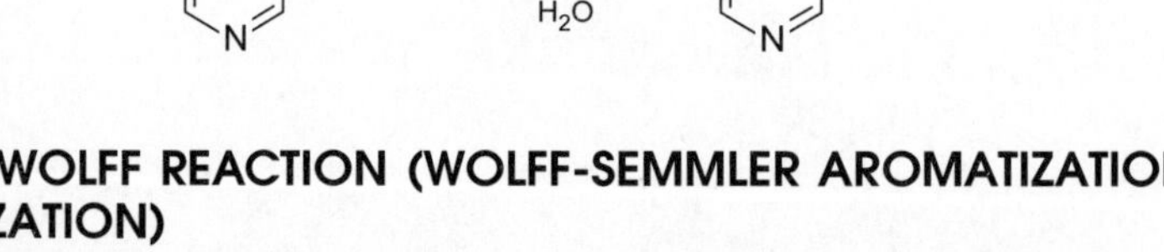

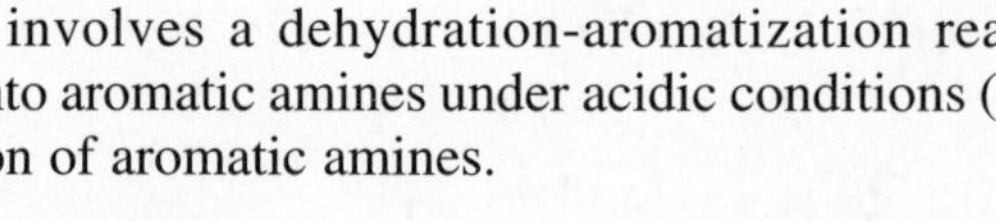

361. SEMMLER-WOLFF REACTION (WOLFF-SEMMLER AROMATIZATION, WOLFF AROMATIZATION)

Semmler-Wolff reaction involves a dehydration-aromatization reaction of α,β-unsaturated cyclohexenyl ketoximes into aromatic amines under acidic conditions (HCl or HBr). This reaction is useful for the preparation of aromatic amines.

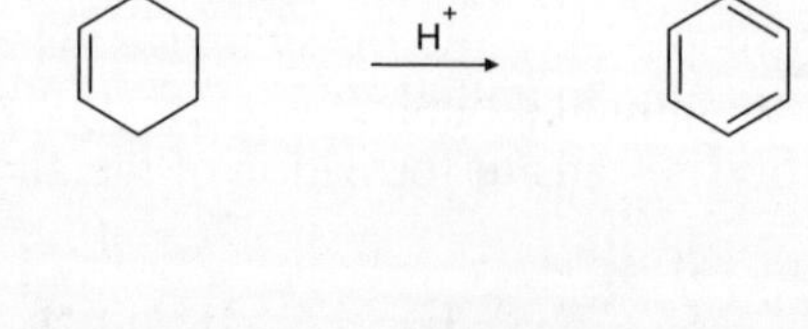

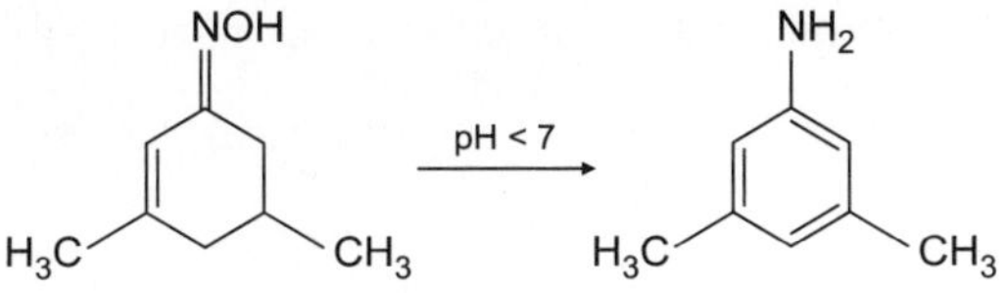

362. SERINI REACTION

The Serini reaction involves zinc-promoted rearrangement of 17-hydroxy-20-acetoxy-steroids into 20-oxo-steroid derivatives. The reaction results in complete inversion of stereochemistry at C-17.

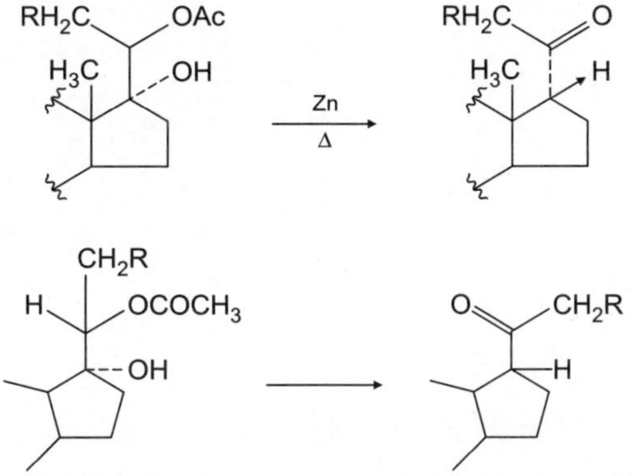

The reaction is found to be extended to other cyclic and open-chain 1,2-diol monoacetates as well as corresponding benzoates and p-nitrobenzoates and has certain application in steroid chemistry.

363. SHARPLESS ASYMMETRIC DIHYDROXYLATION OR SHARPLESS AD

Enantioselective cis-dihydroxylation of olefins using osmium catalyst in the presence of cinchona alkaloid ligands to form a vicinal diol is known as Sharpless asymmetric dihydroxylation (also called the Sharpless bishydroxylation). Therefore, Sharpless AD is used in the enantioselective preparation of 1, 2-diol from prochiral olefins.

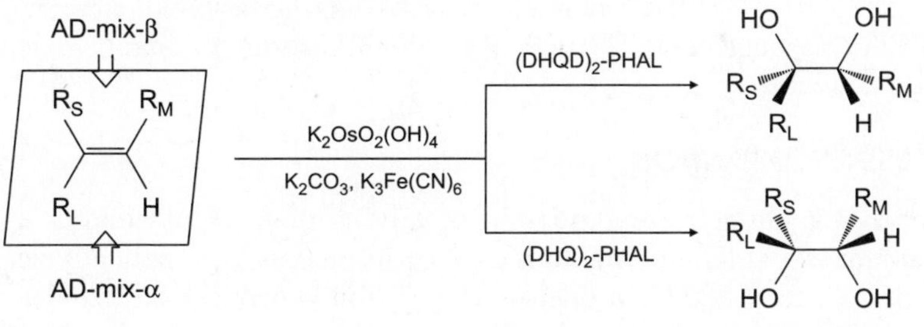

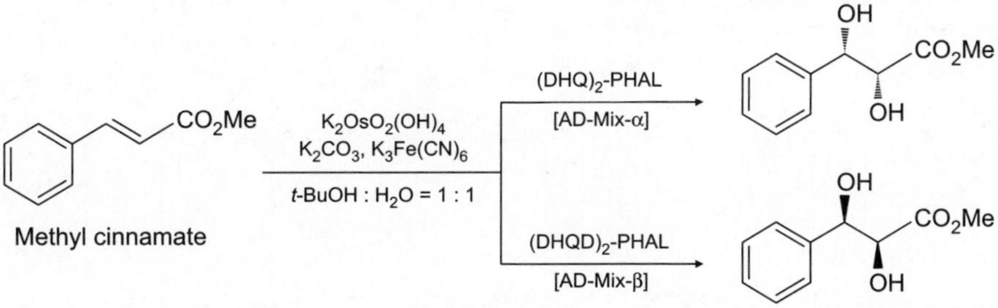

(DHQD)₂-PHAL = 1,4-bis(9-O-dihydroquinidine) phthalazine:
and
(DHQ)₂-PHAL = 1,4-bis(9-O-dihydroquinine) phthalazine

(DHQD)₂PHAL

(DHQ)₂PHAL

Chiral ligands for sharpless asymmetric dihyroxylation

The Sharpless procedure employs catalytic quantities of osmium tetroxide and single-enantiomer chiral amine ligands in conjunction with potassium ferricyanide and potassium carbonate $[K_3Fe(CN)_6–K_2CO_3]$ as stoichiometric secondary oxidant in a biphasic solvent medium. The secondary oxidant effects oxidative hydrolysis of the initially formed cyclic osmate ester in situ, thereby liberating the cis-diol concomitant with regeneration of the osmium tetroxide to complete the catalytic cycle. In presence of a stoichiometric secondary oxidant such as potassium ferricyanide or N-methylmorpholine N-oxide, the Osmium tetroxide is required only in catalytic quantities otherwise the highly toxic and very expensive osmium tetroxide needed in excess amount in the reaction.

The "AD mix β" and "AD mix α" are ready-mixed oxidizing systems containing (DHQD)₂PHAL and (DHQ)₂PHAL respectively. The RS, RM and RL refer to "small", "medium", "large" substituents respectively.

364. SHARPLESS EPOXIDATION

Titanium-catalyzed asymmetric epoxidation of allylic alcohols to yield a 2,3-epoxy alcohol employing titanium alkoxide (e.g. titanium tetra-*iso*-propoxide), an optically active tartrate ester (e.g. diethyl tartrate) and an alkyl hydroperoxide (e.g. *t*-butyl peroxide) referred to as the **Sharpless epoxidation**. A high degree of enantiomeric purity is attainable having predictable absolute stereochemistry:

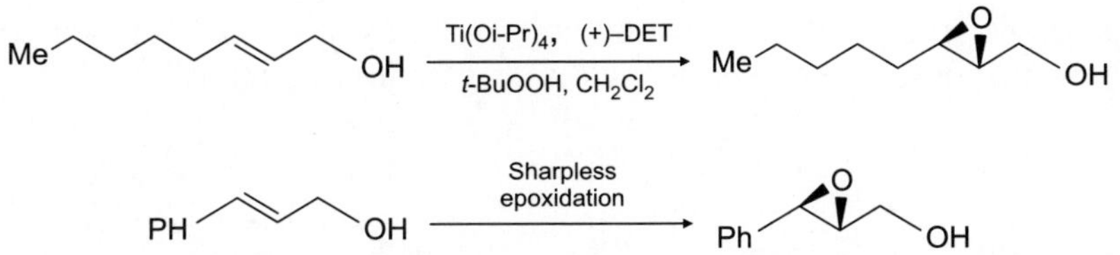

This reaction converts primary and secondary allylic alcohols into 2,3 epoxyalcohols and the reaction is enantioselective (only one enantiomer produced). Enantiomer formed in the reaction depends on stereochemistry of catalyst. The catalyst for the reaction is titanium tetra(isopropoxide) with diethyltartrate. The use of + or − tartrate yield different enantiomers. Tertbutylperoxide is used in the reaction as the oxidizing agent. The reaction is limited to allylic alcohols; other types of alkenes do not efficiently bind to the titanium.

The Reaction

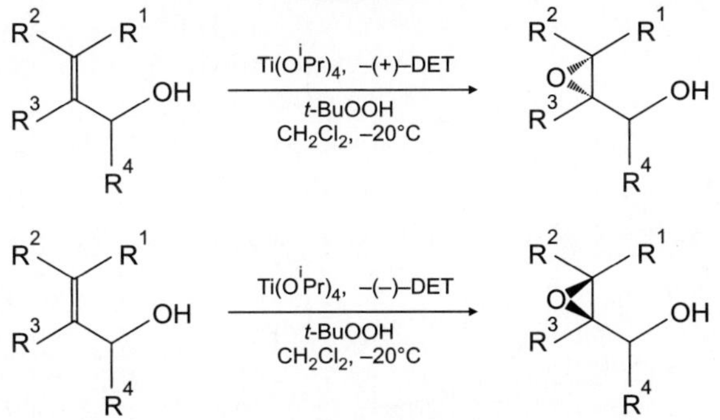

The Catalyst

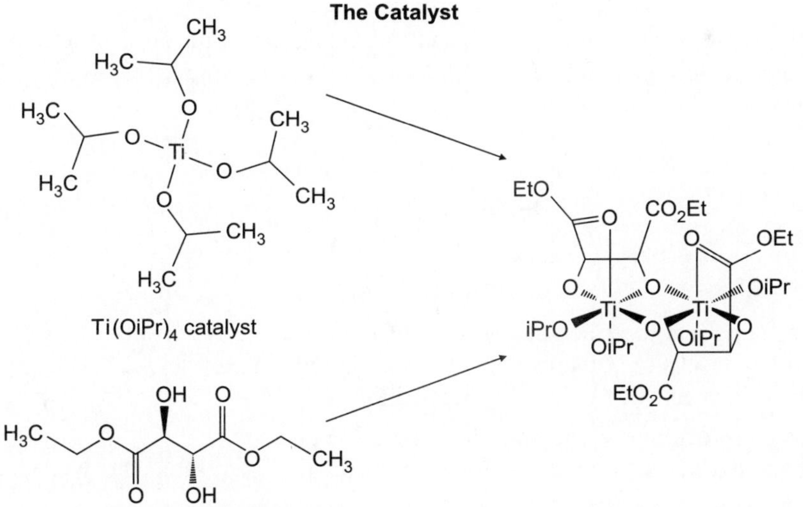

Ti(OiPr)$_4$ catalyst

Diethyl tartrate (DET)
Chirally controls reaction

365. SHARPLESS OXYAMINATION

The Sharpless Aminohydroxylation or Sharpless oxyamination allows the Osmium-mediated syn-selective preparation of 1,2-amino or amido alcohols by reaction of alkenes with salts of

N-halosulfonamides, -amides and -carbamates. Nitrogen and oxygen moieties are added (cis-addition) in this reaction to the substituted olefins to yield vicinal amino or amido alcohols.

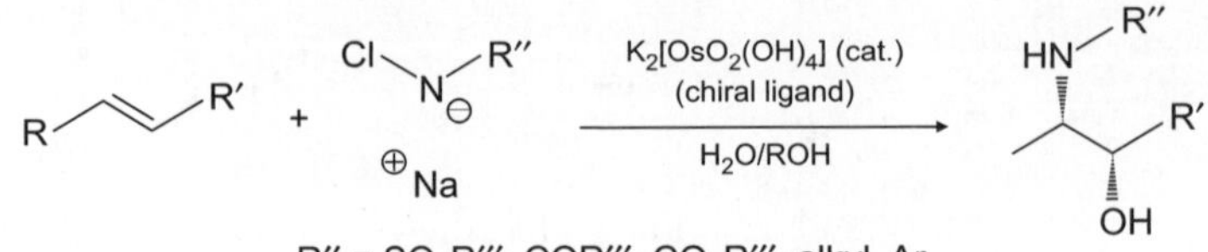

$R'' = SO_2R''', COR''', CO_2R''', alkyl, Ar$

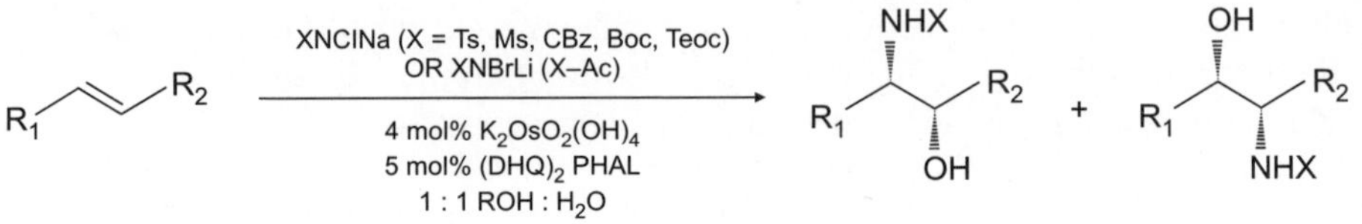

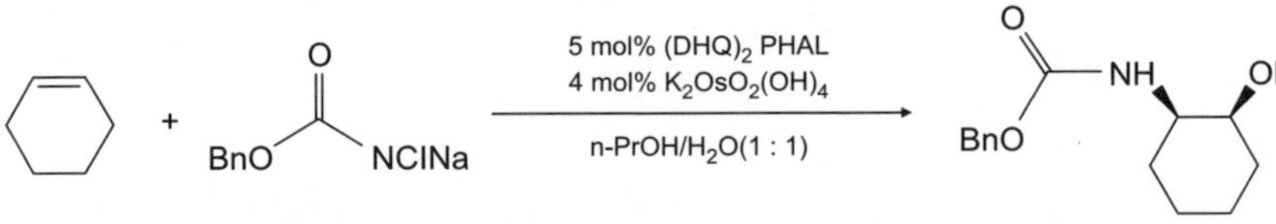

(DHQD)$_2$-PHAL = 1,4-bis(9-O-dihydroquindine)phthalazine

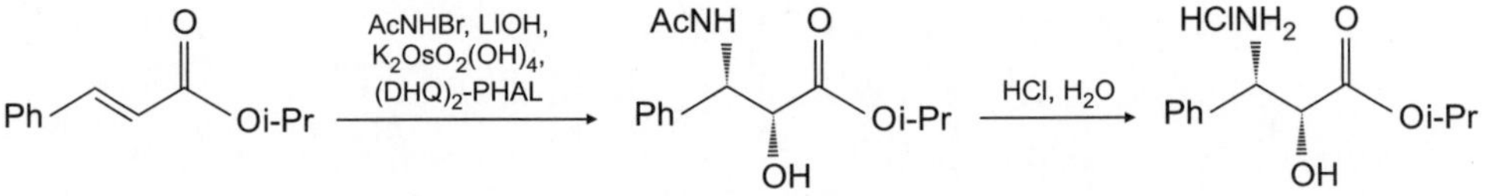

This reaction typically utilizes osmium, a chiral ligand, and a nitrogen source in the generation of 1,2-N-protected amino alcohols from alkenes. As nitrogen sources (X-NClNa), following are usually employed in the reaction:

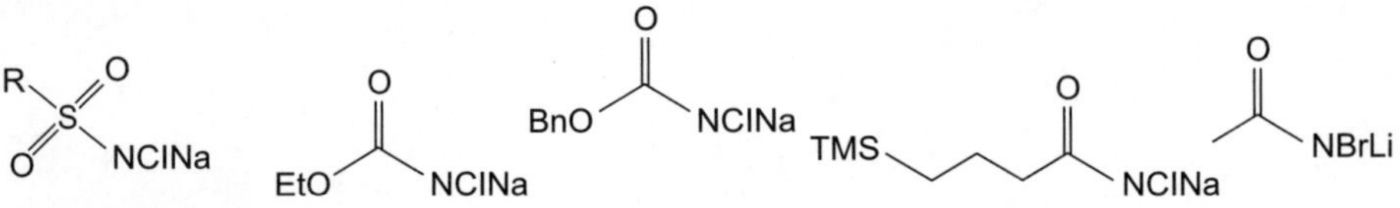

R = p-Tol; Me

It has been found that the reaction works better with smaller N-substituents, (e.g. $MeSO_2^-$), and even better still with salts of N-halocarbamates (NC(O)OR; ease of deprotection!) or N-haloamides. This methodology has two principal drawbacks:

(i) poor regioselectivity with some unsymmetrical alkenes;
(ii) requirement of 3 equivalents of nitrogen source to obtain satisfactory product yields.

366. SIMMONS-SMITH REACTION

The stereospecific transformation of olefins into cyclopropanes by means of the treatment with methylene diiodide and zinc-copper couple is generally known as the Simmons-Smith reaction. The study finds that the cyclopropanation is further facilitated by an intramolecular oxygen atom nearing the olefinic reaction center. All the features of the Simmons-Smith reaction has been discussed. This reaction has a very broad application in organic synthesis.

The reaction was first reported by H. E. Simmons and R. D. Smith in 1958. The reaction affords cyclopropanation of olefins. The **Simmons-Smith reaction** is a stereospecific transformation of cyclopropanes by reaction of olefins with methylene iodide and zinc-copper couple.

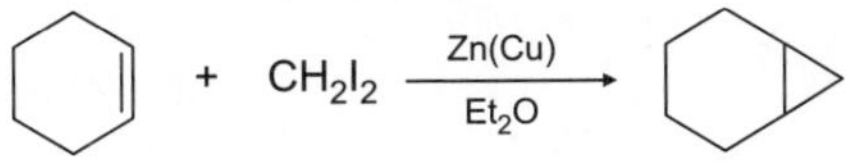

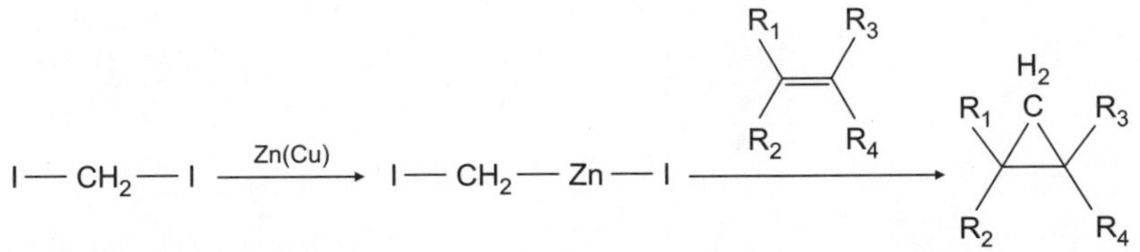

Cyclopropanation is stereospecific: The reaction is stereospecific with respect to to the alkene (mechanism is concerted). Substituents that are trans in the alkene are trans in the cyclopropane etc. (The configuration of the double bond is preserved in the product and the reaction is stereospecific).

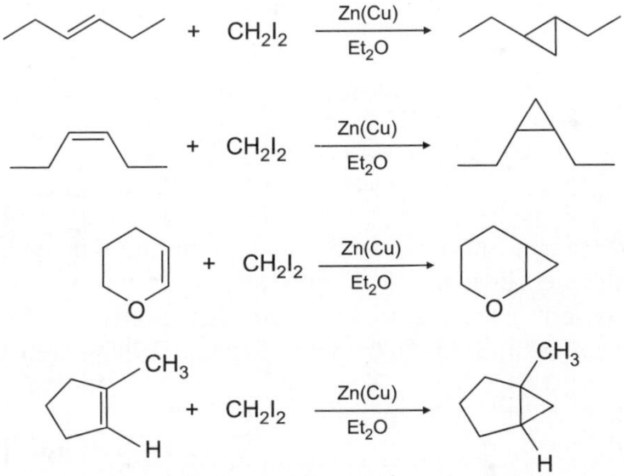

The iodomethyl zinc iodide is usually prepared using Zn activated with Cu.

$$Zn \text{ (dust)} + CuO \xrightarrow[500°C]{H_2} Zn \text{ (Cu)} \approx 90\% \text{ Zn}$$

367. SIMONINI REACTION

The Simonini reaction is an extension of Hunsdiecker reaction and involves the formation of aliphatic ester by the reaction of silver salts of carboxylic acids with iodine. This reaction was named after Angelo Simonini is a useful method for the preparation of esters.

Formation of final product depends greatly on the ratio of the reagents being used in the reaction. For example alkyl iodide is formed when 1:1 ratio of salt and iodine is used and ester RCOOR is obtained with a 2:1 ratio. On the other hand, both the alkyl iodide and ester RCOOR are formed in 3:2 ratio.

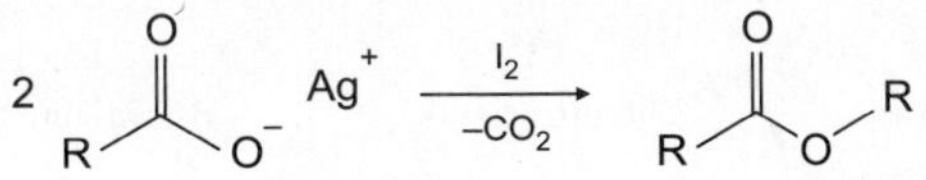

368. SIMONIS CHROMONE CYCLIZATION

Simonis chromone cyclization yields chromones from the reaction of phenols and α-keto esters in the presence of phosphorus pentoxide. This reaction is a variation of the Pechmann condensation reaction.

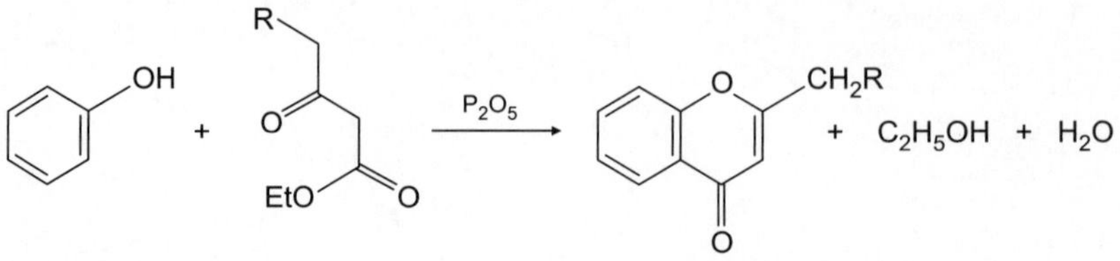

369. SKRAUP REACTION

The synthesis of quinoline derivatives from primary aromatic amines, glycerol, and an oxidizing agent in concentrated sulfuric acid is generally known as the Skraup reaction. In Skraup synthesis aniline, sulphuric acid, glycerol and a mild oxidizing oxidising agent,e.g. nitrobenzene, are heated together to produce quinoline.

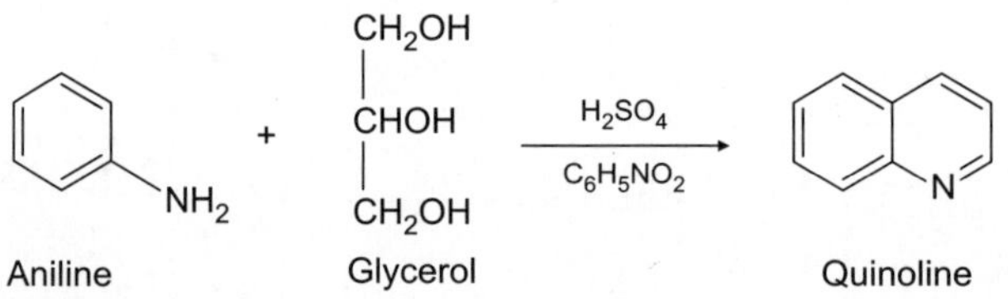

| Aniline | Glycerol | | Quinoline |

In this example, nitrobenzene serves as both the solvent and the oxidizing agent. Apart from nitrobenzene, stannic chloride, ferric salts, oxygen, arsenic or peroxide can also be employed as oxidising agent in the reaction. The acid acts as a dehydrating agent and an acid catalyst. This widely used synthesis for quinoline is named after the Czech chemist Zdenko Hans Skraup.

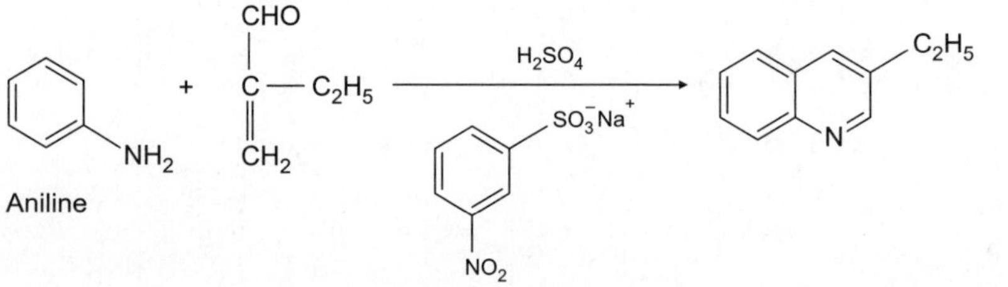

Quinoline with substituent in the benzene ring

In Skraup synthesis, when starts with substituted anilines, quinolines with substituent in the benzene ring are be obtained. o-substituted anilines give 8-substituted quinolines while p-substituted anilines give 6-substituted quinolines.

Ortho-substituted   Glycerol   8-substituted quinoline
aniline

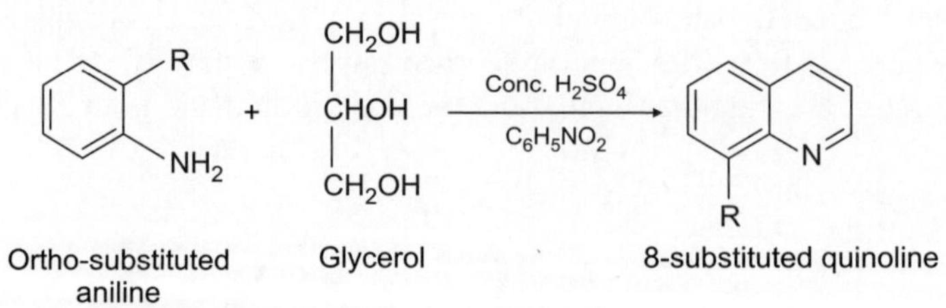

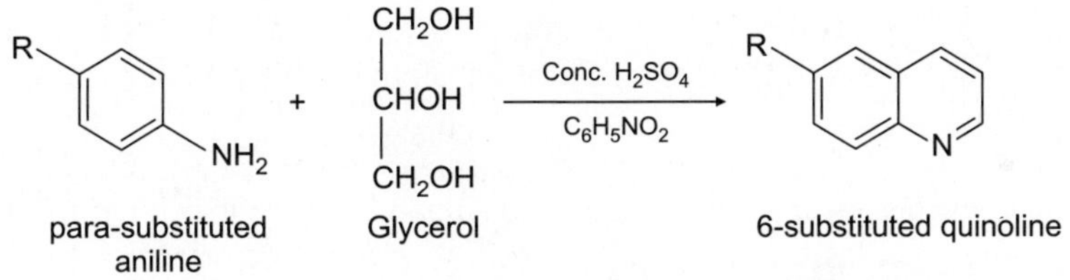

para-substituted aniline + Glycerol $\xrightarrow[C_6H_5NO_2]{\text{Conc. H}_2\text{SO}_4}$ 6-substituted quinoline

m-substituted aniline gives a mixture of 5- and 7-substituted quinolines.

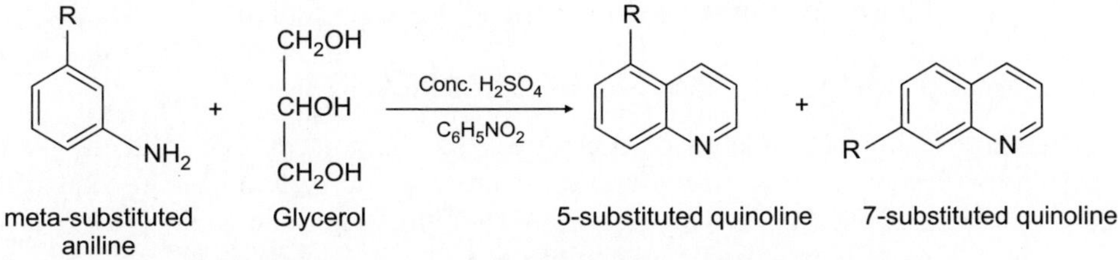

meta-substituted aniline + Glycerol $\xrightarrow[C_6H_5NO_2]{\text{Conc. H}_2\text{SO}_4}$ 5-substituted quinoline + 7-substituted quinoline

Quinoline with substitutent in heterocylic ring (pyridinoid ring)

If the reaction is carried out in presence of substituted α,β-unsaturated aldehyde or ketone, instead of acrolein generated *in situ*, quinoline with substitutent in heterocylic ring is obtained.

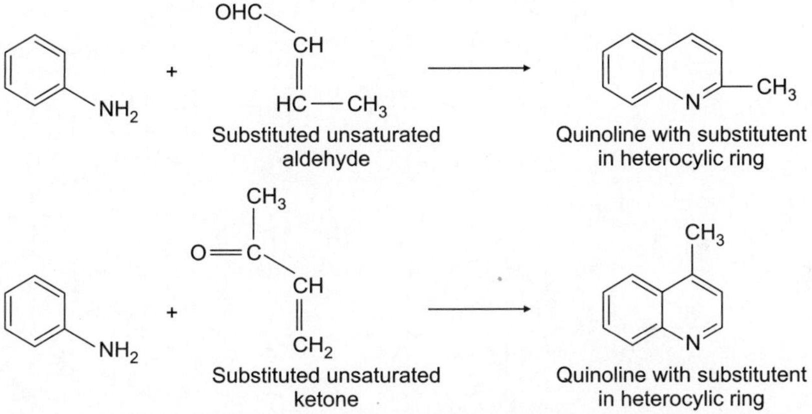

Substituted unsaturated aldehyde → Quinoline with substitutent in heterocylic ring

Substituted unsaturated ketone → Quinoline with substitutent in heterocylic ring

The reaction proceeds in the following manner. Dehydration of glycerol gives acrolein which undergoes Michael addition with aniline followed by electrophilic attack of protonated carbonyl group. The cyclized intermediate undergoes subsequent dehydration and oxidation to give quinoline.

Mechanism

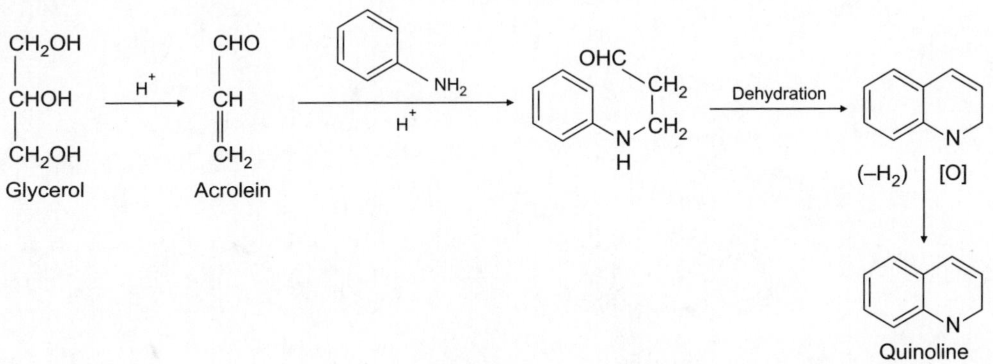

Glycerol $\xrightarrow{H^+}$ Acrolein $\xrightarrow{H^+}$ → Dehydration → $(-H_2)$ [O] → Quinoline

The Skraup reaction is usually vigorous and sometimes violent. The violence of the ordinary Skraup reaction is due to the sudden liberation of acrolein, resulting from the action of sulfuric acid upon the glycerol.

It has been reported that initial Skraup protocol usually gives a very low yield of quinolines. Over the last century, various modifications have been made to improve the yield and reproducibility of the Skraup quinoline synthesis. Various moderators such as acetic or boric acids, ferrous sulfate, thorium, or vanadium or iron oxides have been used to accelerate the reaction and make it higher yielding. Currently, the Skraup reaction can be carried out in protic acid or in the presence of a Lewis acid. In case of unstable aromatic amine, a higher yield of quinoline has been reported.

370. SMILES REARRANGEMENT; TRUCE-SMILES REARRANGEMENT

The Smiles rearrangement is considered to be an intramolecular nucleophilic aromatic substitution in alkaline solution accompanied by the migration of an aromatic system from one heteroatom to a more nucleophilic heteroatom. The two-carbon unit joining X and the nucleophile Y is usually part of an aromatic ring but may also be aliphatic:

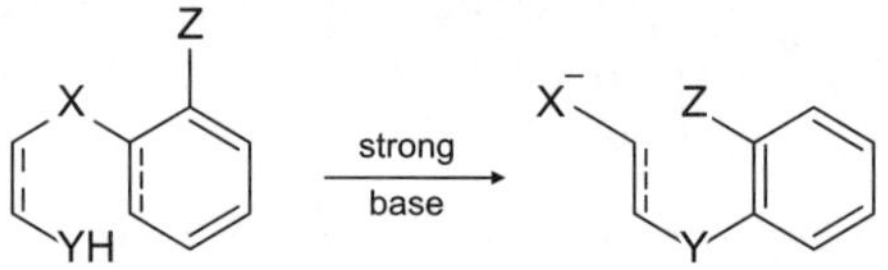

X = S, SO_2, SO, O, COO
Y (the nucleophile) = often the conjugate base of SH, SO_2NHR, SO_2NH_2, NH_2, NHR, OH, OR, CH_2R, CONHR
Z (activating electron-withdrawing group) = NO_2, SO_2R

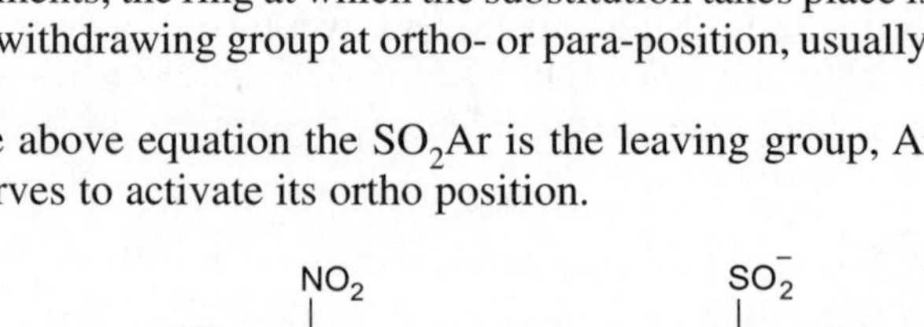

In Smiles rearrangements, the ring at which the substitution takes place is nearly always bearing an activating electron-withdrawing group at ortho- or para-position, usually by ortho- or para-nitro groups.

For example, in the above equation the SO_2Ar is the leaving group, ArO- is the nucleophile, and the nitro group serves to activate its ortho position.

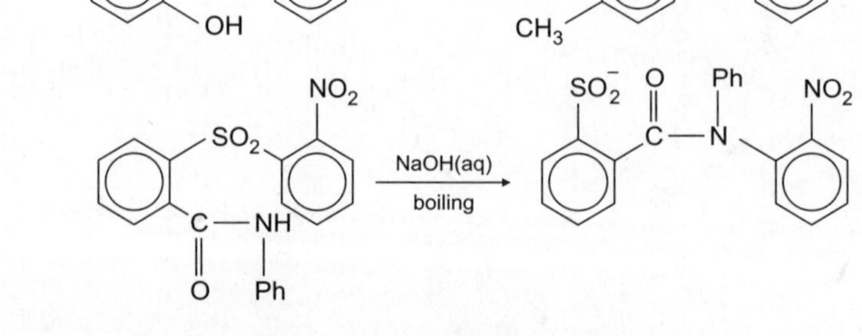

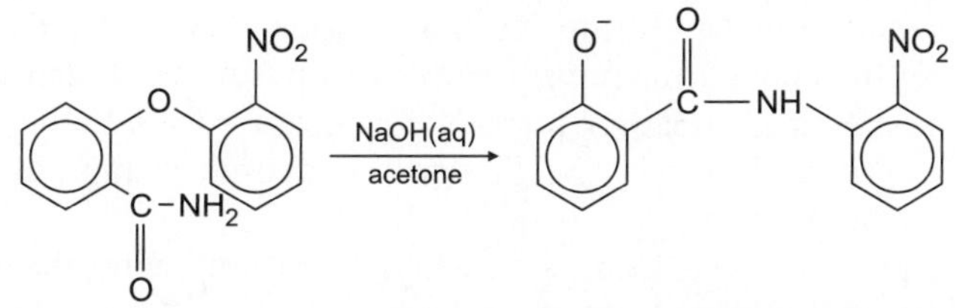

The Truce-Smiles rearrangement involves migration of an aryl group of diaryl sulfones upon treatment with a strong base. It is an extension of the Smiles rearrangement where nucleophile is sufficiently strong that the arene does not require additional activation. for example the conversion of an aryl sulfone into an sulfinic acid by action of n-butyllithium:

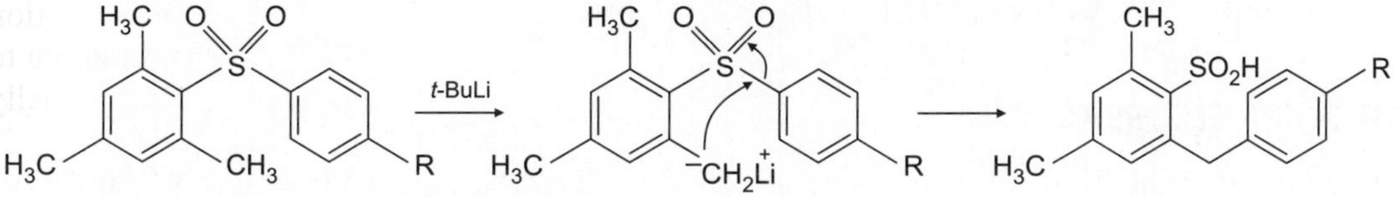

371. SOMMELET-HAUSER REARRANGEMENT

[2,3]-Wittig rearrangement of benzyl quaternary ammonium salts to ortho-substituted benzyl-dialkylamines on treatment with strong base such as alkali metal amides (e.g. sodium amide). An N-dialkyl benzyl amine is produced in the reaction with a new alkyl group in the aromatic ortho position.

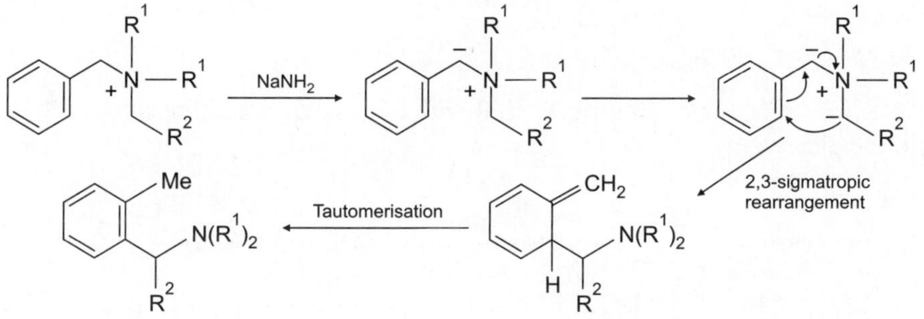

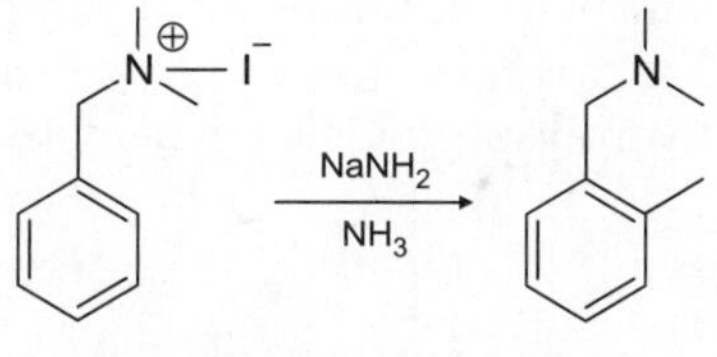

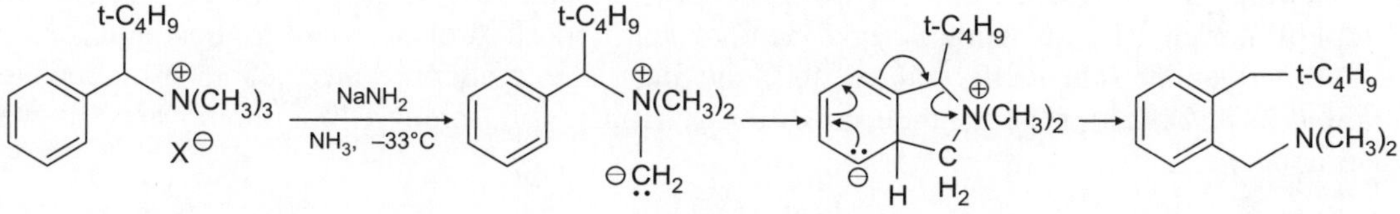

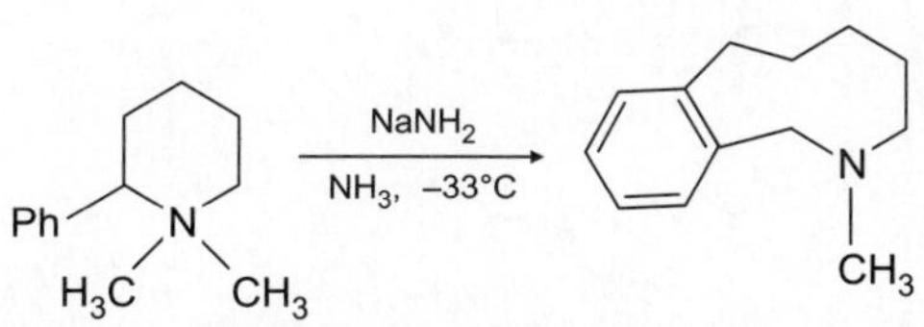

The reaction was named after M. Sommelet and Charles R. Hauser. This rearrangement is often competes with the Stevens rearrangement as well as by the Hofmann elimination. In case where both Stevens and Sommelet-Hauser rearrangement are possible, the selectivity between these depend on various factors such as the Stevens is favored at high temperatures and the Sommelet-Hauser at low temperatures.

In the reaction the deprotonation of benzylic methylene proton takes place to the ylide which undergoes a 2,3-sigmatropic rearrangement in the next step.

372. SOMMELET REACTION

Sommelet reaction is the synthesis of aldehyde from aralkyl halides by treatment with hexamethylenetetramine to yield the quaternary salt, followed by mild acidic hydrolysis.

373. SONN-MÜLLER METHOD

In Sonn-Müller method reaction of ArCONHPh with PCl_5, yield iminium salt, which can then be converted to the aldehyde.

$$ArCONHPh \xrightarrow{PCl_5} \begin{matrix} Ar \\ \diagdown \\ C = NPh \\ \diagup \\ Cl \end{matrix} \xrightarrow[HCl]{SnCl_2} ArCH = NPh \xrightarrow{H_2O} ArCHO$$

Reaction sequence involves treatment of the anilide with phosphorus pentachloride to yield the imidoyl chloride which is reduced with a mixture of stannous chloride and hydrochloric acid to employ to the imine. Subsequent hydrolysis yields the aldehyde.

374. STAUDINGER REACTION

The reaction of an organic azide with a tertiary phosphine (or phosphonite) produce an iminophosphorane (the nucleophilic *aza*-ylide) which upon hydrolysis yield a phosphine oxide and a primary amine is generally referred to as the Staudinger reaction. An azide thus is reduced to an amine in this reaction under mild conditions. Triphenylphosphine is commonly used as the reducing agent.

$$R_3P + N_3X \longrightarrow R_3P = N - N = N - X \xrightarrow{-N_2} R_3P = N - X$$

Phosphazide

$$R_3P = R \overset{R}{\underset{R}{\diagdown}} P$$

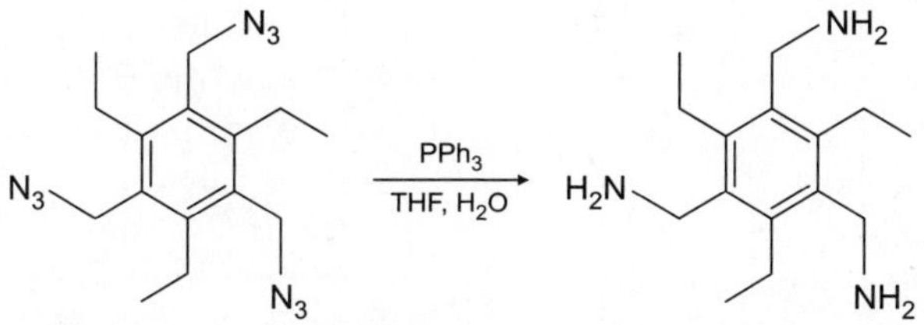

The reaction mechanism consists of formation of an iminophosphorane through nucleophilic addition of the phosphine with the azide and expulsion of nitrogen. Aqueous work up leads to the amine. The reaction was named after its discoverer Hermann Staudinger.

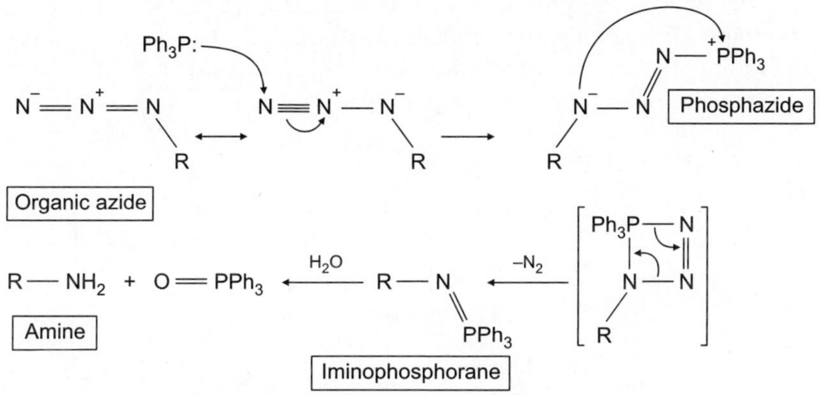

375. STEPHEN ALDEHYDE SYNTHESIS

The reaction was invented by Henry Stephen and hence named. It involves a reaction sequence in which first the treatment of the nitrile with a mixture of anhydrous stannous chloride and dry hydrochloric acid yield the imine salt complex ($[R — CH = NH_2]^+Cl^-$), which is subsequently hydrolyzed with water to the aldehyde. The reaction is applicable for the preparation of aromatic aldehydes.

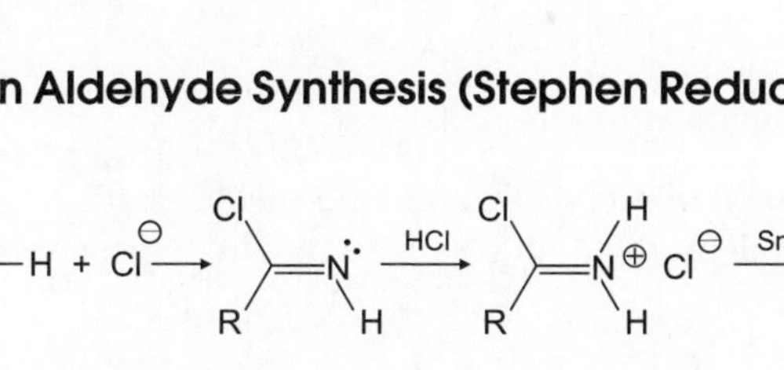

Mechanism for Stephen Aldehyde Synthesis (Stephen Reduction)

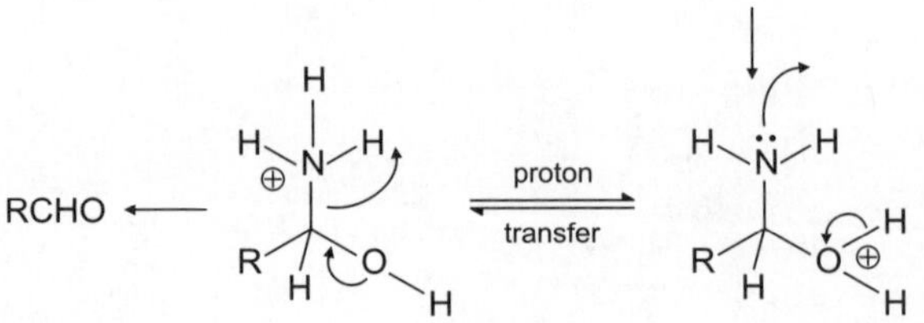

376. STEVENS REARRANGEMENT

Stevens rearrangement involves treatment of a quaternary ammonium salt containing an electron withdrawing group on one of the carbon attached to the nitrogen upon the treatment with a strong base (e.g. $NaNH_2$ or NaOR). The product is a rearranged tertiary amine. In other words, it is an intramolecular thermal [1,2]-electrophilic migration from the heteroatom of an ylide (e.g. quaternary ammonium salt) to the adjacent carbanionic centre upon the treatment with a strong base. The product is a rearranged tertiary amine or sulphide.

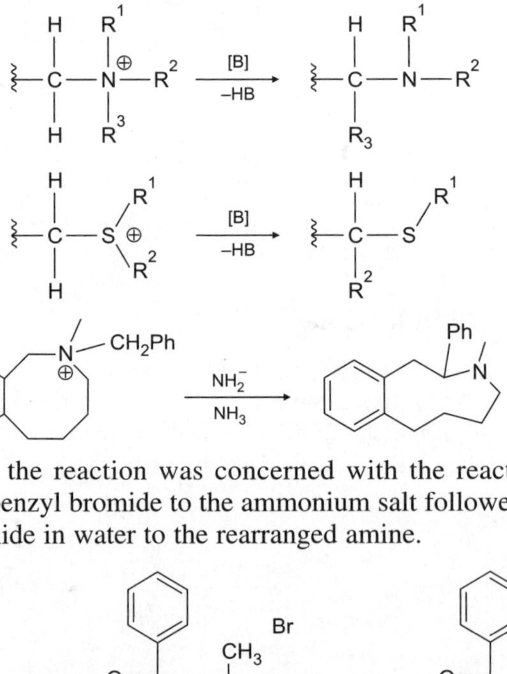

The original protocol of the reaction was concerned with the reaction of 1-phenyl-2-(N, N-dimethyl) ethanone with benzyl bromide to the ammonium salt followed by the rearrangement reaction with sodium hydroxide in water to the rearranged amine.

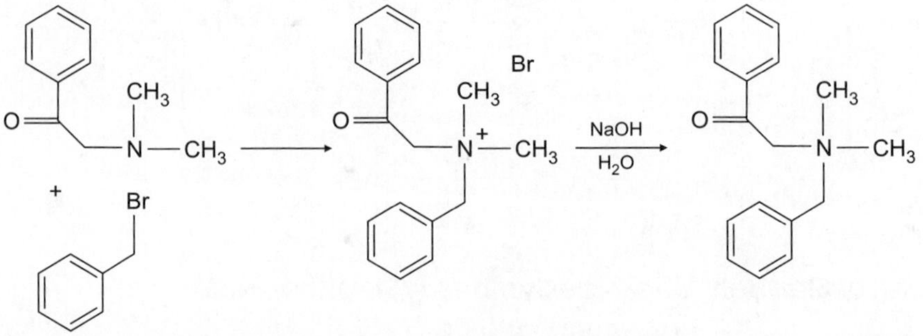

The rearrangement is found to proceed via an intermediate radical-pair or ion-pair. A nitrogen-ylide is formed initially by deprotonation of the ammonium species with a strong base.

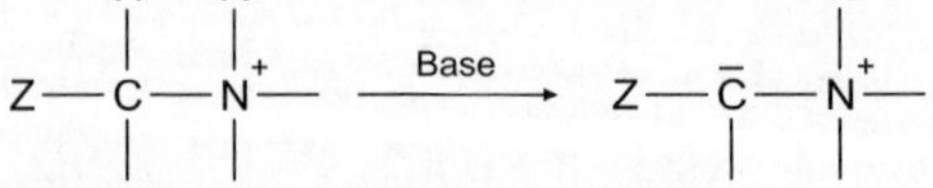

A homolytical cleavage of the N-R bond takes place in next step, which is followed by the rearrangement to yield the tertiary amine via an intermediate radical-pair.

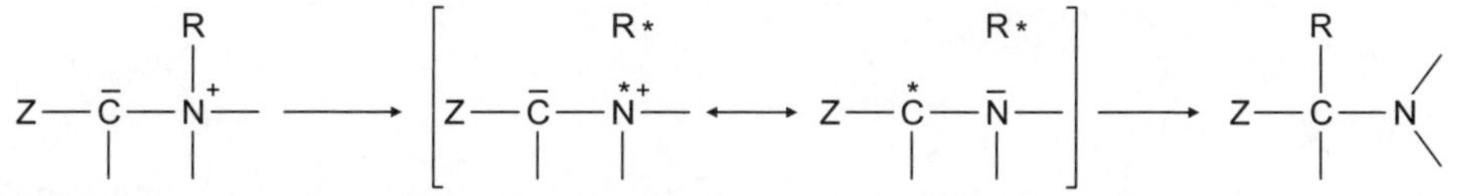

The order of migration is found to be as:

propargyl > allyl >benzyl > alkyl

An ion-pair mechanism is found to be more rational in certain cases than a radical-pair mechanism. A heterolytic cleavage of the N-R bond takes place in this case.

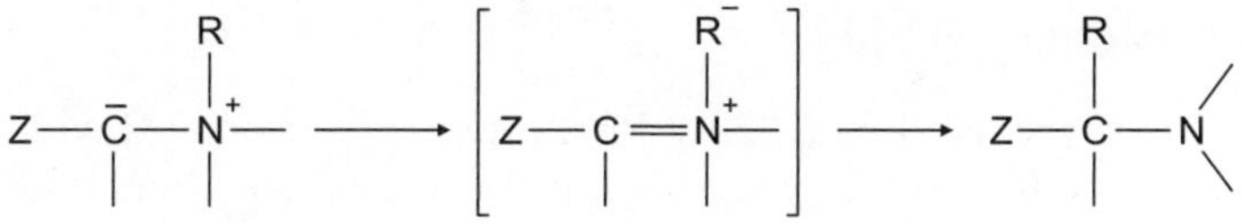

377. STIEGLITZ REARRANGEMENT

Stieglitz rearrangement describes nucleophilic migration from carbon to nitrogen. The migration of a substituent in nitrogen-containing molecules (e.g. hydroxylamines, amines, haloamines, azides) at the tertiary carbon to the adjacent nitrogen atom to form an imine derivative is generally known as the Stieglitz rearrangement. As the reaction involves nucleophilic migration of a substituent from carbon to nitrogen, therefore it is related to the Beckmann rearrangement. Phosphorus pentoxide and boron trifluoride can also promote the Stieglitz rearrangement. This reaction has applications in preparation of Schiff bases and aromatic heterocycles.

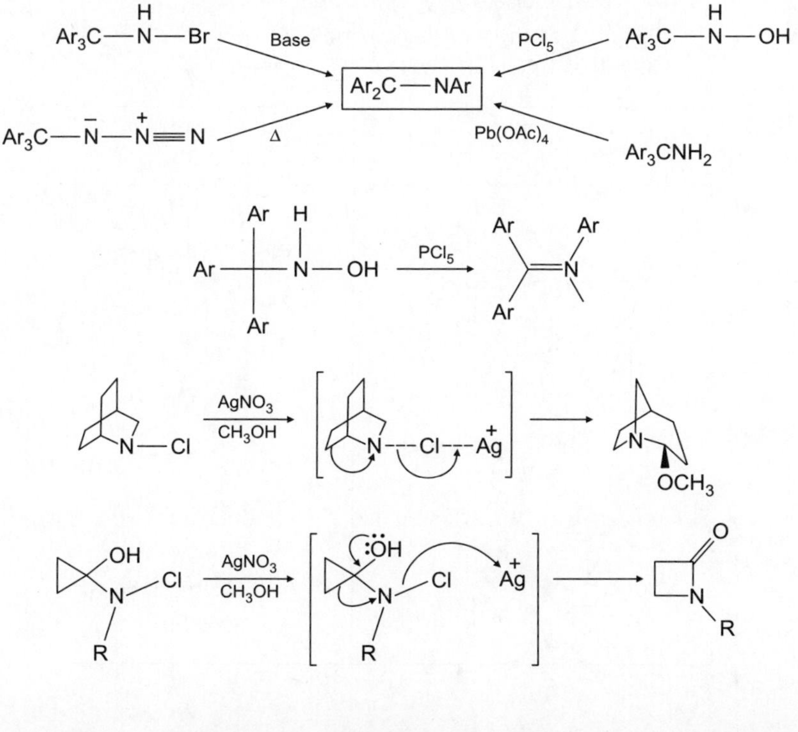

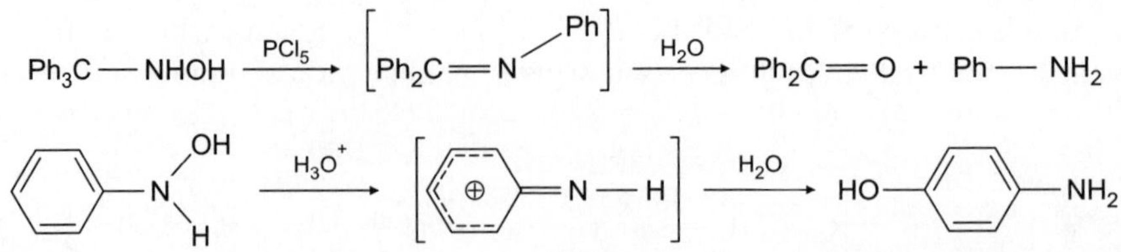

$$Ph_3C-NHOH \xrightarrow{PCl_5} \left[Ph_2C=N\overset{Ph}{}\right] \xrightarrow{H_2O} Ph_2C=O + Ph-NH_2$$

The name Stieglitz rearrangement is generally applied to the rearrangements of trityl N-haloamines and hydroxylamines. These reactions are similar to the rearrangements of alkyl azides, and the name Stieglitz rearrangement is also given to the rearrangement of trityl azides.

$$Ar_3CNHX \xrightarrow{Base} Ar_2CN=NAr$$

$$Ar_3CNHOH \xrightarrow{PCl_5} Ar_2CN=NAr$$

378. STILLE COUPLING

The Stille reaction was discovered in 1977 by John Kenneth Stille and David Milstein. The Stille reaction (also known as Stille coupling) is a chemical reaction, coupling an organotin compound (such as trimethylstannyl or tributylstannyl or tributylstannyl compounds) with carbon electrophiles such as:

- an sp^2 hybridized organic halide (such as Cl, Br, I or a pseudohalide such as a triflate, $CF_3SO_3^-$), or
- an organic acetates, or
- perfluorinated sulfonates lacking a sp^3 hybridized β-hydrogen catalyzed by palladium.

$$R_3SN-R^1 + R^2-X \xrightarrow{[Pd]} R^1-R^2 + R_3SnX$$

[Pd] = Pd(PPh$_3$)$_4$, PhCH$_2$Pd(PPh$_3$)$_2$Cl

R^1 = alkynyl, alkenyl, aryl, allyl, benzyl, alkyl

R^2 = acyl, alkenyl, allyl, benzyl, aryl

X = Cl, Br, I, OCOCH$_3$, OSO$_2$(C$_n$F$_{2n+1}$), n = 0, 1, 4

Electrophile R^1X	Organic reagent $R^2XSn(R^3)_3$
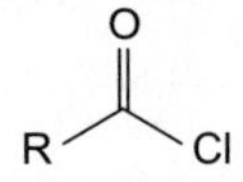	H—SnR$_3$ R'C≡C—SnR$_3$
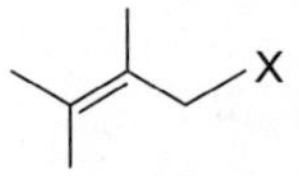 (X = Cl, Br)	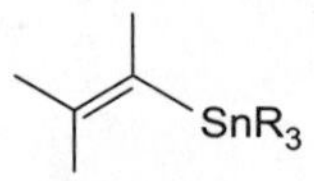
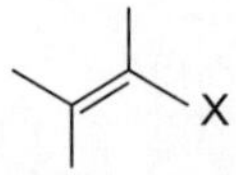 (X = I, OTf)	Aryl—SnR$_3$
Aryl—CH$_2$—X (X = Br, I)	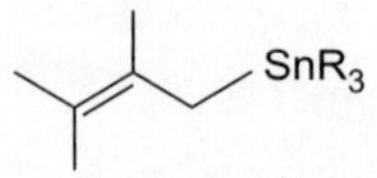
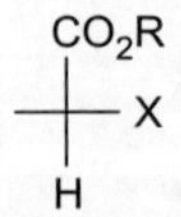 (X = Br, I)	Aryl—CH$_2$—SnR$_3$ R'—SnR$_3$

The reaction is usually performed under inert atmosphere using dehydrated and degassed solvent, as oxygen causes the oxidation of the palladium catalyst and promotes homo-coupling of organic stannyl compounds, and these side reactions lead to a decrease in the yield of the desired cross-coupling reaction.

$$R—X \;+\; R'—SnR_3 \xrightarrow{Pd(0)} R—R' \;+\; XSnR_3$$

Aryl or   Organotin
vinyl halide

The Stille reaction is an extremely versatile alternative to the Suzuki reaction. It replaces the organoboron reagents with organostannanes. As the tin bears four organic functional groups, understanding the rates of transmetallation of each group is important.

Relative rate of transmetallation: Alkynyl > vinyl > aryl > allyl ~ benzyl >> alkyl

The Stille coupling is particularly popular as organostannanes are readily prepared, purified and stored. The reaction also has the advantage that it is run under neutral conditions making it even more tolerant of different functional groups than the Suzuki reaction.

The reaction may also be carried out intramolecularly and with alkynyl stannanes instead of the more usual aryl or vinyl stannanes to form medium-sized rings. For example, the reaction below forms a 10-membered ring containing two alkynes.

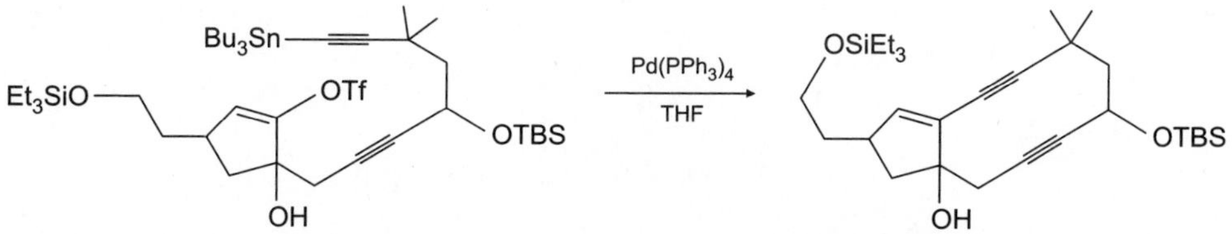

379. STOBBE CONDENSATION

Stobbe condensation describes condensation of diethyl succinate and its derivatives with carbonyl compounds (aldehydes or ketones) in the presence of a strong base such as NaOEt, NaH, or $KOCMe_3$ to form monoesters of α-alkylidene (or arylidene) succinic acids.

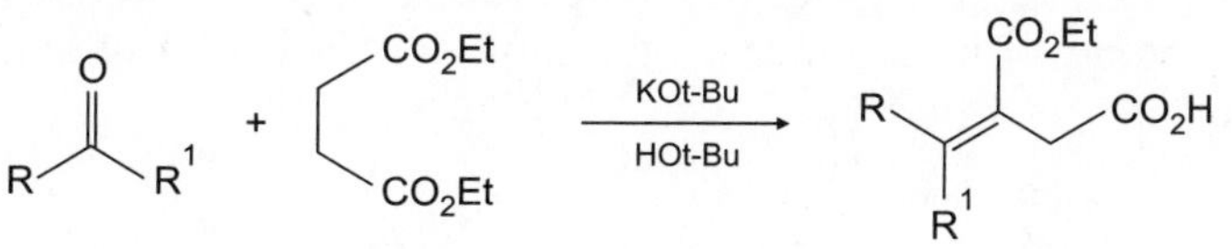

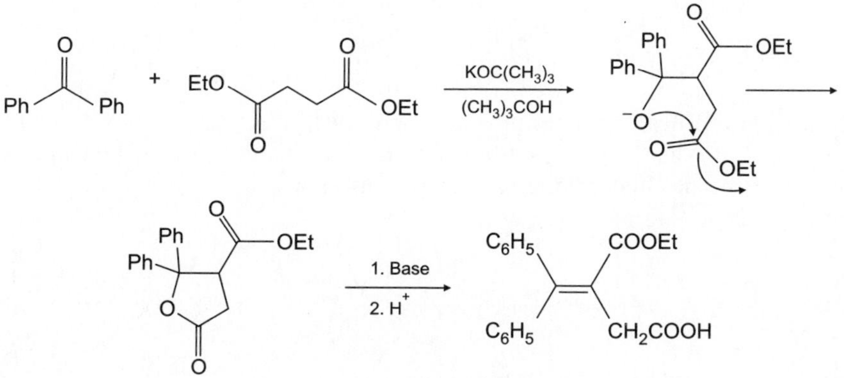

380. STOLLÉ SYNTHESIS

Stollé synthesis involves a series of chemical reactions for the formation of indole derivatives by the reaction of arylamines with α-haloacid chlorides (or oxalyl chloride), followed by cyclization of the resulting amides with aluminum chloride. The first part of the reaction is an amide coupling, while the second part, i.e. cyclization step is a Friedel-Crafts reaction.

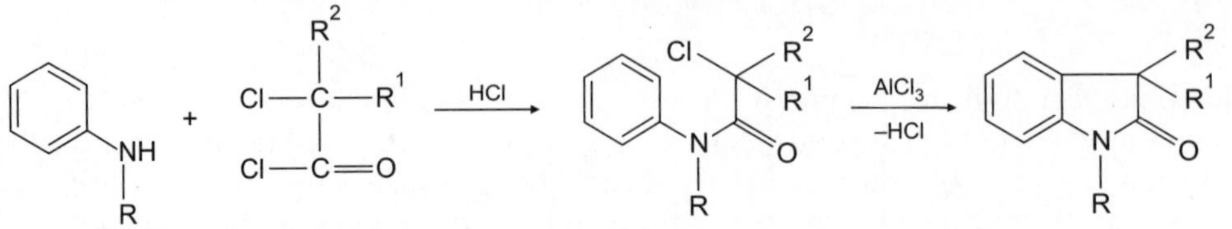

381. STORK ENAMINE REACTION (OR STORK ENAMINE ALKYLATION)

Stork enamine alkylation, also known as the Stork enamine reaction, involves alkylation of enamines. Aldehydes and ketones when react with secondary amines form *enamines*. A cyclic secondary amine like pyrrolidine, piperidine or morpholine is most often used.

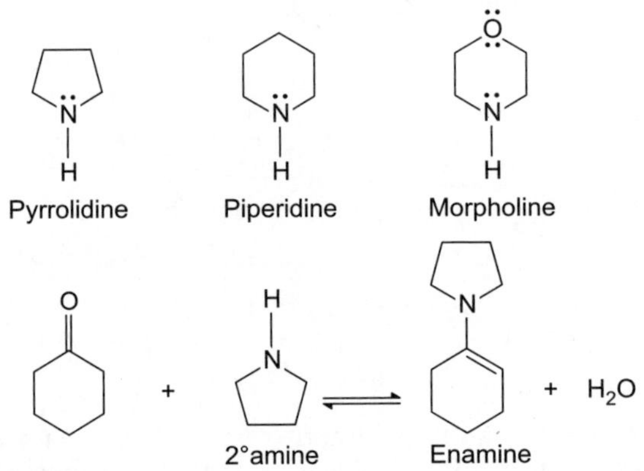

Pyrrolidine   Piperidine   Morpholine

2°amine   Enamine

Gilbert Stork in the year 1954 showed that enamines can be used in the alkylation and acylation of aldehydes and ketones. The usefulness of enamines stems from the fact that they contain nucleophilic carbon.

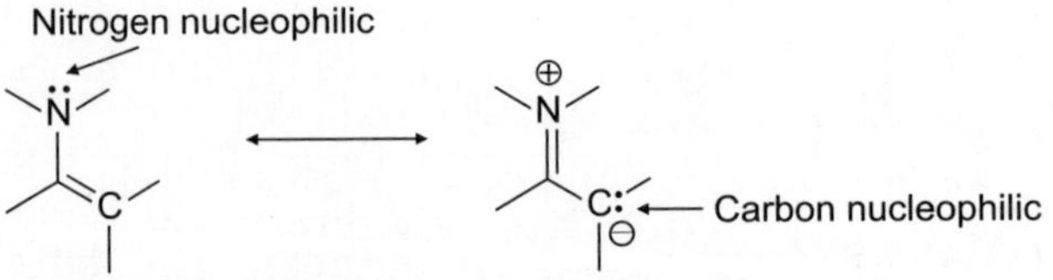

Nitrogen nucleophilic

Carbon nucleophilic

Alkylation of enamines: An iminium ion is produced after alkylation of enamine, which is readily hydrolyzed to regenerate the carbonyl group. Alkylation usually takes place on the less substituted side of the original ketone. The most commonly used amines are the cyclic amines, piperidine, morpholine, and pyrrolidine.

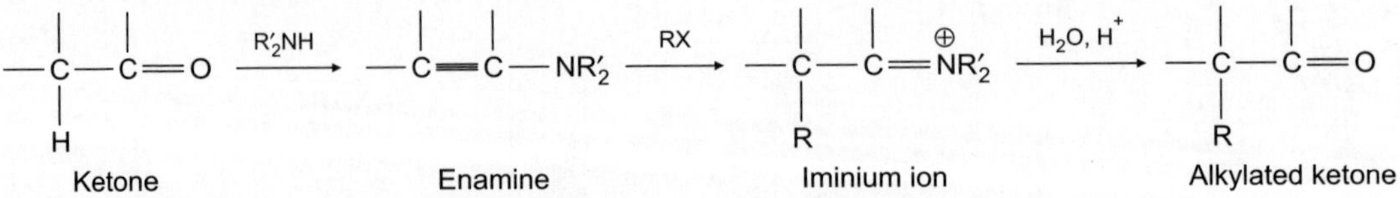

Ketone   Enamine   Iminium ion   Alkylated ketone

In alkylation of enamines, best yields are obtained with reactive halides like allylic, benzylic or propargylic halides or an α-halo ether, α-halo ester or α-halo ketone. For example:

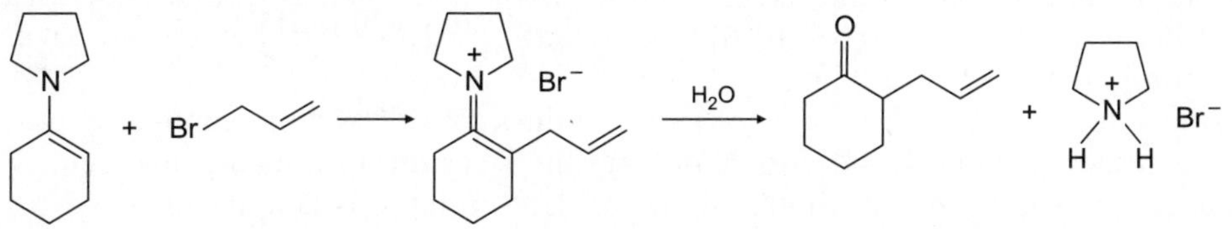

Alkylation of enamines can lead to some N-alkylation, which can be converted to the C-alkylated product by heating. For example:

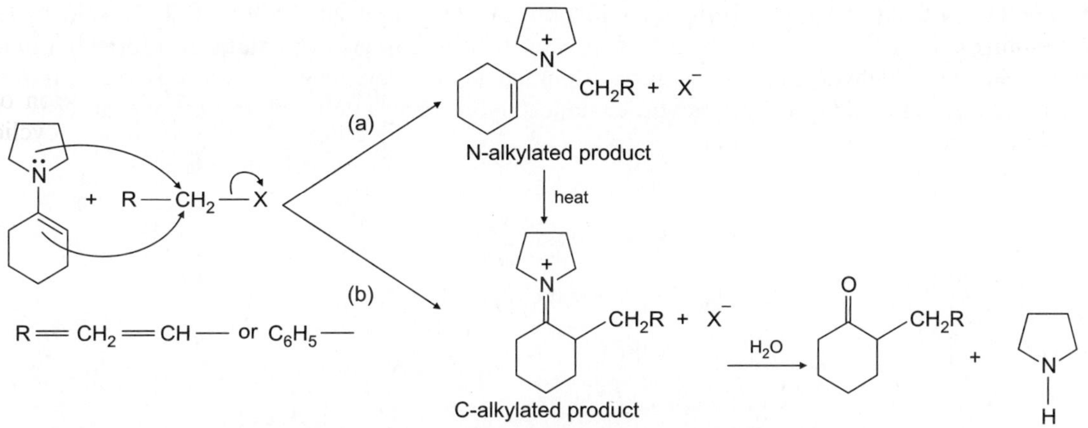

Acylation of enamines is another important application, for example:

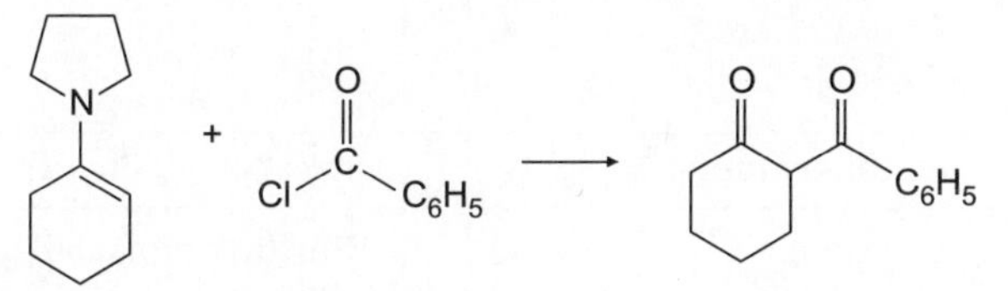

382. STRECKER AMINO ACID SYNTHESIS

Strecker amino acid synthesis affords α-amino nitriles by condensation of an aldehyde, NH_3, and KCN, was first reported in 1850. Subsequent hydrolysis of α-amino nitriles provides α-amino acids. The reaction, as the first multicomponent reaction, provides valuable α-amino nitriles synthons, which are important intermediates for the synthesis of amino acids. This reaction is a special case of the Mannich reaction and generally carried out by addition of cyanide anion to imines.

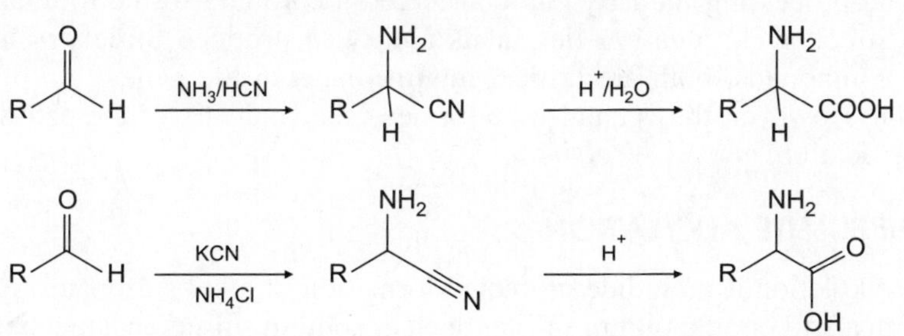

Among various cyanide sources, such as HCN, KCN, Et_2AlCN and Bu_3SnCN which are toxic and require harsh reaction conditions, TMSCN is safer and more efficient. The efficiency of the reaction has been increased by the use of catalyst. Lewis acids such as $Cu(OTf)_2$, $BiCl_3$, $NiCl_2$, $InCl_3$, $RuCl_3$, $Sc(OTf)_3$, $La(NO_3)_3.6H_2O$ or $GdCl_3.6H_2O$ have been used to catalyze homogeneously the Strecker reaction.

It has been found that primary and secondary amines as well as ammonium salts can also be used in the reaction. While usage of ammonium salts gives unsubstituted amino acids, primary and secondary amines give substituted amino acids. Similarly, if ketones are used instead of aldehydes, α,α-disubstituted amino acids are obtained.

383. STRECKER DEGRADATION

The reaction of an α-amino acid with an oxidation reagent to give carbon dioxide and an aldehyde containing one carbon atom less is known as Strecker degradation. The aldehydes formed, often called Strecker aldehydes, can act as food odourants per se. The reaction proceeds via an imine intermediate and can be either catalytic or noncatalytic. Inorganic oxidizing agents can also be used in to bring about the reaction.

Strecker amino acids	Strecker aldehydes
Valine	2-methyl-propanal
Isoleucine	2-methyl-butanal
Leucine	3-methyl-butanal
Methionine	Methional
Phenyl alanine	Phenyl acetaldehyde

The reaction is named after Adolph Strecker, a German chemist. This reaction has been found useful in the food-processing industry and considered a corollary to the Maillard reaction. The importance of Strecker degradation lies in its ability to produce Strecker aldehydes and 2-aminocarbonyl compounds, both are critical intermediates in the generation of aromas during Maillard reaction. However, they can also be formed independently of the pathways established for Strecker degradation.

384. STRECKER SULFITE ALKYLATION

Strecker sulfite alkylation is an iodide promoted formation of alkyl sulfonates by the reaction of alkyl halides with alkali or ammonium sulfites such as sodium sulfite, in aqueous solution. Iodide is used as a catalyst. The reaction was first described by Adolph Strecker in 1868. The general reaction is:

$$RX + M_2SO_3 \xrightarrow{\text{NaI}} RSO_3M + MX$$

385. SUAREZ REACTION (SUAREZ FRAGMENTATION OR SUAREZ CLEAVAGE)

Suarez and co-workers have presented diacetoxyiodobenzene/iodine (DIB/I$_2$) as an effective mixture of reagents for the oxidative decarboxylation of carboxylic acids.

$$R-CO_2H \xrightarrow{\text{DIB/I}_2} R-CO_2I \xrightarrow{-CO_2} RI$$

Suarez fragmentation or Suarez cleavage refers to alkoxy radical mediated photoinduced conversion (photo-fragmentation) of hydroxyl-containing substrates with hypervalent iodine I(III)I$_2$ to the corresponding oxygen-centered radical. This reaction usually works for polyhydroxy cyclic compounds and for the secondary alcohols as well.

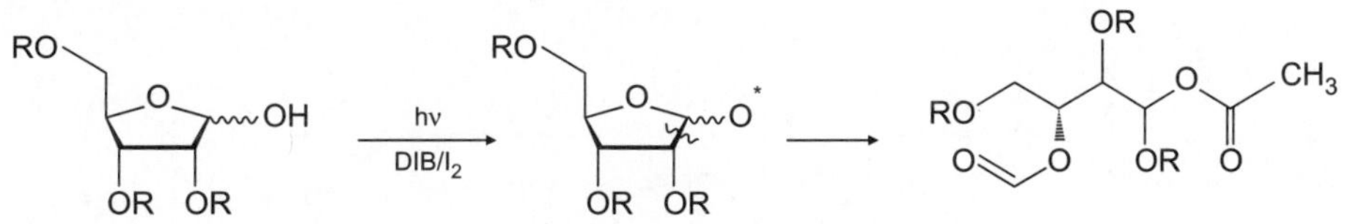

DIB(diacetoxyiodo benzene)

386. SUGASAWA REACTION

Sugasawa reaction refers to ortho acylation of anilines by nitriles as the acylating agents in the presence of BCl$_3$ and an auxillary Lewis acid, e.g. AlCl$_3$. In other words, Sugasawa reaction uses boron trichloride/aluminum chloride to transform anilines and nitriles to 2-aminophenyl ketones.

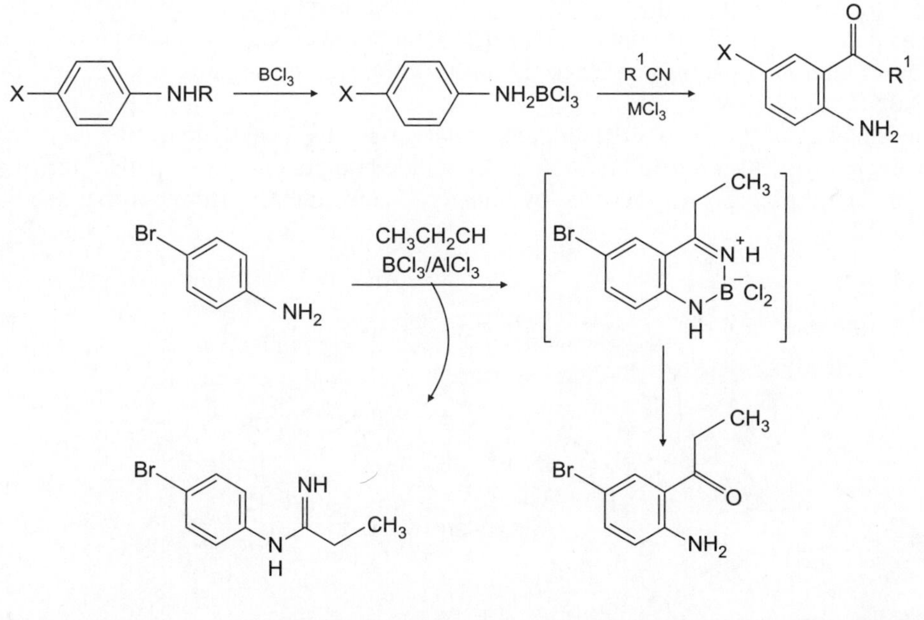

387. SUZUKI COUPLING

The reaction was reported in 1979 by Akira Suzuki and N. Miyaura and commonly referred to as the Suzuki cross-coupling. In broad sense, this reaction involves palladium catalyzed cross-coupling between organoboron compounds and organic halides leading to the formation of carbon-carbon bonds. Palladium-catalyzed carbon-carbon bond forming reactions involves the *Suzuki reaction,* the *Heck reaction* and the *Stille reaction.* The Suzuki reaction is the palladium-catalyzed cross-coupling of organic halides such as an aryl or vinyl halide or triflate or perfluorinated sulfonates with organoborones such as an aryl or vinyl boronic acid using a palladium catalyst in the presence of a base.

$$R^2 - X \ + \ R^1 - B\begin{smallmatrix} R \\ \\ R \end{smallmatrix} \ \xrightarrow[\text{Base}]{\text{Pd(0) (catalytic)}} \ R^2 - R^1 \ + \ \text{Inorganic salts}$$

R^1 = alkyl, allyl, alkenyl, alkynyl, aryl; R = alkyl, OH, O-alkyl; R^2 = alkenyl, aryl; X = Cl, Br, I, OTf

Base = sodium carbonate, sodium hydroxide, M(O-alkyl), potassium phosphate tribasic.

The Suzuki reaction is a powerful cross-coupling method that allows the synthesis of conjugated olefins, styrenes, and biphenyls. Proceeding with high stereo- and regioselectivity is a great advantage of the reaction. Furthermore, boron compounds are generally non-toxic and the reaction can be run under very mild conditions. This has made the reaction popular in the pharmaceutical industry. The stability and weak nucleophilic nature of organoboron compounds has made this reaction very practical. It tolerates a wide range of functional groups and it is highly chemoselective.

The mechanism of the Suzuki reaction closely resembles to that of the Stille coupling reaction. However, in contrast to the Stille coupling reaction, the Suzuki mechanism requires activation of boronic acid (activated species containing a tetravalent boron center), for example with base. This activation of the boron atom enhances the polarisation of the organic ligand, and facilitates transmetallation.

The mechanism of the Suzuki reaction can be described by a catalytic cycle given below. The catalytic cycle invoves *oxidative addition of* the halide component with a palladium(0) complex to give a palladium(II) species followed by transfer of substituent R' from boron to the palladium center (generating a palladium(II) species that contains both the substituent R and R′ that are to be coupled). Finally, *reductive elimination* yields the coupling product and the regenerated catalytically active palladium (0) complex.

388. SWARTS REACTION

The Swarts reaction was discovered and refined by the Belgian chemist F. Swarts. It involves partial fluorination of organic polyhalides (mostly aliphatic) with antimony trifluoride (or zinc and mercury fluorides) in the presence of a trace of a pentavalent antimony salt. The mixture of antimony trifluoride and chlorine ($SbF_3 + Cl_2$) is known as the Swarts reagent.

$$ClHC \equiv CClCHCl_2 \ + \ SbF_3 \ \xrightarrow{SbCl_5} \ ClHC \equiv CClCF_3 \ + \ SbCl_3$$

The following transformation involves a two-step conversion of toluene derivatives into benzotrifluorides through radical chlorination followed by treatment with an inorganic fluoride (e.g. SbF_5) or anhydrous hydrogen fluoride.

This reaction is industrially important in the manufacture of refrigerants (i.e. CFC's/freons).

$$CCl_4 \ \xrightarrow[100\,°C]{HF,\ SbF_5} \ CCl_3F \ + \ CCl_2F_2 \ + \ CClF_3$$

On the industrial scale, trifluoromethylated arenes are mainly produced by the Swarts reaction. The reaction was developed in 1892. The requirement for reactive fluorinating reagents and high temperatures render this strategy incompatible with many common functional groups.

389. SWERN OXIDATION (MOFFATT-SWERN OXIDATION)

The Swern oxidation is one of the most useful methods for the conversion of primary and secondary alcohols to aldehydes and ketones, respectively using oxalyl chloride [$(COCl)_2$], dimethyl sulfoxide (DMSO), and an organic base, such as trimethylamine (Et_3N). The reaction is named after Daniel Swern.

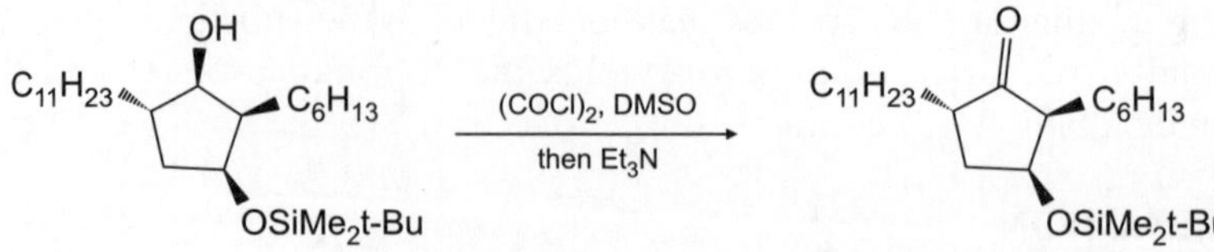

The first step of the Swern oxidation is the generation (*in situ*) of dimethylchlorosulphonium ion by a low-temperature reaction of dimethyl sulfoxide (DMSO) with oxalyl chloride.

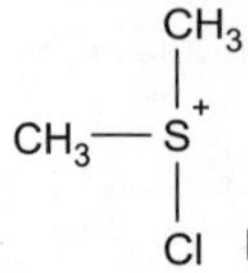

Dimethylchlorosulphonium ion

Then reaction with an alcohol leads to the key alkoxysulfonium ion intermediate.

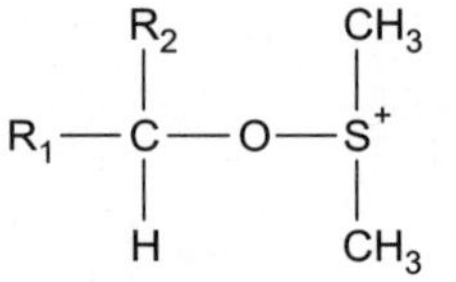

Alkoxysulfonium ion

Deprotonation of this intermediate with at least 2 equivalents of base—typically trimethylamine gives a sulphur ylide, which undergoes intramolecular deprotonation via a five-membered ring transition state and fragmentation to yield the product and dimethyl sulfide.

The combination of dimethylsulfoxide (DMSO) with an electrophilic species to form "activated DMSO" is widely useful for the oxidation of alcohols to their respective carbonyl compounds. Daniel Swern demonstrated that oxalyl chloride was particularly suitable for use as the electrophilic partner. This combination typically affords high conversion of the alcohol to the carbonyl compound and demonstrates good functional group selectivity.

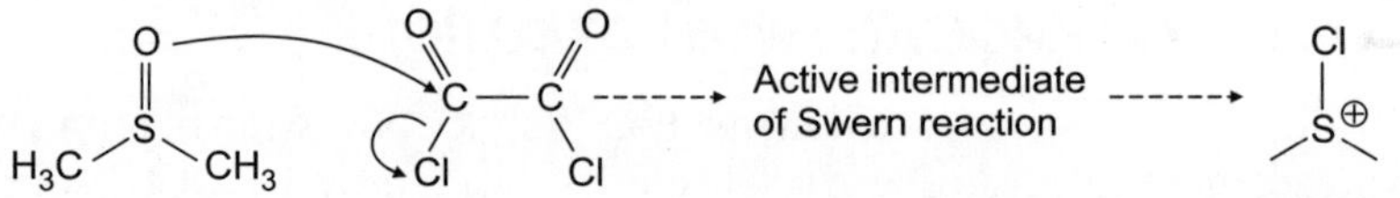

The Swern oxidation has been widely exploited for the oxidation of alcohols to their respective carbonyl compounds, under mild conditions, using activated dimethyl sulfoxide (DMSO) as the oxidizing agent. The reaction is popular for its mild character and wide tolerance of functional groups. However, the cryogenic operating conditions ($< -60°C$) employed for the Swern oxidation limit its utility for scale-up operations.

390. TAFEL REARRANGEMENT

Tafel rearrangement refers to the rearrangement of the carbon skeleton of substituted acetoacetic esters to hydrocarbons with the same number of carbon atoms. Reaction is accompanied by the rearrangement of the alkyl group by electrolytic reduction at a lead cathode in alcoholic sulfuric acid.

$$CH_3COCRR^1COOC_2H_5 \nearrow CH_3CH_2CRR^1CH_3 \quad \text{(normal)}$$
$$\searrow CH_3CH_2CH_2CHRR^1 \quad \text{(rearranged)}$$

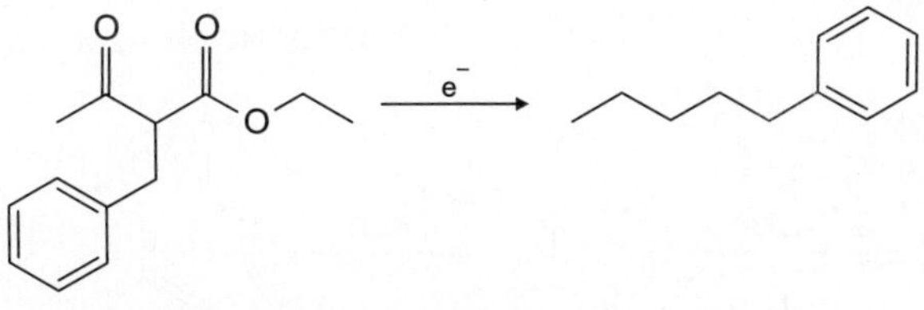

391. TEBBE OLEFINATION (METHYLENATION)

Exchange of the oxygen atom of a carbonyl function for the methylene group of the proposed titanium carbene complex (the Tebbe reagent) to yield terminal alkenes.

The Tebbe reagent is a metal carbenoid prepared from the dimetallomethylene species derived by the reaction of trimethyl aluminium with titanocene dichloride; this reagent exhibits carbenoid behaviour after the addition of a catalytic amount of pyridine. The Tebbe reagent reacts with various carbonyl partners to give the product of methylenation.

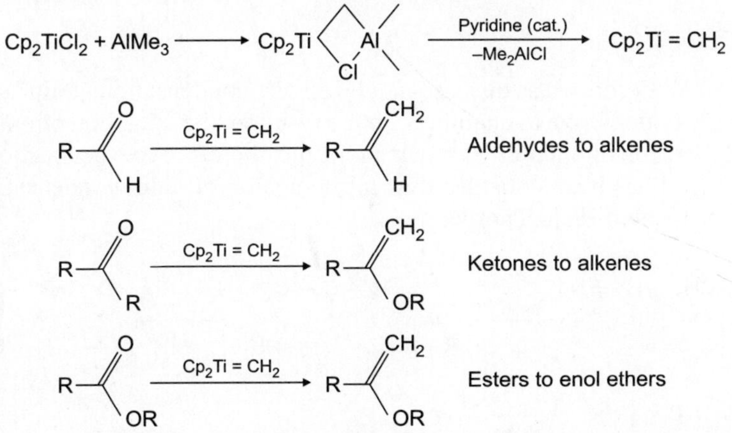

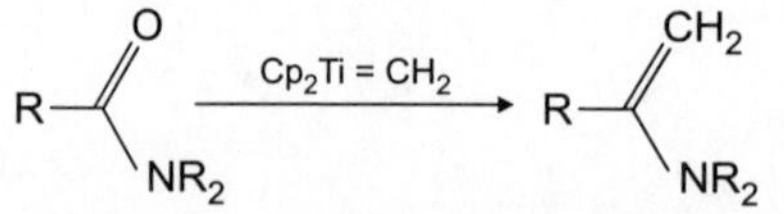

Amides to enamines

392. THIELE REACTION (THIELE-WINTER ACETOXYLATION)

Thiele reaction or Thiele-Winter acetoxylation is a strong acid, e.g. sulfuric acid or boron trifluoride promoted addition of acetic anhydride to quinines to form triacetoxy aromatic compounds. The triacetoxy derivatives can be isolated in fair to excellent yields and can be further hydrolysed either under acidic or basic conditions to the corresponding hydroxyhydroquinone derivatives and found applicability for the preparation of the hydroxyquinone moiety. Therefore, this reaction is useful for the hydroxylation of quinone derivatives. Although best results for the Thiele-Winter acetoxylation are obtained when the acid catalyst is H_2SO_4 or BF_3-etherate, however, CF_3SO_3H, $HClO$, BF_3, CF_3SO_3H and $HClO_4$ were also found to affect the acetoxylation for this reaction effectively.

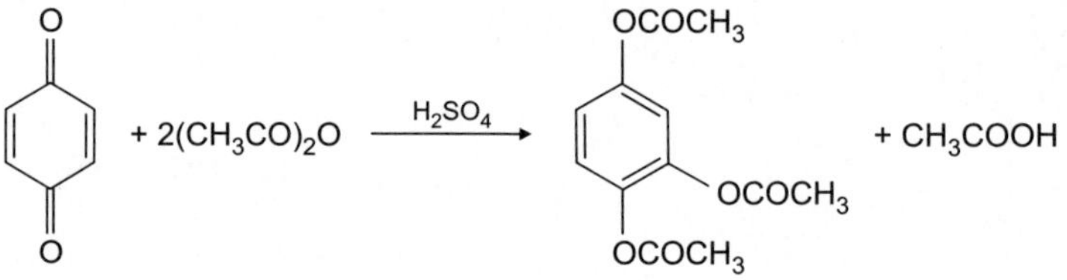

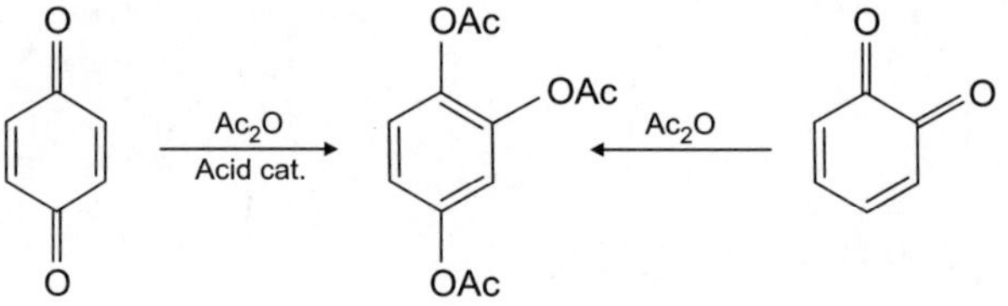

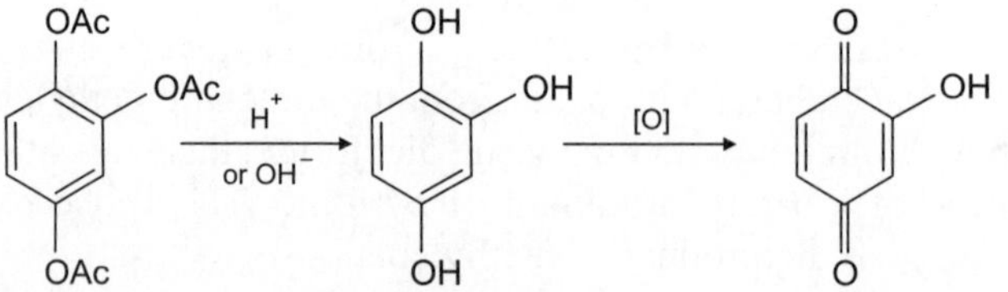

393. THORPE REACTION

The Thorpe reaction is described as the base-catalyzed self-condensation of nitriles to form imines, which subsequently tautomerize to enamines. In the reaction, the α-carbon of one nitrile molecule is added to the CN carbon of other other nitrile molecule, therefore the the reaction is analogous to the aldol reaction and has been considered as nitrile analog of aldol condensation. The reaction was discovered by Jocelyn Field Thorpe.

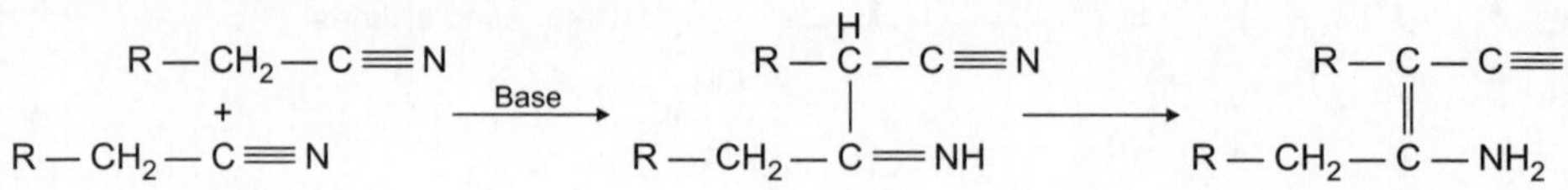

$$C_6H_5-CH_2-C\equiv N$$
$$+$$
$$C_6H_5-CH_2-C\equiv N$$
Phenyl-acetonitrile

$\xrightarrow{C_4H_5OH}$

$$C_6H_5-\overset{H}{\underset{|}{C}}-C\equiv N$$
$$C_6H_5-CH_2-C\equiv NH$$
3-imino-2,4-diphenyl-butyronitrile

$\longrightarrow$

$$C_6H_5-C-C\equiv N$$
$$\parallel$$
$$C_6H_5-CH_2-C\equiv NH_2$$
3-amino-2,4-diphenyl-but-2-enenitrile

$$CH_3CN \xrightarrow[\text{rt. overnight}]{BuOK}$$

(structure: H₂N and H on top; H₂C and CN on bottom)

$$C_2H_5CN \xrightarrow[\text{rt. overnight}]{BuOK}$$

(structure: H₂N and CH₃ on top; C₂H₅ and CN on bottom)

The intermolecular Thorpe reaction and its intramolecular version Thorpe-Zeigler reactions are one of the most promising lines in the synthetic chemistry, e.g. synthesis of five-member amino heterocycles. These are base catalyzed and sodium or potassium alkoxide, sodium hydride, potassium hydroxide, lithium hydroxide and potassium carbonate were employed frequently. Radical alternatives, solvent free strategies as well as iridium hydride complexes also have been applied to intramolecular as well as intermolecular Thorpe reaction. Thorpe cyclization is well known for the formation of synthetically important five-membered heterocycles such as furan, thiophene, pyrazole and many more having adjacent amino and carbethoxy or nitrile functionalities.

394. TIEMANN REARRANGEMENT

In Tiemann rearrangement, amidoximes are rearranged into corresponding substituted ureas by treatment with benzenesulfonyl chloride and hydrolysis with water. Reaction is similar to Beckmann or Lossen type rearrangement.

395. TIFFENEAU-DEMJANOV REARRANGEMENT

The Tiffeneau-Demjanov rearrangement (TDR) is the chemical reaction of β-amino alcohols with nitrous acid to form an enlarged (ring-expanded) cyclic ketone. This reaction provides an easy way to increase amino-substituted cycloalkanes and cycloalkanols in size by one carbon, therefore used to transform a cyclic ketone into a homologue that is one ring size larger.

396. TISHCHENKO REACTION

The Tishchenko reaction was first described in 1906 and entails the Lewis acid mediated condensation of two molar equivalents of an aldehyde to form an ester. When aldehydes, with or without a hydrogen, are treated with aluminum ethoxide (Lewis acis catalyst in Tishchenko reaction), one molecule is oxidized and another reduced.

Besides the aluminum alkoxides and sodium alkoxides that are usually used to affect the Tishchenko reaction, many other molecules or complexes have been found to effectively promote the Tishchenko reaction for example the Tishchenko reaction can also be catalyzed by ruthenium complexes, by Cp_2ZrH_2 or $BuTi(OiPr)_4Li$, and, for aromatic aldehydes, by disodium tetracarbonylferrate, $Na_2Fe(CO)_4$.

In 1990, Evans and Hoveyda reported an important variant of this reaction,2 which has subsequently become known as the Evans-Tishchenko reaction. This transformation involves a preformed β-hydroxyketone and an aldehyde undergoing a Lewis acid catalyzed condensation to generate a 1, 3-*anti* diol monoester. Importantly, the Evans-Tishchenko reaction provides a highly regio-and diastereo-selective route to these ubiquitous structural units; and the reaction itself is remarkably mild and can be carried out in the presence of a number of (often sensitive) functional groups.

The aldol-Tishchenko reaction involves the trimerization of enolizable aldehydes. In an aldol-Tishchenko reaction, two aldehyde molecules undergo addition to form an aldol adduct which is subsequently reduced by a third aldehyde to yield 1,3-diol monoesters, with the occurrence of an intramolecular hydride shift. This step—the intramolecular reduction—has been utilized for highly diastereoselective reduction of β-hydroxyketones.

397. TRAUBE PURINE SYNTHESIS

The Traube purine synthesis, named after Wilhelm Traube in 1900 is considered as the general method for the preparation of purine derivatives. It involves reaction between an amine-substituted pyrimidine and formic acid. The reaction is a multistep synthesis of purine derivatives from 4-amino-6-hydroxypyrimidine or 4,6-diaminopyrimidine involving the nitrosation at 5-position, reduction of nitroso to amino group (by ammonium sulfide) followed by ring closure with formic acid (via 4-amino-5-formamidopyrimidine) or chlorocarbonic ester.

398. TROST ALLYLATION (TSUJI-TROST REACTION)

The Tsuji-Trost reaction (or Trost allylation or allylic alkylation) is a palladium-catalysed substitution reaction of a nucleophile such as active methylenes, alcohols, enolates, amines and phenols with a substrate containing a leaving group in an allylic position such as allyl acetates and allyl bromides. The reaction proceeds in either an S_N2 or S_N2' fashion depending on the catalyst, nucleophile, and substituents on the substrate.

X can be: OAr, $OCOR$, $OCONHR$, OCO_2R, $OP(O)(OR)_2$, Cl, NO_2, SO_2Ph
Nu^- can be $R_1R_2CH_2$, enamines, enolates; R_{1-2} = CO_2R, CN, NO_2, SO_2PH, COR, NC

The reaction was named after Jiro Tsuji who first reported the method in 1965 and named after Barry Trost who in 1973 introduced phosphine ligands in the reaction and introduced an asymmetric version.

The catalytic cycle organized in four key steps:

1. Formation of Pd-olefin complex
2. Ionization/oxidative addition
3. Nucleophilic addition
4. Decomplexation/regeneration of catalyst

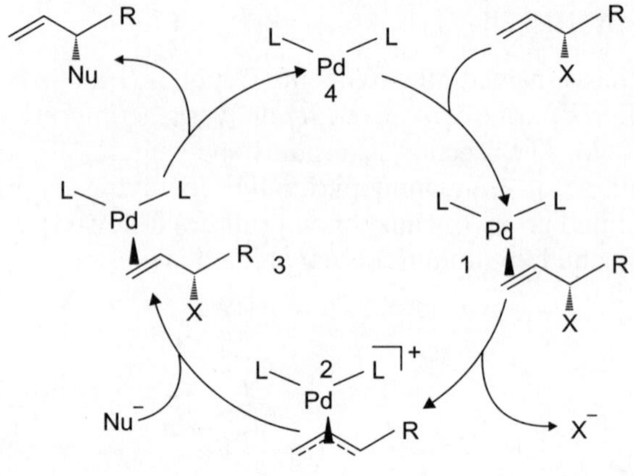

399. TROST DESYMMETRIZATION

Desymmetrization means to remove a symmetric element from the substrate and it may take place in the presence of an enzyme or a nonenzymatic catalyst. Trost desymmetrization involves formation of an enantiomerically pure, azide or amine containing, five or six membered ring by a pallidium catalyzed desymmetrization using a nitrogen nucleophile, where the palladium complex is derived from a chiral ligand and π-allylpalladium chloride.

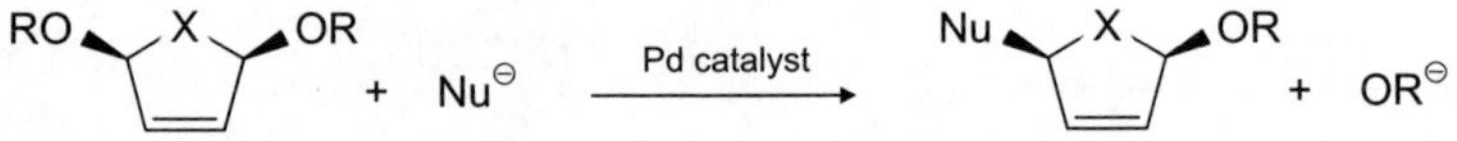

General mechanism of the reaction

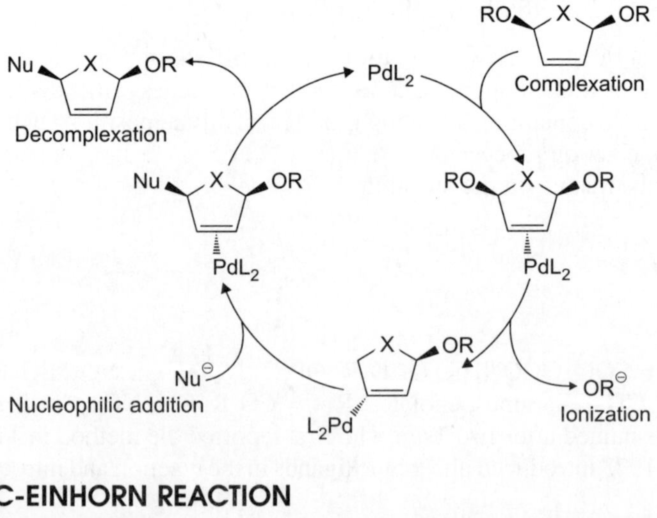

400. TSCHERNIAC-EINHORN REACTION

This reaction was named after Joseph Tscherniac and Alfred Einhorn. It involves aromatic nucleophilic alkylation by using N-hydroxymethyl amide or N-hydroxylmethylimide as the alkylating reagent. The electrophile is a N-methanol derivative of an amide.

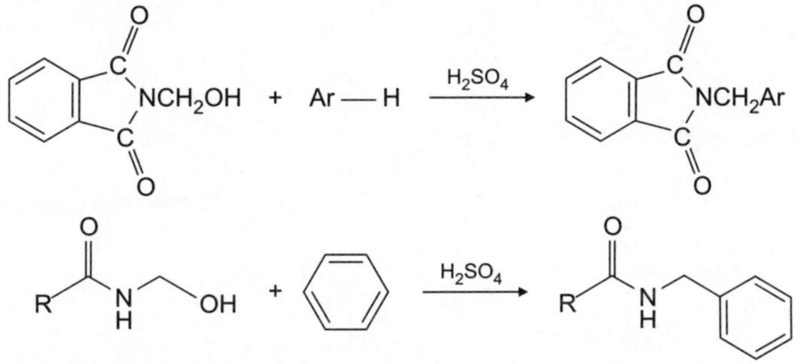

401. TWITCHELL PROCESS

Twitchell process is an industrial process for splitting or hydrolyzing fats and oils into fatty acids and glycerol and fatty acids. In the Twitchell process, the fat is boiled for 20–48 hours in an open tank with steam in a mixture of 25–50% water, 0.5% sulfuric acid and 0.75–1.25% Twitchell reagent (sulfonated petroleum products). The process is often repeated, and the glycerol solution is drawn off after each stage; the fatty acids produced are used for soap, while glycerol is primarily used for explosives (i.e. glycerol dynamite). The process is carried out in a loosely closed wooden, lead-lined or acid-resistant vat.

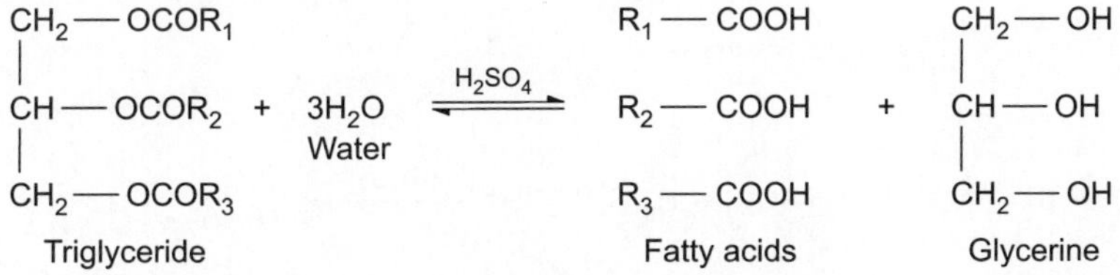

402. UGI REACTION

The Ugi four-component* condensation (U-4CC) between an aldehyde, an amine, a carboxylic acid and an isocyanide allows the rapid preparation (usually complete within minutes) of α-aminoacyl amide derivatives. The reaction involves α-addition of an iminium ion and the conjugate base of a carboxylic acid to an isocyanide, followed by spontaneous rearrangement of the α-adduct to yield an α-aminocarboxamide derivative. Carbonyl compounds and amines, or their condensation products, serve as precursors to the iminium ion. The nature of the product depends primarily on the acid component. This reaction has importance for generating compound libraries for screening purposes. The reaction can also be performed with a preformed imine. This results in an increased yield. The reaction is named after Ivar Karl Ugi, who first published this reaction in 1962.

$$R_1—CHO + R_2—NH_2 + R_3—NC + R_4—COOH \longrightarrow$$

Alpha-acylaminocarboxamide

Generally observed properties of the Ugi reaction:
- Rxn is exothermic and usually complete in seconds minutes at room temperature.
- Aprotic, polar solvents are best, though the low-molecular weight alcohols have been used.
- Can be performed in biphasic media.
- High (0.5–2 M) reactant concentrations are best.
- By virtue of the mechanism, Lewis acids can accelerate the reaction.
- Precondensation of the amine and the carbonyl (preformation of the Schiff base) can increase yields.

*In MCRs, three or more reactants come together in a single reaction vessel to form products that contain portions of all the components.

Concise synthesis of benzodiazepines

Preparation of hydantoins

Ugi reaction   X = O or S

403. ULLMANN REACTION

In the early 1900s, Fritz Ullmann and Irma Goldberg discovered copper-mediated aromatic nucleophilic substitution reactions.

The "classic" Ullmann reaction describes the copper-mediated synthesis of biaryls from aryl halides. It is the oldest known method for C — C coupling of aromatic halides.

X = I, Br

Iodo-benzene   Biphenyl

The Ullmann condensation involves copper-mediated (stoichiometric or catalytic) reaction between an aryl halide and an amine, phenol or thiophenol to synthesize the corresponding aryl-amine, -ether or -thioether compounds, respectively.

X = I, Br   Y = NH, O, S

Biaryl ether synthesis is similarly accomplished with aryl halides and phenols.

Ullmann reaction is a suitable tool for the synthesis of biaryls through dimerisation of aromatic halides. Apart from this, numerous industrial applications, such as synthesis of intermediates in pharmaceutical, agrochemical, fine and polymer chemistry are the other applications were found over the years. The traditional version—the original protocol of the Ullmann reaction has many drawbacks, such as:

1. High reaction temperatures ($\geq 200°C$).
2. Poor solubility of the copper salts consumes stoichiometric copper and generates large amounts of waste. The reaction works best with iodoaryls, which increases the waste problem.
3. Long reaction times.
4. Limited substrate scope (aryl halides for the reaction should be activated/classical Ullmann reaction is limited to electron deficient aryl halides).
5. Excess of copper as promoting agent (high metal loadings or in the presence of stoichiometric amounts of copper reagents).
6. Low functional group tolerance.
7. Moderate and/or erratic yields.

Because of above given harsh reaction conditions, the Ullmann reaction indeed suffered from reduced synthetic scope. However, some improvements and alternative procedures have been introduced. Addition of ligands to the copper catalyst in order to improve the solubility of the copper precursors, leading to the use of milder reaction conditions, i.e. lower reaction temperature and time, lower catalyst loadings, and a widened scope of reactivity. The traditional copper protocols can be improved if the solubility of the copper salts can be increased. Several modifications and improvements of the Ullmann reaction employ copper and its derivatives, but the most frequently used form of copper is activated copper bronze (or bronze powder), a 9:1 mixture of copper with tin. Typically, the copper bronze is activated prior to the coupling reaction using iodine in acetone followed by washing the bronze with concentrated HCl in acetone. Modern variants of the Ullman reaction employing palladium and nickel have widened the substrate scope of the reaction and rendered reaction conditions more mild. Yields are generally still moderate. However, in organic synthesis this reaction is often replaced by palladium coupling reactions such as the Heck reaction, the Hiyama coupling and the Sonogashira coupling.

N-arylamides are valuable compounds, widely employed in the fields of organic synthesis, pharmaceutical chemistry, or biology. One of the most common synthetic protocols for their preparation is the copper-catalyzed Ullmann reaction and the related Goldberg reaction (copper-catalyzed N-arylation of amides).

404. URECH CYANOHYDRIN METHOD; ULTEE CYANOHYDRIN METHOD

The preparation of cyanohydrin by means of the treatment of the carbonyl compounds (aldehydes and ketones) with either alkali cyanide in the presence of acetic acid or anhydrous hydrogen cyanide in the presence of a basic catalyst, the preparation under the former condition is known as the Urech cyanohydrin method, whereas the latter condition with basic catalyst is referred to as the Ultee cyanohydrin method.

The study finds that aldehydes are more reactive than ketones in this reaction. This is a general method for the preparation of cyanohydrin derivatives. The chemist Urech in 1872 was the first to synthesize cyanohydrins from ketones with alkali cyanides and acetic acid.

405. URECH HYDANTOIN SYNTHESIS

The Urech hydantoin synthesis involves formation of hydantoins from α-amino acids by treatment with potassium cyanate in aqueous solution and heating the salt of the intermediate hydantoic acid with 25% hydrochloric acid.

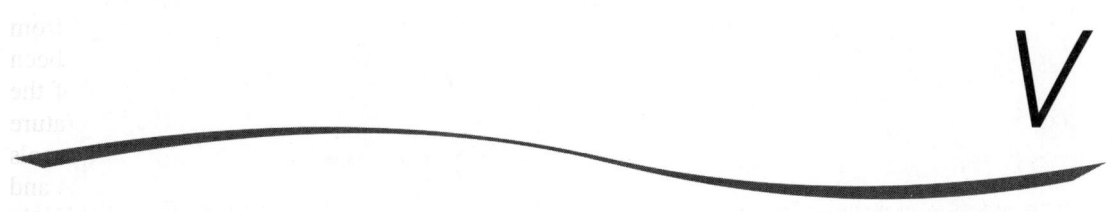

406. VILSMEIER-HAACK REACTION

The reaction is named after Anton Vilsmeier and Albrecht Haack. The Vilsmeier-Haack reaction (also known as the Vilsmeier reaction) allows the formylation of electron-rich arenes (or of activated aromatic or heterocyclic compounds) by the reaction of a N, N-disubstituted amide with phosphorus oxychloride and an electron-rich arene to produce an aryl aldehyde or ketone.

The formylating agent, also known as the Vilsmeier-Haack reagent (a substituted chloroiminium ion), is formed *in situ* from the substituted amide such as DMF and phosphorus oxychloride.

Vilsmeier-Haack reaction is a special type of Friedel-Crafts reaction, which involves electrophilic substitution of an activated aromatic ring with a halomethyleniminium salt. The initial product is an iminium ion, which is rapidly hydrolyzed during workup to give the corresponding aromatic ketone or aldehyde.

The scope of the Vilsmeier reagent is not confined to aromatic formylation reaction alone. A wide variety of alkene derivatives and activated methyl and methylene groups exhibit reactivity towards the Vilsmeier reagent. In addition to the carbon nucleophiles, some oxygen and nitrogen nucleophiles are also reactive towards Vilsmeier reagent. Numerous transformations of the iminium salts into products other than aldehydes have been achieved and these transformations enhance the scope and versatility of the Vilsmeier-Haack reaction.

407. VOIGHT AMINATION

Voigt reaction or Voigt condensation refers to the preparation of alpha-keto amine from condensation of a primary amine (or a secondary amine) with a benzoin in the presence of phosphorus pentoxide or hydrochloric acid.

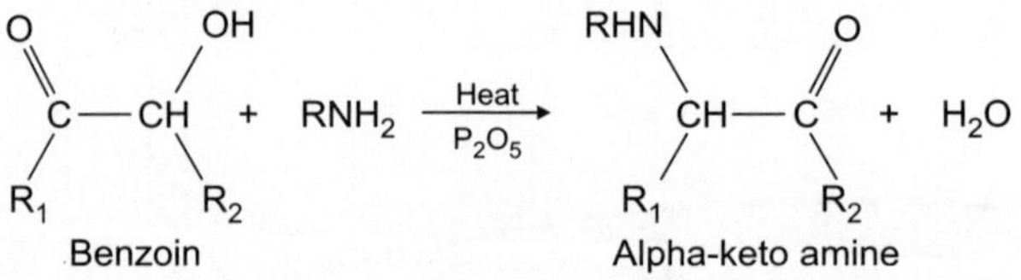

Benzoin   Alpha-keto amine

The reaction generally works well with primary amines and phosphorus pentoxide is used as the condensation reagent in most cases. An electron-donating group on the benzoin slow down the reaction. This reaction has importance for the preparation of α-aminoketones and α-amino alcohols with two phenyl groups.

408. VOLHARD-ERDMANN CYCLIZATION

The reaction was named after Jacob Volhard and Hugo Erdmann. It refers to the synthesis of alkyl and aryl thiophenes by cyclization of disodium succinate or other 1,4-difunctional compounds (γ-oxo acids, 1,4-diketones, chloroacetyl-substituted esters) with phosphorus heptasulfide.

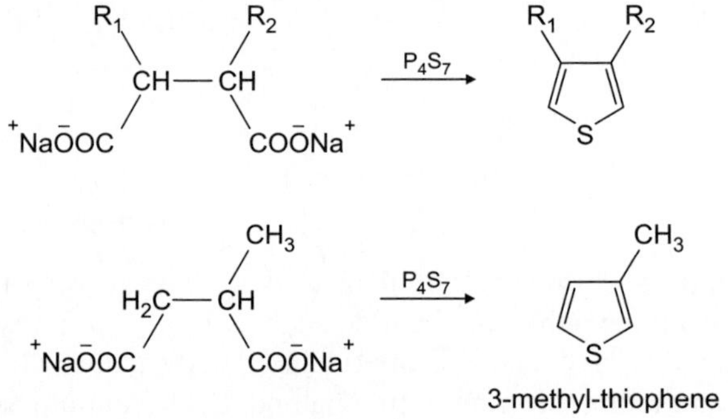

3-methyl-thiophene

409. VON BRAUN AMIDE DEGRADATION

The Von Braun amide degradation refers to the conversion of N-alkyl-substituted amides into nitriles and alkyl chloride by treatment with phosphorus pentachloride (PCl_5).

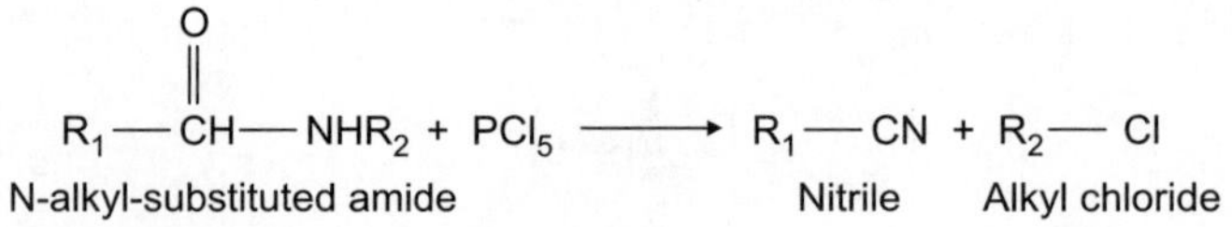

N-alkyl-substituted amide   Nitrile   Alkyl chloride

410. VON BRAUN REACTION

The reaction was first described in 1900 by Julius Von Braun. In Von Braun cyanogen bromide reaction or Von Braun reaction, a tertiary amine reacts with cyanogen bromide to yield an alkyl bromide and a disubstituted cyanamide. The tertiary amine is dealkylated in the reaction by means of the treatment with cyanogen bromide.

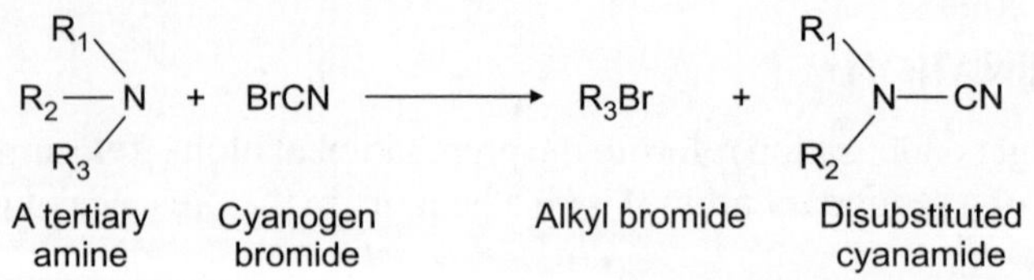

A tertiary   Cyanogen   Alkyl bromide   Disubstituted
amine   bromide   cyanamide

Usually the R group providing most reactive halide cleaves preferentially. An example is given below:

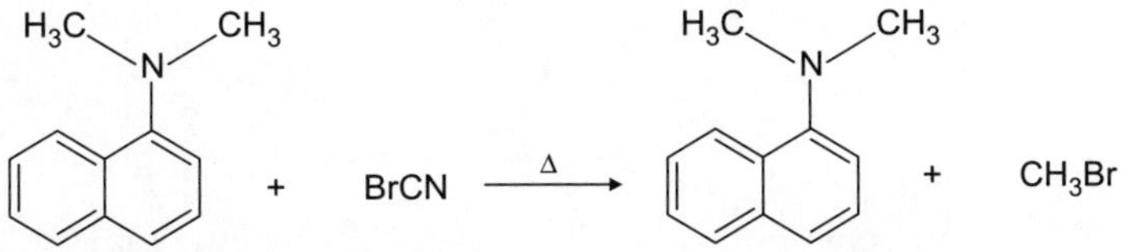

It has been reported that the von Braun reaction can be accelerated by the addition of a Lewis acid. This reaction has been applied for the degradation of alkaloids for the purpose of early structural elucidation as well as the preparation of alkaloid derivatives.

411. VON RICHTER (CINNOLINE) SYNTHESIS

Cinnoline ring was first synthesized by Richter during the diazotization of ortho-aminophenyl-propionic acid and cyclization of the obtained arenediazonium salt. Now, formation of cinnoline derivatives by diazotization of o-aminoarylpropiolic acids or o-aminoarylacetylenes followed by hydration and intramolecular cyclization is known as the Von Richter cinnoline synthesis.

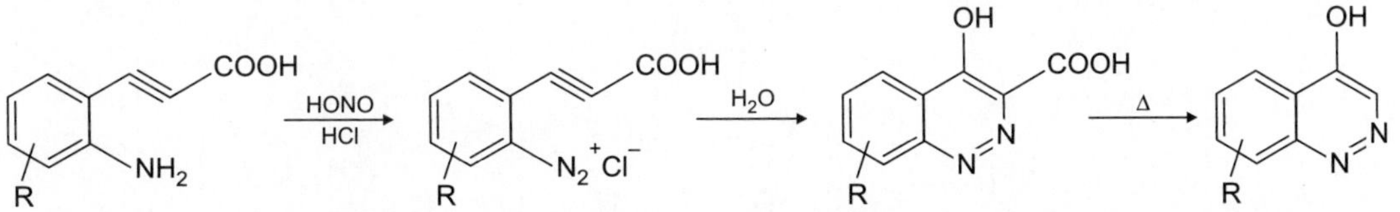

This reaction is still the major reaction for the preparation of cinnoline derivatives. Strongly acidic media is necessary to generate a diazonium salt in a typical Von Richter synthesis.

412. VON RICHTER REARRANGEMENT

The reaction is named after Victor von Richter. It involves carboxylation of aromatic nitro compounds with potassium cyanide. Carboxylation occurs at ortho position of the former nitro group.

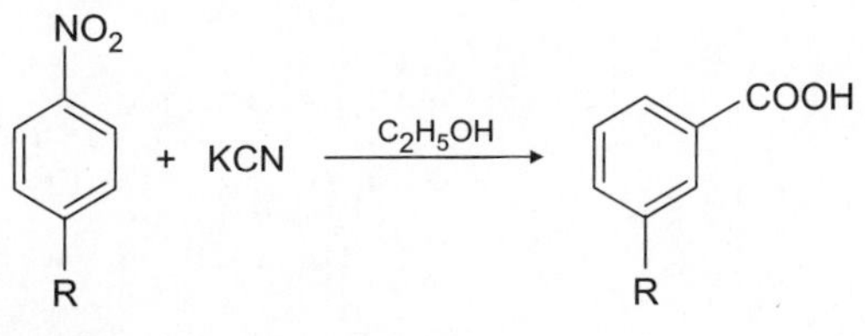

413. VORBRÜGGEN GLYCOSYLATION

The Vorbrüggen glycosylation is the extension of the Hilbert-Johnson reaction for the preparation of N-glycosides by the treatment of a mixture of the peracylated sugar and silylated heterocyclic base with a Lewis acid. The Vorbrüggen glycosylation is usually carried out in MeCN or dichloroethane in the presence of a Lewis acid.

This reaction has been broadly applied in the preparation of nucleosides and nucleotides.

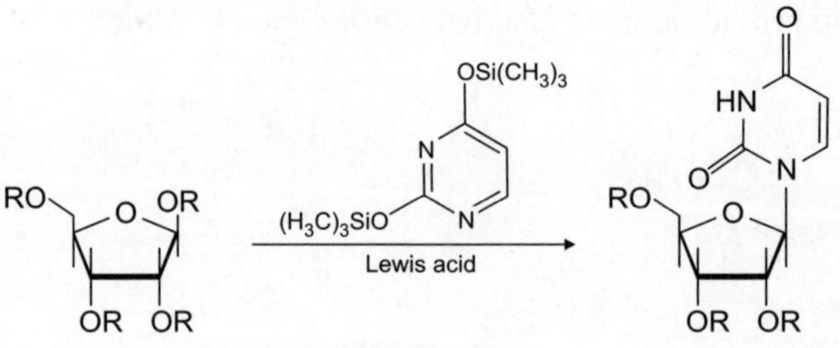

R = COCH$_3$, COPh

Lewis acid = BFa • O(CH$_2$CH$_3$)$_2$, SnCl$_4$, (CH$_3$)$_3$SiOSO$_2$CF$_3$, (CH$_3$)$_3$SiClO$_4$, TiCl$_4$, AlCl$_3$

414. WACKER PROCESS OR WACKER OXIDATION

The Hoechst-Wacker process or simply known as the Wacker process involves the oxidation of ethylene with oxygen in the presence of aqueous acidic solution of palladium chloride and cupric chloride. Copper serves as redox co-catalyst in the reaction. The reaction was initially reported in 1962 and utilized for the industrial production of acetaldehyde from ethylene.

$$H_2C = CH_2 \xrightarrow[\text{CuCl}_2,\ \text{O}_2,\ \text{H}_2\text{O}]{\text{PdCl}_2} H_3C = CHO$$

In its general form, the Wacker oxidation is an aerobic PdCl$_2$-catalyzed, Cu mediated aqueous oxidation of an olefin to a carbonyl compound. It is a versatile reaction that has found broad application in synthetic chemistry for the functionalisation of alkenes. This reaction is generally used to convert the terminal olefins into methyl ketones with oxygen or air as oxidant.

$$R - HC = CH_2 \xrightarrow[\text{CuCl},\ \text{O}_2,\ \text{DMF/H}_2\text{O}]{\text{PdCl}_2} R - \overset{\overset{\displaystyle O}{\|}}{C} - CH_3$$

Terminal alkenes nearly always give methyl ketones, suggesting that hydroxypalladation takes place according to Markovnikov's principles, disubstituted alkenes give products that do not always obey these rules. The σ-bonded (β-hydroxyalkyl) palladium complex has been determined as the key intermediate in this reaction. The Wacker process is an example of homogeneous catalysis.

The Wacker process for oxidation of olefins to ketones has three mechanistically distinct parts:

First, activation of the olefinic double bond toward nucleophilic attack by coordination with Pd(II) and addition of a hydroxide moiety to this electrophilic double bond;

Second, conversion of the resulting 2-hydroxyethylpalladium(II) compound to ketone and a (formally) Pd(0) atom by a series of palladium(II) hydride addition-eliminations involving vinylic alcohol intermediates;

Third, reoxidation of the palladium(0) to palladium(II) by copper(II).

The combination of palladium chloride and cupric chloride is known as the Wacker catalyst.

Wacker-Tsuji Oxidation

The so-called Wacker-Tsuji oxidation is the laboratory scale version of the above reaction, for example the conversion of 1-decene to 2-decanone with palladium (II) chloride and copper(II) chloride in a water/dimethylformamide solvent mixture in the presence of air.

$$nC_8H_{17} - HC = CH_2 \xrightarrow[\text{CuCl}_2,\ \text{O}_2,\ \text{DMF/H}_2\text{O}]{\text{PdCl}_2} nC_8H_{17} - \overset{\overset{\displaystyle O}{\|}}{C} - CH_3$$

415. WAGNER-JAUREGG REACTION

This reaction was named after Theodor Wagner-Jauregg. It describes the addition of maleic anhydride with diarylethylene to form a bis adducts, which after aromatization form higher aromatic analogs. This reaction has certain application in the preparation of polyaromatic compounds.

The reaction is unusual in that the anhydride reacts with the aromatic ring. The presence of the alpha-phenyl group activates the styryl group for a Diels-Alder reaction even at the expense of its aromaticity.

416. WAGNER-MEERWEIN REARRANGEMENTS

A Wagner-Meerwein rearrangement is a class of carbocation 1,2-rearrangement reactions in which a hydrogen, alkyl or aryl group migrates from one carbon to a neighbouring carbon. The Wagner-Meerwein rearrangements are stereospecific and associated with 1° carbonium ions rearranging to more stable 2° and 3° carbonium ion via the 1,2-shift of an adjacent alkyl or aryl function. The driving force of rearrangements is to get more stable carbocations.

The rearrangement was first discovered in bicyclic terpenes, for example the acid-catalysed conversion of isoborneol to camphene.

Isoborneol   Camphene

Rearrangements that occur with elimination of water in the dehydration of an alcohol, of hydrogen halide in the dehydrohalogenation of an alkyl halide, are commonly referred to as Wagner-Meerwein rearrangement (W-M rearrangement). The W-M rearrangement is sometimes named "retrograde pinacol rearrangement" because it is the exact reverse of the pinacol rearrangement discussed below. The most illustrative example of this type of a rearrangement is the formation of tetramethylethylene as the main product of the acid-catalyzed dehydration of methyl-t-butyl carbinol (pinacoyl rearrangement) equation. The rearrangement is in a good agreement with the stability feature of Saytzev olefin.

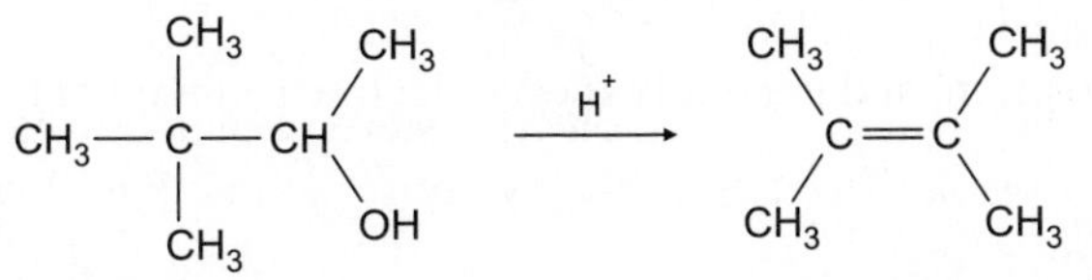

The stability of a carbocation is directly related to the associated substituents on it and the more stable carbocation usually means "more substituted", thus *Wagner-Meerwein shift* leads to the new carbocation being more stabilised than the original. Cations can also be made more stable if they become less strained. So, for example, four-membered rings adjacent to cations readily rearrange to five-membered rings in order to relieve ring strain.

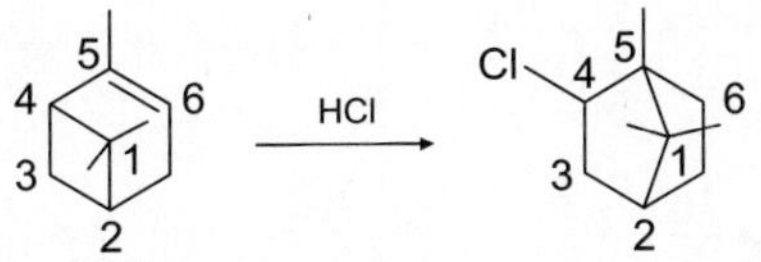

Wagner-Meerwein type cationic rearrangements are usually promoted by acidic reagents or by the presence of the good leaving groups. The presence of an electron-withdrawing group has been found to disfavour the Wagner-Meerwein migration of an adjacent C-C bond. This rearrangement is characterized by a 1,2-shift of an aryl or an alkyl group R (the migrating species) from an α-carbon to the carbonium ion site either in a stepwise mechanism, *a classical ion*, or in a concerted one, *a non-classical ion* as depicted below.

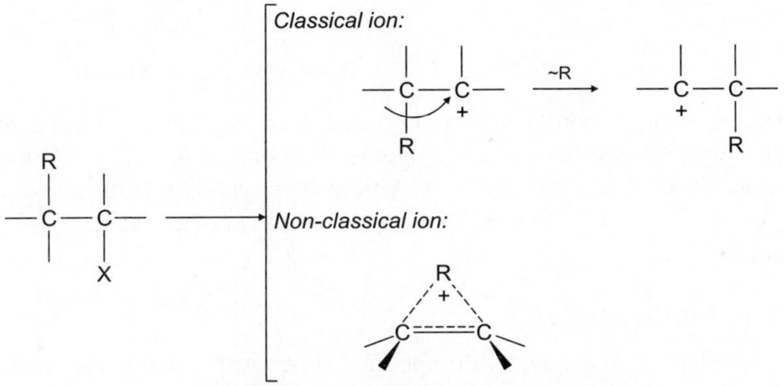

Classical and non-classical carbonium ions

Wagner-Meerwein rearrangements can be used to:

(i) Contract or expand non-aromatic rings: One distinct feature of the W-M rearrangement is that it provides, in some cases, a ring expansion as well as a ring closure which may have a valuable synthetic interest. Indeed, the solvolysis of some cyclic compounds yields the unrearranged products and products from ring enlargement or ring contraction visibly through a 1,2-shift. For example, the hydrolysis of cyclocarbinyl chloride leads to a mixture of the unrearranged alcohol and cyclobutanol, the rearranged alcohol.

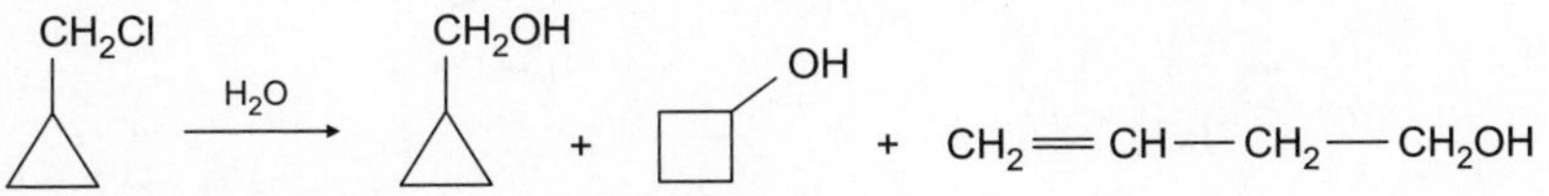

(ii) Make sterically crowded 4° alkyl centres

417. WALDEN INVERSION

The inversion of stereochemical configuration at a stereocenter during a chemical reaction is known as the Walden inversion. It was first observed by chemist Paul Walden in 1896.

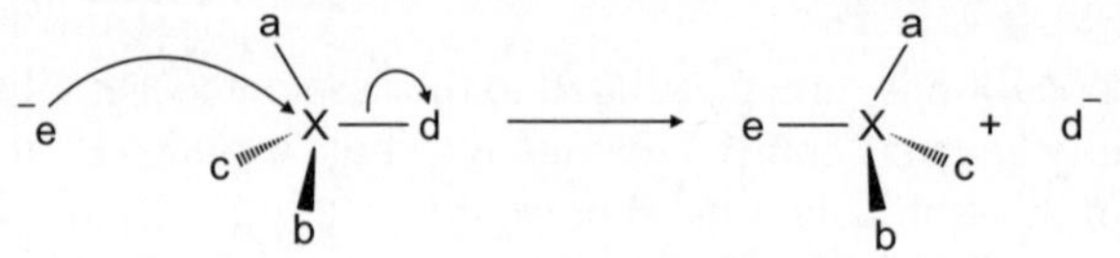

In the S_N2 reaction between iodide ion and (S)-2-chlorobutane, the iodide ion acts as nucleophile and attack on (S)-2-chlorobutane from the backside, which causes Walden inversion. The resulting product is (R)-2-iodobutane.

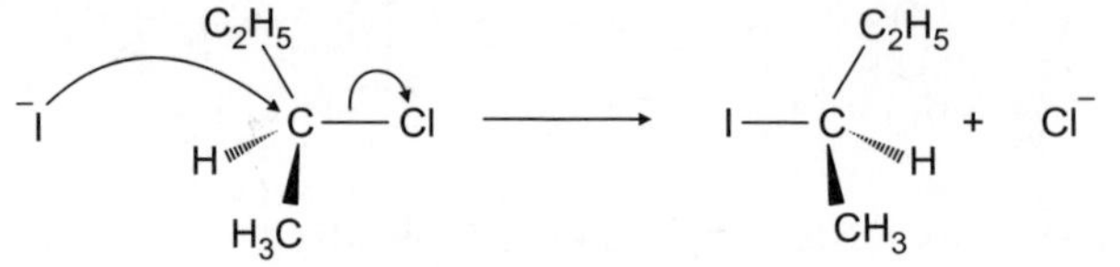

(S)-2-chlorobutane   (R)-2-iodobutane

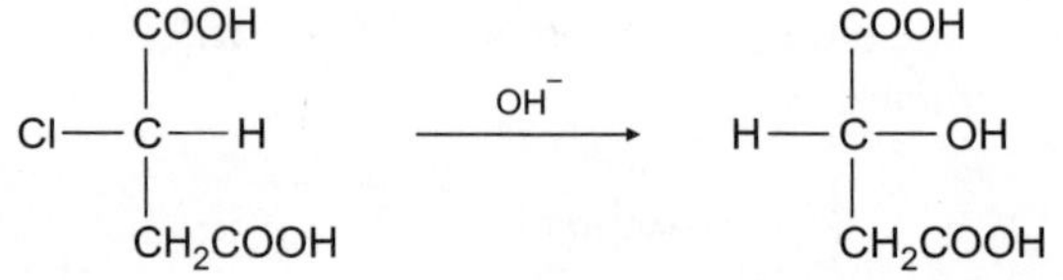

2-chloro-succinic acid   2-hydroxy-succinic acid

A molecule having a chiral center can form two enantiomers around a chiral center, the Walden inversion converts the configuration of the molecule from one enantiomeric form to the other. In a Walden inversion, a chemical species Xabcd, where X is typically carbon having a tetrahedral arrangement of bonds to X, is converted into the chemical species Xabc**e** having the opposite relative configuration.

418. WALLACH REARRANGEMENT

Azoxybenzene and other aromatic azoxy compound* on treatment with strong acids transformed into corresponding hydroxyazobenzene (2- or 4-hydroxyazobenzenes). This rearrangement was discovered by Otto Wallach, in 1880, and hence named.

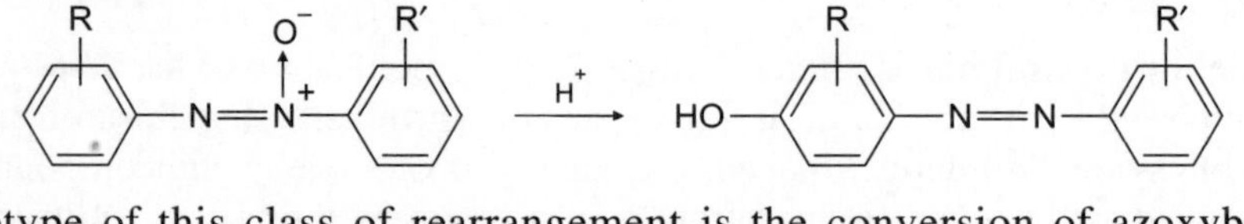

The prototype of this class of rearrangement is the conversion of azoxybenzene to p-hydroxyazobenzene on treatment with concentrated sulfuric acid.

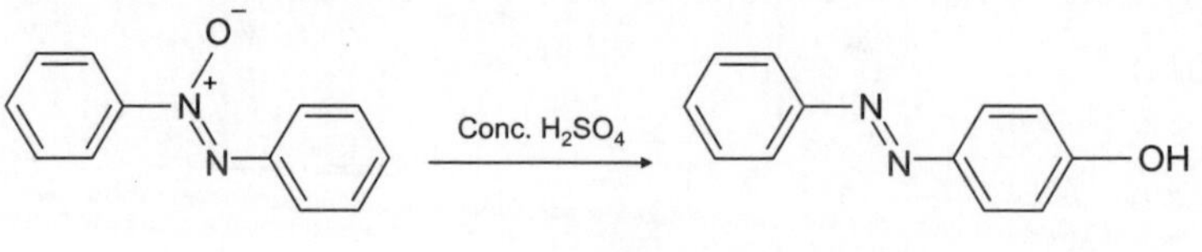

The products of the Wallach transformation have been found to depend on the reaction conditions: the hydroxyl group generally appears in a *para* position, and generally the Wallach rearrangement is referred to as the acid-catalyzed rearrangement of azoxybenzenes to *p*-hydroxyazobenzenes. The *para* rearrangement is found to be intermolecular.

Other reaction conditions such as the application of photochemical or Lewis acid-catalysed reactions or blocking of both *para* positions leads to the formation of *ortho*-hydroxy derivatives. The conversion under photo illumination is referred to as the photo-Wallach rearrangement. The Wallach rearrangement under photochemical conditions is found to be intramolecular in nature forming *o*-hydroxy azo compound.

*Azoxy compounds are also known as diazene oxides are characterized by the $-N = N(O)-$ functional group. They are considered N-oxides of azo compounds sharing a common functional group with the general structure $RN = N^+(O^-)R$.

419. WEERMAN DEGRADATION

Weerman Degradation is mechanistically similar to Hofmann-type reaction, involves the conversion of amides into aldehydes with one less carbon by the reaction of,

- The α-hydroxy amide with sodium hypochlorite to sodium cyanate followed by hydrolysis, or
- The α,β-unsaturated amides in the presence of methanol with sodium hypochlorite followed by acidic hydrolysis of the resulting vinyl urethanes.

This is a general reaction of α-hydroxy carboxylic acids. In carbohydrate chemistry this reaction is used where, an aldonamide (derived from an aldonic acid) is degraded by sodium hypochlorite, forming a new sugar with one less carbon. The reaction is named after R.A. Weerman.

The Weerman reaction is also used for diagnosing the presence of hydroxyl group adjacent to the amido groups i.e. α-hydroxy amides, and this test is referred to as the Weerman test. This reaction has also been used for the direct degradation of carbohydrate, such as the conversion of D-glucose to D-arabinose.

420. WEISS REACTION (WEISS-COOK CONDENSATION, WEISS-COOK REACTION)

Weiss reaction refers to the treatment of 1,2-dicarbonyl compounds (α-dicarbonyl molecule) with 2 equivalents of 3-oxoglutarates to yield *cis*-bicyclo[3.3.0]octane-3,7-dione or [n.3.3] propellanedione (n > 2) tetracarboxylates. Subsequent acid-catalyzed hydrolysis and decarboxylation yield the respective 2,4,6,8-unsubstituted diones. This reaction is also known as the Weiss-Cook condensation or Weiss-Cook reaction.

E = COOCH$_3$;
R^1 or R^2 = H, alkyl, aryl or R^1 — R^2 = (CH$_2$)n, n > 2

The general mechanism for this reaction is thought to involve a series of aldol condensations, dehydrations, and Michael reactions. It has been observed that four carbon–carbon bonds are formed in a one-pot reaction that involves aldol as well as Michael addition reactions.

The Weiss reaction is usually carried out in a buffered solution and good yields are obtained in this reaction if the H^+ concentration is adjusted properly in the reaction mixture.

This reaction has been widely used for the convergent synthesis of polyquinanes and polyquinenes, which are otherwise prepared by much more complicated procedures.

421. WESSELY-MOSER REARRANGEMENT

Wessely-Moser rearrangement refers to an acid- or a base-promoted rearrangement of flavones and flavanones possessing a 5-hydroxyl group, through fission of the heterocyclic ring and reclosure of the of the β-diketone intermediate (diaroylmethanes).

The Wessely-Moser rearrangement occurs frequently in monoflavanoids and is characterized by the reorganization of 5,7,8-subsituents to a 5,6,7-substitution pattern under acidic conditions. In this process, the heterocyclic ring of the monoflavanoid undergoes acid-catalyzed ring opening, followed by recyclization with either of the two ortho-OH groups to give an equilibrium mixture of the two substitution patterns.

It has also been observed that the 5-hydroxy group is necessary for the occurrence of Wessely-Moser rearrangement. This rearrangement is useful for the structural elucidation of the natural products.

422. WESTPHALEN-LETTRÉ REARRANGEMENT

Westphalen-Lettré Rearrangement involves acid-promoted dehydration of 5-hydroxycholesterol derivatives accompanied by C-10 to C-5 methyl migration in compounds with a β-substituent at C-6. In this reaction the 5-hydroxy is removed by acidic dehydration accompanying the migration of the angular methyl group at C-10 to C-5 therefore one equivalent of water is lost and a double bond is formed at C10-C11. In the reaction mechanism the 5-hydroxycholesterol derivatives after being treated with acid forms a carbocation at C5 after which the actual rearrangement takes place.

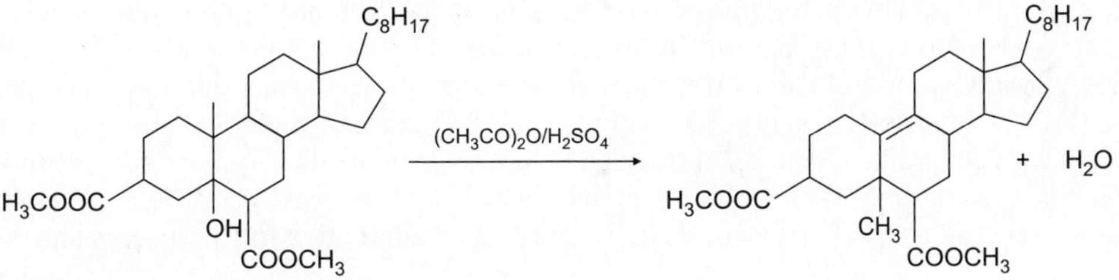

423. WHARTON REACTION

The Wharton reaction or **Wharton oxygen transposition reaction** describes reduction of α,β-epoxy ketones by hydrazine to allylic alcohols. The reaction is named after Peter Stanley Wharton.

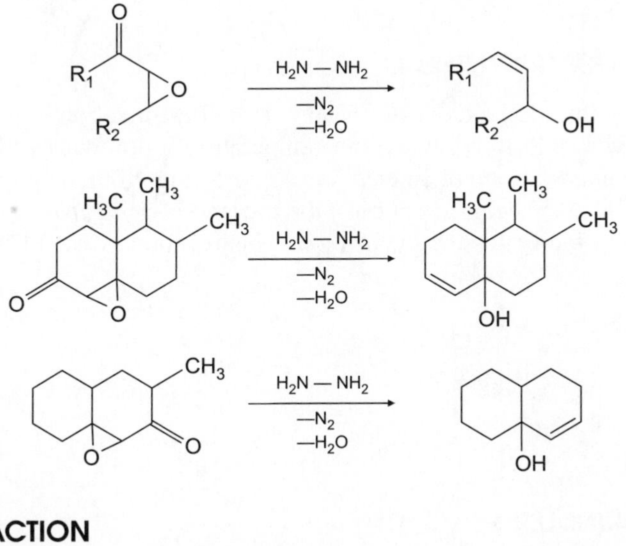

424. WHITING REACTION

The Whiting reaction involves the reduction (in ether or tertiary amines) of alkynyl diols into dienes using lithium aluminium hydride. The reaction proceeds via addition of four hydrogen atoms followed by formal elimination of two equivalents of water.

$$R_1-\underset{}{\overset{OH}{\overset{|}{HC}}}-C\equiv C-\overset{OH}{\overset{|}{CH}}-R_2 \xrightarrow{LiAlH_4} R_1-HC=HC-CH=CH-R_2$$

In general, $LiAlH_4$ does not reduce alkene or alkynyl moieties, but as noted in the Whiting reaction, propargylic alcohol derivatives ($C\equiv C-CH_2-OH$) are important exceptions.

$$H_3C-\underset{\underset{CH_2}{\parallel}}{C}-\underset{\underset{OH}{\mid}}{HC}-C\equiv C-\underset{\underset{OH}{\mid}}{CH}-\underset{\underset{CH_2}{\parallel}}{C}-CH_3 \xrightarrow{\text{LiAlH}_4} H_3C-\underset{\underset{CH_2}{\parallel}}{C}-HC=HC-CH=CH-\underset{\underset{CH_2}{\parallel}}{C}-CH_3$$

425. WICHTERLE REACTION

The Wichterle reaction is a conversion of vinylic chlorides to ketones by sulfuric acid. Most of the applications are transformations of αγ-chlorocrotyl (3-chloro-2-butenyl) derivatives to 3-oxobutyl derivatives. This reaction is a modification of the Robinson annulation in which 1,3-dichloro-cis-2-butene is applied as an equivalent of methyl vinyl ketone to avoid the potentially undesirable condensation and polymerization during the step of the Michael addition, and is generally referred to as the Wichterle reaction. The ketones thus formed may undergo subsequent reactions in compounds containing reactive centers. Thus intramolecular aldol-type condensations occur in compounds containing carbonyl groups, and Friedel-Crafts-type cyclizations in compounds containing aromatic rings. The reaction can be used as an alternative to Robinson's annelation.

426. WIDMAN-STOERMER SYNTHESIS

The synthesis of cinnoline derivatives by means of the diazotization of o-aminostyrenes with sodium nitrite and hydrochloric acid at room temperature is known as the Widman-Stoermer synthesis. In general, an aryl group or an electron-withdrawing group on the β-position of styrene make the reaction difficut to occur; in contrast, the presence of a simple electron-donating group such as methyl or ethyl facilitates the cyclization. This reaction is useful for the preparation of cinnoline derivatives.

427. WILLGERODT-KINDLER REACTION

Willgerodt-Kindler reaction, in its original form, straight- or branched-chain aryl alkyl ketones were found to react with sulfur and secondary amines to give the terminal thioamides as a result of oxidation and rearrangement.

The Willgerodt-Kindler reaction allows the synthesis of amides from aryl alkyl ketones under the influence of a secondary amine and a thiating agent, e.g. aqueous ammonium polysulfide or by sulfur.

This reaction is characterized by the use of alkyl aryl ketones or aldehydes, elemental sulfur, and amines, with morpholine being the most commonly used amine. When ketones are used in this methodology, the carbonyl group is reduced to a methylene group and the terminal methyl group is oxidized to a thiocarbonyl group.

This method has some limitations and has a rather bad reputation for not being very useful for preparative purpose due to the high reaction temperatures and long reaction periods together with poor yields.

Mechanistically, it involves the formation of an enamine which undergoes thiation, and the carbonyl group migrates to the end of the chain via a cascade of thio-substituted iminium-aziridinium rearrangements.

428. WILLIAMSON ETHER SYNTHESIS

The Williamson ether synthesis is a simplest way to synthesize an ether from an organohalide or tosylate and an alkoxide under typical S_N2 conditions. Alkoxides are prepared by the reaction of an alcohol with a strong base such as sodium hydride (NaH). This reaction was developed by Alexander W. Williamson in 1850.

The general reaction mechanism is as follows:

$$R_1\text{---}O^- \quad + \quad R_2\text{---}X \quad \longrightarrow \quad R_1\text{---}O\text{---}R_2 \quad + \quad X^-$$
$$\text{Alkoxide ion} \quad \text{Organo halide} \qquad \text{Ether}$$

The Williamson ether synthesis proceeds via bimolecular nucleophilic substitution (S_N2) mechanism, in which an alkoxide ion displaces a halogen ion. The bond making and breaking occurs simultaneously in the transition state.

$$R_1\text{---}O^- \quad + \quad R_2\text{---}X \quad \longrightarrow [R_1\text{---}O\text{---}R_2\text{---}X] \quad \longrightarrow \quad R_1\text{---}O\text{---}R_2 \quad + \quad X^-$$
$$\text{Alkoxide ion} \quad \text{Organo halide} \qquad \text{Transition state} \qquad \text{Ether}$$

A classical example of Williamson's synthesis involves formation of diethyl ether from the reaction of sodium ethoxide and chloroethane.

$$C_2H_5O^-Na^+ \quad + \quad C_2H_5Cl \quad \longrightarrow \quad C_2H_5\text{---}O\text{---}C_2H_5 \quad + \quad Na^+Cl^-$$
$$\text{Sodium} \qquad \text{Ethyl} \qquad\qquad \text{Diethyl ether} \qquad \text{Sodium}$$
$$\text{ethoxide} \qquad \text{chloride} \qquad\qquad\qquad\qquad \text{chloride}$$

Other examples are

Phenoxide ions can be employed to get aromatic ethers.

An epoxide can be synthesized from a halohydrin using Williamson's reaction.

Since alkoxide ions are highly reactive, they are usually prepared immediately prior to the reaction or generated *in situ* by treating an alcohol with a metal or a strong base. The sodium ethoxide is generated by treating ethyl alcohol with sodium metal.

$$C_2H_5OH \xrightarrow{\text{Na}} C_2H_5O^- \ Na^+$$

Sodium ethoxide

The nature of alkoxide (or aroxide) ion is less important and it may be may be primary, secondary or tertiary. The alkylating agent, on the other hand is most preferably primary. Secondary alkylating agents also react, but tertiary ones are usually too prone to side reactions to be of practical use. This method cannot be used with tertiary alkyl halides, because the competing elimination reaction predominates. The elimination reaction occurs because the rearward approach that is needed for an S_N2 mechanism is impossible due to steric hindrance. An S_N1 mechanism is likewise unfavored, because as the 3° carbon attempts to become a carbocation, the hydrogens on the adjacent carbons become acidic. Under these conditions, the alkoxide ion begins to show less nucleophilic character and, correspondingly, more basic character. This basic character leads to an acid-base reaction, which results in the generation of an elimination product (an alkene).

Finally, the reaction solvent is DMF (*N,N*-dimethylformamide), a classic solvent for S_N2 reactions. DMSO (dimethyl sulfoxide) is also a traditional S_N2 solvent. The Williamson reaction is widely used in both laboratory and industrial synthesis, and remains the simplest and most popular method of preparing ethers. Both symmetrical and asymmetrical ethers are easily prepared.

- Few restrictions regarding the nature of the the the alkoxide
- Works best for methyl- and 1°-halides or tosylates.
- E2 elimination is a competing reaction with 2° -halides or tosylates
- 3° halides undergo E2 elimination
- Vinyl and aryl halides do not react

429. WITTIG REACTION; HORNER REACTION; HORNER-WADSWORTH-EMMONS REACTION

Prof. Georg Wittig, Nobel laureate, invented one of the most important organic reactions in the 1950s, the conversion of a carbonyl compound to an alkene.

The reaction of a phosphorus ylide with an aldehyde or ketone, as first described in 1953 by Wittig and Geissler, is probably the most widely recognized method for carbonyl olefination.

Aldehyde or ketone · The "Wittig reagent" an "ylide" · Alkene

This so-called Wittig reaction has a number of advantages over other olefination methods; in particular, it occurs with total positional selectivity (that is, an alkene always directly replaces a carbonyl group). By comparison, a number of other carbonyl olefination reactions often occur with double-bond rearrangement. In addition, the factors that influence E- and Z-stereoselectivity are well understood and can be readily controlled through careful selection of the phosphorus reagent and reaction conditions. However, the formation of E and Z alkene isomers may be disadvantageous also. The E/Z ratio depends on reaction conditions and the nature of the substituents on the phosphorus.

A wide variety of phosphorus reagents are known to participate in Wittig reactions and the exact nature of these species is commonly used to divide the Wittig reaction into three main groups. Each of these following given reaction types has its own distinct advantages and limitations, and these must be taken into account when selecting the appropriate method for a desired synthesis.

The "classic" Wittig reaction of phosphonium ylides, in broad sense, describes formation of alkene from carbonyl compounds and phosphonium ylides, proceeding primarily through the proposed betaine and/or oxaphosphetane intermediates.

The Horner-Wittig reaction: In 1958, Horner and co-workers described the use of phosphine oxides in Wittigtype reactions (i.e. the ylide is replaced with a phosphine oxide carbanion). This modification allows for the removal of phosphorous as a water-soluble side product.

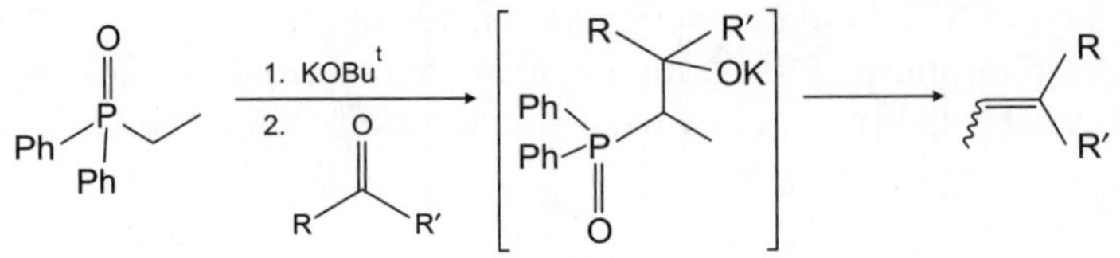

The Horner-Wadsworth-Emmons reaction: In 1961, Wadsworth and Emmons described the increased reactivity of phosphonate-stabilized carbanions with α-electron-withdrawing substituents. This modification subsides one of the limitation of the Wittig reaction of low reactivity of phosphorous ylides (i.e. phosphorous ylides that contain stabilizing groups next to the negatively charged carbon are not reactive enough to undergo the desired reaction with a carbonyl).

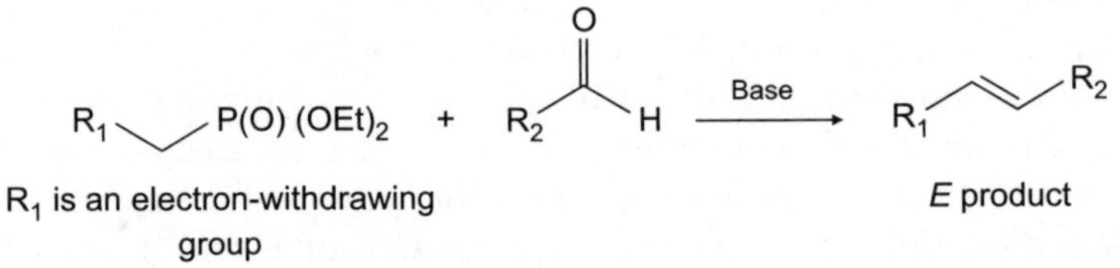

R_1 is an electron-withdrawing group

E product

430. (1,2)-WITTIG REARRANGEMENT

[1,2]-Wittig rearrangement describes base-promoted (e.g. with alkyl lithiums) rearrangement of ethers to yield alcohols (secondary or tertiary alcohols) via a [1,2]-shift. This reaction has limited applications in organic synthesis.

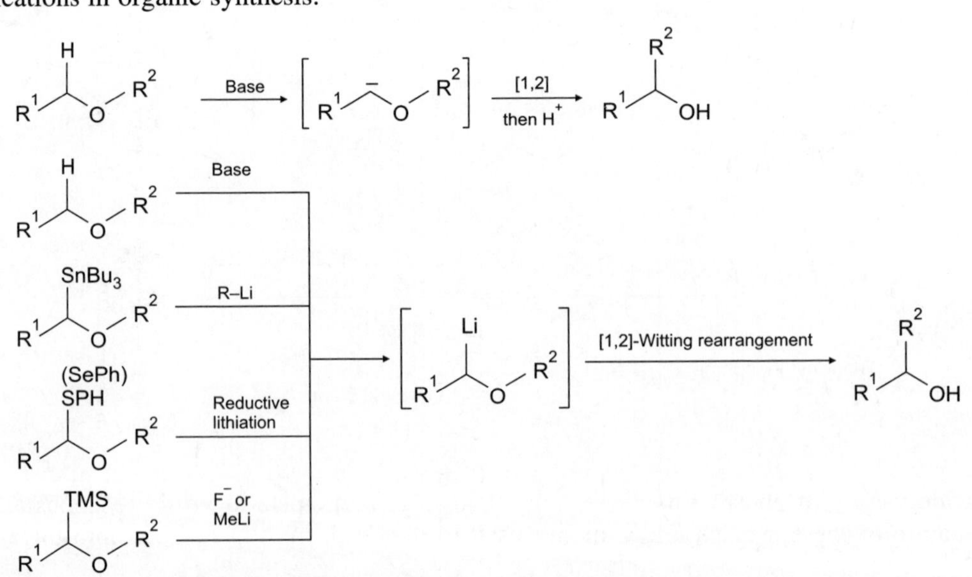

Li-bearing terminus

Migrating carbon

Solvent cage

Similarly, the analogous transformation of thioether into thiol is called the *thio*-Wittig rearrangement, and the corresponding conversion of imine is known as imino 1,2-Wittig rearrangement. The driving force for this reaction is the instability of the α-oxygenated carbanion.

Some important aspects related to this reaction are given below:

- The [1,2]-Wittig rearrangement proceeds via a radical-pair mechanism within a solvent cage.
- Regioselectivity and ease of reaction depend on the substituenets of the substrate.
- Radical recombination with retention of configuration at the migrating carbon.
- Radical recombination with invesion of configuration at the lithium center.
- The [1,2]-Wittig rearrangement competes β-elimination, [2,3]-Wittig rearrangement and [1,4]-shift, giving poor to modest yields of produt.
- Despite its radical character, chirality transfer has been demonstrated within the [1,2]-Wittig rearrangement.

431. (2,3)-WITTIG REARRANGEMENT

The [2,3]-Wittig rearrangement is the [2,3]-Sigmatropic rearrangement or transformation of an allylic ether into a homoallylic alcohol via a concerted, pericyclic process. Because the reaction is concerted, it exhibits a high degree of stereocontrol, and can be employed early in a synthetic route to establish stereochemistry. The stereoselectivity is highly dependent on the nature of the substrate. The Wittig rearrangement requires strongly basic conditions, however, as a carbanion intermediate is essential. [1,2]-Wittig rearrangement is a competitive process. The [2,3]-Wittig Rearrangement allows the synthesis of homoallylic alcohols by the base-induced rearrangement of allyl ethers at low temperatures.

Allyl ether

Homoallylic alcohol

- Y: Anion, heteroatom with lone pairs, ylide
- Base: LDA, n-BuLi, PhLi, ROLi, NaNH$_2$/NH$_3$
- R should be a carbanion-stabilizing group
- Driving force is commonly to quench a charge or to transfer
- Charge to a more stabilizing atom

432. WOHL DEGRADATION; ZEMPLÉN MODIFICATION

The reaction is named after the chemist Alfred Wohl. The Wohl degradation is a chain contraction method for aldoses in carbohydrate chemistry. For example, the conversion of glucose to arabinose. This method converts an aldose into an aldose with one less carbon atom by the reversal of the cyanohydrin synthesis. A nitrile group is eliminated in this reaction by treatment with ammonical silver oxide. Hydroxylamine (NH_2-OH) gives an oxime intermediate used to shorten the length of a monosaccharide by the **Wohl degradation**. The intermediate oxime dehydrates in acetic anhydride to a cyanohydrin acetate that loses H-C°N to regenerate a C=O group. Since the C°N carbon was originally the C of the C(=O)H group, the loss of H-C°N transforms the *CHOH group into an aldehyde group (*C(=O)H). The overall result is an aldose with one less C atom.

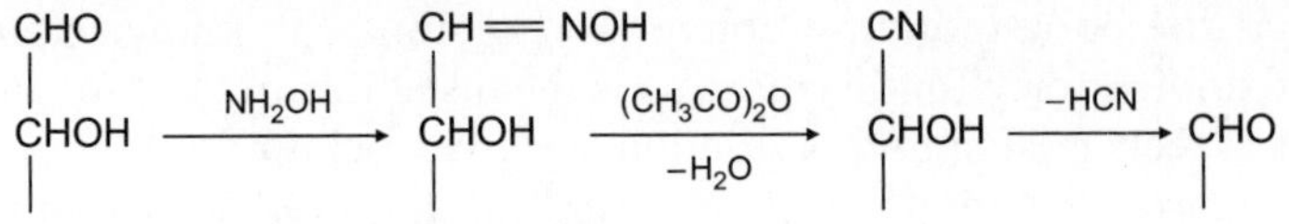

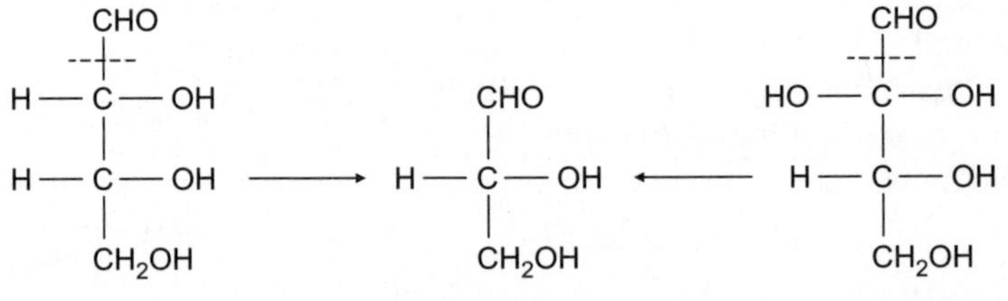

D-glyceraldehyde

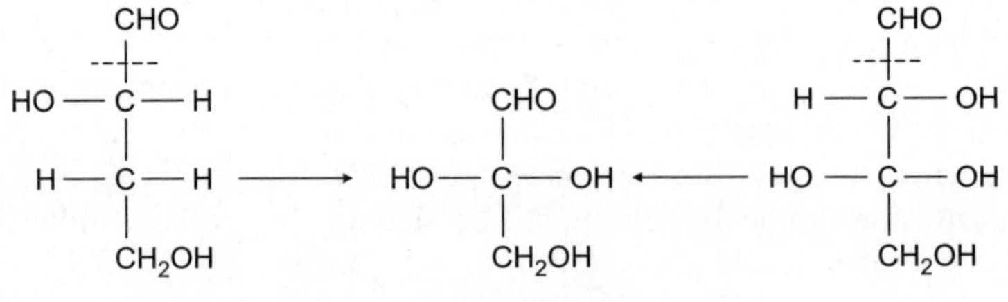

L-glyceraldehyde

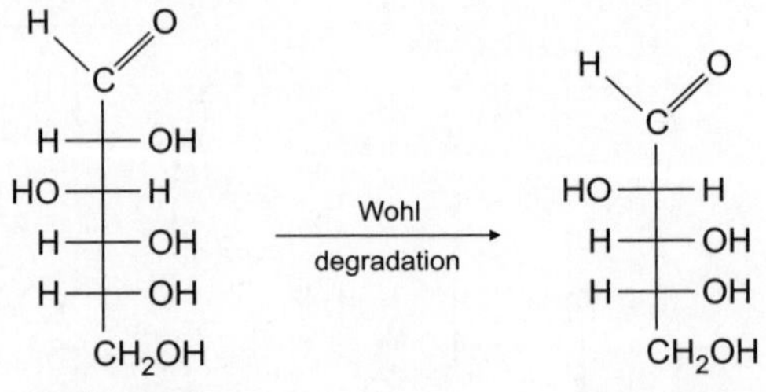

D-Glucose   D-Arabinose

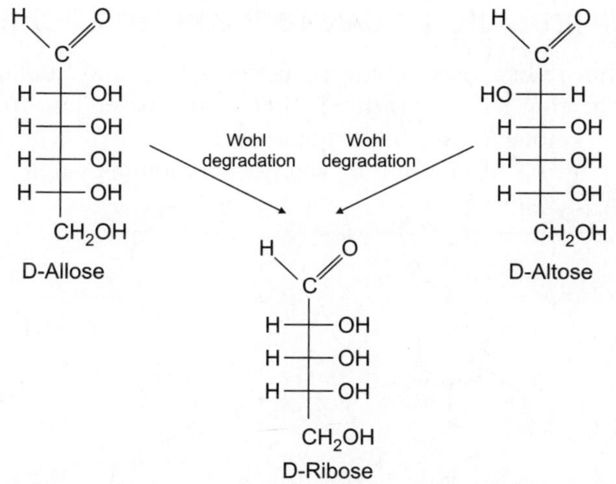

D-Allose

D-Ribose

D-Altose

In the **Zemplén modification**, sodium alkoxide is used in the elimination of the nitrile.

433. WOHL-ZIEGLER REACTION

The bromination of allylic or benzylic position of an unsaturated organic compounds specially olefinic substrate with N-bromosuccinimide (NBS) is known as *Wohl-Ziegler bromination reaction*. This reaction follows a radical pathway. The Wohl-Zeigler reaction has long been used as a general means of brominating unsaturated compounds. This reaction is believed to take place by a free-radical mechanism, is generally conducted in carbon tetrachloride as a solvent, and is catalysed by light and peroxides. N-bromosuccinimide is used as a source of low-concentration bromine, which produces bromine radicals which initiate the reaction.

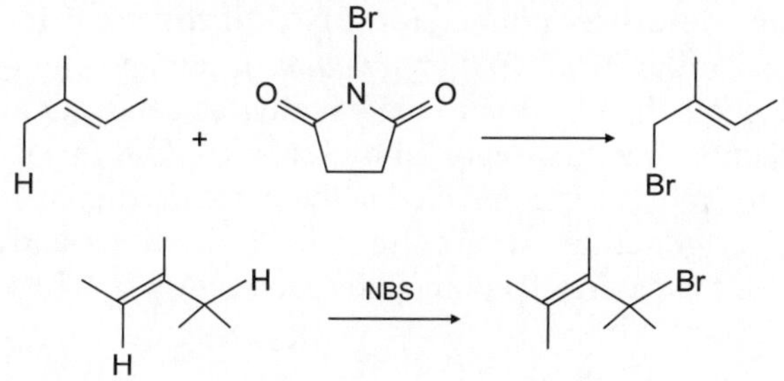

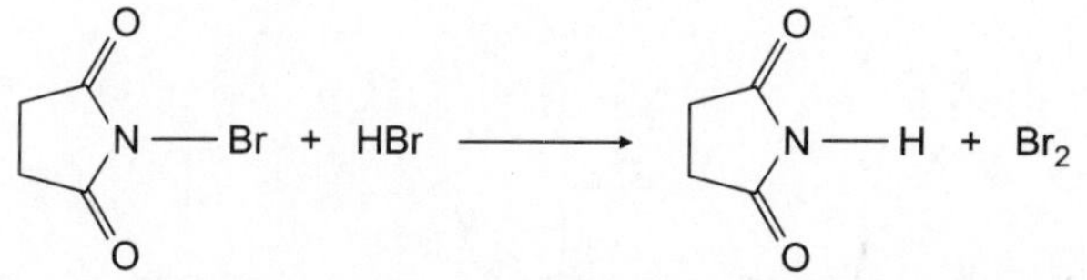

The bromination by N-bromoacetamide is simply referred to as the Wohl reaction, whereas the bromination by NBS is generally known as the Wohl-Ziegler bromination. It has been reported that the Wohl-Ziegler bromination does not work for the conjugated olefins and 1,4-dienes, which are the inhibitors for the Wohl-Ziegler bromination.

434. WOLFF-KISHNER REDUCTION; HUANG-MINLON MODIFICATION

Wolff-Kishner reduction refers to complete reduction of carbonyl compounds to methyl or methylene groups on heating with hydrazine hydrate and a base. The Wolff-Kishner reduction therefore fully reduces a ketone (or aldehyde) to an alkane.

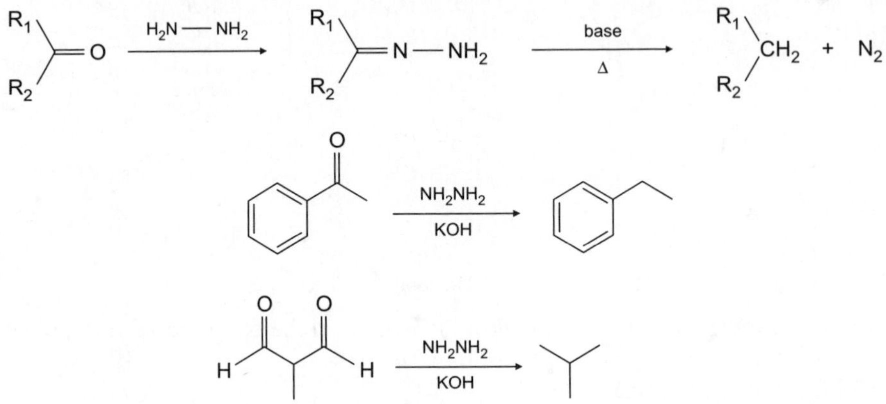

The classical procedure for the Wolff-Kishner reduction, i.e. the decomposition of the hydrazone in an autoclave at 200°C has been replaced almost completely by the modified procedure after *Huang-Minlon*. In the **Huang-Minlon modification**, diethylene glycol is used as a solvent. The Wolff-Kishner reduction is an important alternative method to the *Clemmensen reduction*, and is especially useful for the reduction of acid-labile or high-molecular substrates.

435. WOLFF REARRANGEMENT

The Wolff rearrangement describes conversion of α-diazoketone into a ketene. The *Wolff rearrangement* is one step of the *Arndt-Eistert reaction*. Decomposition of diazo ketone can be accomplished thermally, photochemically or catalytically; as catalyst amorphous silver oxide is commonly used: This reaction was first reported by Ludwig Wolff in 1902.

When the Wolff rearrangements are conducted in the presence of nucleophiles such as alcohols (ROH) or amines (RNH_2) generate carboxylic acid derivatives and when this reaction is carried out in the presence of unsaturated compounds such as A = B, a [2 + 2] cycloadditions takes place.

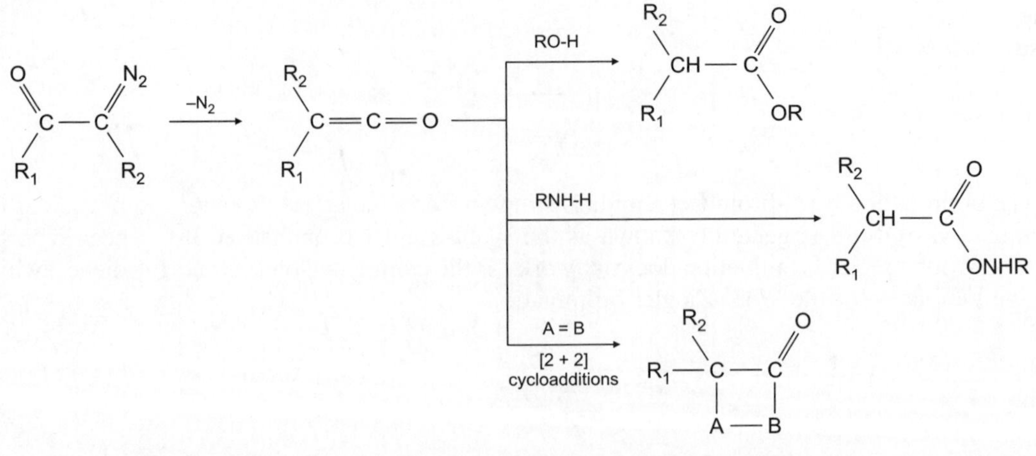

436. WOLFFENSTEIN-BÖTERS REACTION

Wolffenstein-Böters reaction refers to a simultaneous oxidation and nitration of aromatic compounds to form nitrophenols with nitric acid or higher nitrogen oxides in the presence of a mercury salt as catalyst. In this reaction, usually more than one nitro group is introduced to the aromatic nucleus. The nitrobenzene and dinitrobenzene are formed as by-products. This reaction is very useful for the preparation of nitrophenols.

437. WOODWARD CIS-HYDROXYLATION

This reaction is named after its discoverer, Robert Burns Woodward. The Woodward cis-hydroxylation refers to the chemical reaction of alkenes with iodine and silver acetate in wet acetic acid to form cis-diols. This reaction allows the synthesis of syn-diols from alkenes. In the reaction, iodine is added to the alkene followed by nucleophilic displacement with acetate in the presence of water subsequently hydrolysis of the intermediate ester gives the desired diol. This reaction has found application in steroid synthesis. The Prévost reaction gives anti-diols.

438. WURTZ-FITTIG REACTION

The Wurtz-Fittig reaction is named after Charles-Adolphe Wurtz, who discovered in 1855 a similar reaction between two alkyl halides (Wurtz reaction), and Rudolph Fittig, who discovered that also aryl halides undergo this reaction. This reaction allows formation of alkylated aromatic hydrocarbons on coupling of an alkyl and an aryl halide with sodium thus allows the alkylation of aryl halides. The more reactive alkyl halide forms an organosodium first, and this reacts as a nucleophile with an aryl halide as the electrophile. Excess alkyl halide and sodium may be used if the symmetric coupled alkanes formed as a side product may be separated readily. This reaction is useful for production of Biphenyl compounds of type Ph-Ph.

439. WURTZ REACTION

The Wurtz Coupling is one of the oldest organic reactions, named after Charles-Adolphe Wurtz. The Wurtz reaction is a coupling reaction in which two molecules of alkyl halide are coupled in presence of sodium metal in anhydrous ether to form a new carbon-carbon single bond and give a symmetrical alkane.

$$R\!-\!X \;+\; 2Na \;+\; X\!-\!R \xrightarrow{\text{Ether}} R\!-\!R \;+\; 2NaX$$

Alkyl halide ⟶ Symmetrical alkane

Where X = halogen

$$H_3C\!-\!Cl \;+\; 2Na \;+\; Cl\!-\!CH_3 \xrightarrow{\text{Ether}} H_3C\!-\!CH_3 \;+\; 2NaCl$$

Methyl chloride ⟶ Ethane

$$C_2H_5\!-\!Cl \;+\; 2Na \;+\; Cl\!-\!C_2H_5 \xrightarrow{\text{Ether}} C_2H_5\!-\!C_2H_5 \;+\; 2NaCl$$

Eethyl chloride ⟶ Butane

There is a possibility, in the reaction to use different alkyl halides instead of a single halide. If two different or dissimilar alkyl halides (i.e. R — X and R′ — X) are taken in the reaction, then the product is a mixture of alkanes. A particular alkane from this mixture is often difficult to separate.

$$R'\!-\!X \;+\; 2Na \;+\; X\!-\!R \xrightarrow{\text{Ether}} R\!-\!R \;+\; R'\!-\!R \;+\; R'\!-\!R' \;+\; NaX$$

Alkyl halide ⟶ mixture of different alkanes

$$H_3C\!-\!Cl \;+\; 2Na \;+\; Cl\!-\!C_2H_5 \xrightarrow{\text{Ether}} H_3C\!-\!CH_3 \;+\; C_2H_5\!-\!C_2H_5 \;+\; H_3C\!-\!C_2H_5$$

Methyl chloride   Ethyl chloride ⟶ Ethane   Butane   Propane

mixture of different alkanes

The reaction consists of a halogen-metal exchange involving the radical species R•. Anhydrous conditions (e.g. dry ether) are required for the reaction since alkyl free radical are formed during the course of the reaction which are strongly basic in nature and can abstract proton from the water molecule forming alkane thus reduce the yield of the desired product. If alkyl or aryl fluorides or aryl chloride are being taken as reactants, tetrahydrofuran is used as solvent instead of ether. Since the reaction involves free radical species (R·), a side reaction occurs and elimination takes place instead of simple coupling to produce an alkene. This side-reaction becomes more significant when the bulky alkyl halides are involved in the reaction.

The Wurtz reaction is limited to synthesis of symmetrical alkanes with even number of carbon atoms only. The number of carbons in the alkane is double that of alkyl halide. Methane cannot be prepared by the Wurtz reaction.

440. THORPE-ZIEGLER REACTION (ZIEGLER METHOD OR THORPE-ZIEGLER METHOD)

The Thorpe-Ziegler reaction (intramolecular modification of the Thorpe reaction) involves a dinitrile as a reactant and a cyclic ketone as the final reaction product after acidic hydrolysis. In other words reaction between a nitrile and an alkoxide base leads to the formation of a β-iminonitrile from which an α-cyanoketone is obtained by acid hydrolysis. This reaction is known as the Thorpe reaction when intermolecular, and called the Thorpe-Ziegler reaction when intramolecular.

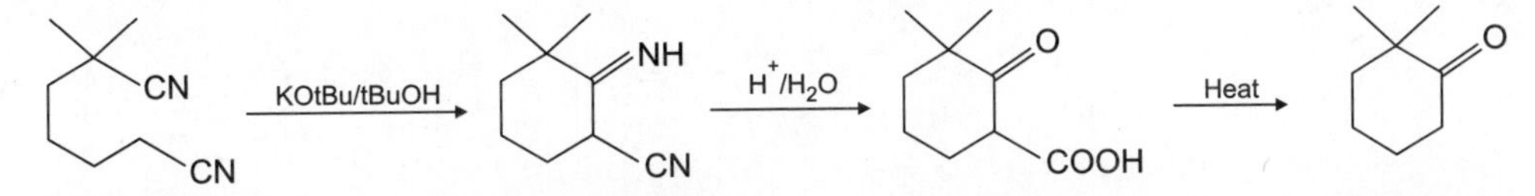

2,2-dimethyl-
heptanedinitrile

2-imino-3,3-dimethyl
cyclohexanecarbonitrile
(An imine)

3,3-dimethyle-2-oxo
cyclohexanecarboxylic
acid

2,2-dimethyl
cyclohexanone

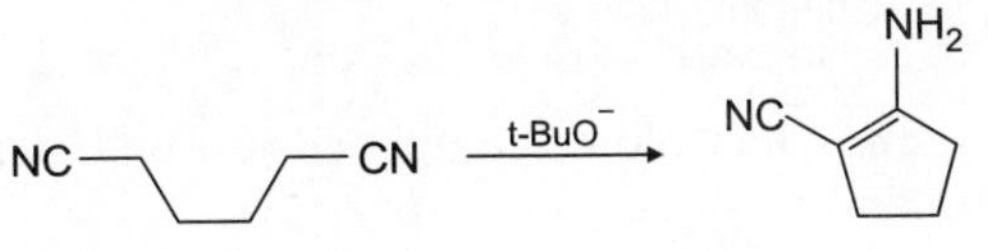

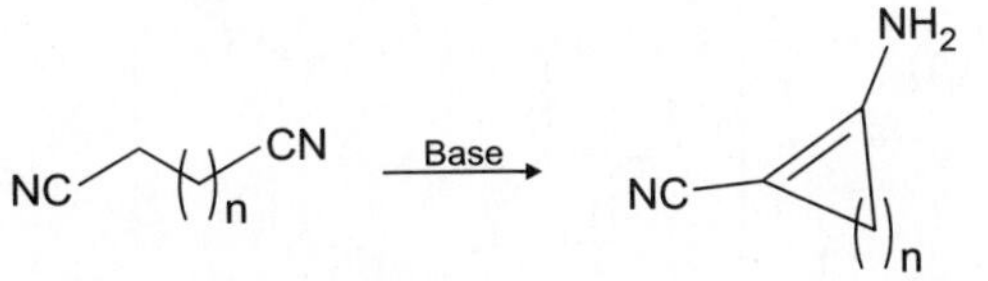

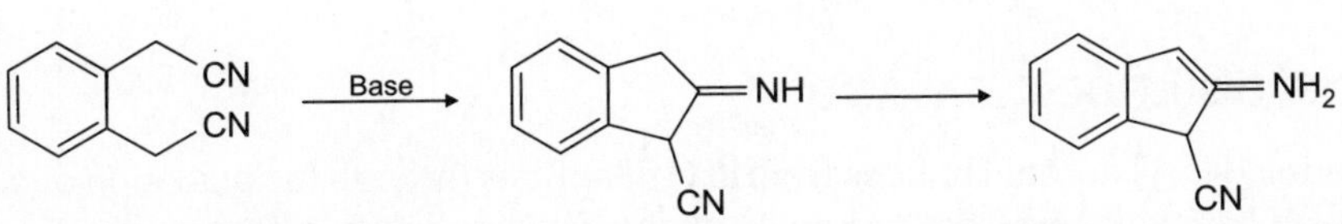

The reaction was named after Jocelyn Field Thorpe and Karl Ziegler. The Thorpe-Ziegler reaction is especially useful in the formation of five- to eight-membered rings and for rings with more than thirteen members, although it fails for nine- to twelve-membered rings.

441. ZIEGLER-NATTA POLYMERIZATION

Polymerization of vinyl monomers under mild conditions using Lewis acid catalysts to give a stereoregulated, or tactic, polymer.

A Ziegler-Natta catalyst, named after Karl Ziegler and Giulio Natta, is a catalyst used in the synthesis of polymers of 1-alkenes (α-olefins). Two broad classes of Ziegler-Natta catalysts are employed, distinguished by their solubility.

Heterogeneous supported catalysts based on titanium compounds are used in polymerization reactions in combination with cocatalysts, organoaluminum compounds such as triethylaluminium, $Al(C_2H_5)_3$. This class of catalyst dominates the industry.

Homogeneous catalysts usually based on complexes of Ti, Zr or Hf. They are usually used in combination with a different organoaluminum cocatalyst, methylaluminoxane (or methylalumoxane, MAO). These catalysts traditionally include metallocenes but also feature multidentate oxygen- and nitrogen-based ligands.

Ziegler-Natta catalysts are used to polymerize terminal 1-alkenes (ethylene and alkenes with the vinyl double bond):

$$n\ CH_2 = CHR \longrightarrow -[CH_2 - CHR]n -$$

442. ZIMMERMANN REACTION

Zimmermann reaction occurs between methylene ketones and aromatic polynitro compounds in the presence of alkali.

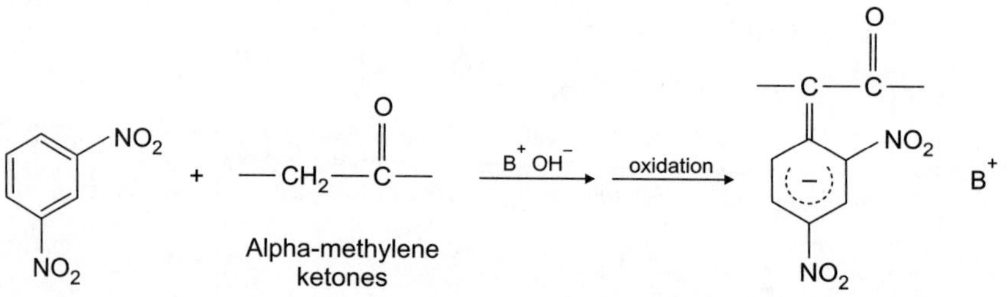

When applied to 17-oxosteroids, the colored compounds formed can be used for the quantitative determination of 17-oxosteroids:

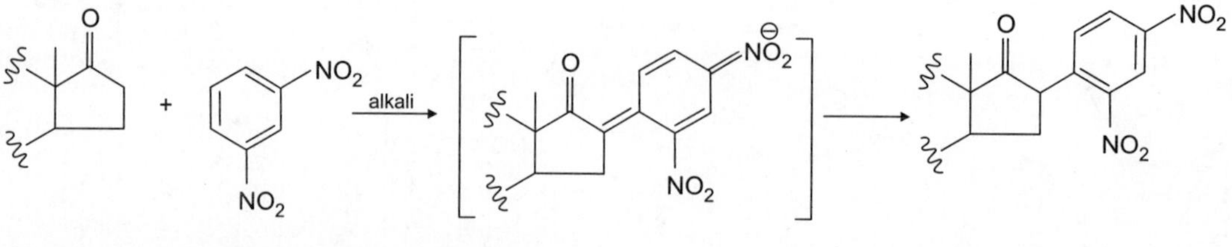

443. ZINCKE DISULFIDE CLEAVAGE

The preparation of arylsulfenyl halides from aryl disulfides through the oxidation of corresponding disulfides with halogens is generally known as the Zincke disulfide cleavage.

$$Ar-S-S-Ar \xrightarrow{X_2} Ar-S-X\ +\ X-S-Ar$$

The Zincke disulfide cleavage is usually performed at low temperature without light and in an anhydrous solvent to minimize possible aromatic halogenation.

Action of chlorine or bromine on thiophenols or arylbenzyl sulfides also result in formation of arylsulfenyl halides.

$$Ar-S-H \xrightarrow{X_2} Ar-S-X$$

$$Ar-S-CH_2-\underset{}{\bigcirc} \xrightarrow{2X_2} Ar-S-X \quad + \quad X_2HC-\underset{}{\bigcirc} \quad + \quad HX$$

444. ZINCKE NITRATION

Replacement of ortho- or para-bromine or iodine atoms in phenols by a nitro group on treatment with nitrous acid or a nitrite in acetic acid is generally referred to as the Zincke method or Zincke nitration. The reaction was named after Theodor Zincke. This reaction is a manifestation of nucleophilic aromatic substitution.

Under the standard reaction conditions, only bromo and iodo groups are replaced by a nitro group, although chloro and fluoro groups can also be substituted under special conditions. This reaction has been used for the preparation of nitrophenols.

445. ZINCKE-SUHL REACTION

Zincke-Suhl reaction refers to phenol-dienone rearrangement of p-cresols by addition of carbon tetrachloride in the presence of aluminum chloride with formation of 4-methyl-4-trichloromethylcyclohexa-2,5-dienone. This reaction is considered as a special case of a Friedel-Crafts alkylation and was first described by Theodor Zincke and Suhl.

Abbreviations and Acronyms

ee	Enantiomeric excess
(DHQ)2-PHAL	1,4-bis(9-O-dihydroquinine)-phthalazine
(DHQD)2-PHAL	1,4-bis(9-O-dihydroquinidine)-phthalazine
⬤—	polymer support
[bimim]Cl•2AlCl$_3$	1-butyl-3-methylimidazolium chloroaluminuminate (a Lewis acid ionic liquid)
°C	Temperature in degrees Centigrade
Ⓟ—	Polymeric backbone
1,5-HD	1,5-hexadienyl
9-BBN	9-borabicyclo[3.3.1]nonane
Ac	Acetyl
AIBN	2,2′-azobisisobutyronitrile
Alpine-borane®	B-isopinocampheyl-9-borabicyclo[3.3.1]-nonane
aq.	Aqueous
Ar	aryl
B:	generic base
BER	Borohydride exchange resin
BINAP	(2R,3S),2,20-bis(diphenylphosphino)-1,10-binapthyl
BINAP	2,2′-bis(diphenylphosphino)-1,1′-binaphthyl
Bn	Benzyl
BOC	tert-butoxycarbonyl
Boc	tert-butyloxycarbonyl
bpy (bipy)	2,20-bipyridyl Ot-Bu
Bu	n-Butyl
Bz	Benzoyl
c-	Cyclo
CAM	Carboxamidomethyl
CAN	Ceric ammonium nitrate
cat.	Catalytic
Cbz	Carbobenzyloxy
Chirald	(2S,3R)-(þ)-4-dimethylamino-1,2-diphenyl-3-methylbutan-2-o1
Cod	1,5-Cyclooctadiene (ligand)
Cot	1,3,5,7-Cyclooctatetraene (ligand)
Cp	Cyclopentadienyl
CSA	Camphorsulfonic acid
CTAB	Cetyltrimethylammonium bromide
CuTC	copper thiophene-2-carboxylate

Cy	Cyclohexyl
DABCO	1,4-diazobicyclo[2.2.2]octane
dba	Dibenzylidene acetone
DBE	1,2-dibromoethane
DBN	1,5-diazabicyclo[4.3.0]non-5-ene
DBU	1,8-diazabicyclo[5.4.0]undec-7-ene
DCC	1,3-dicyclohexylcarbodiimide
DCE	1,2-dichloroethane
DDQ	2,3-dichloro-5,6-dicyano-1,4-benzoquinone
DEA	Diethylamine
DEAD	Diethylazodicarboxylate
DIAD	diisopropyl azodicarboxylate
DIBAL	diisobutylaluminium hydride
Dibal-H	Diisobutylaluminium hydride
DIPEA	diisopropylethylamine
Diphos (dppe)	1,2-bis(Diphenylphosphino)ethane
Diphos-4 (dppb)	1,4-bis(Diphenylphosphino)butane
DMA	N,N-dimethylacetamide
DMAP	4-N,N-dimethylaminopyridine
DME	1,2-Dimethoxyethane
DMF	N,N-dimethylformamide
DMFDMA	N,N-dimethylformamide dimethyl acetal
dmp	bis-[1,3-Di(p-methoxyphenyl)-1,3-propanedionato]
DMS	dimethylsulfide
DMSO	Dimethyl sulfoxide
DMSY	dimethylsulfoxonium methylide
DMT	dimethoxytrityl
dpm	Dipivaloylmethanato
dppb	1,4-bis(diphenylphosphino)butane
dppe	1,2-bis(diphenylphosphino)ethane
dppf	1,1′-bis(diphenylphosphino)ferrocene
dppp	1,3-bis(diphenylphosphino)propane
DTBAD	di-tert-butylazodicarbonate
DTBMP	2,6-di-tert-butyl-4-methylpyridine
dvb	Divinylbenzene
E1	unimolecular elimination
E1cB	2-step, base-induced β-elimination via carbanion
E2	bimolecular elimination
EAN	ethylammonium nitrate
EDA	Ethylenediamine $H_2NCH_2CH_2NH_2$
EDDA	ethylenediamine diacetate
EDTA	Ethylenediaminetetraacetic acid
EE 1-	Ethoxyethyl
Ei	two groups leave at about the same time and bond to each other as they are doing so.
Eq	equivalent
Et	ethyl
EtOAc	ethyl acetate
FMN	Flavin mononucleotide

fod	tris-(6,6,7,7,8,8,8)-heptafluoro-2,2-dimethyl-3,5-octanedionate
Fp	Cyclopentadienyl-bis(carbonyl iron)
FVP	Flash vacuum pyrolysis
h	Hour (hours)
HMDS	hexamethyldisilazane
HMPA	Hexamethylphosphoramide
HMPT	Hexamethylphorous triamide
HMTTA	1,1,4,7,10,10-hexamethyltriethylenetetramine
hn	Irradiation with light
Imd	imidazole
iPr	Isopropyl
IR	Infrared
KHMDS	potassium hexamethyldisilazide
LAH	lithium aluminium hydride
LDA	Lithium diisopropylamide
LHMDS	Lithium hexamethyl disilazide
LICA (LIPCA)	Lithium cyclohexylisopropylamide
LTMP	Lithium 2,2,6,6-tetramethylpiperidide
M	metal
MABR	Methylaluminium bis(4-bromo-2,6-di-tert-butylphenoxide)
MAD	bis(2,6-Di-tert-butyl-4-methylphenoxy)methyl aluminium
m-CPBA	m-chloroperoxybenzoic acid
Me	Methyl
MEM	β-methoxyethoxymethyl
Mes	Mesityl
MOM	Methoxymethyl
Ms	Methanesulfonyl
MTM	Methylthiomethyl
MVK	methyl vinyl ketone
NAD	Nicotinamide adenine dinucleotide
NADP	Sodium triphosphopyridine nucleotide
Napth	Naphthyl
NBD	Norbornadiene
NBS	N-bromosuccinimide
NCS	N-chlorosuccinimide
Ni(R)	Raney nickel
NIS	N-iodosuccinimide
NMP	N-methyl-2-pyrrolidinone
NMR	Nuclear magnetic resonance
Nos	nosylate (4-nitrobenzenesulfonyl)
N-PSP	N-phenylselenophthalimide
N-PSS	N-phenylselenosuccinimide
Nu	nucleophile
Oxone	2 $KHSO_5.KHSO_4.K_2SO_4$
PCC	Pyridinium chlorochromate
PDC	Pyridinium dichromate
PEG	Polyethylene glycol
Ph	Phenyl
PhH	Benzene

PhMe	Toluene
Phth	Phthaloyl
pic	2-pyridinecarboxylate
Pip	Piperidyl N
Piv	pivaloyl
PMB	para-methoxybenzyl
PMP	4-methoxyphenyl
PPA	polyphosphoric acid
PPTS	pyridinium p-toluenesulfonate
Pr	n-propyl
Py	Pyridine N
PyPh2P	diphenyl 2-pyridylphosphine
Pyr	pyridine
quant.	Quantitative yield
Red-Al	sodium bis(methoxy-ethoxy)aluminium hydride (SMEAH)
Salen	N,N′-disalicylidene-ethylenediamine
sBu	sec-butyl
sBuLi	sec-butyllithium
SET	single electron transfer
Siamyl	Diisoamyl
SM	starting material
SMEAH	sodium bis(methoxy-ethoxy)aluminium hydride (Red-Al)
S_N1	unimolecular nucleophilic substitution
S_N2	bimolecular nucleophilic substitution
SNAr	nucleophilic substitution on an aromatic ring
TADDOL	a,a,a0a0-tetraaryl-4,5-dimethoxy-1,3-dioxolane
TASF	tris-(diethylamino)sulfonium difluorotrimethyl silicate
TBABB	tetra-n-butylammonium bibenzoate
TBAF	tetra-n-butylammonium fluoride
TBDMS	tert-butyldimethylsilyl
TBDPS	tert-butyldiphenylsilyl
TBHP	tert-butylhydroperoxide
TBS	tert-butyldimethylsilyl
t-Bu	tert-butyl
TEA	triethylamine
TEBA	Triethylbenzylammonium
TEMPO	Tetramethylpiperdinyloxy free radical
TEOC	trimethylsilylethoxycarbonyl
Tf	trifluoromethanesulfonyl (triflyl)
Tf (OTf)	Triflate
TFA	Trifluoroacetic acid
TFAA	Trifluoroacetic anhydride
TFP	tri-2-furylphosphine
THF	Tetrahydrofuran
THP	Tetrahydropyran
TIPS	triisopropylsilyl
TMEDA	N,N,N ′,N′-tetramethylethylenediamine
TMG	1,1,3,3-tetramethylguanidine

TMP	2,2,6,6-tetramethylpiperidine
TMS	Trimethylsilyl
TMSCl	trimethylsilyl chloride
TMSCN	trimethylsilyl cyanide
TMSI	trimethylsilyl iodide
TMSOTf	trimethylsilyl triflate
Tol	toluene or tolyl
Tol-BINAP	2,2'-bis(di-p-tolylphosphino)-1,1'-binaphthyl
TosMIC	(p-tolylsulfonyl)methyl isocyanide
TPAP	tetra-n-propylammonium perruthenate
Tr	Trityl
TRIS	Triisopropylphenylsulfonyl
Ts(Tos)	Tosyl
TsO	tosylate
UHP	urea-hydrogen peroxide
UV	Ultraviolet
Xc	Chiral auxiliary
Δ	Solvent heated under reflux